Nanoscale Science and Engineering Forum 2013

AA003380

Core Programming Area at the 2013 AIChE Annual Meeting: Global Challenges for Engineering a Sustainable Future

San Francisco, California, USA
3 – 8 November 2013

ISBN: 978-1-63439-047-7

Printed from e-media with permission by:

Curran Associates, Inc.
57 Morehouse Lane
Red Hook, NY 12571

proceedings.com

Some format issues inherent in the e-media version may also appear in this print version.

Printed by Curran Associates, Inc. (2014)

For permission requests, please contact AIChE
at the address below.

AIChE
3 Park Avenue
New York, NY 10016-5991

Phone: (203) 702-7660
Fax: (203) 775-5177

www.aiche.org

Additional copies of this publication are available from:

Curran Associates, Inc.
57 Morehouse Lane
Red Hook, NY 12571 USA
Phone: 845-758-0400
Fax: 845-758-2634
Email: curran@proceedings.com
Web: www.proceedings.com

TABLE OF CONTENTS

(8a) Enhancing in Vitro and in Vivo Gene Delivery of Adeno-Associated Viruses Type 2 By Cell-Permeable Peptides

Liu, Y., University of Southern California
Wang, P., University of Southern California
Originally presented on: 11/3/2013 15:30:00 - 15:50:00

Adeno-assocated virus type 2 (AAV2) is considered a promising gene delivery vector and has been extensively applied in several disease models; however, inefficient transduction in various cells and tissues has limited its widespread application in many areas of gene therapy. In this study, we have developed a general but efficient strategy to enhance viral transduction both *in vitro* and *in vivo*, by incubating viral particles with cell-permeable peptides (CPPs). We show that CPPs increase internalization of viral particles into cells by facilitating both energy-independent and energy-dependent endocytosis. Moreover, CPPs can significantly enhance the endosomal escape process of viral particles, thus enhancing viral transduction to those cells that have exhibited very low permissiveness to AAV2 infection due to impaired intracellular processing of virus. Thus, the membrane-penetrating ability of CPPs enables us to generate an efficient method for enhanced gene delivery of AAV2 vectors, potentially facilitating its applicability to human gene therapy.

(8b) Optimization of Mucus Penetrating Particles and Vehicle Composition for Improved Mucosal Surface Drug Delivery

Ensign, L., Johns Hopkins University
Xu, Q., Johns Hopkins University
Tang, B.,
Wang, Y. Y.,
Cone, R., Johns Hopkins University
Hanes, J., Johns Hopkins University
Originally presented on: 11/3/2013 15:50:00 - 16:08:00

Introduction

Local mucosal drug delivery has many applications, including treatment of various cancers, inflammatory conditions such as inflammatory bowel disease and chronic obstructive pulmonary disease, and prevention of sexually transmitted infections. However, achieving effective drug delivery to mucosal epithelia can be challenging. Epithelial surfaces of the female reproductive tract, gastrointestinal tract, and lung have numerous folds and/or topography, making significant portions of the epithelial surface difficult to access. Additionally, mucosal epithelia are highly absorptive and permeable, which can limit the duration of delivery of soluble drug systems. Nanoparticles can be used as controlled-release delivery systems to slow the absorption of drugs, but the viscoelastic, adhesive mucus layers lining mucosal epithelia trap conventional nanoparticles (CP), limiting distribution and facilitating rapid clearance. We recently discovered that densely coating the surface of nanoparticles with low molecular weight polyethylene glycol (PEG) effectively shields the nanoparticle core from adhesive interactions with human cervicovaginal mucus (CVM) and human respiratory mucus (1,2). These so-called mucus penetrating particles (MPP) rapidly diffuse through the fluid-filled pores in the mucus mesh, whereas CP are adhesively immobilized. However, diffusion alone may not be rapid enough for MPP to distribute throughout mucosal surfaces, including penetrating the mucus blanket. Thus, we have found that using osmotic gradients to induce fluid absorption and subsequent advective delivery to the epithelial surface is very effective; MPP delivered in a hypotonic vehicle rapidly distributed throughout the vaginal tract of mice, including reaching into the deep folds within 10 minutes (3). By reaching the more slowly cleared mucus layers in the rugae, MPP were well-retained in the vagina, whereas mucoadhesive CP were rapidly cleared with the luminal mucus layers (3). We anticipate that similar improvements in distribution and retention with hypotonically-administered MPP can be achieved at various mucosal surfaces, potentially improving drug delivery for a wide range of conditions. Here, we optimize nanoparticle surface PEG density and vehicle tonicity for enhanced distribution, facilitating effective drug delivery.

Materials and Methods

Non-biodegradable probe MPP for testing various vehicles, capable of diffusing through human CVM, were prepared as previously reported (4). Biodegradable nanoparticles composed of poly(lactic-co-glycolic acid) (PLGA) and PEG copolymer (PLGA-PEG) were synthesized with various PEG surface densities using a self-assembly emulsification method and varying ratios of PLGA:PLGA-PEG. PEG surface density was quantified using two NMR methods: the first method involved dissolving the nanoparticles and determining total PEG content, while the second method measured surface PEG content of intact particles. The PEG density (Γ) was calculated as the number of PEG molecules on the nanoparticle surface per 100 nm^2. The number of unconstrained molecules per 100 nm^2 (Γ^*) was calculated by assuming PEG behaves as a flexible polymer in a random-walk conformation. The ratio Γ/Γ^* represents the relative steric constraint on the surface PEG; as Γ/Γ^* increases, the PEG conformation progresses from the mushroom to the brush regime (5). Transport rates of MPP and CP were quantified using multiple particle tracking (6). Vehicles of varying osmolality were prepared with various amounts of water and saline. Vaginal distribution was assessed by qualitatively (cross-sectional images) and quantitatively (surface area covered on flattened tissue).

Results

We found that increasing PEG content more effectively shielded the nanoparticle core, reducing mucoadhesion and increasing transport rates in human mucus. Increasing PEG content reflected an increase in the relative ratio of PEG chains per surface area compared to the number of unconstrained PEG chains that would occupy a given surface area, indicating a more constrained surface conformation and close packing in the ?brush? regime. Improvement in mucus penetration correlated well with improved vaginal distribution and surface coverage in mice. Administering MPP in hypotonic vehicles led to rapid fluid absorption and distribution throughout the vaginal tract. We found that even minimally hypotonic vehicles (220 mOsm/kg) facilitated favorable distribution over the entire epithelial surface, including the deep folds. In contrast, MPP administered in an isotonic vehicle were found diffusely spread throughout the vaginal lumen, rather than forming the uniform coating on the vaginal epithelium seen with hypotonically-administered MPP.

Discussion and Conclusion

Many conventional vehicles for vaginal and rectal delivery are hypertonic, which causes fluid secretion that facilitates clearance (7). Hypertonicity has also been demonstrated to cause inflammation that increases susceptibility to sexually transmitted infections (8,9). Although purely hypotonic vehicles appear to be safe after daily vaginal administration for 1 week (3), we anticipate that minimally hypotonic vehicles are less likely to cause any toxicity or epithelial distress, particularly with repetitive administration. As such, we demonstrated here that improved epithelial distribution can be achieved with a minimally hypotonic vehicle. Although osmotically-induced fluid absorption appears to be advantageous for nanoparticle delivery at mucosal surfaces, it is vital that the nanoparticle be sufficiently mucoinert to penetrate through the mucus barrier. Although we previously demonstrated that nanoparticles sufficiently coated with PEG can rapidly penetrate human CVM, here we more thoroughly characterize what is a ?sufficient? coating. We found that increasing PEG content, such that the conformation was in the ?brush? regime, correlated with increased diffusion rates in human CVM and improved distribution in the mouse vagina. We anticipate that the PEG surface density can be similarly optimized for penetration in mucus at other mucosal surfaces, including the GI tract and the lungs. Additionally, mucosal surfaces such as the GI tract and the lung are also absorptive, such that hypotonic vehicles will also be beneficial for distribution and retention. Combining MPP with and hypotonic vehicles has promise for more efficacious drug and gene delivery to various mucosal epithelia, potentially improving prevention and treatment of a wide array of mucosal diseases and conditions.

(1) Lai SK, et al. Rapid transport of large polymeric nanoparticles in fresh undiluted human mucus. Proc Natl Acad Sci U S A. 2007;104(5):1482-7.

(2) Schuster BS, et al. Nanoparticle diffusion in respiratory mucus from humans without lung disease. Biomaterials. 2013; 34(13): 3439-46.

(3) Ensign LM, et al. Mucus-penetrating nanoparticles for vaginal drug delivery protect against herpes simplex virus. Sci Transl Med. 2012;4(138):138ra79.

(4) Nance EA, et al. A dense poly(ethylene glycol) coating improves penetration of large polymeric nanoparticles within brain tissue. Sci Transl Med. 2012;4(149):149ra19.

(5) Boylan NJ, et al. Highly compacted DNA nanoparticles with low MW PEG coatings: in vitro, ex vivo, and in vivo evaluation. J Control Release. 2012; 157(1): 72-79.

(6) Suh J, et al. Real-time multiple-particle tracking: applications to drug and gene delivery. Adv Drug Deliv Rev. 2005; 57(1):63-78.

(7) Zeitlin L, et al. Leakage of three commercial vaginal gels in women. Abstract. Contraception. 2003 2003;68:139-55.

(8) Moench TR, et al. Microbicide excipients can greatly increase susceptibility to genital herpes transmission in the mouse. BMC Infect Dis. 2010;10:331.

(9) Fuchs EJ, et al. Hyperosmolar sexual lubricant causes epithelial damage in the distal colon: potential implication for HIV transmission. J Infect Dis. 2007;195(5):703-10.

(8c) Modeling Bioresponsive Hydrogels for Targeted Homeostasis in Drug Delivery

Wilson, A. N.,
Guiseppi-Elie, A., Clemson University
Originally presented on: 11/3/2013 16:08:00 - 16:26:00

Modeling Bioresponsive Hydrogels for Targeted Homeostasis in Drug Delivery

A. Nolan Wilson and Anthony Guiseppi-Elie

Center for Bioelectronics, Biosensors and Biochips (C3B), Clemson University Advanced Materials Center, 100 Technology Drive, Anderson, South Carolina 29625, USA

Department of Chemical and Biomolecular Engineering, Department of Bioengineering, Department of Electrical and Computer Engineering, Clemson University, Clemson, South Carolina 29634, USA

ABTECH Scientific, Inc., Biotechnology Research Park, 800 East Leigh Street, Richmond, Virginia 23219, USA.

The use of bioresponsive hydrogels as drug delivery platforms has enabled functionality and versatility far beyond what is achievable by classical tablet formulations or bolus injection. Key features of these materials leading too much of their success are high biocompatibility, adaptability, tuneability, and relative ease of manufacturing. Bioresponsive hydrogels have proliferated into a wide range of applications including drug delivery, biosensing, and tissue engineering. Initially, hydrogels began as passive materials with an objective to remain invisible to the host tissue; however, as engineering capabilities grew, intrinsic stimuli responsiveness of hydrogels were coupled with the biospecificity of biological moieties to create a new class of materials, bioresponsive hydrogels, who actively interact with the surrounding environment[1]. These materials have distinct advantages over their predecessors as they can sense and respond to a specific metabolic stimuli. Further advancement of bioresponsive hydrogels, in the realm of drug delivery, has been to engineer self-regulating control loops into the materials such that metabolic activity dictates the release rate and establishes a pseudo-steady state, or homeostatic conditions, over the delivery period. Homeostatic drug delivery is akin to the intrinsic controls existing in natural metabolic pathways as they both exert spatiotemporal control to dampen dynamic inputs from the environment to maintain target metabolic activities[2].

Here we have performed an *in silico* investigation into a bioresponsive hydrogel which controls the relative activity of the bathing matrix metalleoprotease (MMP). The goal is to regulate MMP-9 activity to levels associated with normal wound repair over a delivery period that is time consistent with wound remodeling.

Immobilized to the hydrogel is a peptide-drug conjugate which upon exposure to the specific enzyme matrix metalleoprotease-9 (MMP-9), a protease involved in tissue remodeling, is cleaved and the pendant inhibitor (drug) released. After release, the drug, an inhibitor of MMP-9, is able to modulate the activity of the very protease which cleaved it from the hydrogel matrix. Key engineering factors were illuminated by perturbing system constants such as component diffusion coefficients, co-polymer content, and enzyme kinetic parameters; it was determined that transport path length (i.e. microsphere radius) and pendant inhibitor concentration (i.e. drug loading) offered two key design parameters for engineering the control set point, from 99% to 7% relative activity ($R = 5 - 200\ \mu m$), and performance time, from 3 to 24 hours ($[I]_{loaded} = 0.050 - 0.300$ mM), respectively. The model was used to demonstrate enzyme activity could be modulated to normal, <1ng/mL, levels under static and dynamic conditions. By calculation of the Thiele modulus, it was determined the system operated under reaction limited conditions for short length scales and that the material could reduce metabolic activity to a desired set point under dynamic enzyme activities in the surrounding environment. Additionally, the relation between radii ($R = 300 - 2500\ \mu m$) and effective diffusivity ($D_{eff} = 7.71\times10^{-10} - 4.90\times10^{-8}\ cm^2s^{-1}$) was interrogated to show operation in diffusion limited and reaction limited domains. Self-regulated delivery offers advanced functionality and increases the range of applications of bioresponsive hydrogels as the focus turns toward integrating the material's response into native biochemical pathways and away from drug release rates.

1. Wilson, A.N., R. Salas, and A. Guiseppi-Elie, *Bioactive hydrogels demonstrate mediated release of a chromophore by chymotrypsin.*Journal of Controlled Release, 2012.

2. Wilson, A.N. and A. Guiseppi-Elie, *Bioresponsive Hydrogels.* Advanced healthcare materials, 2012.

(8d) Minimal "Self" Peptides That Inhibit Phagocytic Clearance and Enhance Delivery of Nanoparticles

Rodriguez, P. L., University of Pennsylvania
Harada, T., University of Pennsylvania
Christian, D. A., University of Pennsylvania
Pantano, D. A., University of Pennsylvania
Tsai, R., University of Pennsylvania
Discher, D. E., University of Pennsylvania
Originally presented on: 11/3/2013 16:26:00 - 16:46:00

Foreign particles and cells are rapidly cleared from the body by professional phagocytes that more often encounter and recognize "self" cells. From studies of mice, CD47 appears to be a factor in multiple glycoprotein complexes that impede phagocytosis of self cells, with signaling of 'self' occurring through a highly polymorphic, species-specific receptor, SIRPα. Here, short peptides were computationally designed from human-CD47, synthesized with anchoring groups, and attached to virus-size nanoparticles for injection into NOD.SCID (NSG) mice that are known to exhibit uniquely high compatibility with human cells. The hCD47-peptides delay splenic clearance of particles by macrophages with an exponential advantage in persistent circulation, which we exploit in enhancing by >10-fold the near-infrared imaging of human tumor xenografts. The affinity of hCD47 for NSG-SIRPα proves weak but within the broad range (0.1~5 mM) measured for ten constructed variants of human-SIRPα; several versions of hCD47-peptide are likewise shown to bind and potently inhibit nanoparticle uptake by a surprising cytoskeletal mechanism. The reductionist approach reveals the importance and utility as well as some limits of a human 'Marker of Self'.

(8e) Enhanced Intracellular Concentration and Endosomal Release of Gold Nanorods Through Tannic Acid Functionalization: Implications for Advancement in Nano-Based Drug Delivery Mechanisms

Comfort, D., University of Dayton
Untener, E., University of Dayton
Comfort, K., University of Dayton
Hussain, S., Air Force Research Laboratory
Originally presented on: 11/3/2013 16:46:00 - 17:04:00

Due to their unique properties, including a general biocompatibility, enhanced surface area, and a strong plasmonic response, gold nanorods (GNRs) have shown tremendous potential for applications including bio-imaging and drug delivery. However, one primary challenge associated with nanoparticle-dependent delivery systems is the frequent occurs of endosomal entrapment; which severely limits the desired targetability and functionality of nano-constructs. This study sought to overcome that challenge through a systematic approach to GNR functionalization for targeted cellular internalization and delivery potential, evaluating the influence of both aspect ratio and surface properties. We identified that longer and positively charged GNRs demonstrated a higher rate of cellular internalization compared to their shorter and negatively charged counterparts. Most interestingly, tannic acid (TA) functionalized GNRs were taken up independently of endocytosis, as assessed through the application of multiple endocytosis inhibitors. This distinct manner of internalization resulted in a unique orientation of TA GNRs within the cells, as viewed through TEM. To verify this phenomenon, GNRs were further functionalized with the fluorescent drug Doxorubicin and through confocal imaging the successful cytosolic delivery of the GNRs was ascertained, but only in conjunction with TA surface chemistry. Taken together, these results identify TA functionalized GNRs as a strong candidate for future development of nano-based application in the areas of delivery mechanism and therapeutics.

(8f) Histidine-Mediated Hydrogen Bonding Enhances Stability and Silencing Activity of Peptide siRNA Polyplexes

Chou, S. T., University of Maryland
Hom, K., University of Maryland Baltimore
Zhang, D., University of Maryland
Tricoli, L., University of Maryland
Hustedt, J., University of Maryland
Lee, A., University of Maryland
Leng, Q., University of Maryland Baltimore
Shapiro, M. J., Pifzer
Seog, J., Univeresity of Maryland
Kahn, J., University of Maryland
Mixson, A. J., University of Maryland Baltimore
Originally presented on: 11/3/2013 17:04:00 - 17:22:00

Branched peptides containing histidines and lysines (HK) have been determined to be effective carriers for DNA or siRNA. We anticipate that elucidation of the binding mechanism of HK with siRNA will provide greater insight into the self-assembly and delivery of the polyplex. In siRNA silencing experiments with mammalian cells, a four-branched HK peptide siRNA polyplex maintained silencing activity even with prolonged pre-incubation with serum. In contrast, siRNA in complex with four-branched N3K4b peptide, in which histidines were substituted with asparagines, showed a marked decreased in silencing activity with pre-exposure to serum. Consequently, we explored the hypothesis that histidines might form non-covalent bonds with nucleic acids to enhance the stability of siRNA polyplexes. To accomplish this, we initially compared the biophysical properties of H3K(+H)4b with N3K4b and polylysine. Consistent with siRNA silencing experiments, gel electrophoresis analysis demonstrated that the HK siRNA polyplex maintained its integrity for more than 24 h incubation in 50% serum, whereas siRNA in complex to N3K4b or polylysine was degraded in a time-dependent manner. We next studied the thermodynamic profiles of various peptides binding to siRNA at pH 7.3 with isothermal titration calorimetry. While four-branched polylysine and linear lysine-alanine peptide (K4b and A2K) interacted with siRNA resulting in an endothermic reaction, branched and linear lysine-histidine peptides (H3K(+H)4b and H2K) both exhibited an exothermic reaction at the initial binding step. This indicates that an important part of the exothermic interaction was based on non-ionic bond formation of histidines with siRNA. To investigate the type of non-ionic bond, we studied the protonation state of the imidazole ring of a selectively N15 labeled H3K(+H)4b upon siRNA binding with heteronuclear single quantum coherence (HSQC) NMR. The peak of $N^{\delta 1}$-protonated tautomers of imidazole (type α, neutral form) shifted downfield (in the direction of deprotonation) by 0.5 to 1.0 ppm with addition of siRNA at pH 7.3, providing direct evidence that uncharged histidines formed hydrogen bonds with siRNA at physiological pH. Moreover, unshifted $N^{\varepsilon 1}$ tautomers described the directionality of hydrogen bonds. Taking together, these results indicate that histidine-rich peptides form non-ionic bonds, and in particular hydrogen bonds, with siRNA, thereby enhancing the stability and biological activity of the polyplex *in vitro* and *in vivo*.

(8g) A Tale of Two Targets - Increasing Tumor-Specific Gene Expression By Targeting Overexpressed $\alpha 5\beta 1$ Integrin and Upregulated Transcriptional Activity of NF-κB

Adil, M., University of Minnesota
Kokkoli, E., University of Minnesota
Originally presented on: 11/3/2013 17:22:00 - 17:42:00

Non-viral gene delivery vehicles targeted to cancer surface receptors *in vivo* often non-specifically deliver to off target tissues expressing the same receptors, resulting in undesirable side effects. To address this and increase the specificity of gene expression in cancer tissues, we designed a modular delivery system multi-targeted to the overexpressed cancer surface receptor $\alpha_5\beta_1$ integrin and the upregulated transcriptional activity of NF-κB. We therefore condensed a luciferase gene under the control of a NF-κB responsive element (pNF-κB-Luc) with polyethyleneimine and encapsulated it in PR_b peptide functionalized stealth liposomes that specifically target cancer cells that overexpress the $\alpha_5\beta_1$ integrin. The multi-targeted system achieved increased gene expression in DLD-1 colon cancer cells compared to BJ-fibroblast healthy cells and was also able to differentiate between cancer cells and healthy cells better than either of the individually targeted systems. In addition, we constructed a novel cancer therapeutic plasmid by cloning a highly potent diphtheria toxin fragment A (DTA) expressing gene under the control of an NF-κB responsive element (pNF-κB-DTA). We obtained a dose dependent reduction of cellular protein expression and a dose dependent increase in cytotoxicity in cancer cells transfected with the pNF-κB-DTA plasmid. When pNF-κB-DTA was delivered in PR_b functionalized stealth liposomes to colon cancer cells and healthy cells *in vitro*, cancer cells were specifically eradicated by more than 60% while healthy cells were minimally affected. The modular nature of our design allows targeting novel pairs of extracellular receptors and upregulated transcription factors for increased specific delivery to a variety of applications.

(8h) Alteration of Nanoparticle Pharmacokinetics By Tumor-Induced Modulation of the Immune Response

Kai, M. P., North Carolina State University
Keeler, A., University of North Carolina at Chapel Hill
Luft, J. C., University of North Carolina at Chapel Hill
O'Neal, S., University of North Carolina at Chapel Hill
Zamboni, W.,
DeSimone, J. M., University of North Carolina at Chapel Hill
Originally presented on: 11/3/2013 17:42:00 - 18:00:00

Circulation persistence has been established as a key characteristic of nanoparticles for passively targeting tumors, an important facet of chemotherapeutic delivery. However, many drugs are hindered by toxicity issues. Notably, Cisplatin—approved for use in a wide variety of cancers—has significant renal and neurological toxicities. Incorporating the drug into a nanoparticle increases tumor accumulation while simultaneously reducing side effects caused by systemic exposure.

We have formulated and optimized PEGylated, polymeric hydrogels using the PRINT platform, and complexed cisplatin into the particles. The maximum tolerated dose of the particles is being determined and compared to free drug. Cytotoxicity in several lung carcinoma lines is being investigated, and efficacy studies in mice will be carried out in both subcutaneous and orthotopic non-small cell lung xenografts (A549). Renal toxicity and neuropathy will also be assayed. We anticipate reduced side effects and higher survival probability compared to the current standard of care.

(28a) Synthesis of Hierarchical Carbon Structures With Controlled Pore Size Using Polymer Blends

Peer, M., Pennsylvania State University
Qajar, A., PennState University
Rajagopalan, R., Penn State University
Foley, H. C., Pennsylvania State University
Originally presented on: 11/3/2013 15:35:00 - 15:50:00

Hierarchical porous carbon materials with well connected micro and mesoporosity have shown high performance in different applications including catalysis, adsorption and energy storage. Synthesis of carbonaceous materials with bimodal porosity using a one-pot approach is of great interest due to process simplicity. Using soft templating approach and pluronic F-127 as the structure directing agent, the use of an expensive hard template can be avoided. This will also eliminate the post-synthesis treatments needed o remove the template. In this study, simultaneous polymerization of furfuryl alcohol and phloroglucinol was used to synthesize a polymer blend which resulted in a hierarchical carbon structure after pyrolysis. Polyfurfuryl alcohol is a thermosetting polymer which is typically used to produce microporous carbon with a narrow pore size distribution (~0.5 nm). On the other hand, phloroglucinol polymerization in the presence of HCl and formaldehyde forms a polymeric network which results in an ordered mesoporous carbon structure with the average pore size of 8-9 nm, after pyrolysis. The miscible polymeric blend synthesized by simultaneous polymerization of these two monomers was used as the hierarchical carbon source. It was shown that, pore volume and pore size of the resultant carbon material can be controlled by adjusting the relative monomers composition, HCl (polymerization catalyst) and formaldehyde concentration. Carbons with different mesopore size ranging from 3 nm to 17 nm were synthesized using this simple approach.

(28b) A Self-Consistent Field Study of Diblock Copolymer/Charged Particle System Morphologies for Nanofiltration Membranes

Zhang, B., University of Tennessee
Ye, X., University of Tennessee
Edwards, B. J., University of Tennessee
Originally presented on: 11/3/2013 15:50:00 - 16:05:00

A combination of self-consistent field theory and density functional theory was used to examine the stable, equilibrium morphologies of thin films formed by diblock copolymers with a tethered nanoparticle attached either between the two blocks or at the end of one of the blocks. Both neutral and charged particles were examined, with and without interaction potentials between the particles and the blocks. The phase diagrams of the various systems were constructed, with the aim of determining the most appropriate morphologies for application to nanofiltration membrane technology. In particular, we examined the conditions under which periodic, hexagonal arrays of cylinders could be generated wherein the tethered particles would primarily be located within the interface between the two blocks of the copolymer, thus allowing chemical or thermal degradation to tailor a thin membrane composed of hollow channels lined with charged particles to facilitate both charge and steric exclusion in nanofiltration applications. Key factors influencing these properties were determined to be the particle position along the diblock chain, the interaction potentials of the blocks and particles, the volume fraction of the blocks, and the radii of gyration of the blocks.

(28c) Phase Behavior of Tethered Nanoparticle Telechelics

Marson, R. L., University of Michigan
Phillips, C. L., University of Michigan
Anderson, J. A., University of Michigan
Glotzer, S. C., University of Michigan
Originally presented on: 11/3/2013 16:05:00 - 16:20:00

Tethered Nanoparticles (TNPs) are a unique class of building blocks with assembly behavior analogous to copolymers and surfactants. In all such polymeric systems, complex network morphologies, such as the gyroid, appear for suitable choices of building block architecture. These morphologies are highly sought after for applications ranging from photonics, where band gaps can be tuned based on the length-scales of the network, to microfluidics, for their potential to subdivide space into non-interacting regions. We discuss the presence of such complex morphologies in TNP Telechelics - two nanoparticles of different chemical specificity joined by a polymer tether - simulated via Langevin Dynamics at varying nanoparticle size ratios, polymer tether lengths, temperatures, and packing fractions. By drawing comparisons to other types of systems, particular focus is placed on how introducing a polymer tether as an assembly strategy affects the types of ordered structures that emerge. We further discuss how TNP systems can be adapted to create structures with symmetries not traditionally observed in polymeric systems.

Research supported by the U.S. Department of Energy, Office of Basic Energy Sciences, Division of Materials Sciences and Engineering under Award #DE-FG02-02ER46000.

(28d) How Particle Geometry Controls the Properties and Assembly of Nanoobjects

Walker, D. A., Northwestern University
Grzybowski, B., Northwestern University
Originally presented on: 11/3/2013 16:20:00 - 16:35:00

While there has been significant progress in controlling the size, shape and properties of nanomaterials, their controlled assembly into nanostrucutres remains a significant challenge. To date, many nanoparticles (NPs) of unusual shapes have been reported - but the influence of geometry on the surface chemistry of these particles remains a mystery. We recently found that the pK_a of carboxylic acids bound to the surface of NPs deviates drastically compared to the values for free ligands in solution. This deviation can be described by the coupling the acid/base chemical equilibrium taking place at the particle surface with the surrounding electrostatic environment. Interestingly, by controlling the NP geometry, one can control how charges are distributed on the surface of NPs as well as other chemical properties (such as shifts in redox potentials). The gradients in chemical properties across particle surfaces can be exploited to create higher order structures which would not be possible without the influence of externals forces (such as the use of electronic or magnetic fields). A mixture of experimental and theoretical results will be presented.

(28e) Serum Protein Interactions With Nanoscale Osteogenic Thin Films

Samuel, R. E., Massachusetts Institute of Technology
Johnson, S.,
Cowan, D., HarvBoston Children's Hospital
Originally presented on: 11/3/2013 16:35:00 - 16:50:00

The coating of biomaterials enhances their biocompatibility as implants. Binding serum proteins to the surface greatly facilitates the adhesion and proliferation of hosts cells on implanted biomaterials. We have previously shown that protamine-based layer-by-layer (LbL) polyelectrolyte multilayer (PEM) nanoscale thin film surface coatings of biomaterials support osteoblast adhesion, proliferation, and differentiation during in vitro culture in the presence of fetal bovine serum (FBS). The primary objective of the current studies is to determine the effect of binding specific protein components of FBS to protamine-based PEMs on the adhesion, proliferation, and differentiation of osteoblast cells. LbL nanoscale PEM thin film coatings using charged polyelectrolyte (protamine sulfate and SPS) were made using the spraying method. The linear growth of these PEM thin films was monitored by UV-Vis spectroscopy and the adsorption of bovine serum protein (BSA) and mixed protein constituents of FBS to the surfaces of negatively-charged outer surface (PrS/SPS)n and positively-charged outer surface (PrS/SPS)n (PrS)1 PEMs was evaluated. The net negative outer charge (PrS/SPS)n PEMs demonstrated a higher BSA adsorption compared to (PrS/SPS)n (PrS)1 PEMs. The total UV-Vis absorbance decreased similarly when both formulations of PEMs were incubated in the mixed protein FBS solution. These results suggest that negatively charged BSA may exchange with negatively charged SPS at the (PrS/SPS)n PEM surface. In contrast, incubation of both PEM formulations in FBS may result in the diffusion of both PrS and SPS from the PEM into the bulk FBS solution resulting in alteration in the surface topography and enhanced cell adhesion.

(28f) Chiral Templating of Self-Assembling Nanostructures With Circularly Polarized Light

Yeom, J., University of Michigan
Kotov, N. A., University of Michigan
Originally presented on: 11/3/2013 16:50:00 - 17:05:00

Chiral Templating of Self-Assembling Nanostructures with Circularly Polarized Light

Jihyeon Yeom, Nicholas A. Kotov

ABSTRACT: Chirality is a typical property for natural compounds such as DNA, RNA and proteins existing predominantly as levorotatory enantiomers. However, enantiomeric preference for artificial inorganic nanomaterials was difficult to control without the use of templates based on biological compounds. The fabrication of complex nanostructures with spatial and sequential assembly of desired components via the self-assembly approach requires directed control of system properties. One potential route to this capability is the incorporation of self-assembly induced and controlled by light. Here we show that racemic cadmium telluride nanoparticles (CdTe NPs) selectively self-assemble into chiral nanoribbons under circularly polarized light illuminations. Depending on the helical directions of laser, different light adsorption is induced in CdTe NPs which lead to greater reactivity of a certain chirality. Chiral twisted nanowires can open the door to chiooptical devices including display technologies, physiological properties and pharmacological effects using polarized light.

(28g) Toward Dynamic Control of Magnetic Nanoparticle Monolayers Fabricated By Electrophoretic Deposition: A New Path to Ordered Superlattices?

Dickerson, J. H. II, Brookhaven National Laboratory
Krejci, A. J.,
Darbandi, M.,
Mandal, J.,
Thomas, C. G. W.,
Originally presented on: 11/3/2013 17:05:00 - 17:20:00

Studies of nanoparticle monolayer formation have been motivated by potential applications of the films in magnetic storage, surface-enhanced Raman scattering coatings, ceramic joining, and biosensors, among others. Potential applications of nanoparticle monolayers rely on fabrication techniques that are facile, rapid, and site selective. Three of the most common techniques used to create monolayer films are Langmuir-Blodgett, ligand-mediated self-assembly, and evaporative self-assembly. Electrophoretic deposition (EPD) of nanoparticles s from colloidal suspensions stands as an equally powerful and versatile liquid-phase process to fabricate inexpensively films and coatings of materials. A wide assortment of materials, from macroscopic crystals to nanoparticles and including luminescent, magnetic, and ceramics materials, can be cast by EPD onto field-emanating conducting, semiconducting, and insulating substrates. Many potential commercial and industrial applications of nanoparticle ensembles rely on fabrication techniques that are facile, rapid, and site-selective which create homogenous, densely packed, defect-free thin films. EPD is a technique that meets all of these criteria.

In this presentation, we discuss the development of a new technique for fabricating locally-ordered arrays of NPs, electrophoretic deposition (EPD). EPD is a rapid, safe, and facile method for depositing suspended nanomaterials on a large scale. In one minute of deposition, we can fabricate homogeneous NP monolayers on macroscopic surfaces as large as square centimeters. Colloidal suspensions of magnetic nanoparticles [iron oxide (FeO_x/Fe_3O_4 core/shell, Fe_3O_4, and $CoFe_2O_4$ NPs] have been employed to fabricate nanoparticle monolayers by EPD. The rate of deposition and the total particle deposition were controlled by varying the concentration of nanoparticles and the deposition time, respectively. Using scanning electron microscopy and transmission electron microscopy, we observed tight, hexagonal packing of the nanoparticles, indicating strong particle-particle interactions. Multilayer growth also was achieved and assessed using scanning electron microscopy and atomic force microscopy, revealing a monolayer-by-monolayer growth process. Evidence of ordered monolayer films and the ease of creating multilayers based on said monolayers provide substantial promise for future device applications, ones for which the facile assembly of macroscopic, tightly packed, ordered casts of nanoparticles, i.e., a crystal of nanoparticles, could be transformative.

(28h) Shape-Selective Synthesis of Au Nanoparticles: The Role of PVP

Liu, S. H., Penn State University
Saidi, W. A., University of Pittsburgh
Fichthorn, K., The Pennsylvania State University
Originally presented on: 11/3/2013 17:20:00 - 17:35:00

Structure-directing agents, such as polyvinylpyrrolidone (PVP), play an important role in determining the shapes of metal nanostructures in solution-phase syntheses. It is usually hypothesized that structure-directing molecules bind more strongly to certain crystal facets, which grow at the expense of facets on which they are less strongly bound. In this study, we use dispersion-corrected density functional theory to resolve the role of PVP in the shape-selective synthesis of {111}-faceted Au nanostructures in the solution-phase polyol process. Interestingly, Ag nanostructures grown via the polyol process are {100}- faceted and we discern the differences between the two metals. We calculate binding energies for various conformers of the 2-pyrrolidone ring of PVP on Au(111) and (5x1) Au(100)-hex – the energetically favored reconstruction of the (100) surface. We find a wide variety of binding configurations on the two surfaces, with binding energies that vary from 0.92 to 1.14 eV. On Au(111), there is high and uniform density of PVP binding sites, all with binding energies that fall in the middle of the binding-energy range. On (5 x 1) Au(100)-hex, the binding sites form an interesting striped pattern, with alternating stripes of high, medium, and low binding energies. We discuss the origins of these binding-energy stripes. The stripes of high binding energies are the thinnest and they are widely spaced, making it impossible that a single polymer could have many segments contacting these sites. We also analyze the surface energies of the two surfaces with adsorbed PVP and find that the surface energy of Au(111) is lower than (5x1) Au(100)-hex. Therefore, PVP is more inclined to bind to Au(111) rather than (5x1) Au(100)-hex, causing the formation of nanostructures with (111) facets that is seen experimentally. Differences between the experimentally observed performances of PVP for forming Au and Ag nanostructures stem from the propensity of Au(100) to reconstruct to lower its surface energy, while Ag(100) retains the bulk truncation.

(28i) Mixed-SAM of Aliphatic Thiols On Gold Nanocolloids: Collodial Stability and Reactivity

Kizil, R., Istanbul Technical University

Originally presented on: 11/3/2013 17:35:00 - 17:50:00

Gold nanoparticles with diameters below 20 nm are extensively employed to numerous fields in nanotechnology, spanning from fabrication of nanodevices to photovoltaics and catalysis. The unique optical and electronic properties, and chemical functionality of Au nanoparticles can be tailored by changing the size, shape or aggregation state and most importantly surface chemistry. Chemisorption of alkanethiols on gold surface has long been exploited to form self-assembled monolayes (SAM) on the nanocolloid surface. Chemical functionalization of Au nanocolloids significantly changes the rates of flocculation of the gold dispersion depending on the pH, the chain length and terminal functional group of the alkanethiol.

For the formation of SAM on colloidal gold, in general, a single type of alkanethiol previously dissolved in alcohol is chemisorbed in aqueous gold colloid dispersion. The most widely used molecules for SAM formation are 16-mercaptohexadecaonic acid (MHDA) and 11-mercaptoundecanoic acid (MUDA) mainly due to good association characteristics in water. It is inevitable that formation of SAM from mixed molecules will provide an additional functionality to gold colloids with apparent advantages of precluding steric effects and enabling the particles possess multifunctional reactive terminal groups. However, a potential risk of preparing mixed-SAM on nano-sized colloidal gold is the loss of colloidal stability or fast agglomeration due to loss of surface charge.

This study investigates formation of bifunctional colloidal gold dispersions with a good colloidal stability and pH dependent agglomeration behavior. Citrate reduced Au nanoparticles in aqueous solutions (particle size of nearly 13 nm) was mixed with an equimolar mixture of two different alkanethiols dissolved in ethanol. Before the chemisorption of alkanethiols, Au nanoparticles were treated with tween solution for physical adsorption of the surfactant molecules onto the metallic surface. The first mixed-SAM on Au nanoparticles was prepared by chemisorption of 0.5 mM MHDA and decanthiol (DT) via Au-S covalent link with exposed carboxylic acid and alkyl ends. The second mixed-SAM on gold was prepared using a mixture 0.5 mM MHDA and butyl 3-mercaptopropionate (BMP). A UV-Vis spectrometer was used to monitor capping of the surface of colloidal gold with mixed-SAM. A red shift in the spectrum of bare Au nanoparticles indicates the surface binding of mixed alkanethiols as a consequence of size extension. The colloidal stability was tested for a pH range of 6.0 to 9.0 by 0.5 increments. While the best colloidal stability was observed at pH 7.5 for MHDA-BMP mixed SAM, a different pH response was recorded for the other type of mixed-SAM. The talk will cover the discussion of agglomeration kinetics of multifunctionalized gold colloids and chemical binding of amine containing biotin molecules with the functional groups on the colloids.

(44a) PLGA-Peg-Based pH Sensitive Nanoparticles for Targeted Drug Delivery
Zhou, Z.,
Originally presented on: 11/4/2013 8:30:00 - 8:50:00

PLGA-PEG-Based pH Sensitive Nanoparticles for Targeted Drug Delivery

Zilan Zhou and Joo-Youp Lee

Chemical Engineering Program, School of Energy, Environmental, Biological, and Medical Engineering, University of Cincinnati, Cincinnati, OH 45221-0012

Histidine-poly(lactic-co-glycolic acid)-poly(ethylene glycol)-azide triblock copolymers (His-PLGA-PEG-N_3) were prepared for the development of biocompatible and biodegradable nanoparticles with a controlled drug release feature to be used for cancer therapeutics. This triblock copolymer contains two features: an imidazole ring in the histidine part can be protonated or deprotonated which allows for pH sensitivity, and an azide group at the other end can be coupled with an antibody through linker conjugation. This nanoparticle system is expected to have two features of targeted drug delivery and controlled drug release. ^{1}H-NMR was used to characterize the synthesized polymer and confirm the conjugations between these components. Dynamic light scattering (DLS) was used to measure the nanoparticle sizes and the results show size changes in the nanoparticles by responding to pH. Paclitaxel was employed as a model anticancer drug. An In vitro drug release study results for drug release kinetics will be presented when pH changes from 7 to acidic pH.

(44b) Polyelectrolyte Complex Micelles As Vehicles for miRNA Delivery

Leon Gibbons, L. F., University of Chicago
Perry, S. L., University of Illinois at Urbana-Champaign
Priftis, D., University of Chicago
Kade, M. J., Argonne National Lab
Wong, D., University of Chicago
Tirrell, M., University of Chicago
Originally presented on: 11/4/2013 8:50:00 - 9:10:00

Electrostatically driven micellization is a useful way of assembling functional nanostructures because of the ability to tune intermolecular interactions using pH and salt. Studies have shown that the macroscopic phenomenon of complex coacervation can be achieved at the nanoscale by coupling the polyelectrolyte to a neutral yet hydrophilic block, forming micelles with a coacervate core and a hydrophilic corona. Polypeptides have been used as a model system to study both complex coacervates and coacervate core micelles. Interestingly, the ability of polypeptides to participate in secondary structure formation impedes complex coacervation and affects the structure of the corresponding polyelectrolyte complex micelles. In this work, we characterize the structure and stability of these polypeptide based model micellar systems using static and dynamic light scattering, electron microscopy, circular dichroism and small angle x-ray scattering. Additionally, we create polyelectrolyte complex micelles that contain therapeutically relevant charged molecules such as miRNA and peptides, specifically for the treatment of atherosclerotic lesions and cancer. To increase the efficacy of delivering miRNA, we use triblock copolymers consisting of a polyelectrolyte block, a polyethylene oxide block, and a targeting ligand, creating polyelectrolyte complex micelles with a targeting ligand outside the corona of the micelle. Initial cell studies conducted with the different miRNA containing micelles will be discussed.

(44c) Stable Hydrogel-Anchored Liposomes With Double Stranded DNA Linkers

Dayani, Y., University of Southern California
Malmstadt, N., University of Southern California
Originally presented on: 11/4/2013 9:10:00 - 9:28:00

Due to similarity of liposomes to cell membranes, they are valuable biocompatible nanostructures for handling biological molecules. They have various biomedical applications, and they can afford targeted delivery of therapeutic components, but they usually show insufficient stability, limiting their utility as drug delivery vehicles. Although several approaches have been used for enhancing liposome stability, most of them suffer multi-step processing and lack of biocompatibility.

To improve liposome stability, we have developed a biomimetic lipid-based nanostructure which has potential application in gene and drug delivery. In a self-assembly procedure, the lipid bilayer of a liposome was covalently anchored to a biocompatible poly(ethylene) glycol (PEG) hydrogel core using programmable double-stranded DNA (dsDNA) linkers. These dsDNA linkers are composed of two 15-mer complementary single-stranded DNAs (ssDNAs) modified with cholesterol and vinyl groups. Hybridized dsDNA linkers were added to the lipid mixture used for formation of liposomes, and the lipid mixture was dried. Dry lipid film was rehydrated with PEG-diacrylate (PEG-DA) monomer solution, and small unilamellar liposomes were formed by extrusion. We thermally controlled the formation of hydrogels inside liposomes while preventing bulk gel formation. During the liposome formation process, the DNA linkers spontaneously insert into the lipid bilayer via hydrophilic cholesterol and link the bilayer to the hydrogel by direct incorporation of the vinyl group in a free radical polymerization process.

Size exclusion chromatography (SEC) of intact and surfactant-treated liposomes confirmed the formation of nanogels anchored to the lipid membrane. Polymerized particles had the same column retention time regardless of whether or not they were treated with surfactant. This confirms the formation of liposome-templated nanogels which have same hydrodynamic radius as liposomes. Conjugation of dsDNAs to the nanogel surface was also confirmed by SEC of liposomes containing dye-labeled dsDNAs. Transmission electron microscopy (TEM) showed ~100 nm nanoparticles before and after removal of unanchored lipids confirming the formation of PEG nanogels in liposomes. Dynamic light scattering (DLS) also showed ~120 nm size nanoparticles corresponding to hydrogel-anchored liposomes. The size of surfactant-treated liposomes decreased by ~20 nm due to the removal of unanchored lipids and dsDNA molecules attached to the outer leaflet of the lipid bilayer.

A dsDNA sensitive dye, Hoechst 33342, verified the location of dsDNA linkers on the surface of nanogels. The change in quantum yield of this dye upon specifically binding dsDNA was used to show that DNA was accessible on the surface of nanogels from which the non-anchored lipids had been stripped away by surfactant.

A dye dequenching assay was used to demonstrate the vastly improved stability of hydrogel-anchored nanostructures versus both standard liposomes and liposomes containing a hydrogel core without anchoring. In this assay, a pair of well-known dye and quencher (ANTS/QSY) were co-encapsulated in liposomes incubated in 20% serum at 37°C. Hydrogel-anchored liposomes showed ~22% release over 184 h incubation in serum, while standard liposomes and hydrogel-liposomes with no dsDNA anchors showed faster release of dye with 100% release obtained within 184 h. Although hydogel-liposomes with no anchor showed slower dye release in first 90 h, their overall stability was not comparable with anchored liposomes.

The fabricated lipid-polymer nanostructures can provide robust drug delivery vehicles with advantages of sustained drug release and the potential for targeted delivery. We have deployed these particles in the intracellular delivery of a peptide with anticancer properties. SEC confirmed the successful and efficient encapsulation of fluorescently labeled 17-amino acid peptide within the PEG hydrogel core. We included a pH-sensitive lipid in the membrane to trigger release in and disruption of late-stage endosomes. Release kinetics studies of these pH-sensitive liposomes showed that liposomes are stable at neutral pH while rapidly releasing cargo at acid pH. We traced intracellular delivery of the nanoparticles and their cargo by fluorescent confocal microscopy. In this pathway, cholesterol-dsDNAs arranged on surface of nanogels can facilitate the intracellular delivery of peptide-encapsulated nanogels by disrupting late-stage endosomes.

These stable nanostructures represent a unique platform for protected delivery of oligonocleotides and therapeutic compounds and allow formation of templated nanogels with programmable molecular recognition sites.

(44d) Carbon Nanotubes-Poly(lactide-co-glycolide) Conjugates for the Delivery of Pro-Apoptotic Transcription Factors in Osteosarcoma Cells

Cheng, Q., University of South Carolina Columbia
Jabbarzadeh, E., University of South Carolina
Originally presented on: 11/4/2013 9:28:00 - 9:46:00

Gene therapy approaches have offered a new avenue to treat diseases. Such approaches rely on transferring proteins or genes to program cells into desired fate. Traditionally, liposomes, synthetic polymers and viral vectors have been used as gene carriers. However, the enthusiasm to use liposomes and synthetic polymers has been hampered by the low specificity and transfection efficiency. The application of viral vectors has also been limited due to host immune response. Recently, carbon nanotubes have received significant attention for their promising potential in penetrating cell membranes and transfecting cells effectively. However, major concerns still remain with the toxicity side-effects of carbon nanotubes. The present study introduces a new generation of CNT-based gene vectors based on the use of PLGA to reduce the intrinsic toxicity of carbon nanotubes, and to control the temporal release of genes within cells. Carbon nanotube-PLGA conjugates were fabricated through an atom transfer radical polymerization process. Briefly, oxalyl chloride activated carboxyl carbon nanotubes were linked to PLGA. Caspase-3 (CP3), as a pro-apoptosis protein, was conjugated to PLGA by carbodiimide crosslinking chemistry that included activation of carboxyl group in PLGA by N-hydroxysuccinimide (NHS) and ethyl-dimethylaminopropyl carbodiimide (EDC). Our chemical characterization results demonstrated PLGA wrapping over individual carbon nanotube as confirmed by TEM, UV and XRD. Cellular uptake of CNT-PLGA-protein conjugates was examined through fluorescent microscopy. Images showed CNT-PLGA conjugates were able to penetrate osteosarcoma cell membrane and release caspase-3 with a transfection efficiency of close to 100%. Strikingly, upon the introduction of CNT-PLGA-CP3 conjugates, cell survival significantly diminished to ~20% on day 3, and remained suppressed till day 7. Our results confirmed this effect on cell survival was solely due to the release of CP3 and not CNT-PLGA conjugates (negative control). In contrast, liposomes (positive control) demonstrated an inhibition of only ~50% on day 1 with cells continuing to grow back to fully confluent on day 3. Our results all together demonstrate the successful development and application of CNT-PLGA-protein conjugates to deliver transcription factors within cells. This approach has a broad applicability to a number of research areas including cancer therapeutics, regenerative medicine, cellular imaging and tissue engineering.

(44e) Enhanced Therapeutic Index in Acute Myeloid Leukemia Models By Gold Nanoparticles Directly Passivated With Chemotherapeutics

Egusa, S., The Cleveland Clinic Foundation
Ebrahem, Q., The Cleveland Clinic Foundation
Mahfouz, R. Z., The Cleveland Clinic Foundation
Sauntharajah, Y., The Cleveland Clinic Foundation
Originally presented on: 11/4/2013 9:46:00 - 10:04:00

An enduring, fundamental issue in cancer medicine is that of poor therapeutic index. Treatments typically destroy normal as well as cancer cells, causing toxicity that limits the safety and efficacy. Enabling cancer-selective drug delivery would substantially address this problem, by identifying and exploiting the difference between normal and cancer cells. In this study, we utilize the higher redox stress of cancer cells, and consequently the higher glutathione (an endogenous anti-oxidant) expression, as the key to realize selective delivery. We have developed Au nanoparticles directly passivated with amine-containing drug molecules as drug vehicles, where drug release would be triggered by glutathione via ligand exchange. Au nanoparticles are commonly passivated by thiol-containing molecules via strong thiol-Au bond. Our amine drug-passivated Au nanoparticles relies on amine-Au interaction that is of lower affinity than thiol-Au interaction, and hence drug delivery mediated by *in situ* ligand exchange with bioavailable thiols *e.g.* glutathione could be facilitated compared to the conventional thiol-pasivated Au nanoparticles. Using Au nanoparticles directly passivated with methotrexate (Au:MTX), the mechanism of drug delivery is elucidated *in vitro* and *in vivo*. Enhanced therapeutic index of Au:MTX is demonstrated using a murine xenotransplant model of primary human acute myeloid leukemia (AML), a cancer which is disseminated without large tumor masses: therefore the improved efficacy suggests a mechanism involving cancer-selective drug delivery at the cellular level, as opposed to *e.g.* the enhanced permeability and retention (EPR) effect via leaky neovasculature in solid tumor models. The structural simplicity of the amine drug-Au nanoparticles moreover allows for facile functionalizations and scale-up, and could positively impact a breadth of therapeutics.

(44f) Chromonics Nanoparticles As Novel Drug Carriers: Size Distribution and Control Release Studies

Misra, R., Indian Institute of Technology, New Delhi

Originally presented on: 11/4/2013 10:04:00 - 10:24:00

Chromonic liquid crystals (or chromonics) are formed by the self-organization of aromatic compounds with ionic or hydrophilic groups in aqueous solutions. Due to their self-assembly, they aggregate and stack onto each other forming an inner core region which can be loaded with drug and which in turn can be released on specific time periods in controlled manner. This study reports the process of designing the chromonic nanoparticles and drug release in sustained and controlled manner from them. Folic acid has been considered as a chromonic material in the present study and was used to develop nanoparticles with the help of HPMC polymer, which in turn crosslinked with the help of polyvalent cationic salt. Characterization of these nanoparticles were carried out with help of DLS and SEM technique. It was observed that HPMC polymer dosage and concentration can be used to manipulate and control the size of these nanoparticles.In the respect to release the drug in sustainable manner for prolonged cytotoxicity and constant therapeutic index, folic acid release studies from the nanoparticles were carried out in 0.8% NaCl solution and DMEM growth media. It was observed that both nanoparticle's size and release medium affects the release of folic acid from the nanoparticles. Therefore, chromonic nanoparticles can prove to be a potent drug nanocarrier for delivery of chemotherapeutics drugs, proteins and other compounds.

(44g) Kinetic Analysis of Cytotoxicity With Chromonic Nanoparticles On Cancer Cell Lines: Comparative Analysis With Free Drug

Misra, R., Indian Institute of Technology, New Delhi
Upadhyay, M., Indian Institute of Technology, New Delhi
Originally presented on: 11/4/2013 10:24:00 - 10:42:00

Methotrexate (an anticancer drug) and folic acid (chromonic material) bears structural similarity, folic acid nanoparticles have been designed as mimicking drug molecules.The aim of this study is to compare the cytotoxic effect of nanoparticle-based anticancer drug with that in free drug form on cancer cell lines HepG2, Huh7 and HeLa. Folic acid. We investigated the effect of methotrexate and folic acid using different cancer cell lines. The proliferation of the cell line was found to be inhibited in a dose dependent manner which was detected using MTT assay. It was observed that the cytotoxic effect of the methotrexate and folic acid was higher and in controlled fashion in nanoparticle form than in the case of direct dosage. Kinetic studies revealed the sustained and controlled release patterns of the anticancer drug. Based on the results obtained, it can be envisaged that the in vivo use of nanoparticle based drug delivery and targeting is a promising approach in chemotherapy, overcoming the toxic effects of the known aqueous drugs.

(44h) Designing of Drug Nanocarriers for Chemotherapeutics Delivery: Effect of Size, Shape, Charge, & Surface Chemistry

Upadhyay, M., Indian Institute of Technology, New Delhi
Misra, R., Indian Institute of Technology, New Delhi
Originally presented on: 11/4/2013 10:42:00 - 11:00:00

Nanocarriers have seen a growing interest in drug delivery technologies and the success of the carrier depends on the size, shape and the surface structure and chemistry of the carrier in the context of the delivery pathway. This article develops a map of guildelines for design of nanoparticle based chemotherapy. It reviews the mechanism of delivery in different pathways, physiology and chemistries involved and barriers to transport and delivery of nanocarrier based drugs, specifically for chemotherapeutic drugs. The microenvironment and physiology of a tumor site and its chemical environment is also reviewed, focusing on the impact on delivery. For each of these primary pathways including transdermal, parenteral, mucosal, and oral, nanoparticle size, method of formation and functional characteristics are defined based on the hurdles to transport and mechanisms that aid delivery. Targeting and release mechanisms with these different kinds of nanocarriers are also reviewed. Successful lab and commercially available solutions are compared to these design parameters and a map of nano-carrier design guidelines are developed and presented.

(50a) Crosslinked Anion Exchange Membranes for Redox Flow Batteries

Jung, M. S., Illinois Institute of Technology
Parrondo, J., Illinois Institute of Technology
Arges, C., Illinois Institute of Technology
Ramani, V., Illinois Institute of Technology
Originally presented on: 11/4/2013 8:30:00 - 8:55:00

All-vanadium redox flow batteries (VRFB) have been very popular as they use the same electrochemical species in both electrode chambers and therefore they do not suffer from electrolyte cross contamination. A core component of any RFB is the electrolyte membrane (PEM or AEM) used to separate the positive and negative electrolytes. The key requirements for the separator are good chemical and electrochemical stability, good mechanical stability, high ionic conductivity, and low permeability for the vanadium ions. Nafion® membranes have been the most popular separators in RFBs because of their excellent oxidative stability and high proton conductivity, however, they exhibit high vanadium crossover. Anion exchange membranes (AEMs) are ideal candidates as RFB separators as they have much lower vanadium ion crossover than proton exchange membranes (10-100 times less) due to Donnan exclusion. Moreover, the reduced penetration of VO_2^+ into the AEM minimizes chemical degradation.

We selected poly (2,6-dimethyl 1,4-phenylene) oxide (PPO) based AEMs as battery separators as they possess good conductivity, low water uptake and good mechanical stability. PPO is commercially available, relatively inexpensive, and can be readily modified by attaching bromine to the backbone to induce a phase-separated morphology. PPO can be easily functionalized into an AEM by bromination followed by reaction with a tertiary amine. In this talk, we will discuss the effect of polymer crosslinking on the conductivity, water uptake, vanadium (IV) permeability, ultimate tensile stress, glass transition temperature and storage modulus of PPO- based AEMs. Polymer crosslinking has been performed by the reaction of brominated PPO (Br-PPO) with tetramethylhexanediamine (TMHDA) in liquid phase. Non crosslinked AEMs have been prepared by reaction of the Br-PPO with quinuclidine (PPO-$ABCO^+$) and trimethylamine (PPO-TMA^+). Nuclear Magnetic Resonance techniques [^{1}H-NMR, ^{13}C-NMR, correlation spectroscopy (COSY), and ^{1}H-^{13}C heteronuclear multiple quantum correlation (HMQC) spectroscopy] have been used to investigate the polymer structure before and after exposure to VO_2^+ solution (over time) and to identify any possible degradation products and mechanisms. Crosslinked membranes will be tested (and compared against Nafion® and non crosslinked AEMs) in a vanadium redox flow battery to evaluate its performance and durability. Postmortem analysis of the membranes via NMR, conductivity/IEC measurements, and mechanical testing will be performed to unearth key degradation modes.

(50b) Challenges to Implementation of High-Rate Energy Storage in Military Applications

Heinzel, J., Naval Sea Systems Command - Philadelphia
Ostanek, J.,
Hoffman, D., Office of Naval Research
Peek, I., Naval Sea Systems Command - Philadelphia
Originally presented on: 11/4/2013 8:55:00 - 9:20:00

Future military platforms will require more extensive use of electronic power systems to achieve required performance levels. Energy storage is becoming a key enabler for power systems operating with high levels of transient power requirements and supporting fuel-efficiency operations while maintaining capability. High density energy storage which operates continuously at high charge and discharge rates to support these loads under the combination of finite space, high power and energy requirements, stringent safety characteristics and wide upper temperature ranges drive the requirement for advancement in a multiphysical mix of electrical, chemical, thermal, safety and controls elements, which must be considered and have development performed to enable widespread implementation. These will be discussed from the perspective of requirements and considerations for system design and engineering for applications associated with high rate support of transient loads, power quality and optimization of generator and electrical architecture dynamics.

(50c) Novel Pt/TixM1-xO2 Nanocatalysts for Fuel Cells

Hwang, B. J., National Taiwan University of Science and Technology
Ho, V. T. T.,
Pan, C. J., National Taiwan University of Science and Technology
Su, W. N.,
Originally presented on: 11/4/2013 9:20:00 - 9:45:00

This work presents a new approach by exploring robust non-carbon $Ti_xM_{1-x}O_2$ as a novel functionalized co-catalytic support for Pt. This new approach is based on the novel nanostructure of the $Ti_xM_{1-x}O_2$ supports which holds strong metal-support interactions (SMSI) between $Ti_xM_{1-x}O_2$ and Pt that can not only modify surface electronic structure of Pt, evidenced from a shift in the d-band centre of the surface Pt atoms but also provide dual reaction sites for methanol oxidation. Furthermore, another benefit of $Ti_xM_{1-x}O_2$ is the extremely high stability of Pt/ v during potential cycling, which is attributable to the SMSI between Pt and $Ti_xM_{1-x}O_2$. Interestingly, $Ti_xM_{1-x}O_2$ can be fabricated as a much thinner catalyst layer, resulting in significantly improved the mass transport kinetics and performance of the resulting membrane-electrode-assembly (MEA).The new approach presented in this work opens a reliable path to the discovery of advanced concepts that may lead to the design of new catalyst materials that can replace the traditional catalytic structures and motivate further research in this field.

(50d) A Tunable Hierarchical Porous Carbon Structure for Studying Electrochemical Energy Conversion

Chan, K. Y., The University of Hong Kong

Yang, C.,

Originally presented on: 11/4/2013 9:45:00 - 10:10:00

A Tunable Hierarchical Porous Carbon Structure for Studying Electrochemical Energy Conversion

Chunzhen YANG and Kwong-Yu CHAN*, Department of Chemistry, The University of Hong Kong, Pokfulam Road, Hong Kong

Abstract

A well-defined porous carbon structure with tunable parameters provides a platform for systematic study of transport and confinement effects on electrochemical conversion. Synthesized via silica template techniques, the hierarchical structure has a hollow core within a mesoporous shell (HCMS) with uniform pore size, shell thickness, hollow core diameter, as shown in Fig. 1.

By varying the synthesis parameters, the structural parameters can be tuned to study

1) combined effects of surface area, mesopore size, macropore size, and hierarchy of porosity on ionic transport and adsorption in electrochemical capacitance behavior;

2) how distribution of platinum or other electrocatalyst nanoparticles in the porous structure affect catalytic faradaic reactions;

3) how distribution of mixed metal nanoparticles such as PtRu and varying of Pt and Ru content in the porous structure affects faradaic reactions such as methanol oxidation; and

4) the effectiveness of a mathematic model in analyzing electrochemical conversion in porous structures.

Electrochemical capacitance is studied with a series with meso-shell thickness stepwise increased from 0, 25, 50 to 100 nm while keeping an identical 330 nm hollow core and mesopore of 3.9 nm.[1] A thicker shell has a higher surface area with a proportional increase of electrochemical capacitance which however, can only be fully realized at low scan rates/currents. At high currents, ionic transport limits the electrochemical capacitance of a thick mesoporous shell. Electrochemical impedance spectra (EIS) obtained on the family of HCMS carbon structures were fitted to equivalence circuits described by models of network structures.[2] The results reveal the need to match the AC frequency with the characteristic time constant of a structure for optimum active and reactive power. For fuel-cell electrode conversion, the distribution of Pt and PtRu nanoparticles in a mesoporous carbon can vary in different synthetic protocol. Uniform Pt particle distribution can be achieved with a CPDP method [3] as shown in Fig. 2. Different electrocatalysts with varying Pt distribution can be compared for activity and utilization in methanol oxidation. The particles dispersed into a thicker shell are more stable and give better performance over time, as shown in voltage cycling and extended methanol oxidation.

SHAPE * MERGEFORMAT

Fig. 1. A hollow core mesoporous shell (HCMS) carbon structure with three varying geometrical parameters: pore diameter (d_p), shell thickness (S), and diameter of core (C).

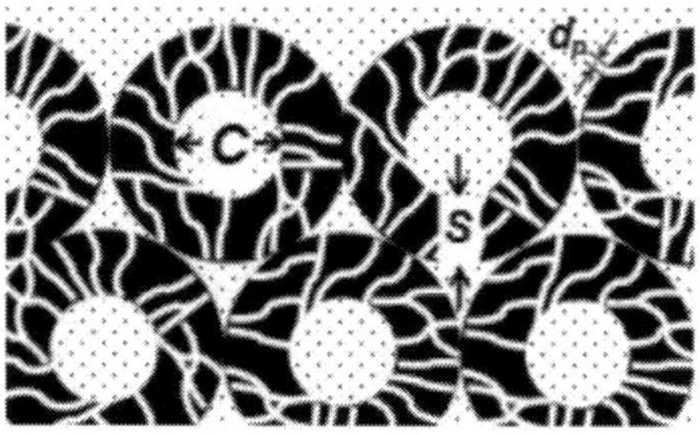

Fig. 1. A hollow core mesoporous shell (HCMS) carbon structure with three varying geometrical parameters: pore diameter (d_p), shell thickness (S), and diameter of core (C).

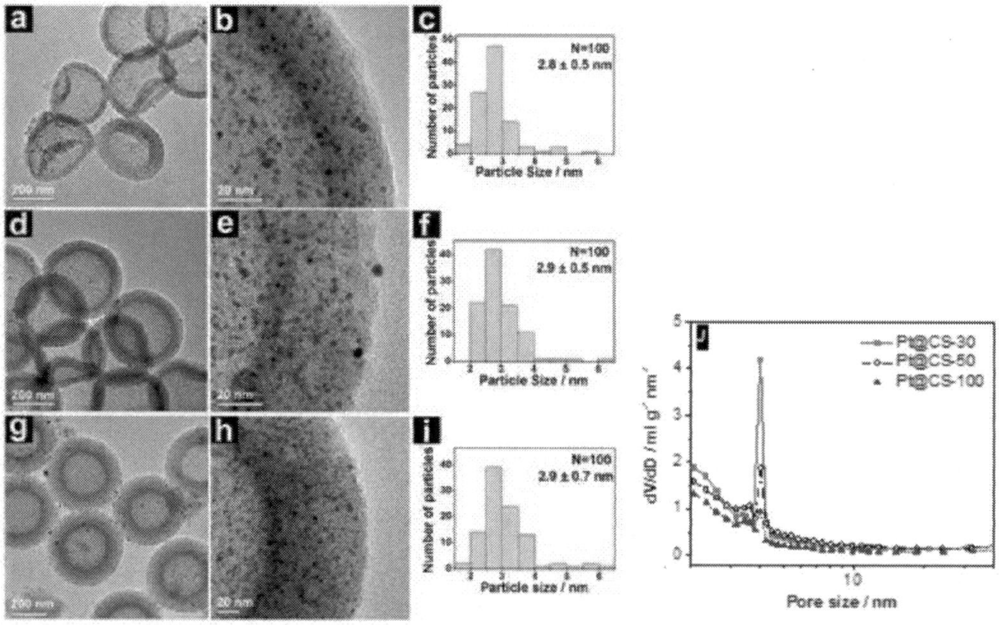

SHAPE * MERGEFORMAT

Fig. 2 TEM images of Pt nanoparticles loaded in mesoporous carbons of same hollow core diameter, *C* = 330 nm but different Shell thickness (a)-(c) Pt@CS-30, *S*=30 nm ; (d)-(f) Pt@CS-50, *S*=50 nm ; (g)-(i) Pt@CS-100, *S*=100 nm; (j) pore size distribution of the porous structures loaded with Pt.

Fig. 2 TEM images of Pt nanoparticles loaded in mesoporous carbons of same hollow core diameter, *C*=330 nm but different Shell thickness (a)-(c) Pt@CS-30, *S*=30 nm ; (d)-(f) Pt@CS-50, *S*=50 nm ; (g)-(i) Pt@CS-100, *S*=100 nm; (j) pore size distribution of the porous structures loaded with Pt.

References:

[1]. F. Li, M. Morris and K.Y. Chan*, ?Electrochemical Capacitance and Ionic Transport in the Mesoporous Shell of a Hierarchical Porous Core-Shell Carbon Structure?, *J. Mater. Chem.* 21 (2011) 8880-8886.

[2] Chunzhen Yang, Chi-Ying Vanessa Li, Fujun Li, Kwong-Yu Chan*, ?Complex Impedance with Transmission Line Model and Complex Capacitance Analysis of Ion Transport and Accumulation in Hierarchical Core-Shell Porous Carbons?, *J. Electrochem. Soc.* (2013) to appear.

[3] F. Li, K.Y. Chan*, and H. Yung, ?Carbonization over PFA-Protected Dispersed Platinum: An Effective Route to Synthesize High Performance Mesoporous-Carbon Supported Platinum?, *J. Mater. Chem.* 21 (2011) 12139-12144.

(50e) High-Performance Electrodes for Hydroxide Exchange Membrane Fuel Cells

Kaspar, R. B., University of Delaware
Letterio, M., University of Delaware
Jensen, K., University of Delaware
Gu, S., University of Delaware
Yan, Y., University of Delaware
Originally presented on: 11/4/2013 10:10:00 - 10:35:00

Hydroxide exchange membrane fuel cells (HEMFCs) are an attractive alternative to conventional proton exchange membrane fuel cells (PEMFCs) because they are compatible with non-precious-metal catalysts. However, HEMFC performance is lower than the PEMFC benchmark, in part because HEMFCs continue to borrow electrode architectures that were originally designed for PEMFCs. In PEMFCs, water is produced at the cathode, so hydrophobic gas diffusion layers (GDLs) are employed to control flooding. But hydrophobic GDLs may not be necessary for HEMFC cathodes, in which water is consumed instead of produced, reversing the direction of water transport in the cell. This work demonstrates that the performance and durability of membrane-electrode assemblies (MEAs) can be improved by tailoring electrode design to reflect the unique kinetics and water transport behavior of HEMFCs.

(50f) Capacitive Mixing in Porous Electrodes for Energy Production

Gabitto, J. F., PVAMU
Tsouris, C., Oak Ridge National Laboratory
Originally presented on: 11/4/2013 10:35:00 - 11:00:00

Energy obtained by salinity-gradients (blue energy) is a promising renewable energy source. Mixing of a high-concentration saline solution, such as seawater, with fresh water (river water) to produce a brackish solution dissipates more than 2.2 MJ of free energy per m^3 of fresh water. The charging-discharging cycle inside the pores of the solid electrodes is strongly influenced by the pore size distribution. Macroporous materials behave like standard plate capacitors, while mesoporous materials behave like cylindrical capacitors. The goal of this project was to optimize the recovery efficiency of free energy change by using tailored-design carbon electrodes. Morphology, pore size, and carbon electrode composition were chosen to optimize the charge-discharge steps of the reversible cycle. The physicochemical processes inside the electrode and the corresponding effects on the reversible energy cycle were investigated. A proposed energy cycle comprises a charging step, a discharging step and two switching steps that replace a saltwater solution inside the pores by a fresh water solution and vice versa. Equations derived using the volume averaging method have been employed to simulate the charging-discharging cycle in homogeneous porous materials of a macropore size distribution. Ionic transport by diffusion and electrical double layer expansion processes were considered to simulate the two alternating steps. Improved energy production was achieved by optimizing the duration of each step and variation of the operating variables.

(66a) Biomimetic Enzyme Entrapment Strategies

Naik, R. R., Air Force Research Laboratory
Originally presented on: 11/4/2013 8:30:00 - 8:55:00

Silks are fibrous proteins produced by insects or spiders, and provide structural and protective functions. Silkworm silk from *Bombyx mori* is one of the most characterized silks and consists of the heavy and light chains that form fibers with impressive mechanical strength. Aqueous processing conditions, biocompatibility, and biodegradability of silk fibroin, along with facile chemical modifications, are attractive features of silk that have led its exploitation for a variety of medical and non-medical applications. More recently, renewed emphasis has been place on silk fibroin as a matrix for the entrapment of enzymes. We will present our studies on the entrapment of enzymes using silk, and describe how the unique structure of silk fibroin protein creates an environment that is conducive to the stabilization of entrapped proteins. The water-insoluble films maintain the activity of the entrapped enzymes under a variety of temperature, pH and storage conditions. Silk is robust and malleable, thus increasing the range of processing possibilities. Silk was also used to modify paper substrates for the immobilization of enzymes for paper-based analytical devices. Based on our and other studies, silk is an intriguing material for the immobilization of biomolecules and expands the possibilities for generating multifunctional materials.

(66b) New Enzyme Nanoconjugates for Green Synthesis of Chemicals

Ge, J., Tsinghua University
Lu, D., Tsinghua University
Liu, Z., Tsinghua University
Originally presented on: 11/4/2013 8:55:00 - 9:20:00

Enzyme catalysis with high chemo-, regio-, and stereo-selectivity is appealing for the green synthesis of chemicals, particularly those of complex structures. However and unfortunately, most of the native enzymes that offer high catalytic performances in vivo, display very limited activities in chemical environments, particularly in the presence of organic solvents or at high temperatures. Nanostructured enzyme catalysts that take the advantageous properties of nanomaterial have thus been pursued to overcome above mentioned problems. This presentation summarizes our recent efforts of developing enzyme catalysts for green synthesis of chemicals. Examples include one-step synthesis of chloramphenicol palmitate with substrate-imprinted lipase nanogel, lipase-catalyzed one-step and regioselective synthesis of clindamycin palmitate. One-step synthesis of chloramphenicol palmitate with the imprinted lipase nanogel gave a yield of ~99% and a purity of ~99% within 9 hours. Lipase-catalyzed one-step synthesis of clindamycin palmitate reached a conversion above 90% in 12 hours with a highly regioselective acylation (~100%) at the 2-hydroxyl of clindamycin. Novel nanostructures such as magnetic enzyme nanoparticles and temperature-responsive enzyme nanoconjugates, which facilitate the operation and the recovery of the enzyme catalysts, will also be included.

(66c) PCNA-Mediated Gelation and Immobilization of Multi-Enzyme Complex

Nagamune, T., The University of Tokyo
Watanabe, H.,
Hirakawa, H., The University of Tokyo
Originally presented on: 11/4/2013 9:20:00 - 9:45:00

In bacterial P450 systems, FdXs act as shuttle molecules between FdRs and P450s. Therefore, co-immobilization of FdRs, FdXs and P450s using carries or cross-linking reagents inhibits electron transfer to P450s from NADH through FdRs and FdXs. Here, we overcome the problem by supermolecular assembly of the Proliferating cell nuclear antigen (PCNA), which is a trimeric ring-shaped protein that encircles dsDNA. FdRs, FdXs and P450s were fused at the C-terminus of each PCNA subunit [1] and the three fusion proteins were homodimerized by genetically linking to a homodimeric phosphite dehydrogenase (PTDH) at the N-terminus. An equimolar mixture of the three fusion proteins, PTDH-PCNA1-FdR, PTDH-PCNA2-FdX and PTDH-PCNA3-P450 immediately gelated because the three fusion proteins were connected by the PCNA heterotrimerization. The gel was stabilized by introduction of selective disulfide bonds between PCNA subunits, which was achieved using the Cys mutants of the PCNA subunits [2]. The gel of FdX-FdR-P450 protein complex showed the monooxygenase activity and was available for a repeated batch reaction. Furthermore, because PTDH can regenerate NADH with phosphite, the gel catalyzed the monooxygenation reaction consuming phosphite in the presence of NAD^{+}. This PCNA-mediated gelation, that is, insolubilization is a promising immobilization method for single enzymes as well as multienzyme complexes because each enzyme can completely retain its activity.

References

[1] H. Hirakawa and T. Nagamune, (2010) *ChemBiochem* **11**: 1517-1520.

[2] H. Hirakawa, A. Kakitani and T. Nagamune (2013) *Biotechnol. Bioeng. in press.*

(66d) Engineered Nanomaterials for Environmental and Energy Related Applications
Rayalu, S.,
Originally presented on: 11/4/2013 9:50:00 - 10:15:00

The continuing, rapid worldwide progress in material science and catalysis promises "innovative &cutting edge processes" for improving quality of life and increasing the efficiency of harvesting and utilizing energy resources.Thus to ensure nation's economic competitiveness catalytic chemistry is proving to be an essential prime requisite for making cleaner environment a reality and enhancing sustainability and carbon balance. In order to meet the challenges of these research areas, engineered nanomaterials and biocatalysts are being developed for environmental remediation and hybrid energy systems with novel cross-disciplinary paradigm. Also the tool of biocatalysis is being used for re-engineering and biomimicking the natural processes, preferably, in nature's way. In this connection, wide array of materials including engineered bio-polymers, stabilized enzyme nanoparticles are being synthesized for applications like i) Simultaneous carbon capture and sequestration and ii)CO_2capture and valorization.

i) Simulataneous carbon capture ,valorization and sequestration : CA as ubiquitously found enzyme is involved in gaseous CO_2 sequestration and is being looked as a promising candidate for combating global warming. Efforts are also being made to mimic the reaction for fixation of anthropogenic CO_2 into bi-carbonate using carbonic anhydrase (CA) as a biocatalyst. CA is being employed to accelerate the rate of hydration of CO_2 to form carbonate ions and proton. Presently, carbonate is being precipitated from aqueous solution as calcium carbonate given a suitable saturation of calcium and carbonate ions by addition of appropriate buffer. A unique protocol for the synthesis of stabilized enzyme nanoparticles (SENs) has been developed for CA to stabilize the enzyme activity by encapsulating each enzyme molecule with a hybrid organic/inorganic polymer network using chitosan and silane. Since, SENs are still in nanometer scale and soluble in aqueous buffer, they can be further immobilized into nanoporous materials. As such, this hierarchical approach allows separate control over stabilization and immobilization in a nanoporous host. Also new molecule and functionalized bio-materials are being developed for capture of carbon dioxide with sorption capacity in the range of 100-200mgof CO_2/g of material or molecule. The regeneration of these new molecules are being attempted by using carbonic anhydrase with very encouraging results by exploiting the reversible reaction of CA with bicarbonate to release CO_2.

ii) CO_2 capture and valorization: New hybrid process has been developed for generation of chemical and fuels including syngas, formic acid , methane and hythane by coupling enzyme driven biomimetic carbonation with photocatalysis based chemical looping reaction (CLR). Carbonic anhydrase is being tested for its catalytic interaction with carbon dioxide in presence of suitable catalysts including Titania and appropriate co-catalysts. An innovative approach to generate syngas and other chemicals has been identified by photo catalytically reducing the bicarbonate/carbonate rich stream to carbon monoxide and formaldehyde . The process developed is clean and green by virtue of the fact that it ensures i) Pt and organic donor free system ii) sustained generation of CO, hydrogen and formic acid iii) ensures stability of carbonic anhydrase by flux engineering iv) concomitant avoidance of buffering of the system and v) usage of carbonates as sacrificial donor . The rate of hydrogen and CO evolution in syngas is 124.16µmols/4h and 1.949µmols/4h, respectively .The utilization of CO_2 as a building block in a Pt and donor free system using this new approach of PC-CLR may represent an interesting approach to synthetic methodologies less intensive in carbon and energy. Also, the production of chemicals and fuels instead of concentrated CO_2stream /carbonates as end products during biomimetic carbonation may make the process commercially viable to be adopted by industries emitting carbon dioxide. The engineered methanation process is being developed for generation of methane from calcite as well as bicarbonate with methane evolution rate of 4.3µmol/h .

Engineered materials and processes ensures cleaner and greener and sustainable environment.

References

1. Renu Yadav, Meenal Joshi, Snehal Wanjari, Chandan Prabhu, Swati Kotwal, T. Satyanarayanan, Sadhana Rayalu. Immobilization of Carbonic Anhydrase on Chitosan Stabilized Iron Nanoparticles for the Carbonation Reaction. Water, Air, & Soil Pollution Journal 223(2012) 5345-5352.
2. Renu Yadav, Snehal Wanjari, Chandan Prabhu, Vivek Kumar, Nitin Labhsetwar, T Satyanarayanan, Swati Kotwal, Sadhana Rayalu. Immobilized carbonic anhydrase for the biomimetic carbonation reaction. Energy Fuels 24 (2010) 6198 - 6207.
3. Renu Yadav, Nitin Labhsetwar, Swati Kotwal, Sadhana Rayalu. Single enzyme nanoparticle for biomimetic CO_2 sequestration. Journal of Nanoparticle Research 13 (2011) 263 - 271.
4. Renu Yadav, T Satyanarayanan, Swati Kotwal, Sadhana Rayalu. Enhanced carbonation reaction using chitosan based carbonic anhydrase nanoparticles. Current Science 100 (3) (2011) 520 - 524.
5. S. Wanjari, C. Prabhu, T. Satyanarayana, A. Vinu, S Rayalu. Immobilization of carbonic anhydrase on mesoporous alumina silicate Microporous Mesoporous Mater 160 (2012)151.
6. S. Wanjari, C. Prabhu, N. Labhsetwar, S. Rayalu Biomimetic carbon dioxide sequestration using immobilized biocomposite material J Mol Catal B Enzym 93 (2013)15-22.
7. C. Prabhu, A. Valechha, S. Wanjari, N. Labhsetwar, S. Kotwal, T. Satyanarayanan, S. Rayalu. Carbon compositre beads for immobilization of carbonic anhydrase ,J Mol Catal B Enzym 71(2011) 71-78.
8. Snehal Wanjari, Chandan Prabhu, Renu Yadav, Nitin Labhsetwar, T Satyanarayanan, Sadhana Rayalu. Immobilization of carbonic anhydrase on chitosan beads for carbonation reaction. Journal of Process Biochemistry 46 (2011) 1010-1018.
9. Chandan Prabhu, Snehal Wanjari, Adarsh Puri, Abhishek Bhattacharya, Rajesh Pujari, Renu Yadav, Nitin Labhsetwar, Anjana Sharma, T Satyanarayanan, Sadhana Rayalu. Region specific bacterial carbonic anhydrase for the biomimetic sequestration of carbon dioxide. Energy Fuels 25 (2011) 1327-1332.
10. P. Mirjafari; K. Asghari; N. Mahinpey. Investigating the Application of Enzyme Carbonic Anhydrase for CO_2 Sequestration Purposes. Ind. Eng. Chem. Res. 46 (2007) 921-926.
11. N. Liu; G.M. Bond; A. Abel; B.J. McPherson; J. Stringer. Biomimetic sequestration of CO_2 in carbonate form: Role of produced waters and other brines. Fuel Processing Technology 86 (2005) 1615– 1625.
12. C. Prabhu; S.Wanjari; S. Gawande; S. Das; N. Labhsetwar; S. Kotwal; A.K. Puri; T. Satyanarayana; S. Rayalu. Immobilization of carbonic anhydrase enriched microorganism on biopolymer based materials. J Mol. Catal B: Enz. 60 (2009) 13-21.
13. S. Hosseinkhani; M. Nemat-Gorgani. Partial unfolding of carbonic anhydrase provides a method for its immobilization on hydrophobic adsorbents and protects it against irreversible thermoinactivation.

Enz. Microb. Technol. 33 (2-3) (2003) 179–184.
14. E. Ozdemir. Biomimetic CO_2 sequestration: immobilization of carbonic anhydrase within polyurethane foam. Energy Fuels. 23 (2009) 5725–5730.
15. S. Wanjari; C. Prabhu; R. Yadav; T. Satyanarayana; N. Labhsetwar; S. Rayalu. Immobilization of carbonic anhydrase on chitosan beads for enhanced carbonation reaction. Process Biochem. 46 (2011) 1010-1018.
16. G.M. Bond; G. Egeland; D.K. Brandvold; M.G. Medina; F.A. Simsek; J. Stringer. Enzymatic catalysis and CO_2 sequestration. World Resource Review 11 (1999) 603-619.
17. G.M. Bond; J. Stringer; D.K. Brandvold; F.A. Simsek; G. Medina; G. Egeland. Development of Integrated System for Biomimetic CO_2 Sequestration. Using the Enzyme Carbonic Anhydrase. Energy & Fuels 15 (2001) 309-316.
18. S. Bhattacharya; A. Nayak; M. Schiavone; S.K. Bhattacharya. Solubilization and Concentration of Carbon dioxide: Novel Spray reactors with immobilized Carbonic anhydrase. Biotechnol. Bioeng. 86 (2004) 37-46.
19. G.M. Bond; M.G. Medina; J. Stringer; F.A. Simsek. CO_2 Capture from Coal-Fired Utility Generation Plant Exhausts, and Sequestration by a Biomimetic Route Based on Enzymatic Catalysis (2008).
20. M.C. Trachtenberg. Methods, apparatuses and reactors for gas separation. US Patent Appl. Publ. No. 2007/0004023 (2007).
21. M.C. Trachtenberg. **Enzymes systems for gas processing**. Canadian Patent 2,222,030 (2007).

(66e) Engineering the Self-Assembly of Ultrastable Protein Filaments Into 2D and 3D Multifunctional Nanostructures

Glover, D., University of California
Giger, L., University of California
Reese, H., University of Californi
Clark, D. S., University of California
Originally presented on: 11/4/2013 10:15:00 - 10:30:00

Self-assembling protein templates have enormous potential as biomaterials for the fabrication of multifunctional nanostructures that require precise positioning of individual molecules, such as enzymes, in regular patterns. We are endeavouring to expand the realm of multidimensional protein assembly by generating new proteins that assemble into 2D and 3D shapes of controllable size and symmetry for template-based construction of advanced biomaterials. One of the central protein building blocks enabling these efforts is the g-prefoldin (g-PFD), a chaperone protein isolated from the hyperthermophilic archaeon *Methanocaldococcus jannaschii*. The g-PFD oligomerizes into filaments several microns in length and exhibit remarkable stability up to at least 94°C. The malleable and distinct modular architecture of the g-PFD filament provide an ideal starting point to construct novel protein architectures. Our current aim is to develop a standardized biomolecular construction set comprised of g-PFD filaments of specific size, and two- and three-way connectors that can be assembled into diverse structures of controllable size and shape.

We have demonstrated the creation of a three-way connector containing the g-PFD fused with a trimerization domain that is capable of assembling with filaments into 2D "pinwheel-like" structures, and self-closing loops using two-way connectors. Furthermore, we are able to position nanoparticles at specific locations upon the pinwheel assembly. Based on our initial success to incorporate a two- and three-way connector into the γ-PFD subunit we are working to create tunable binding partners to control the assembly and dimensions of our biomoleculer construction kit. In particular, we used simple but well-understood structural elements such as coiled-coils to create strong hydrophobic interactions between two self-associating helical bundles. As a consequence, two g-PFD subunits containing opposing α-helical stretches (so-called "bait" and "prey" sequences) form a tight heterodimeric coil-coiled bundle. These bait-prey proteins are specific for each other, and are capable of capping filament length. Subsequently, the bait-prey pair will be used to control *de novo* assembly of the three-way connector into various 3D structures. Ultimately, the ability to control the position and orientation of functional molecules at the nanoscale using self-assembling supramolecular protein templates, such as the g-PFD, will greatly expand opportunities for the template-based construction of biomaterials.

(66f) Fabrication of Active Enzyme Fibers Via Electrospinning

Tran, D., University of Texas at Dallas
Balkus, K. J. Jr., The University of Texas at Dallas
Originally presented on: 11/4/2013 10:30:00 - 10:45:00

The use of enzymes for chemical conversion is of great interest when chemo, regio, and stereo specific products are of vital importance. Enzymatic conversions also allows for catalysis at moderate conditions however the major limiting factor in the use of enzymes for large scale operation is the cost of enzymes. This disadvantage can be minimized by extending the catalytic lifetime of the typically short lived enzyme enzymes, usually by immobilization. Electrospinning has been used to immobilize enzymes within nanofiber polymer matrices either through encapsulation or covalent attachment to the fibers. Through the use of concentric electrospinning, we have reported the first enzymatically active enzyme fibers of cellulase. Composites were also generated using the same technique with a coaxial configuration with cellulase in the inner core and a polymer (Nafion) in the outer shell. Fibers provide a higher surface area when compared to protein immobilized on films translating to high active site accessibility and activity while allowing for ease of handling and reuse. The outer shell Nafion helps hydrolyze cellulose which then is broken down to glucose by enzymes in the inner core. These fibers can potentially be incorporated as reactive membranes for the production of biofuels using solid substrates under mild conditions.

(66g) Multi-Enzyme Co-Localization On Nanoparticles Facilitated By DNA Hybridization

Jia, F., Iowa State University
Narasimhan, B., Iowa State University
Mallapragada, S., Iowa State University
Originally presented on: 11/4/2013 10:45:00 - 11:00:00

Multi-enzyme complexes (MECs) in Nature exhibit highly efficient catalytic mechanism(s) in reaction cascades. Researchers have developed several strategies to co-localize multiple enzymes on nano-carriers to improve multi-enzyme catalytic efficiency by mimicking the structure and function of MECs. Numerous studies have indicated that the spatial arrangement and orientation of multiple enzymes in confined space are of paramount importance in facilitating cooperative enzymatic activity in multi-enzyme co-localization. Biomolecule scaffolds based on DNA hybridization are very attractive for multi-enzyme co-localization because of their ability to enable unique effective control of the relative positions of different enzymes. In this work, glucose oxidase (GOX) and horseradish peroxidase (HRP) were co-localized onto polystyrene nanoparticles via DNA hybridization. The rationale behind selecting the particulate geometry in contrast to a planar surface is that the particulate geometry is more suitable for co-localization of multiple layers of enzymes due to lower steric hindrance. Free DNA hybridization and co-localization efficiency were studied using Förster resonance energy transfer (FRET) techniques. Various capture DNA concentrations were compared and optimized. The immobilization of one enzyme was studied initially to investigate the steric hindrance effect. Evidence of co-localization of GOX and HRP was obtained by FRET studies of dye-labeled tag DNAs. Finally, it was demonstrated that co-localizing GOX and HRP via DNA hybridization significantly improved the overall reaction efficiency in comparison to a mixture of immobilized single enzymes. In summary, DNA-directed co-localization of multiple enzymes provides excellent control of the relative positions of the enzymes, thereby mimicking MECs, and leading to high reaction efficiency.

(67a) Invited Paper: From Nanotechnology to Picotechnology: Revolutionizing Medicine

Webster, T. J., Northeastern University

Originally presented on: 11/4/2013 8:30:00 - 9:05:00

Inspired from biological systems, nanotechnology is beginning to revolutionize medicine including improved prevention, diagnosis, and treatment of numerous diseases. This talk will summarize efforts over the past decade that have synthesized novel nanoparticles, nanotubes, and other nanomaterials to improve medicine. Efforts focused on the use of nanomaterials to minimize immune cell interactions, inhibit infection, and increase tissue growth will be especially emphasized. Tissue systems covered will include the nervous system, orthopedics, bladder, cardiovascular, vascular, and the bladder. Due to complications translating in vitro to in vivo results, only in vivo studies will be emphasized here. Materials to be covered will include ceramics, metals, polymers, and composites thereof. Self-assembled nano-chemistries will also be emphasized. As the FDA has now approved several nanomaterials for medical applications, recent results from FDA trials will also be discussed. In summary, this talk will provide the latest information concerning the design and use of numerous nanomaterials in regenerative medicine while highlighting what is necessary for this field to continue to grow.

(67b) Invited Talk: Biodegradable Nanomaterials for Future Medical Implants

Liu, H., University of California at Riverside

Originally presented on: 11/4/2013 9:05:00 - 9:30:00

In today's medical world, biomaterials are widely used in medical implants from coronary stents to orthopedic devices. Current medical implants and devices are mostly made of titanium alloys or stainless steel – permanent materials that often require secondary surgeries for removal. This presentation will highlight recent advances in development of biodegradable implant materials and nanostructured interfaces for medical applications. Biodegradable materials as scaffolds are advantageous for regenerative medicine applications because they can be rationally designed to promote cellular functions when needed and naturally degrade with tissue healing. Results showed that biodegradable nanocomposites significantly prolonged the release of a functional peptide and enhanced human mesenchymal stem cell functions.

(67c) Photothermal Welding of Ruptured Intestinal Tissue Using Plasmonic Nanocomposites

Huang, H. C., Massachusetts General Hospital, Harvard Medical School
Walker, C. R., Arizona State University
Nanda, A., Arizona State University
Ramos, J., Arizona State University
Pushpavanam, K. S., Arizona State University
Rege, K., Arizona State University
Originally presented on: 11/4/2013 9:30:00 - 9:45:00

Each year, over 600,000 patients undergo surgical operations to treat colorectal diseases in the United States (www.sages.org). Anastomotic leakage following standard colorectal anastomosis is commonly associated with life-threatening bacterial infections (incidence rate: 4-17%). The purpose of this study is to demonstrate that plasmonic nanocomposites, generated from gold nanorods and elastin-like polypeptides, can facilitate the photothermal welding of ruptured colorectal tissues. We hypothesized that heat transfer from the light-activated nanocomposite will facilitate the binding of ruptured tissue and result in a fluid-tight seal. We test our hypothesis using *ex-vivo* porcine intestines.

Gold nanorods (GNRs) and cysteine-containing elastin-like polypeptides (CELPs) were synthesized by adapting procedures from literature. Nanocomposites (1 mm diameter, ~250 μm thick) were prepared by two steps: (i) self-assembly of CELPs (2 mg/ml) on GNRs (9.5-190 μg/mL) via gold-thiol bonds to form dispersed GNR-CELP nanoassemblies at 4°C and (ii) irreversible phase separation of nanoassemblies to form solid-phase nanocomposites at 37°C. The optical and rheological properties of the nanocomposites were measured by UV-Vis spectroscopy and rheometer, respectively. The biological properties of the nanocomposites were evaluated using murine fibroblast cells. A titanium sapphire laser pumped by a solid-state laser (Spectra-Physics, Millennia) was employed for welding (800 nm, 20 W/cm^2, 0.4-2.8 kJ) of the ruptured porcine small intestines (Animal Technologies Inc.) in presence and absence of nanocomposites. Tissue tensile strength, leakage pressure, bursting pressure and bacteria (E. coli DH5-alfa) leakage were measured immediately post-photothermal welding.

We have delivered reproducible, stable, light-activable, and biocompatible GNR-CELP nanocomposites; the swelling ratio and the stiffness of the nanocomposite can be modulated by adjusting the GNR content (0-8.7 wt%). We observed that fibroblast cells could proliferate with minimal cytotoxicity (<10% cell death) on top of the nanocomposites up to 72h of culture. Photo-activation of nanocomposites resulted in a tissue temperature increase up to ~60°C, which enhanced the tensile strength and leaking pressure of the ruptured-tissue up to 47% and 78% of its original intact form, respectively.

The plasmonic nanocomposites possess tremendous translational potential in the repair of intestinal and colorectal tissues in several diseases, including cancer. The use of therapeutically relevant cells, including stem cells, and encapsulation of drugs within the nanocomposite to further enhance healing and repair (while preventing infection) will be further investigated and discussed.

*** Formatting Issue - Best Available Paper/Graphic ***

(67d) Nanostructured 3D PLGA Scaffolds For Skin-Tissue Engineering Applications

Karahaliloglu, Z.,

Ercan, B., Northeastern University

Denkbas, E. B.,

Webster, T. J., Northeastern University

Originally presented on: 11/4/2013 9:45:00 - 10:00:00

NANOSTRUCTURED 3D PLGA SCAFFOLDS FOR SKIN-TISSUE ENGINEERING APPLICATIONS

Z. Karahaliloğlu[1], B. Ercan[2], E.B. Denkbaş[3], T. Webster[2]

[1]Hacettepe University, Nanotechnology and Nanomedicine Division, 06800, Beytepe, ANKARA, Turkey

[2]Northeastern University, Chemical Engineering Department, 02115, Boston, MA, USA

[3]Hacettepe University, Chemistry Department, Biochemistry Division, 06800, Beytepe, ANKARA, Turkey

Background and aims: Skin is the largest organ in the body and performs various functions (i.e., maintaining temperature, waste exchange, etc.) critical for our survival. When skin-tissue is damaged due to a burn or a trauma, natural skin grafts in the form of a xenograft, allograft or autograft have been providing optimal results for wound healing applications. However, the use of graft tissue comes with numerous challenges including limited donor site for severe skin trauma patients, potential for immune response for allographs and xenographs, requirement of a surgical procedure, and high cost[1,2]. Although some of the commercially available synthetic skin substitutes regenerate some of the functions of damaged epidermal tissue, there is no currently-used material that can replace the functional complexity of natural skin. To address this challenge, we are proposing the use of nanofeatured poly(lactic-*co*-glycolic acid) (PLGA) as an alternative skin substitute. PLGA is an FDA approved biodegradable and a biocompatible copolymer used in various medical applications such as resorbable sutures, surgical clips and drug delivery systems[3]. Furthermore, there is plenty of in vitro and vivo study showing that nanostructured PLGA enhances cellular adhesion, proliferation and long-term functions[4,5]. In this research study, nanostructured 3D PLGA scaffolds were created and investigated as an alternative material for skin-tissue engineering and its regenerative performance was compared to a nanosmooth PLGA scaffold in vitro.

Methods: In this work, PLGA scaffolds were prepared by solvent casting and salt leaching process as reported previously[4]. Briefly, 0.5 g of PLGA (PLGA; 50:50 wt%; Polysciences, Inc.) were dissolved in 6 ml of chloroform in 45 °C for 30 min. 4.5 g of sodium chloride salt (NaCl, up to 250 μm, Sigma) were used to introduce pores into the substrate. NaCl particles were added to the dissolved polymer solution. After mixing and sonicating for 20 min, the mixture was poured into a Teflon petri dish and dried overnight at room temperature. To evaporate of all chloroform, the dried polymer was placed into a vacuum for 48 h. The scaffold was soaked in deionized water for 3 days. To understand the effect of PLGA nanorough surfaces on human epidermal keratinocyte (İnvitrogen C0015C) and fibroblast cell lines (ATCC CCL-110), PLGA scaffolds were soaked in a NaOH solution prepared at different concentrations (0.1, 0.5 and 1 N) and for various times (5, 10, 15 and 20 minutes). Characterization of the treated- and untreated PLGA scaffolds was performed by using Scanning Electron Microscopy (SEM). Cell adhesion to the scaffold and cell proliferation was assessed using human fibroblast and human keratinocyte cell lines. Experiments were completed in triplicate and repeated at least three times.

Results: SEM images of untreated and treated PLGA scaffolds exhibited a significant change in surface roughness. Treated PLGA had nano topographical surface properties while untreated PLGA scaffolds showed a smooth surface. Our preliminary results (not shown here) indicate that nanofeatured PLGA surfaces promoted more skin cell adhesion compared to its nanosmooth counterpart.

Conclusion: This study confirmed that PLGA scaffolds altered to have a nanorough surface topography by chemical etching technique should be further studied for improved skin-tissue engineering applications.

a

b

d

c

e

f

g

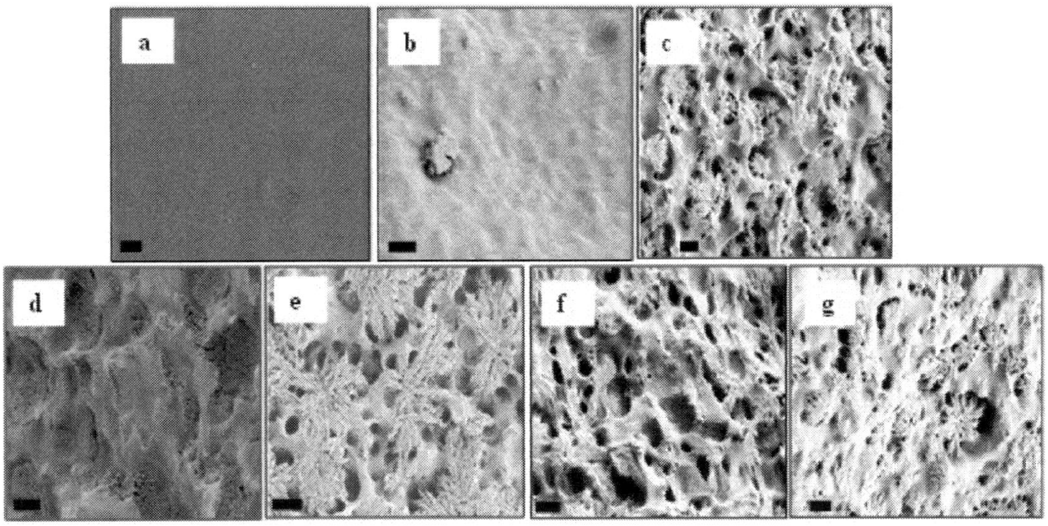

Figure 1. SEM images of (a) untreated and (b) 0.1 N NaOH-treated for 10 min, (c) 0.5 N NaOH-treated for 10 min, (d) 1 N NaOH-treated for 5 min, (e) 1 N NaOH-treated for 10 min, (f) 1 N NaOH-treated for 15 min

*** Formatting Issue - Best Available Paper/Graphic ***

and (g) 1 N NaOH-treated for 20 min PLGA scaffolds. Scale bars= 200 nm.

Acknowledgements: The authors would like to thank Daniel Hickey and Linlin Sun for their assistance. The authors would also like to thank Northeastern University and Hacettepe University for funding.

References

1. Sheila, M. Nature, 2007, 445, 874¨C880.

2. Marler, J. J.; Upton, J.; Langer, R.; Vacanti, J. P. Advance Drug Delivery Rev. 1998, 33, 165-182.

3. Kumbar, S.G.; Nukavarapu, S.P.; James, R;, Nair, L.S.; Laurencin, C.T. Biomaterials, 2008, 29, 4100¨C4107.

4. Smith, L.L.; Niziolek, P.J.; Haberstroh, K.M.; Nauman, E.A.; Webster, T.J. Int J Nanomedicine, 2007, 2(3), 383¨C388.

5. Park, G.E.; Pattison, M.A.; Park, K.; Webster, T.J. Biomaterials, 2005, 26(16), 3075.

(67e) Using Magnesium Oxide Nanoparticles to Improve Inhomogeneous Soft and Hard Tissue Growth

Hickey, D. J.,

Sun, L., Northeastern University

Ercan, B., Northeastern University

Webster, T. J., Northeastern University

Originally presented on: 11/4/2013 10:00:00 - 10:15:00

Introduction: Where ligaments meet bone, there exists a transitional region of inhomogeneous tissue that is graded from hard, highly mineralized fibrocartilage at the bone interface to un-mineralized soft tissue at the ligament. This structure, called the enthesis, disperses stress concentrations that arise due to the vastly different mechanical properties of bone and ligaments. However, in the event that a ligament is injured and requires surgery, the tendon-to-bone insertion site (TBI) must be destroyed and cannot easily be restored due to avascularity in the region and because of consequent slow healing times within the inhomogeneous tissue compared to healing within uniform tissues. It is believed that this loss of enthesis functionality following joint reconstructive surgery is a leading cause of high failure rates for such surgeries (5-25% failure rates for ACL surgery) [1]. Therefore, there is considerable interest in the development of a nanostructured biomaterial that is capable of regenerating the TBI.

The current study developed and characterized for the first time poly (l-lactic acid) (PLLA) mineralized with nanoparticles of hydroxyapatite (HA) and magnesium oxide (MgO) as a tissue scaffold to direct the regeneration of the enthesis. HA constitutes about 60% of the dry mass of bone and has been used previously as a successful material for bone grafts. Magnesium is an essential mineral that regulates HA crystal size and density, and Weng and Webster have shown that nano-rough magnesium increased bone cell density by up to three-fold when compared to bulk magnesium [2]. Presently, the ability of materials to promote tissue growth at the TBI was characterized via cell adhesion and proliferation experiments with fibroblasts and osteoblasts. Further characterization of materials was performed using scanning electron microscopy (SEM), contact angle tests, and mechanical load testing.

Methods: Hydroxyapatite (HA) precipitates were synthesized by a wet chemistry process and then treated hydrothermally to give nanoparticles of sizes consistent with HA crystals observed in natural bone, as reported by Zhang *et al.* [3]. Poly (l-lactic acid) (PLLA) (MW=50,000), (Polysciences, Warrington, PA) was placed in 20 mL scintillation vials and dissolved in 10 mL of chloroform (Sigma Aldrich, St. Louis, MO) to reach 3 wt% PLLA in chloroform. HA nanoparticles and MgO nanoparticles (particle diameters of 20 nm, US Research Nanomaterials Inc., Houston, TX) were added to separate vials in concentrations ranging from 5-20 wt% in PLLA. Samples were heated to 55 °C and sonicated for 1 hour, then poured into 60 mm diameter pyrex petri dishes (Sigma Aldrich, St. Louis, MO). Samples were heated at 60 °C for 1.5 hours to evaporate excess chloroform, producing polymer films about 0.2 mm thick which were then allowed to sit overnight before being cut into strips for further study.

Samples were cut into 1 cm x 3 cm rectangular strips for tensile testing with a uniaxial tensile tester equipped with material analysis software (ADMET, Norwood, MA). A 10 lb. load cell was used and small pieces of paper towel were attached to the tensile grips to prevent the polymer composites from breaking from the force of the grips. This set-up was used to obtain the elastic modulus, material elongation, and maximum load endured for each sample. Cell adhesion and proliferation tests were performed by seeding

3500 cells/cm^2 of fibroblasts and osteoblasts (American Type Culture Collection, Manassas, VA) onto 1-cm^2 samples and culturing for 4, 24, 72, and 120 hours under standard conditions. Cell numbers were quantified using MTT assays (Sigma Aldrich, St. Louis, MO). Experiments were conducted in quadruplet and repeated three times. Data was analyzed using Student's t-tests.

Results: Results indicated for the first time that MgO nanoparticles in plain PLLA or PLLA/HA composites significantly increased osteoblast and fibroblast adhesion on PLLA (Figure 1). SEM images showed considerable differences in nanoscale surface topography between samples. However, preliminary contact angle results demonstrated no differences in wettability among all samples. Mechanical tensile testing revealed that the addition of HA nanoparticles to plain PLLA hardened the polymer, reducing the material elongation and increasing its elastic modulus. These results were exaggerated for PLLA mineralized with MgO nanoparticles. The maximum load that the polymer was capable of withstanding was slightly increased for MgO mineralized samples.

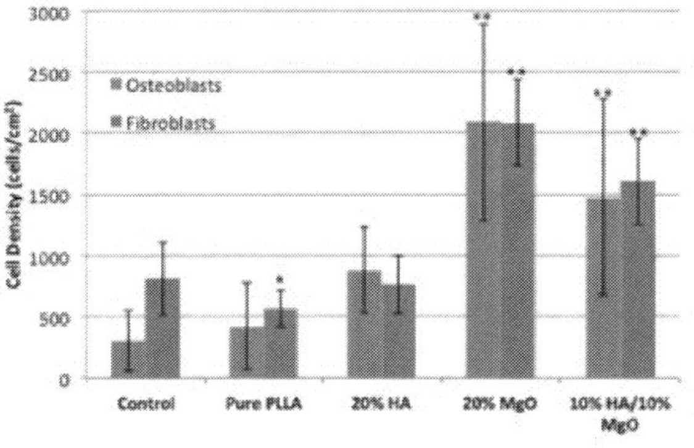

Figure 1. Adhesion of fibroblasts and osteoblasts on pure PLLA and PLLA mineralized with nanoparticles (20% HA, 20% MgO, and 10% HA/10% MgO). Control is empty cell culture well. Data represents mean ± SD. n = 4. % = weight%. *$P<0.05$, **$P<0.005$ compared to controls.

Conclusions: Here, MgO nanoparticles have been shown to increase osteoblast and fibroblast cell adhesion on PLLA, and should therefore be further investigated as a material to promote bone tissue growth at one end of the tendon-to-bone insertion site, and fibrous tissue growth at the other end of the TBI.

Acknowledgements: The authors thank Robert Eagan, William Fowle, Scott McNamara, David McKee, the Northeastern University Department of Chemical Engineering for facilities and funding, and the NSF-IGERT grant #0965843 for funding.

References

[1] Smith L, Thomopoulos S, *US Muscoskel. Rev.*, 6 (2011), 11-5.

[2] Weng L, Webster TJ, *Nanotechnology.* 23 (2012), 485105.

[3] Zhang *et al.*, *Int. Journal of Nanomedicine,* 3 (2008), 323-34.

(67f) Tunable Assembly of Biomaterials for Bone Tissue Engineering

Shah, N., David H. Koch Institute for Integrative Cancer Research at MIT
Hyder, M. N., Massachusetts Institute of Technology
Seeherman, H., Restituo LLC.
Padera, R., Brigham and Women's Hospital
Spector, M., Massachusetts Institute of Technology
Hammond, P. T., Massachusetts Institute of Technology
Originally presented on: 11/4/2013 10:15:00 - 10:30:00

Biomedical devices such as implants for total joint replacements and grafts to fill large bone defects replace native tissue, restore mobility, and alleviate morbidity. However, in some cases, these devices may fail by post-surgical loosening or stress shielding due to a modulus mismatch between the bone and implant. To attenuate these incidences, endogenous progenitor cells can be recruited to regenerate tissue to (i) promote the direct integration of the implant with the host bone tissue and (ii) repair critical size defects. We have developed modular nanostructured coatings containing osteogenic materials, including multiple biological growth factors and inorganic bone minerals, to mimic aspects of the natural bone wound healing cascade. This enabled the formation of new, functional bone tissue which bonded a permanent prosthesis to the native tissue with superior strength and rapidly filled large defects and restored tissue strength.

We used a layer-by-layer approach to coat implant surfaces with individual layers of materials with nanoscale control over their deposition. Angiogenic growth factor VEGF and osteoinductive growth factor BMP-2 were incorporated in gradient degradable coatings, such that VEGF released rapidly during the early acute phase of wound healing, along with slower releasing BMP-2 for longer term osteogenic effects. In vivo, coated polycaprolactone/β-tricalcium phosphate scaffolds were implanted in the rat quadriceps muscle pocket. Dual growth factor from the scaffolds surface resulted in superior de novo ectopic bone formation compared to single growth factor release as measured by μCT and histology. New bone formation was restricted to the scaffold. In a complementary study, surrogate orthopedic implants were biologically fixed to the bone tissue by recapitulating the bone matrix environment with films containing a combination of hydroxyapatite (HAP) and BMP-2. The films contained a permanent osteoconductive HAP base layer and a degradable BMP-2 releasing top layer. In vivo release of BMP-2 was monitored using fluorescence and ELISA of the marrow aspirates. In combination, HAP and BMP-2 accelerated the differentiation of human mesenchymal stem cells in vitro with concurrent upregulation of osteogenic markers. We observed that the mode and rate of growth factor release significantly influenced bone regeneration. When implanted in the rat tibia, implants bonded to the parent bone via maturing trabecular bone, with a temporal increase in the pull-out force when coated with a combination of these bone matrix materials. Long term studies indicated that the implant was stable over at least 18 months with minimal foreign body response. The interfacial tesile strength was 2-3 higher than clinically used bioactive bone cements.

These results demonstrate that surface modification alone can be used to induce tissue regeneration. The local availability of precise amounts of regulators is as a potent tool in tissue engineering and controlling the amount and duration is critical. The layer-by-layer technique is scalable, adaptable and highly tunable which makes it promising for clinical use. Precisely assembling materials using this technique in modular nanoscale assemblies can result in a synergistic response which accelerates bone regeneration and enables implants to last their natural lifetime and scaffolds to introduce rapid bone healing.

This work was funded by the NIH (R01 AG029601) and supported in part by the NCI (P30 CA014051) and the U.S. Army Research Office (under contract W911NF-07-D-0004).

(67g) Sibling-Collagen Scaffolds As a Novel Bone Tissue Engineering Material

Zurick, K., University of Missouri - Columbia
Bernards, M. T., University of Missouri, Columbia
Originally presented on: 11/4/2013 10:30:00 - 10:45:00

Native bone tissue is comprised of a mixture of collagen, non-collagenous proteins, and hydroxyapatite (HA). The SIBLING (small integrin binding, N-linked glycoprotein) family of proteins is the primary group of non-collagenous proteins found in bone tissue, several of which have been identified ahead of the mineralized front of developing bone. By replicating the native binding interactions between collagen and the SIBLING family of proteins, it is believed that a new type of acellular bone tissue engineering scaffold can be developed. In this work, we show preliminary studies which suggest that both mineralization and cell growth can be enhanced by the addition of SIBLINGs to a collagen scaffold.

The mineralization induction capabilities of three of the SIBLING members, bone sialoprotein (BSP), osteopontin (OPN), and the calcium binding subdomain of dentin sialophosphoprotein, dentin phosphoprotein (DPP), were directly compared on a biomimetic collagen substrate. ^{125}I radiolabeled adsorption isotherms were constructed for each of the SIBLING proteins on the collagen substrate and were then used to ensure equivalent amounts of were bound to each substrate in subsequent experiments. Collagen substrates were then prepared with either BSP, DPP, or OPN and exposed to a simulated body fluid for 5, 10, and 24 hours. The resulting mineral morphology was assessed with both atomic force microscopy and colorimetric assays for Ca^{2+} and PO_4^{3-}. Mineralization was seen in the presence of all three proteins, but only minerals formed in the presence of DPP were morphologically similar to those seen in developing bone tissue.

The accessibility of the cell binding RGD domain of DPP when bound to collagen was explored and compared to previous studies with BSP and OPN. ^{125}I adsorption isotherms were developed for DPP adsorbing to both tissue culture polystyrene (TCPS) and collagen coated TCPS to ensure equal amounts of DPP were bound to each substrate. An MC3T3-E1 cell adhesion assay was performed on TCPS and collagen-coated TCPS in the presence of identical amounts of adsorbed DPP. A statistically significant difference was seen between cell binding on collagen coated TCPS and TCPS with bound DPP, and the cells exhibited a preference for plain TCPS over collagen coated TCPS. This suggests that the RGD sequence of DPP is less accessible when it is bound to collagen. A cell inhibition assay was performed to ensure that the results seen were due to interactions between the RGD domain of DPP and the MC3T3-E1 cell integrins.

The effects of SIBLINGs on the early stages of collagen fibrillogenesis were also investigated. Fibrillogenesis assays were performed in a 96-well plate using a solution of collagen, a fibrillogenesis buffer, and one of three SIBLINGs. Absorbance measurements were performed at 400 nm to quantify both the rate and extent of collagen fibrillogenesis in the presence of either BSP, DPP, or OPN.

This work has shown that a collagen-SIBLING scaffold is capable of both promoting cell growth and mineralization that is morphologically similar to that found in *in vivo* developing bone tissue, allowing for an acellular bone tissue scaffold. By replicating the native binding interactions between collagen and the SIBLING family of proteins it is thought that a novel type of biomimetic bone scaffold can be developed without the need of cells.

(67h) Reduced Adhesion of Macrophages On Nanotubular PDMS Molds for Catheter Applications

Liu, L., Northeastern University

Ercan, B., Northeastern University

Sun, L., Northeastern University

Webster, T. J., Northeastern University

Originally presented on: 11/4/2013 10:45:00 - 11:00:00

Introduction: Polydimethylsiloxane (PDMS) has been used for short- and long-term catheters. However, PDMS shunt tubing has been scrutinized recently because of their extremely high failure rates. While there are many reasons why PDMS shunt systems fail, one is tissue occlusion of the ventricular catheter, most often caused by macrophages. It is hypothesized that nanotextured and nanotubular surfaces can be carefully manipulated to inhibit immune cell (specifically, macrophages) and improve the fibroblast adhesion due to their unique surface energy properties which have the ability to control initial protein absorption and subsequent cell behavior. The objective of this in vitro study was to create polydimethylsiloxane (PDMS) molds of titanium (Ti) anodized to possess nanotubes and test inflammatory cell and fibroblast responses on such substrates.

Materials and Methods: 99.2% pure titanium foils (Alfa Aesar, Ward Hills, MA) were cut into 1cm×1cm squares and cleaned with acetone, 70% ethanol, deionized water. The samples were etched for 1min with a solution of 1.5% nitric acid and 1.5% hydrofluoric acid (HF) to remove the thin oxidized layer. The cleaned Ti sample was used as an anode, while a platinum (Pt) mesh served as a cathode. Both were immersed in an electrolyte solution consisting of 1.5% HF. The Ti and Pt samples were connected to a DC power supply. A constant voltage of 20V was applied for 10 minutes and the distance between Ti anode and Pt cathode was kept constant at 1cm. After anodization Ti samples were rinsed with large amounts of deionized water immediately and air dried.

Next, the PDMS monomer and cross-linking agents (Sylgard 184, Dow Chemical Co.) were mixed at 10:1 (weight ratios) for 15min and were then placed in a vacuum chamber for 30min to remove air bubbles. The mixture was cast onto the nanotubular Ti master mold and then was placed into a vacuum chamber for another 1h. The PDMS was cured at 60 °C for 2h followed by cooling and was gently peeled away from the Ti template. To improve the peeling off ability, a gold film was sputtered onto the top surface of the PDMS template and the PDMS slurry was again poured onto the first PDMS. The desired PDMS replica was obtained after curing.

The IC-21 macrophage cell line was used (TIB-186; ATCC, Manassas, VA) for determining an immune response. Macrophages were cultured according to ATCC instructions in RPMI-1640 medium (Invitrogen, Carlsbad, CA) with 10% fetal bovine serum and 1% penicillin/streptomycin (HyClone Laboratories Inc, Logan, UT) under standard incubator conditions (37°C, humidified, 5% CO_2/95% air environment). Macrophages were seeded at 3500 cells/cm^2 per substrate and were cultured in a humidified 5% CO_2/95% air environment at 37°C for 24 hours. After 24 hours, nonadherent macrophages were removed by rinsing in phosphate-buffered saline (PBS). Macrophages adherent on the substrates were fixed with 4% formalin in PBS and stained with both rhodamine phalloidin (R415) (Molecular Probes, Eugene, OR) and DAPI (4', 6-diamidino-2-phenylindole) (33258) (Sigma-Aldrich). F-actin filaments and cell nuclei were visualized using tetramethyl rhodamine iso-thiocyanate and DAPI, respectively.

Also, to determine the adhesion and proliferation of skin fibroblasts (ATCC, CCL-110), a cell proliferation assay (CellTiter 96, Promega) was used. Briefly, for cell adhesion, cells were seeded at 3500 cells/cm^2 in standard cell culture media and were incubated for 4 hours. For proliferation studies, cells were seeded at 3500 cells/cm^2 for 1, 3 and 5 days. The dye solution were added to the cells after the end of the prescribed period for 4h, then the stop solution was added and incubated overnight. A plate reader was used to test cell density. Numerical data were analyzed with a Student's t-test to make pair-wise comparisons. Statistical significance was considered at $p<0.05$.

Results and Discussion: As expected, compared with unmodified Ti (Figure 1. a), nano-sized tubes were distributed uniformly on the Ti surface after anodization (Figure 1, c). The uniform pores, as observed by scanning electron microscopy, were estimated to have a diameter of 60-70nm and a depth of 200nm. After pouring the PDMS slurry onto the surface of the nanotubular Ti and peeling off the first PDMS template, a nanopatterned structure was observed (Figure 1, d). Most importantly, the PDMS replica showed similar nanostructures as anodized Ti.

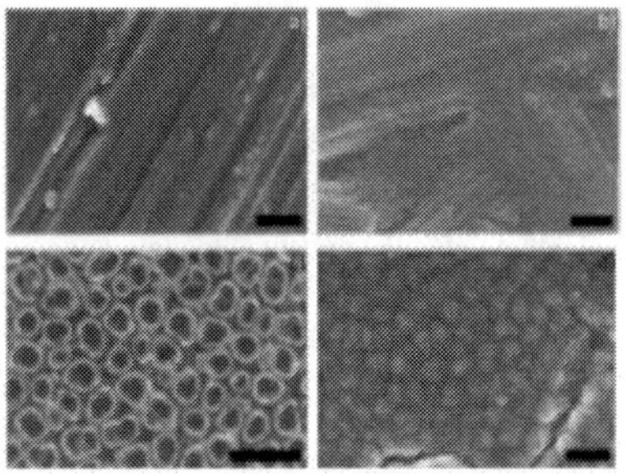

Figure 1. SEM images of

a) Unanodized Ti b) PDMS based on unanodized Ti

c) Anodized Ti d) PDMS based on anodized Ti

Scale bars are 200nm.

The results also demonstrated decreased numbers of macrophages adhered to nanotubular Ti and PDMS replica surfaces. Most importantly, the increased fibroblast adhesion, proliferation, and long-term functions on nanotubular substrates indicated the nanoscale topographies of polymers should be strongly considered for catheter applications.

Conclusions: A PDMS replica was obtained from anodized nanotubular Ti and demonstrated a promising ability to control macrophage and fibroblast functions.

Acknowledgements: The authors would like to thank Northeastern University for funding.

(81a) A Lattice Model of Silica Polymerization

Khan, M. N., University of Massachusetts Amherst
Auerbach, S. M., University of Massachusetts
Monson, P. A., University of Massachusetts
Originally presented on: 11/4/2013 8:30:00 - 8:55:00

We have developed a lattice model to describe the gelation transition in sol-gel synthesis and other process for silica material synthesis. In this model, we have used an atomic representation of silica monomer where the silica tetrahedra occupies a unit cell in the lattice with silicon atom located at the center and hydroxyl groups on the vertices. This lattice model has proved useful in understanding a great variety of materials formation processes such as silica polymerization at low pH, silica crystallization into microporous materials, and formation of surfactant-templated mesoporous materials. However, the transition from silica gel at low pH to nanoparticles at high pH remains poorly understood. To address this, we have adopted a coarse grained representation of the charge balancing species (cation) which occupy sites on the lattice with excluded volume. Silica condensation/hydrolysis is represented as the overlap of hydroxyl groups, from two different silica monomers. We have also included a simplified version of the electrostatic interaction of the cation with the silica molecule. Such an approach has enabled us to study silica polymerization over the pH spectrum. At zero concentration of the cation, which represents the iso-electric point of silica (pH=2), we have obtain a gel. In this region large cluster percolates the system. Whereas at high cation concentration we have observed nanoparticles that have a core-shell structure with silica in the core surrounded by a shell of cations. At intermediate values of the cation concentration, we have observed a gelation transition. This region is marked by decrease plateau regime in the average cluster size of the particles around 3 nm, in good agreement with experiment findings. During the transition, the cations start populating the surface of the gel, prohibiting further polymerization. As the concentration of the cations is increased further, larger particles break up into smaller particles. We are in the process of developing a structural descriptor that would be useful in studying the transition. Future work is required to determine the detailed structure of these core-shell nanoparticles, and how they transform to zeolites.

(81b) Computational Mechanics of Aerogels and Aggregation-Based Nanocomposite Materials

Gelb, L. D., University of Texas at Dallas
Ferreiro-Rangel, C. A., University of Texas at Dallas
Originally presented on: 11/4/2013 8:55:00 - 9:20:00

Aerogels are materials with very low densities and high surface areas prepared through sol-gel processing. Aerogels are amorphous materials formed by in-solution aggregation processes, with fractal structure at scales from near-atomic to greater than the mean pore size. They can have extremely high surface area and porosity, low thermal conductivity, and impressive specific strength. We have previously introduced (J. Phys. Chem. C **111** (2007) 15792) a coarse-grained flexible model where Langevin dynamics are used to mimic the gelation process of the aerogel primary particles; this model also naturally includes the effects of relaxation and fluctuation on the gel structure. The work described in this study consists of two parts. In the first, simulation methods for determination of the Young's modulus and Poisson ratio of model aerogels are introduced, and results obtained in good agreement with experiment. In a second area, we study a selection of novel material structures obtainable from sol-gel-like aggregation processes. These include (a) bidisperse and nanocomposite aerogel-like materials, consisting of a mixture of small and large particles; (b) interpenetrated aerogels, consisting of multiple independent gel networks; and (c) hierarchically-structured materials, formed by the (random) aggregation of regular nanoparticle assemblies. These material models are mechanically and structurally characterized, and the results analyzed in order to establish structure-property-synthesis correlations for nanocomposite aerogels and related aggregation-based materials.

(81c) Fabrication of Nanoporous Silicon Oxycarbide Materials Using Layered Double Hydroxide As a Sacrificial Template

Yan, X., University of Southern California
Deng, W., University of Southern California
Khawaji, M., University of Southern California
Tsotsis, T., University of Southern California
Sahimi, M., University of Southern California

Originally presented on: 11/4/2013 9:20:00 - 9:45:00

Nanoporous silicon oxycarbide is considered to be a promising material for a variety of applications. Highly porous silicon oxycarbide materials with high surface area and a hierarchical pore size distribution were successfully synthesized using tri-isopropylsilane (TPS) as a precursor and a layered double-hydroxide (LDH) as a sacrificial template via chemical vapor deposition (CVD) at the relatively low temperatures of 700 °C – 800 °C. Techniques such as X-ray diffraction (XRD), X-ray photoelectron spectroscopy (XPS) and transmission and scanning electron microscopies (TEM or SEM) were used to characterize the resulting materials and showed them to be hollow spheres with layered structures. Nitrogen adsorption-desorption (BET) measurements also validate the fact that the synthesized porous $Si_xO_yC_z$ materials have hierarchical structures containing both mesopores and micropores. The materials exhibit high BET surface areas ranging from 360 m^2/g to 540 m^2/g and total pore volumes from 0.69 cm^3/g to 0.91cm^3/g. The $Si_xO_yC_z$ materials fabricated in this research, using low cost templates and precursors and convenient synthesis procedures, have desirable structural properties, and show promise for applications as catalysis support, sorbents, battery anodes, and membrane films for water filtration and gas separations applications.

(81d) Worm-Like Micelle Assisted Synthesis of Iron Oxide Nanorods

Duttagupta, S., Indian Institute of Technology Bombay
Thaokar, R., Indian Institute of Technology, Bombay
Mehra, A., Indian Institute of Technology, Bombay (IITB)
Originally presented on: 11/4/2013 9:45:00 - 10:10:00

For synthesis of nanoparticles, when assisted with a soft template, the morphologies of the micelles play an important role in evolution of shape of the nanoparticles. Depending on the concentrations of surfactant and counter-ions as well as that of the solvent, micelles can exhibit range of morphologies. Here we discuss about anisotropic iron oxide nanoparticles being synthesized by the direct reaction of iron chloride salt ($FeCl_3.6H_2O$) with Ammonium Hydroxide (NH_4OH) in a cetyl trimethyl ammonium chloride (CTAC) – sodium salicylate (NaSal) worm-like micellar system. It has been observed that the presence of a worm-like micellar phase is critical in the formation of these anisotropic nanoparticles. When NaSal is below the critical concentration required for the formation of worm-like micellar phase, only spherical particles were obtained. Addition of precursors to the surfactant system leads to formation of spherical nanoparticles at short times, which then possibly coagulate and consolidate along a surfactant backbone to form nanorods. Interestingly, when pre-formed spherical nanoparticles are added to a worm-like micellar system, similar nanorods are observed.

Even when the precursors are made to react in the worm-like micellar phase, the system needs to be sheared continuously as it ages. Spherical particles are obtained if the system does not experience any shear as it ages. The lack of continuous shear during the aging process may be hindering with the particle - particle interactions in order for the spherical nanoparticle to grow into nanorods. The initial study of the formation process using transmission electron microscopy indicates that the nanorods are formed by anisotropic coagulation and consolidation of spherical nanoparticles. We are now conducting transient studies of the system and analysing the same with a cryo-TEM in order to understand as well as capture the growth mechanism.

The system has also been subjected to controlled shear ($0 – 32\ s^{-1}$) in a Linkam shear cell in order to examine the effect of shear on particle morphology. We observe a change in number yield of anisotropic particles as well as their morphology with a shift from lower to higher shear rate. The cross over frequency seems to plays a significant parameter with respect to effect of shear on the morphology. The controlled shear based experiments provide us with data that can help us explore scale-up for synthesis of anisotropic particles using soft templates.

We have further analysed the system with varying concentration of precursor in the system and notice there is no drastic change in the size. Our group has attempted simulation to observe the effect on particle size with increase in initial loading but experimentally above a critical concentration, the nanoparticles precipitate out of the micellar system. We are exploring the effect of WLM concentration in the system in order to increase the particle loading capacity of the system as well.

(81e) Preparation of Mesoporous Materials: Approaching to Industrial Viability

Santamaria, E.,

Maestro, A., Universitat de Barcelona

Porras, M., Universitat de Barcelona

Gutierrez, J. M., Universitat de Barcelona

Gonzalez, C., Universitat de Barcelona

Originally presented on: 11/4/2013 10:10:00 - 10:35:00

Since the discovery of the ordered mesoporous materials [1-2] many papers have been published about preparation of these kind of materials, but most of the proposed methods are very far of being applicable at industrial scale due to cost of precursors, catalysts, structure directing surfactants, and complex processes which are proposed.

In most of the processes alcoxides of the corresponding metal are proposed as precursors, acid or basic catalysts are used, and surfactants used as structure directing agents are not normally recovered.

In this work, meso and meso-macroporous materials are prepared using metasilicate as precursor, experiments using acid resins as acidifying agent are presented, and experiments using recovered surfactant are also presented. Influence of concentration of precursor and surfactant on properties of obtained materials is presented.

In case of using acid resin the preparation method is the following. First, water and surfactant are and kept under stirring at 50°C to ease the melt of the surfactant until a clear solution was obtained. Then ion exchange resin is added to the water and surfactant solution. Later on, a sodium silicate solution is added drop by drop. The resulting slurry is stirred, so the Na^+ of the sodium silicate solution substituted the H^+ on the exchange sites of cation resin. The filtered liquid, which did not have condensed silica yet, is placed in an oven at 100 °C during 24 h in order to allow the silicate condensation producing a white precipitate that corresponds to the ordered mesoporous material. In case of using macroemulsions as templates for macroporous materials, HCl is used as catalysts.

Surfactant is extracted from material by ethanol in a soxhlet, and recovered by evaporating the ethanol in a rotavapor.

The material is finally dried and calcined in order to eliminate the residual surfactant

Materials are characterized by Small-angle X-ray diffraction scattering (SAXS), Scanning electron microscopy (SEM), Transmission electron microscopy (TEM), Hg intrusion porosimetry and N_2-sorption analysis.

In figure 1, examples of macromesoporous material and the ordered mesoporous structure are presented.

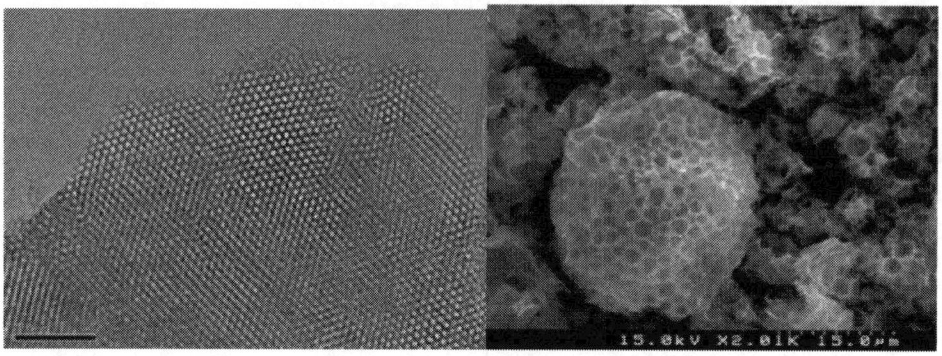

Figure 1. SEM and TEM images showing macroporous and mesoporous structures obtained

Mesostructure was confirmed by SAXS. High specific surfaces were obtained depending on the values of variables studied, the less silicate or surfactant concentration the more specific surface. No significant differences in specific surface were observed when recovered surfactant is used.

As main conclusion of this work, remark that very good materials with high specific surfaces and ordered mesostructure can be obtained using inorganic precursor, acid resin as catalyst, and recovered surfactant, which can make much less expensive the fabrication of this kind of materials.

[1] C.T. Kresge, M.E. Leonowicz, W.J. Roth, J.S. Beck. Ordered mesoporous molecular sieves synthetized by a liquid-crystal template mechanism. *Nature* 359, **1992**, 710

[2] J.S. Beck, J.C. Vartuli, W.J. Roth, M.E. Leonowicz, C.T. Kresge,, K.D Schmitt, C.T.W Chu, D.H. Olson, E.W. Sheppard, S.B. McCulle, J.B. Higgins, J.L. Schlender. Anew family of mesoporous molecular sieves prepared with liquid crystal templates, J.*Am. Chem. Soc.* 114, **1992**, 10834

Keywords: Silica, emulsion, meso-macroporous, ordered mesopores, cooperative templating mechanism

(81f) Engineering the Architectural Diversity of Heterogeneous Metallic Nanocrystals

Yu, Y., National University of Singapore
Zhang, Q., National University of Singapore
Xie, J., National University of Singapore
Lee, J. Y., National University of Singapore

Originally presented on: 11/4/2013 10:35:00 - 11:00:00

The development of a diversity of materials with the desired functionalities and performance from a limited number of elements provided by nature has always been a challenge for chemists and materials scientists. To this end the organic chemists have done very well – for they have crafted numerous molecules from basic elements such as carbon, hydrogen and oxygen by changing the number and the type of atoms and their spatial relationship. Taking a lesson from such an approach, we developed a synthesis strategy capable of engineering the architecture of heterogeneous metallic nanocrystals (HMNCs) through rational and independent programming of every architectural-determining element of a HMNC, i.e. the shape and size of the component nanocrystals (NCs) and their spatial arrangement. Architectural engineering of HMNCs can generate metallic nanostructures with unprecedented diversity and structural complexity, and hence increases the possibility to discover or create new and more varied properties for both basic and applied research.

Our synthesis strategy takes advantage of the geometry-dependent distribution of electrons in a chemical reaction system to direct the deposition of satellite NCs onto specific regions of a polyhedral central NC. The electron distribution is guided by the curvature of the polyhedral NC surface, the defining feature of any polyhedron. Consequently the strategy can be used to position satellite NCs exclusively on the corners and along the edges of different polyhedral central NCs. The satellite NCs can develop their own shape and size through the control of crystal growth kinetics. Furthermore, this strategy can produce HMNCs consisting of two or more types of component NCs, and of different metals in various combinations.

(110a) Single Phase Reaction-Precipitation Systems for the Synthesis of Poly (β-Amino Esters) Nanogels

Gupta, P., University of Kentucky
Dziubla, D. T., University of Kentucky
Hilt, D. J. Z., University of Kentucky
Originally presented on: 11/4/2013 12:30:00 - 12:50:00

Poly (β-Amino Esters) (PBAE) are a a class of versatile, pH sensitive, hydrolytically biodegradable polymers with tunable degradation and release properties. These polymers have shown a great potential as a matrix for gene and small molecule drug delivery via classic drug encapsulation/formulation approaches. Moreover, PBAE can be polymeric prodrugs, wherein active agents (e.g., curcumin, quercetin) can be acrylated and subsequently conjugated into the polymeric backbone of the resulting hydrogel, greatly enhancing the release rate and providing a mechanism for stabilizing highly labile drugs.

However, owing to the high reactivity of the monomers and stability concerns of the ensuing particles, standard oil-water nano-emulsion methods are not amendable to their formation. Further, nanoprecipitation, when done in two immiscible organic solvent suspension systems, can lead to aggregation and instability during the purification processes. In this work, a novel approach to formulate anti-oxidant conjugated nanogels in a single organic phase system is explored where two multifunctional monomers, when added in appropriate concentrations in an organic solvent, react with each other and form stable nanogel/nanoparticle suspension. These nanogels are then subsequently coated with PEG to prevent post purification (via centrifugation) instability. In the present research, quercetin was used as the molecule of interest as it possesses strong anti-oxidative, anti-inflammatory and anti-cancerous properties. To formulate nanoparticles (NPs), quercetin multiacrylate (QMA) was reacted with a secondary diamine (DA), (N,N' dimethyl-1-3 propanediamine) in acetonitrile via a Michael addition reaction to form a crosslinked polymeric nanoparticle suspension. Different monomer concentrations were used to obtain variable particle sizes. The obtained NP suspension was then reacted with polyethylene glycol methyl ether methacrylate (PEGMEMA Mn=5000) for one hour to cap unreacted amines with PEG and increase the particle stability. Purification of NPs was done via centrifugation followed by re-suspension in fresh acetonitrile.

The yield of the reaction was between 83-97% as determined from UV-Vis spectroscopy. The size of synthesized NPs ranged from 200-400nm as observed using SEM and DLS. Upon hydrolysis in PBS with pH 7.4 at 37°C, NPs showed prolonged and uniform release of active quercetin for at least 36 hours, which was studied using UV-Vis spectroscopy.

(110b) Responsive Nanoscale Hydrogels for the Oral Delivery of Chemotherapeutics

Puranik, A., The University of Texas at Austin
Peppas, N., The University of Texas at Austin
Originally presented on: 11/4/2013 12:50:00 - 13:10:00

Responsive Nanoscale Hydrogels for the Oral Delivery of Chemotherapeutics

Amey S. Puranik[1] and Nicholas A. Peppas[1,2,3]

[1]Department of Chemical Engineering, [2]Department of Biomedical Engineering and [3]College of Pharmacy, The University of Texas at Austin, Austin, TX, 78712

Small-molecule chemotherapeutics, which are currently delivered intravenously, are predominantly associated with a large number of side effects owing to their inherent hydrophobicity, the resultant poor systemic pharmacokinetic profiles and an undifferentiating ability to kill cells. We hypothesize that the use of a nanoscale hydrogel-based delivery system could help overcome drug solubility and dissolution-related bottlenecks in the delivery of hydrophobic chemotherapeutics. A nanoscale drug delivery system was devised to offer an enhanced opportunity for orally delivering hydrophobic cancer therapeutics by selectively localizing the drug within the lipophilic core of a nanoscale hydrogel, which in turn, can respond to changing environmental conditions in the gastrointestinal tract. Hydrogel nanoparticles (size: 100-120 nm) comprising of methacrylic acid co-polymerized with a hydrophobic monomer, and colloidally stabilized by poly(ethylene glycol) tethers, were synthesized using an UV-initiated emulsion polymerization technique. Doxorubicin (DOX) was chosen to be a model hydrophobic chemotherapeutic to evaluate the loading and release capabilities of these nanoparticles. Nanoparticle formulations with varying crosslinking density and core hydrophobicity were able to load doxorubicin with efficiencies ranging from 36-55% within 2 hours of incubation with the drug solution. As the percentage of crosslinking density of the nanoparticles was increased, their ability to swell as well as their therapeutic agent loading and release capability, decreased. The physically encapsulated doxorubicin was completely released within a period of 4-6 hours, at pH 7.4, while still being retained at pH values representative of conditions in the stomach. The nanocarriers were also studied for their cytotoxicity, mucoadhesion and ability to enhance uptake of doxorubicin by cancer cells. The pH-responsive nanocarriers displayed amenable physicochemical characteristics that are considered crucial to hydrophobic drug delivery and achieved improved uptake of DOX by cancer cells.

Acknowledgement: This work was supported by a grant from the National Institutes of Health (EB 1R01-000246-18).

(110c) Chemotactic Live Autonomous Drug Delivery Agents With Different Body Geometries

Sahari, A., Virginia Tech
Traore, M. A., Virginia Tech
Behkam, B., Virginia Tech
Originally presented on: 11/4/2013 13:10:00 - 13:30:00

Chemotactic live autonomous drug delivery agents with different body geometries

Ali Sahari[1], Mahama A. Traore[2], and Bahareh Behkam[1, 2]

[1]School of Biomedical Engineering and Sciences, Virginia Tech, Blacksburg, VA 24061

[2]Mechanical Engineering Department, Virginia Tech, Blacksburg, VA 24061

INTRODUCTION

Bacteria-based drug delivery carriers are envisioned to significantly improve the targeted delivery of drug molecules and imaging agents to the regions of interest. These autonomous drug carriers if directed by chemical cues, normally present in diseased areas such as tumors, can offer novel solutions for targeted drug delivery. DrugBots, the potential drug carriers developed by the authors, are constructed by interfacing various shapes micro/nano-particles with an ensemble of live engineered bacteria with the purpose of using the bacteria for actuation, sensing, communication and control. Non-spherical micro/nanoparticles are recognized to have advantages over their spherical counterparts such as higher margination and adhesion as well as prolonged circulation which make them attractive for drug delivery purposes. Gradients of chemical attractants (such as naturally occurring gradients of disease biomarkers) can be utilized to control the motion of the bacteria (and hence DrugBots with spherical and non-spherical body shapes) towards higher concentrations of chemo-effectors (i.e. disease site). In the work presented here, the chemotactic response of DrugBots is quantitatively characterized in the presence of a stable and well-controlled concentration gradient of a chemo-attractant, established within a flow-free hydrogel-based microfluidic device. This study focuses on investigating the role of body shape and chemotaxis in the motile behavior of DrugBots. We established a linear gradient of L-Aspartic acid for bacteria in a hydrogel-based microfluidic device in order to characterize the DrugBot transport for two different morphologies of the DrugBot, namely sphere and elliptical disk. Disks were fabricated here using a casting and mechanical stretching technique [1]. This non-spherical shape was selected as it has been shown to exhibit longer half-life circulation and higher targeting specificity in mouse models [2] which makes them a good candidate, along with their spherical counterparts, for chemotaxis experiments.

MATERIALS AND METHODS

PEG-DA based Microfluidic device fabrication: A diffusion-based microfluidic device using 700 Da polyethylene glycol diacrylate (PEG-DA) hydrogel was fabricated through a one-step UV polymerization process. The three parallel channels separated by the two porous hydrogel walls enabled us to establish concentration gradients of chemo-effectors for bacteria (and DrugBots) (see Fig. 1). This was accomplished by continuous flowing of a chemo-attractant and a buffer solution in the two outer channels and lateral diffusion of the chemo-attractant through the hydrogel wall and into the flow-free buffer filled center channel after assembling the device. The quasi-steady linear gradient in the center channel was maintained throughout the entire experiment.

DrugBot Construction: Elliptical disk-shaped (ED) polystyrene (PS) particles were fabricated using a high throughput particle casting and stretching technique [1]. Motile sub-populations of *Escherichia coli (E. coli)* MG1655 transformed with pHC60 (GFP-expressing plasmid) were grown in Tryptone Broth to OD_{600}=0.5 at 32 °C and incubated with goat polyclonal anti-lipid A LPS antibody labeled with biotin (10 µg/ml). The two collections of 6 Âµm microparticles (spheres and EDs) were coated with Cy3-streptavidin and were added to the bacterial solution which produced the DrugBots (see Fig. 2) through self-assembly by utilizing the high affinity between biotin and streptavidin.

L-Aspartic acid, a known chemoattractant for *E. coli* bacteria, was selected to control the movement of the DrugBots in the device. Spherical and ED-shaped DrugBots were injected into the chemotaxis observation channel (see Fig. 3) after a concentration gradient of the chemo-effector (0.1 mM/mm) was established. Motile DrugBots were observed using a Zeiss AxioObserver D1 inverted microscope equipped with a motorized stage and an AxioCam HSm camera plus a 20Ã— objective and the captured videos were analyzed using the ImageJ software (NIH, Bethesda, MD). For data, analysis, the width of the observation channel is divided into two partitions and the spatial distribution of DrugBots is calculated across the width of the channels in order to quantify the migration profile of the DrugBots.

RESULTS AND DISCUSSION

Quantitative investigation of DrugBot chemotaxis in response to concentration gradients of L-Aspartic acid as well as screening the effect of particle geometry on this chemotactic response is the main purpose of this study. The fabricated PEG-DA hydrogel microfluidic device generates a well-controlled chemical gradient across the chemotaxis observation channel and allows us to measure the chemotactic response quantitatively. The chemo-attractant gradient reached a steady state after around 35 min and it was maintained by a slow flow through the channels at 5 Âµl min^{-1} flow rate. The fabricated spherical and ED-shaped DrugBots were then introduced into the center channel and every 5 min, 10 second videos were recorded at 20 frames/sec frame rate. In order to quantify the migration profile of the DrugBots, the spatial distribution of the DrugBots in the two partitions across the chemotaxis chamber was calculated using ImageJ software. The results showed that the number of the DrugBots at the chemo-attractant side increases during the first 20 min of the experiment implying that there was a distinct bias towards the high concentration of L-Aspartic acid for both types of DrugBots. Preliminary data analysis shows that spherical DrugBots have a slightly higher chemotaxis partition coefficient (CPC), compared to ED-shaped DrugBots after 40 minutes of running the experiment. CPC is a measure of the strength of the response of the DrugBots to the chemoattractant gradient. Further statistical analysis is being performed to determine the statistical significance of the observed trends. Moreover, in order to ensure that the DrugBot movement is because of the chemo-attractant, a control experiment was conducted with no chemo-attractant and as expected, the distribution of the DrugBots remained unchanged (data not shown). The gradient steepness can also influence the migration of the DrugBots. A steeper gradient is expected to generally yield a stronger chemotaxis response. However, an inhibitory effect might be observed if the high chemo-attractant concentration saturates the bacterial chemotaxis receptors. A thorough study of the effect of a varying gradient steepness on the chemotactic response of the DrugBots is currently being performed.

CONCLUSIONS

Bacteria-based cargo carriers with spherical and ED shapes along with *E. coli* strain MG1655 were constructed here. Chemotaxis was exploited to control the movement of these autonomous drug delivery vehicles introduced into a PEG-DA hydrogel based microfluidic device. Both types of these carriers, called DrugBots, have shown a biased movement towards higher concentrations of L-Aspartic acid. These controllable DrugBots have a great potential to autonomously navigate and carry/deploy a wide variety of cargos such as therapeutics and contrast agents to hard-to-reach diseased areas.

REFERENCES

[1] Champion, J.A. et al., Making polymeric micro- and nanoparticles of complex shapes, Proceedings of the National Academy of Sciences, 104:11901 (2007).

[2] Muro, S. et al. Control of endothelial targeting and intracellular delivery of therapeutic enzymes by modulating the size and shape of ICAM-1-targeted carriers. Molecular Therapy, 16:1450?1458 (2008).

(110d) Redox-Adaptive Protein Intracellular Delivery For Targeted p53 Therapy

Zhao, M., UCLA
Tang, Y., University of California, Los Angeles
Originally presented on: 11/4/2013 13:30:00 - 13:50:00

Specific induction of cell death in tumors is considered one of the most desired and effective anticancer therapies. A potent chemotherapy option is directly arming the cancer cells with executioner proteins or apoptotic-inducing proteins that are not targeted by anti-apoptotic maneuvers found in many tumors. We have developed a new method to treat breast cancer by using Redox-Adaptive Protein Intracellular Delivery (RAPID). RAPID is a platform to deliver proteins in native forms into cells. The key design feature of RAPID is to first encapsulate protein molecules in a thin layer of water soluble, positively charged, protease degradable polymer to form nanometer-sized nanocapsules. The nanocapsule shell facilitates uptake of the protein content into cells, and protects the protein both during in vivo circulation and endocytosis. To endow the nanocapsules biodegradability once entered the target cells, the polymer shell is crosslinked with a disulfide containing, redox-sensitive crosslinkers that can be degraded once encountering the reducing intracellular environment. In this presentation, we will describe using RAPID to deliver the cellular guardian transcription factor p53. It controls the normal cell homeostasis and is mutated in a large number of breast cancers, including those hard to treat triple negative breast cancers. We hypothesized that delivering functional forms of p53 to tumor cells defective in p53-dependent apoptosis can resurrect the pathway and lead to tumor clearance. This presentation will cover three topics: 1) demonstration that recombinant p53 can indeed be delivered to breast cancer cells and lead to either senescence or apoptosis of cancer cells; 2) establishment of a new nanocapsule synthesis strategy that allows modification of the particle surface with targeting ligands using "Click" chemistry; and 3) the use of p53 nanocapsules on tumors grafted onto nude mice. THE RAPID method will be general applicable to a variety of therapeutic proteins and human diseases.

(110e) Cancer Treatment Using Drug Delivery Nanoparticles

Shah, P.,
Shaqfeh, E. S. G., Stanford University
Originally presented on: 11/4/2013 13:50:00 - 14:10:00

In tumors, pores of size on the order of 100nm develop in the microvasculature wall due to abnormal endothelial growth. Drug delivery to these tumors via nanoparticles involves two physical processes: migration of nanoparticles to the endothelial wall (macroscale) and extravasation of these particles through the vasculature pores (microscale). Recent studies show that the treatment efficiency depends on nanoparticle shape and size specific to each tumor type, characterized by microvasculature geometry, pore size distribution and associated flow properties such as shear rates and oncotic pressure. We attack the two scales of the problem separately. As our preliminary model for the micro-scale, we develop a singular perturbation theory as well as develop a Brownian dynamics simulation to evaluate the pore-extravasation (flux) rate of point Brownian particles through a circular pore when they are in the presence of shear flow and pressure-driven suction flow (Sampson flow). Predictions from the Brownian dynamics simulation are found to agree well with our theory at the physiologically relevant range of Peclet numbers and suction strength. We then extend the validated Brownian dynamics simulation from point particles to finite sized spherical and rod shaped particles, and interface these simulations with a large-scale computational model for blood and nanoparticle transport in a model tumor microvasculature. With this, we demonstrate the capability to detemine nanoparticle leakage rates and how it varies depending on tumor type. We identify Brownian motion as a major deteminant in treatment efficiency and we characterize its effect, coupled with other flow parameters on the treatment efficiency for different tumor types.

(582c) Calibrating Schema-Guided Computational Design of Adeno-Associated Virus Chimeras

Ho, M. L., Rice University
Adler, B., Rice University
Torre, M., Rice University
Silberg, J. J., Rice University
Suh, J., Rice University
Originally presented on: 11/4/2013 14:10:00 - 14:30:00

Adeno-associated virus (AAV) is a non-pathogenic member of the *Parvoviridae* virus family that is increasingly being exploited as a gene delivery vehicle for human gene therapy. While there is great promise in virus nanoparticle (VNP) gene therapy, it has yet to be established as the leading treatment platform due to cell targeting inefficiencies. Researchers have been slow to create VNPs with new functions, such as specific cell targeting, because 1) the effects of even single amino acid substitutions on virus capsid structures cannot yet be predicted and 2) the current throughput of combinatorial screening is a limiting factor in mining chimeric libraries for new structures and functions. To overcome these challenges, we are examining whether the SCHEMA algorithm can predict structural disruption in AAV capsids when highly divergent serotypes are recombined. Homologous recombination has contributed to AAV evolution, but the extent to which AAV capsid proteins (VP1, VP2, and VP3) tolerate amino acid substitutions created through this process remains poorly understood. To evaluate SCHEMA predictions, we created 17 chimeras by recombining divergent (~58% sequence identity) AAV serotypes 2 and 4 and analyzed the structures and functions of these chimeras. Interestingly, chimeric capsids that differed by more than one hundred residues from natural capsids still retained their ability to package viral genomes, a characteristic of properly assembled 60-subunit virus capsids. Even within our small chimera population, we observed structural and functional diversity. For example, chimeric capsids varied in their ability to assemble and protect genomes from nucleases. We also found that chimera structural disruption, calculated using the Schema algorithm, tended to correlate with capsid disruption: only the chimera with the lowest structural disruption completely protected its genome from nuclease digestion and retained the ability to transduce Cos-7 cells like wild-type AAV2 and AAV4. These results suggest that the sequence space accessed by recombination of highly divergent AAV serotypes is rich in genetically-encoded VNPs with diverse sequences and binding properties. Our findings indicate that Schema-guided recombination should be useful for creating libraries enriched in capsid-forming proteins with a range of functions for future gene therapy applications.

(110g) Multi-Targeted siRNA Delivery Liposomes Inhibit Metastatic Breast Cancer Cell Migration

Guo, P., Children Hospital Boston
You, J. O., Harvard University
Auguste, D., Harvard University
Originally presented on: 11/4/2013 14:30:00 - 14:50:00

Metastatic breast cancer (MBC) is the second leading cause of cancer-related fatalities in women, accounting for more than 40,000 deaths each year. MBC has a five-year relative survival rate of 23% compared with 99% for patients with non-metastatic breast cancer. These statistics suggest an urgent and significant need for developing novel and effective therapeutics for the treatment of MBC. In this study, we engineered a liposomal drug delivery vehicle designed to inhibit both the C-X-C chemokine receptor type 4 (CXCR4) and Lipocalin-2 (Lcn2) mediated migratory pathways. CXCR4 is a G protein-coupled receptor that is responsible for cell migration along chemokine gradients. Lcn2 is a secreted protein that mediates breast cancer cell EMT. pH-responsive liposomes were designed to protect and trigger the release of Lcn2 siRNA. Liposomes were modified with anti-CXCR4 to target MBC cells and block migration along the CXCR4-CXCL12 axis. This synergistic approach - coupling the CXCR4 axis blockade with Lcn2 silencing - significantly reduced migration of triple-negative human breast cancer cells MDA-MB-436 (-92%) and MDA-MB-231 (-88%). These results suggest that drug delivery vehicles engineered to target multiple migratory pathways may effectively slow progression of MBC.

(135a) Biocatalytic Nanocomposites That Selectively Target Uncontrolled Pathogens

Dordick, J. S., Rensselaer Polytechnic Institute

Originally presented on: 11/4/2013 12:30:00 - 12:55:00

We have taken cues from nature to design materials with unique structural and functional properties, along with new process technologies with the ability to produce a wide range of biomimetic structures. Specifically, we have focused on the generation of nanostructures that are functionalized with and in some cases constructed from biological molecules, complete with tailored selectivities and biocatalytic activities. These nanostructures have been exploited in the generation of biocatalytically functional polymeric films, coatings, and paints that kill bacteria, prevent biofilm formation, and reduce fouling by bioorganic molecules. In this talk I will highlight our recent efforts to exploit the interface of biology with materials science, enhancing enzyme function along the way. Both fundamental advances and applications will be discussed, the latter focused on enzyme-nanomaterial composites with a wide range of activities that endow surfaces with decontaminating properties that target selectively key pathogens in our societal infrastructure, including MRSA, Listeria, and bacillus spores. Such activity provides a safe and potentially broadly applicable route to eliminating toxic compounds and pathogenic microorganisms from common surfaces.

(135b) Engineering of Quorum Sensing Molecules for Biodegradation, Bioelectricity Generation, and Nano-Material Biosynthesis By Whole-Cell Biocatalysts

Zhong, J. J., Shanghai Jiao Tong University
Yong, Y. C.,
Song, H.,
Originally presented on: 11/4/2013 12:55:00 - 13:20:00

Quorum sensing (QS) is widely conserved in bacterial cell-cell communication systems and plays an important role in many biological processes. This work demonstrates that engineering of QS molecules contributes to bioelectricity generation and nanomaterial synthesis by *Pseudomonas aeruginosa*, in addition to its impact on toxic aromatics biodegradation. In microbial fuel cells (MFCs) system under different QS expression patterns, we found that *P. aeruginosa* could use different electron shuttles. By overexpressing rhl QS system, more electricity generation with broader potential range from -0.40 V to 0 V was reached compared to non-engineered whole-cell biocatalysts. The electrochemical and high performance liquid chromatography analyses indicated that the electron shuttle produced by the rhl overexpression cells was phenazines - pyocyanin and phenazine-1-carboxylate, a type of efficient electron shuttle. The rhl QS system was also found significant to carbon nanomaterial biosynthesis by *P. aeruginosa*. By overexpressing rhl QS, carbon nanomaterial was synthesized faster by the engineered biocatalysts. More interestingly, the rhlIR double mutant could not synthesize the nanomaterial even with addition of N-butyryl-L-homoserine lactone (BHL, a signaling molecule of rhl QS system); but the rhlI single mutant could synthesize under BHL addition, which will be discussed in this presentation. The information obtained shows good potential of *P. aeruginosa* as whole-cell biocatalysts for bioelectricity production and nanomaterial biosynthesis by QS engineering.

(135c) Performance Characterizations of Nano-Bio-Catalysts for Enzymatic Biofuel Cells

Ha, S., Washington State University
Kim, J., Korea University
Wee, Y., Korea University
Garcia-Perez, T., Washington State University
Originally presented on: 11/4/2013 13:20:00 - 13:45:00

Carbon-based nanomaterials as a support for enzyme immobilization hold a great potential to fabricate the enzyme electrode that meet both high power density and long-term stability requirements for the successful applications of enzymatic biofuel cells. In this paper, we will immobilize our model anode and cathode enzymes over various carbon nano-materials to form high performing enzyme electrodes via nano-confinement and/or multipoint crosslinking approaches. In terms of achieving the high power density output, we will explore a large surface area of carbon nano-materials for improving the enzyme loading per a unit geometrical surface area of electrodes for enzymatic biofuel cells. By increasing the enzyme loading, the available active sites for electron generation also increases in case of the enzyme anode. As the more number of electrons are generated, a probability of "collecting" them at the anode current collector would increase as well. Furthermore, carbon-based nanomaterials possess a high electronic conductivity, which can enhance the electron transfer rate of enzyme electrodes by decreasing its ohmic overpotential. In terms of achieving the high stability, we will explore the nano-confinement and/or multipoint crosslinking effect of enzymes. The electrochemical activities and long-term stability of these enzyme electrodes will be evaluated and correlated to their enzyme immobilization process parameters and their structures. Based on this performance-structure relationship, we will further optimize their electrochemical performances toward the biofuel cell applications.

(135d) Biocatalyzed Artificial Photosynthesis: Visible Light-Driven Cofactor Regeneration Coupled With Redox Biocatalysis

Kim, J. H.,
Lee, S. H., Korea Advanced Institute of Science and Technology
Nam, D. H.,
Park, C. B., Korea Advanced Institute of Science and Technology
Originally presented on: 11/4/2013 13:55:00 - 14:20:00

Photosynthesis is a complex process in which green plants produce essential biological fuels using solar energy. Its unique features, such as environmental compatibility and near-unity quantum yield, have fascinated scientists and provided inspiration to improve the efficiency of solar cells and photoelectrochemical hydrogen production systems. On the other hand, recent interest has been expressed in redox enzymes that can catalyze complex synthesis reactions under mild conditions; conventional catalysts rarely accomplish this task. With recent advances in genomics and genetic engineering, the use of biocatalysis for industrial synthetic chemistry is growing rapidly. Most biocatalysts in current use, however, are limited to cofactor-independent enzymes that perform relatively simple chemistry. In comparison, cofactor-dependent enzymes are capable of accelerating much more complex reactions and catalyzing a large number of synthetically useful reactions. In this talk, I will discuss about how to regenerate cofactors using visible light to couple with redox enzymatic reactions. Many challenges are expected for the development of nanobiocatalytic assemblies to mimic natural photosynthesis, such as the formulation of light harvesting system suitable for photobioreactor application under solar light, the design of photosensitizer-redox enzyme coupled system. Biocatalyzed artificial photosynthesis will create a technological foundation for the efficient synthesis and production of fine chemicals utilizing solar energy.

(135e) Light-Triggered Biocatalysis Using Thermophilic Enzyme-Gold Nanoparticle Complexes

Wong, M. S., Rice University
Blankschien, M. D., Rice University
Pretzer, L. A., Rice University
Huschka, R., Rice University
Halas, N. J., Rice University
Gonzalez, R., Rice University
Originally presented on: 11/4/2013 14:20:00 - 14:45:00

The use of plasmonic nanoparticle complexes for biomedical applications such as imaging, gene therapy, and cancer treatment is a rapidly emerging field expected to significantly improve conventional medical practices. In contrast, the use of these types of nanoparticles to noninvasively trigger biochemical pathways has been largely unexplored. Here we report the light-induced activation of the thermophilic enzyme Aeropyrum pernix glucokinase, a key enzyme for the decomposition of glucose via the glycolysis pathway, increasing its rate of reaction 60% with light by conjugating the enzyme onto Au nanorods. The observed increase in enzyme activity corresponded to a local temperature increase within a calcium alginate encapsulate of ~20 °C when compared to the bulk medium maintained at standard, nonthermophilic temperatures. The encapsulated nanocomplexes were reusable and stable for several days, making them potentially useful in industrial applications.

(135f) One-Step Regioselective Synthesis of Clindamycin Palmitate By Immobilized Lipase

Li, Z., Department of Chemical Engineering
Ge, J., Tsinghua University
Liu, Z., Tsinghua University
Originally presented on: 11/4/2013 14:45:00 - 15:00:00

Clindamycin is an established antibiotic with proven effectiveness against Gram-positive organisms and Gram-negative anaerobes and thus has been widely applied to treat diseases including infections of the respiratory tract, skin and soft-tissue. Clindamycin could be used as a first-line treatment option for the methicillin-resistant Staphylococcus aureus (MRSA).

However, the synthesis of clindamycin palmitate involves laborious steps of protection and deprotection and gives an overall yield below 50%. Here we report the first example of the one-step synthesis of clindamycin palmitate with high regioselectivity using immobilized *Candida antarctica* lipase B (Novozym 435) as the catalyst. The lipase catalyzed synthesis reached a conversion above 90% in 12 hours in an organic solvent. Characterized by ^{1}H and ^{13}C NMR, DEPT, HMQC, ESI-MS and FT-IR spectra, a nearly 100% regioselective acylation was confirmed occurred at the 2-hydroxyl of clindamycin. The enzymatic acylation of clindamycin at the solvent-free condition catalyzed by Novozym 435 and other types of novel immobilized lipase such as the lipase-inorganic crystal hybrid nanoflower will also be introduced in this presentation. In summary, the one-step regioselective synthesis of clindamycin palmitate with a high yield (>90%) and a high regioselectivity (~100%) greatly simplifies the synthesis by excluding the group protection/deprotection procedures and downstream purifications, which makes this enzymatic process very attractive for the manufacturing of clindamycin ester derivatives in pharmaceutical industry.

(191a) Engineering of Multilayered Nanoparticles Structures for Stretchable Conductors With Self-Organized Conductive Pathways

Kim, Y., University of Michigan, Ann Arbor
Zhu, J., University of Michigan
Yeom, B., University of Michigan
Prima, M. D.,
Su, X., University of Michigan
Kim, J. G.,
Yoo, S. J.,
Uher, C., University of Michigan
Kotov, N. A., University of Michigan
Originally presented on: 11/4/2013 15:15:00 - 15:31:00

Stretchable conductors are essential components of next generation electronic devices that can facilitate integration of human-machine interfaces. Despite the fundamental difficulties of combining stretchability and conductivity tremendous research and development efforts have been put into exploring possibilities envisioned from nanoscale building blocks with various polymeric matrixes. Carbon nanotubes (CNTs) have been chosen mainly because of its high aspect ratio that enabled low percolation threshold and good electrical properties during mechanical deformation. Typical stretchable conductors from CNTs, however, have several problems; high quality of composite is not reproducible, metallic conductivity is unlikely, and properties are anisotropic. To overcome those problems, we present a systematic approach making stretchable conductors from extreme content of nanoparticles (NPs) with polymer matrix by layer-by-layer (LBL) assembly technique. To highlight the structural influences on the properties of composites, layered composites by vacuum-assisted flocculation (VAF) method were also prepared. NPs, in general, might be an inadequate choice as fillers because they have minimal aspect ratio, however specially synthesized NPs with judiciously chosen polymer matrix are expected to have better electromechanical properties. Indeed, unexpectedly high electrical and mechanical properties with metallic charge transport mechanism were observed and these were originated from dynamic self-assembly of NPs under stress. Theoretical study about dynamic conducting percolation showed excellent match with experimental data. Also, stretchable nanoparticle conductor composites showed a new approach of electro-tuning mechanical properties. Scanning electron microscope, transmittance electron microscope, atomic force microscopy, electrical properties measurement system, dynamic mechanical analysis, small-angle X-ray scattering were performed to analyze properties of nanocomposites.

(191b) Formation of High-Aspect Ratio Helical Nanorods Via Peptide Driven Self-Assembly of Fullerodendrimers

Hilmer, A. J., Massachusetts Institute of Technology
Shimizu, S., Massachusetts Institute of Technology
McNicholas, T. P., Massachusetts Institute of Technology
Strano, M. S., Massachusetts Institute of Technology
Originally presented on: 11/4/2013 15:31:00 - 15:47:00

The ability to selectively form contiguous, high aspect-ratio electron acceptor domains could aid in the control of active layer morphologies in polymer photovoltaics. We have synthesized two novel, asymmetric methanofullerenes, which self-assemble into one-dimensional supramolecular structures in cyclohexane. We show that, through the introduction of a dipeptide sequence to one terminus of the dendritic methanofullerene, it is possible to transform the assembly behavior of these molecules from poorly formed aggregates to high aspect ratio nanorods. These nanorods have diameters of 3.76 ± 0.52 nm, and possess lengths which can exceed 10 μm. Both direct AFM observations and circular dichroism (CD) measurements indicate that these structures are comprised of helical assemblies of dendritic fullerenes. Temperature-dependent absorbance and CD measurements indicate that self-assembly is initiated by intermolecular peptide interactions, which simultaneously drive self-assembly and impart chirality to the resulting supramolecular structures. Here, we present optical and morphological characterization of these assembled structures.

(191c) Engineering Localized Surface Plasmon Resonances in Si Nanowires

Chou, L. W., Georgia Institute of Technology
Filler, M. A., Georgia Institute of Technology
Originally presented on: 11/4/2013 15:47:00 - 16:03:00

Nanoscale semiconductors are rapidly emerging as an alternate materials platform with which to engineer localized surface plasmon resonances (LSPRs). These quantized excitations, which result from the collective oscillation of free carriers at a metallodielectric interface, promise new opportunities for ultrafast information transfer and efficient energy conversion. Here, we show how phosphorus-doped regions positioned along the length of Si nanowires permit user-programmable spectral responses in the mid-infrared. Highly aligned Si nanowire arrays are synthesized via the vapor-liquid-solid (VLS) technique with a combination of Si_2H_6 and PCl_3 precursors. All phosphorus-doped sections are sandwiched between intrinsic base and tip regions to deconvolute substrate and catalyst effects, respectively. *In-situ* infrared absorption spectroscopy measurements, which mitigate the complications that arise from surface oxidation, reveal intense absorption bands between 800 and 2200 cm^{-1} that exhibit dopant concentration and shape-dependent spectral shifts consistent with longitudinal LSPRs. Discrete dipole approximation calculations confirm the observed spectral response results from resonant absorption and that free carrier densities are on the order of 10^{20} cm^{-3}. Multimodal spectral responses, achieved by introducing multiple doped regions in each Si nanowire, are also possible. Our findings open the door to surface plasmon engineering in a ubiquitous semiconductor and also highlight the utility of VLS synthesis for this purpose.

(191d) Impurities in Semiconductor Nanostructures

Holmberg, V. C., University of Texas at Austin
Sahu, A.,
Norris, D. J., Swiss Federal Institute of Technology (ETH ZÃ¼rich)
Korgel, B. A., The University of Texas at Austin
Originally presented on: 11/4/2013 16:03:00 - 16:19:00

As in bulk semiconductors, impurities play an important role in semiconductor nanostructures. Solid-state reactions such as silicide formation are used to make low-resistance electrical contacts in transistor applications, and cation-exchange reactions can be used to completely (or partially) transform a nanostructure from one material to another. Likewise, impurities are used to store electrochemical energy in devices such as Li-ion batteries. In confined semiconductor nanostructures, even single atomic impurities can have dramatic effects on optical, electrical, and magnetic properties. For example, the addition of only 2–3 Ag atoms to a CdSe nanocrystal can lead to an order-of-magnitude increase in fluorescence efficiency. Controlled impurity incorporation and characterization pose significant challenges in these systems. Here, we will discuss the effects of impurity incorporation in semiconductor nanostructures, as well as how *in situ* transmission electron microscopy (TEM) can be used to study the dynamics of solid-state reactions and impurity diffusion on the nanoscale. Dramatic morphological changes can be observed due to the volume change associated with silicide formation in Si nanowires, and in some systems, the rate of impurity diffusion can be orders of magnitude different than what is observed in the bulk.

(191e) Dense Semiconducting Nanowire Arrays Grown Directly On Graphene

Alper, J.,
Gutes, A.,
Carraro, C., University of California Berkeley
Maboudian, R., University of California, Berkeley
Originally presented on: 11/4/2013 16:35:00 - 16:51:00

Semiconducting nanowires have been demonstrated as promising in a number of devices due to novel properties which are intrinsic to their nanoscale. Silicon nanowires are promising for lithium battery electrodes, where the nanowire geometry allows for relaxation of mechanical stress associated with lithium insertion. Silicon nanowires are also attractive for thin film solar cells, where the nanowire geometry results in enhanced broadband absorption and enables thinner cells with increased efficiency. As well as silicon, the properties of silicon carbide nanowires including large specific surface area and high aspect ratio lend them to application as supercapacitor electrodes and field emitters respectively. Typically these materials are grown via vapor deposition techniques at high temperatures in harsh chemical environments. These growth conditions require refractory, rigid substrates. Yet in order to integrate these materials into devices, they must be accurately transferred and electrically contacted on the array scale. Previous efforts in the transfer step have utilized the Langmuir-Blodgett technique or polymer stamping. The arrays are then electrically contacted through metal evaporation methods. Here we demonstrate an improvement over these previous techniques by growing the nanowires directly on a conductive, flexible, and mechanically robust graphene. The arrays of wires may then be transferred in a facile manner while maintaining good electrical contact throughout the array. The wire-graphene hybrids are structurally characterized via Raman spectroscopy, scanning electron microscopy and x-ray diffraction confirming their chemical identity and crystalline nature. As well, Raman spectroscopy is used to confirm that high quality graphene is maintained during the nanowire growth process. Transfer of the hybrids is demonstrated to a number of arbitrary substrates while maintaining electrical contact. Silicon carbide nanowire – graphene hybrids are also applied as a supercapacitor electrodes with performance indicating good electrical contact throughout the array. Lifetime cycling studies on these electrodes indicate that the hybrid is a robust material with over 95% capacitance retention after 10,000 charge discharge cycles in aqueous electrolyte.

(191f) Inorganic Organic Ligand Exchange On the Surface of Zinc Sulfide Nanocrytals

Lawal, Q., Stanford University
Herron, S., Stanford University
Bent, S. F., Stanford University
Originally presented on: 11/4/2013 16:51:00 - 17:07:00

Inorganic ligand exchange on the surface of zinc sulfide nanocrystals

Qudus Lawal, Steve M. Herron and Stacey F. Bent

Department of Chemical Engineering, Stanford University, Stanford, CA 94305

Nanocrystals (NCs) have been the subject of intense interest because of their current and potential applications in optical and opto-electronic devices, sensing, and medical therapeutics. Control over the surface chemistry of NCs is crucial to the performance of the NCs in opto-electronic devices, and the NC surface properties play a major role in thin-film formation from nanocrystal inks. Specifically, small inorganic species such as ammonium sulfide can provide colloidal stability in polar solvents and may facilitate charge transport between individual nanocrystals, creating more opportunities for device integration of these materials. In these studies, zinc sulfide nanocrystals were synthesized with long-chain hydrophobic ligands including oleylamine. Subsequently, a series of ligand exchanges were performed on these nanocrystals to replace the nascent nonpolar ligands with a metal-free inorganic species, ammonium sulfide. This was done in a two-step process, in which the first step was to reduce the length of the bulky oleylamine ligand to butyl amine, to facilitate transport of ammonium sulfide to the nanocrystal surface. The NC's were dispersed in the polar solvent formamide to stabilize this matrix before finally extracting excess hydrophobic ligand with hexanes. Different mass ratios (0.1:1, 1:1, 1:10) of NC to sulfide ion were used to determine the optimal ratio for a desirable exchange. The NCs before and after ligand exchange were characterized using FTIR spectroscopy to determine the efficacy of ligand exchange by monitoring the change in surface chemical environment, and UV-Vis spectroscopy to examine the optical characteristics of the NCs before and after ligand exchange. The effect of ligand exchange on the NC properties will be discussed.

(191g) Leveraging Molecular and Quantum Confinement in Arrayed Nanostructures for Energy Technologies

Crosser, L. A., Mainstream Engineering Corporation
Nicol, K. T., Mainstream Engineering Corporation
Chester, G. E., Mainstream Engineering Corporation
Zastrow, D. J., Mainstream Engineering Corporation
Schwartz, N. R., Mainstream Engineering Corporation
Wong, J. C., University of Florida
Chew, J. O., University of Florida
Ziegler, K. J., University of Florida
Hill, J. J., Mainstream Engineering Corporation
Reeves, R., Mainstream Engineering
Originally presented on: 11/4/2013 17:07:00 - 17:23:00

Structural- and energy-based anisotropy in nanomaterials can enhance the material's ability to store, transport or convert energy. As a result, researchers have demonstrated that nanomaterials can have advantageous properties relative to their bulk material. For example, transport mechanisms can be selectively tuned by altering nanostructure geometry. Many novel lab-scale experiments have demonstrated this phenomenon and how it can be used to enhance device performance. However, assembly of nanostructures at the device-scale while maintaining the property enhancement is the dominant engineering challenge. Mainstream Engineering is currently developing several nanomaterial-based energy conversion and storage technologies with a focus on device manufacturability. We will highlight our current efforts to not only fabricate nanomaterials with unique and advantageous properties, but also assemble these nanostructures in a configuration that maintains the properties at macro-scale. Using template-directed assembly of nanoparticles and wires, we will discuss their application to thermoelectric composites, hydrogen, oxygen and thermal energy storage, photovoltaics, as well as the development of electrode materials for electrochemical double layer capacitors.

(191h) Development and Application of Tin/Indium (Sn/In) Nanoparticles As Low Melting Temperature Nanosolder Materials

Shu, Y., University of Massachusetts Lowell
Rajathurai, K., University of Massachusetts Lowell
Gao, F., University of Massachusetts Lowell
Cui, Q., University of Massachusetts Lowell
Gu, Z., University of Massachusetts Lowell
Originally presented on: 11/4/2013 17:23:00 - 17:39:00

Low melting temperature lead-free solder materials are attractive for many electronic assembly and packaging processes. There are several advantages for low temperature applications, for example, applicable for flexible electronics, environmental friendly, and higher energy efficiency. Herein we report the synthesis and development of tin/indium nanosolder particles to serve as a new nanosolder material for low temperature processing and applications. The nanosolder particles are synthesized through a surfactant-assisted chemical reduction method in an aqueous phase. Parameters that affect the size and shape of the nanosolder particles have been investigated. SEM, TEM, XRD and EDS have been used to determine the structure and composition of the nanosolders. DSC measurement has been employed to determine the thermal property of the nanosolder particles. Furthermore, the characterizations of melting behavior and intermetallic formation are provided from the reflow and soldering tests. The results indicate that the tin/indium nanosolder particles have good potential for low temperature soldering applications.

(193a) Nanostructured Optoelectronics: Using New Energy States and High-Energy Photons

Singh, V., University of Colorado Boulder
Goodman, S.,
Nagpal, P., University of Colorado Boulder
Originally presented on: 11/4/2013 15:15:00 - 15:33:00

Semiconductor nanocrystals or quantum dots are currently being explored for solution processable optoelectronic devices, and as candidates for developing third-generation photovoltaics. While the size-tunable bandgap and solution chemistry affords attractive opportunities, discrete quantum-confined electronic states have been researched for utilizing photons with energy higher than the nominal nanocrystal bandgap. Recently, we discovered new energy states in non-toxic, earth abundant semiconductor nanocrystals, which can be utilized for development of efficient optoelectronic devices like light emitting diodes (LEDs), and extremely-thin photovoltaics and photodetectors for absorption of incident solar radiation. I will summarize the recent efforts in my group on proof-of-concept demonstrations, in single nanoparticle studies and their ensemble counterpart nanoscaled devices, of these new energy states, and advances made towards utilizing high-energy photons and hot-carriers in nanostructured films.

(193b) Plasma Synthesis of Metal-Sulfide Nanocrystals

Thimsen, E., Washington University in St. Louis
Kortshagen, U. R., University of Minnesota
Aydil, E. S., University of Minnesota
Originally presented on: 11/4/2013 15:33:00 - 15:51:00

Metal-sulfide nanocrystals (e.g. CdS, PbS, Cu_2ZnSnS_4) are typically produced through liquid solution synthesis in solvents that may pose challenges in achieving high purity materials. Moreover, solution synthesis generates large volumes of hazardous waste and the commonly used hot injection method is difficult to scale up to achieve high throughput. Gas-phase approaches, such as the continuous nonthermal plasma processes we have been developing, do not require the use of solvents and can offer very high purity and very high throughput. The benefits of the plasma approach come without sacrificing control over materials properties. In many cases the materials have excellent quality and achieve high performance in applications. For example, intrinsic and doped silicon nanoparticles have been produced and already commercialized for use in solar cells. However, to date no plasma synthesis method has been developed for metal-sulfides. Starting with a model material, zinc sulfide (ZnS), we have developed a gas phase approach to making binary sulfide nanocrystals. ZnS is a wide band gap semiconductor (E_g = 3.7 eV) that has been extensively used as a phosphor, buffer layer in thin film solar cells and surface passivation shell for semiconductor quantum dots. Crystalline ZnS particles with 2-10 nm diameters were synthesized from diethyl zinc and elemental sulfur vapor. Particle diameter and composition could be controlled through the gas flow rate, feed gas composition and pressure. Even without optimization, production rates were as high as 250 mg/hr with 25% mass yield in our bench scale 8 cm^3 reactor. The nanocrystals could be deposited on substrates directly or formed into stable nanocrystal dispersions (inks) by capping their surfaces with organic ligands such as oleylamine. This, to our knowledge, is the first report of a metal-sulfide nanocrystalline material made using a nonthermal plasma process. Beyond ZnS, we will also present findings on the Fe, Sn and Cu sulfides.

(193c) Layer-By-Layer/M13 Virus Assembled Porous Photoanodes for Efficient Electron Collection in Dye-Sensitized Solar Cells

Chen, P. Y., David H. Koch Institute for Integrative Cancer Research
Belcher, A. M., Massachusetts Institute of Technology
Hammond, P. T., Massachusetts Institute of Technology
Originally presented on: 11/4/2013 15:51:00 - 16:09:00

Layer-by-Layer/M13 Virus Assembled Porous Photoanodes Facilitating Electron Collection in Dye-Sensitized Solar Cells

Presenter: Po-Yen Chen

Principal Authors: Prof. Angela M. Belcher and Prof. Paula T. Hammond

Session Selection: Thin Film and Organic Photovoltaics

Keywords: Layer-by-layer assembly; M13 bacteriophage; Electron transport; Electron diffusion length; Dye-sensitized solar cells.

Abstract

Dye-sensitized solar cells (DSSCs) represent a promising solar technology because of their ease of fabrication and high power conversion efficiency (PCE). Two important factors determining the device performance, light harvesting and electron collection, are influenced by the architecture of the titanium dioxide (TiO_2) photoanode. However, the widely-used architecture of randomly-packed nanoparticles, made via screen printing or doctor blading, are optimized for light harvesting but not necessarily for electron collection. Controlling the nanoscale morphology of the TiO_2 photoanode is required to minimize electron recombination; however, this control is not necessarily accessible via doctor blading or screen printing. To address this challenge, strategies have been developed to create oriented nanostructures and process them into photoanodes to form directional electron pathways and to improve the electron collection. While these methods improve the electron collection of DSSC, complicated processes limit their application in mass production. Layer-by-layer (LbL) assembly is a widely-used thin film deposition method, which offers several advantages: it is a simple and cheap process with strong potential for scalability and a high degree of nanoscale control over thickness, morphology, and chemical composition of films. Nano- and micro-porous LbL assemblies have been formed using a simple pH post-treatment developed by Rubner et al, and these porous films have been used to template DSSC photoanodes.

Herein, we present a facile process to generate an oriented porous TiO_2 photoanode via dip and spray LbL assemblies, using alternating layers of positively charged linear poly(ethylenimine) (LPEI), negatively charged poly(acrylic acid) (PAA) and phage. In this study, bilayer architectures without phage $(LPEI/PAA)_n$ and tetralayer architectures with phage $(LPEI/PAA/LPEI/phage)_m$ are both assembled via dip and spray LbL. A porous transition occurs to the nonporous $(LPEI/PAA)_n$ or $(LPEI/PAA/LPEI/phage)_m$ films when they are exposed to a low pH environment. A liquid phase deposition reaction is then used to achieve a conformal surface coating of TiO_2. The subsequent calcination step removes the sacrificial polymer and phage template to generate an anatase TiO_2 photoanode. Moreover, the presence of the high aspect ratio M13 bacteriophage (phage) further optimizes electron collection by incorporating into the LbL-assembled porous films to generate interconnections among the TiO_2 crystallites inside the oriented photoanode.

The DSSCs employing the LbL-templated photoanodes show excellent photocurrent and high PCE with relatively low dye loading compared to nanoparticle DSSC. The dip and spray LbL-templated DSSC show ~98% and ~106% of photocurrent with only ~52% and ~64% of the adsorbed dye compared to the nanoparticle device, respectively. This is due to the efficient electron collection in both interconnected porous LbL-templated photoanodes, while the randomly packed nanoparticle photoanode is designed for sufficient light harvesting but is not ideal for efficient electron collection. To further optimize the electron pathway inside the TiO_2 pore structure, the high aspect ratio phage nanowire as a biomineralization template is incorporated within the LbL templates. The phage templates form a percolative network to generate interconnected TiO_2 crystallite chains throughout the photoanode. To realize the effect of the interconnected TiO_2 wires embedded in the architecture, the phage-containing LbL-templated device performance was investigated. Both phage-containing dip and spray LbL-templated devices can further achieve 114% of the photocurrent compared to the nanoparticle device, while the dye loadings (and thus light harvesting) almost remain the same. Also, the inclusion of phage-templated nanowires further enhances the electron diffusion length (L_n) of LbL-templated devices. The L_n between the LbL-templated devices without and with phage templates was improved by a factor of from ~4-5 and ~7-8, respectively. These results shows that the inclusion of phage-templated nanowires in LbL-templated photoanodes can help the crystallite interconnection and optimize the electron pathway, which further reduces interfacial recombination and thus further enhances electron collection.

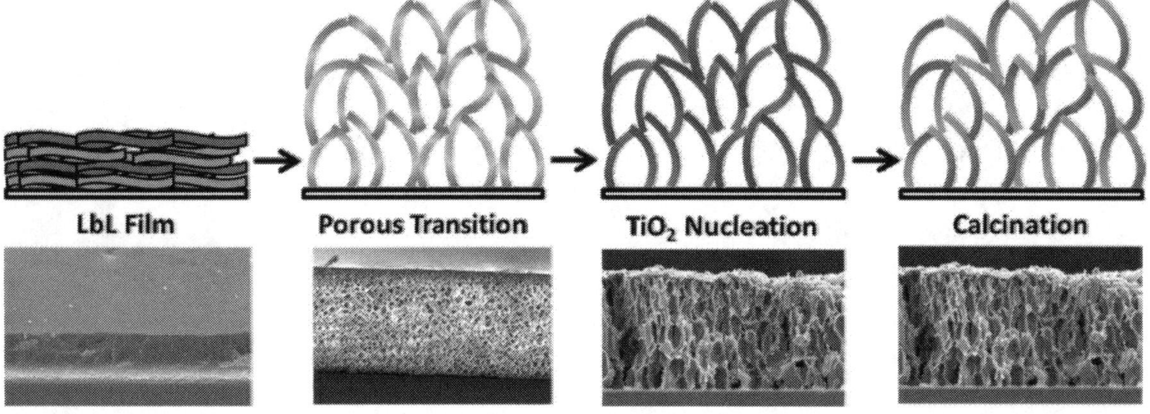

Figure 1.

(193d) Interface Engineering in Solid-State Quantum Dot-Sensitized Solar Cells: Strategies to Improve Charge Collection

Roelofs, K. E., Princeton Unversity
Brennan, T. P., Stanford University
Yang, T. Q., Stanford University
Bent, S. F., Stanford University
Originally presented on: 11/4/2013 16:09:00 - 16:27:00

Solid-state dye-sensitized solar cells (ss-DSSCs) are composed of cheap, abundant materials, and have record power conversion efficiencies now at 6.1%.[1] In this work, the dye in ss-DSSCs was replaced with inorganic quantum dots as the light-absorbing material, creating solid-state quantum dot-sensitized solar cells (ss-QDSSCs). Quantum dots (QDs) show favorable absorption properties due to their size-dependent band gap, as well as higher molar extinction coefficients than commonly-used dyes. Increased absorption is important as the active layer thickness in the devices is limited by charge recombination.

Despite the promise of quantum dots as a light absorbing material, QDSSC efficiencies remain low – especially in solid-state devices – due to high rates of recombination and low surface coverages of the photoanode by QDs. A recent study places QD surface coverages below 14% of the photoanode surface area.[2] The record efficiency for liquid QDSSCs is 5.6%,[3] and only 1.5% for QDSSCs with a solid-state hole-transport material (HTM).[4] In this work, we examine two strategies for improving efficiency in QDSSCs: inserting barrier layers to reduce recombination, and increasing QD surface coverage.

We first demonstrate that ultra-thin recombination barrier layers of Al_2O_3 deposited by atomic layer deposition (ALD) can improve the performance of cadmium sulfide (CdS) quantum dot-sensitized solar cells (QDSSCs) with spiro-OMeTAD as the solid-state HTM. CdS QDs were grown directly on nanoporous TiO_2 substrates by successive ion layer absorption and reaction (SILAR). The deposition of CdS quantum dots was confirmed with transmission electron microscopy (TEM), X-ray photoelectron spectroscopy (XPS), and UV-vis spectroscopy. We explored depositing the Al_2O_3 barrier layers by ALD either before or after the QDs, resulting in TiO_2/Al_2O_3/QD and TiO_2/QD/Al_2O_3 configurations. The effects of barrier layer configuration and thickness were tracked through current-voltage (*J*-V) measurements of device performance and transient photovoltage measurements (TPV) of electron lifetimes.

The Al_2O_3 barrier layers were found to suppress dark current and increase electron lifetimes with increasing Al_2O_3 thickness in both barrier layer configurations in CdS ss-QDSSCs. For thin barrier layers, gains in open-circuit voltage (V_{OC}) and concomitant increases in efficiency were observed, although at greater thicknesses, losses in short-circuit current (J_{SC}) caused net decreases in efficiency. Devices with 1 Al_2O_3 ALD cycle deposited before the QDs had the highest average device parameters, with an efficiency of 0.35%, J_{SC} of 0.83 mA/cm^2, V_{OC} of 0.70 V, and fill factors of 0.60.

A close comparison of the electron lifetimes in TiO_2 in the TiO_2/Al_2O_3/QD and TiO_2/QD/Al_2O_3 configurations suggests that electron transfer from TiO_2 to spiro-OMeTAD is a major source of recombination in ss-QDSSCs, although recombination of TiO_2 electrons with oxidized QDs can also limit electron lifetimes, particularly if the regeneration of oxidized QDs is hindered by a too-thick coating of the barrier layer.

However, the low efficiencies of CdS ss-QDSSCs, due in part to the large, non-ideal band gap of CdS, motivated the development of a lead sulfide (PbS) ss-QDSSC system. PbS QDs were grown *in situ* on nanoporous TiO_2 by SILAR as well as by ALD. With the ultimate goal of increasing QD surface coverage, we compare the impact of these two synthetic routes on the absorption and device electrical properties. Thus far, the highest efficiencies were observed in devices with ALD-grown QDs: 0.30 % at 10 ALD cycles, although the devices with SILAR-grown QDs had the highest J_{SC}'s observed in our work to date, of 0.98 mA/cm^2, and efficiencies of 0.23 %. We hypothesize that the higher J_{SC}'s and the higher reverse-bias photocurrent observed in the case of SILAR-grown QDs is due to a higher number of defect states in these solution-deposited QDs, as compared to the vapor-deposited QDs grown by ALD.

To understand the effects of QD surface coverage on device performance, particularly interfacial recombination, TPV measurements were carried out to determine electron lifetimes, and external quantum efficiency (EQE) measurements were used to track charge collection efficiencies. These measurements were taken for varying QD deposition cycles. Electron lifetimes were found to decrease with increasing SILAR cycles. We hypothesize that recombination to oxidized QDs is the critical source of recombination in these devices. We are currently investigating methods to increase QD surface coverage and improve electron injection by pre-deposition chemical treatments of the QD surface.

1. Chung, I.; Lee, B.; He, J.; Chang, R. P. H.; Kanatzidis, M. G., All-Solid-State Dye-Sensitized Solar Cells with High Efficiency, *Nature* **2012,** *485* (7399), 486-489.

2. Guijarro, N. S.; Lana-Villarreal, T.; Mora-SerÃ³, I. N.; Bisquert, J.; GÃ³mez, R., CdSe Quantum Dot-Sensitized TiO_2 Electrodes: Effect of Quantum Dot Coverage and Mode of Attachment, *J. Phys. Chem. C* **2009,** *113* (10), 4208-4214.

3. Lee, J.-W.; Son, D.-Y.; Ahn, T. K.; Shin, H.-W.; Kim, I. Y.; Hwang, S.-J.; Ko, M. J.; Sul, S.; Han, H.; Park, N.-G., Quantum-Dot-Sensitized Solar Cell with Unprecedentedly High Photocurrent, *Scientific Reports* **2013,** *3.*

4. Lee, H.; Leventis, H. C.; Moon, S. J.; Chen, P.; Ito, S.; Haque, S. A.; Torres, T.; Nuesch, F.; Geiger, T.; Zakeeruddin, S. M., et al., PbS and CdS Quantum Dot-Sensitized Solid-State Solar Cells: ''Old Concepts, New Results,'' *Adv. Func. Mater.* **2009,** *19*, 2735-2742.

(193e) Virus-Templated Three-Dimensional Photoanodes for Dye-Sensitized Solar Cells With Efficient Electron Collection and Plasmon-Enhanced Light Absorption

Chen, P. Y., David H. Koch Institute for Integrative Cancer Research
Hammond, P. T., Massachusetts Institute of Technology
Belcher, A. M., Massachusetts Institute of Technology
Originally presented on: 11/4/2013 16:45:00 - 17:03:00

Virus-Templated Three-Dimensional Photoanodes for Dye-Sensitized Solar Cells with Efficient Electron Collection and Plasmon-Enhanced Light Absorption

Presenter: Po-Yen Chen

Principal Authors: Prof. Paula T. Hammond and Prof. Angela M. Belcher

Session Selection: Nanostructured Photovoltaics

Keywords: M13 bacteriophage; Three-dimensional network; Localized surface plasmon; Dye-sensitized solar cells.

Abstract

M13 virus is a filamentous bacteriophage of 880 nm in length and 6.5 nm in diameter, and has been demonstrated as a versatile bio-template to assemble and to nucleate a broad range of materials. The targeted materials can be assembled along the phage surface, and used to assemble functional materials into two- and three-dimensional structures for various device applications, including photovoltaics. Dye-sensitized solar cells (DSSCs) are a promising solution-processed photovoltaic technology. The power conversion efficiency (PCE) of DSSCs is mainly determined by light harvesting and electron collection. In general, to efficiently harvest light and collect electrons, different design criteria must be considered. For efficient electron collection, nanostructures of titanium dioxide (TiO_2) with high aspect ratio have been proven to be effective, compared to the commonly used architecture of photoanodes composed of randomly-packed nanoparticles (NPs). However, the reduced surface area of one-dimensional structures leads to insufficient dye adsorption and thus restricts light harvesting. As a result, achieving sufficient light harvesting requires developing strong-absorbing dye-molecules or various mechanisms to increase optical absorption, *e.g.* localized surface plasmon (LSP)-enhanced optical absorption by using metal NPs. Therefore, different functional materials are required for electron collection and light harvesting, which makes it attractive to develop an approach to assembling different functional components together into a homogeneously distributed composite with a three-dimensional (3-D) structure.

Here, we report a novel 3-D viral network as a versatile platform for DSSCs which can template TiO_2 nanowires for improved electron transport, and can also bind Au NPs and then incorporate them uniformly into the photoanodes for the LSP-enhanced light harvesting of dye-molecules. The virus-templated TiO_2 nanowires promote electron transport from the photo-excited dyes to the current collector (**Figure 1**) in the photoanodes, resulting in an increased electron diffusion length compared to conventional NP-based photoanodes. In addition, the M13 virus can bind Au NPs prior to the templated synthesis of TiO_2; the resulting TiO_2-coated gold nanoparticles generate LSP effects that improve the photo-absorption of dye-molecules adsorbed on the Au@TiO_2 nanocomposites (**Figure 1**). The 3-D thin film is composed of tightly packed TiO_2 crystallites forming interconnected nanowires (scanning electron microscopy image in **Figure 1**). Compared to the size of individual viral particles, the virus-templated TiO_2 nanowires are approximately 100 nm in diameter and 2-3 mm in length, which suggests that each nanowire in the photoanode is templated by a bundle of viruses. In addition, the mesoporous structure of the interconnected nanowire network with interconnected pores maximizes interfacial contact with the electrolyte, which increases the mobility of redox couples and also minimizes back recombination.

The ability of M13 viruses to bind with Au NPs can fabricate the virus-templated Au@TiO_2 photoanodes to improve the optical absorption. The resulting Au@TiO_2 nanocomposite results in the LSP-improved light harvesting of dye molecules in the DSSC photoanodes. The device performance of the virus-templated Au@TiO_2 photoanodes is optimized by varying the concentration of the plasmonic Au NPs and the thickness of the photoanodes (**Figure 1**). The virus-templated Au@TiO_2 DSSCs (concentrations of Au NPs, ~0.24 wt.% and ~0.8 wt.%) always show higher PCE than the virus-templated TiO_2-only DSSCs with similar thicknesses of the photoanodes. However, PCE decreases when the concentration of Au NPs is increased to 2.4 wt.%, probably due to the increased trapping of photo-excited electrons by Au NPs and competing light absorption of Au NPs with dye-molecules. Moreover, the LSP from Au NPs in nanowire-based photoanodes has the ability to increase the light harvesting without affecting the high diffusion length achieved by the virus-templated nanowire network, and thus maintain the efficiency of electron collection at 99.4%. Thus, the optimal concentration of incorporated Au NPs was found to be 0.8 wt.%, and the best virus-templated plasmon-enhanced DSSC achieves an increased efficiency of 8.46% due to the efficient electron collection and the LSP-improved light harvesting simultaneously.

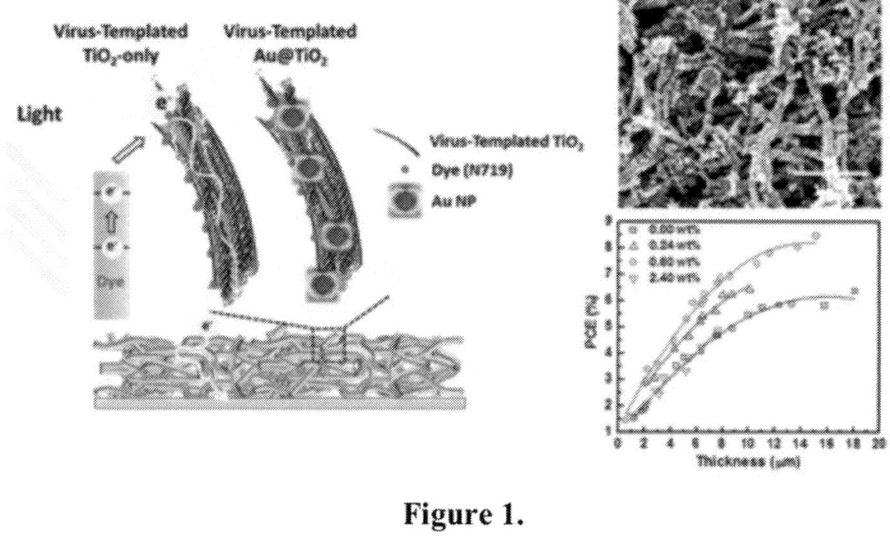

Figure 1.

(193f) Effect of Oxygen Flow Rate On the Propeties of RF-Sputtered SnO2 Protective Layer and the Performance of DSSC

Lu, H. C., Chang Gung University
Chen, Y. H.,
Originally presented on: 11/4/2013 17:03:00 - 17:21:00

RF-sputtered tin oxide (SnO_2) thin films were attempted to be used as the protection layers for the transparent conductive oxide (TCO) electrodes of dye-sensitized solar cells (DSSC). The effect of oxygen flow rate during the sputtering process on the properties of SnO_2 thin films and the performance of DSSC were also investigated. Tin oxide thin films with rutile structure, high average visible transmittance (>85%), and moderately low resistivity (Â □`10^{-1} Ωcm) were obtained when deposited on glass substrates at 373 K. By acting as protection layers for the TCO electrodes, not only the changes in the electrical properties of the TCO electrodes during the fabrication process of DSSC could be suppressed, but the performance of the DSSC was also found to be enhanced.

(193g) Improved Performance of Polymer Solar Cells By Silica Nanoparticles

Shen, H.,
Mackay, M. E., Michigan State University
Originally presented on: 11/4/2013 17:21:00 - 17:39:00

We have found an improvement in the performance of polymer-based solar cells by filling the bulk-heterojunction active layer consisting of polymer and fullerene with 133 nm diameter silica nanoparticles. These silica particles are electrical insulators and weak light scatters in the visible spectra due to their small sizes and similar refractive index to the polymer and fullerene. With a particle volume fraction of about 10% in the active layers, the performance of the devices was not hurt but instead slightly improved. In the devices made with regioregular poly(3-hexylthiophene) (P3HT) and phenyl-C61-butyric acid methyl ester (PCBM), the improvement was evident across a wide range of mixing ratios between PCBM and P3HT. In particular, the power conversion efficiency at AM 1.5G spectrum was increased from 2.75±0.06% to 3.35±0.17% in the device with 60% of PCBM by weight. The improvement mainly comes from the larger fill factors, indicating enhanced charge transport properties in the active layer. We investigated the mechanism of the improvement by using small angle neutron scattering and other microscopy techniques. This study suggests that when nanoparticles are implemented into the active layer, their physical presence and its influence on the structure of the bulk-heterojunction should be considered in addition to other electrical and optical properties they possess.

(233a) Invited: Harnessing Instabilities Of Oil/Water Interfaces To Tailor Structure and Functionality Of Amphiphilic Polymer Assemblies

Hayward, R. C., University of Massachusetts, Amherst
Originally presented on: 11/5/2013 8:30:00 - 9:20:00

Micellar and microparticulate assemblies of amphiphilic polymers hold promise for a variety of applications in delivery of therapeutic compounds and imaging contrast agents. However, as such materials typically face large barriers against structural reorganization in aqueous environments, their assemblies are usually far from equilibrium, yielding great sensitivity of their structures to the route by which they were processed. Our group has studied the formation of micelles and microparticles through interfacial instabilities during the removal of volatile solvents from oil-in-water emulsions containing amphiphilic block copolymers. We have shown that this phenomenon is consistent with a nearly vanishing solvent/water interfacial tension that is expected upon a transition from a macrophase-separated emulsion to a microemulsion phase. Further, the morphologies developed through this process can be tuned by tailoring the composition of the copolymer as well as the concentration of small-molecule surfactants. Finally, we have studied the opportunities that this approach offers for the preparation of multi-functional micelles, containing several different types of inorganic nanoparticles, and of textured particles with interesting immunogenic properties.

(233b) Invited: Creating Therapeutic Nanomaterials From Proteins

Champion, J. A., Georgia Institute of Technology

Originally presented on: 11/5/2013 9:20:00 - 10:10:00

Protein drugs can provide a key advantage over small molecule drugs; they evolved to perform their function, while small molecules are often selected for "best" function compared to a pool of candidates. However, proteins can present challenges in delivery that must be overcome in order to be used as therapeutic drugs. Though their folded structure and specific sequence are critical to their biological function and makes them sensitive and difficult to package, but also provides an opportunity to create materials from proteins that is not possible for small molecules. The main goal of this work is to engineer nanomaterials made from therapeutic proteins, and control the interactions between these materials and cells for applications in cancer, inflammation and vaccines. Similar to polymeric materials, we use self-assembly and covalent chemistry approaches to build nanomaterials from proteins. However, we take advantage of proteins' unique properties of bioactivity, tertiary structure, affinity interactions, and defined and controllable sequence. By controlling the amino acid sequence and nanomaterial fabrication conditions, we engineer materials that protect and deliver proteins, and control their presentation to cells. I will show some of the materials we have made from biologically active proteins and discuss our progress in using these as therapies.

(233c) Invited: Yin/Yang Innate Immune Effectors Are The Key To Understanding Human Neurodegenerative and Plaque Diseases

Barron, A. E., Stanford University
Karande, P., Rensselaer Polytechnic Institute
Originally presented on: 11/5/2013 10:10:00 - 11:00:00

My laboratory has been interested, for the past 10 years, in developing stable, non-natural mimics of natural antimicrobial peptides, the latter of which are better termed host defense peptides (HDPs). For both HDPs and novel peptoid mimics we created and studied side by side, we found that the activity and selectivity of these antibiotic peptides depend on relative extents of net positive charge, hydrophobicity, amphipathicity, and in some cases, their self-association into meta-stable assemblies that dissemble upon encountering electronegative cell membranes. We completed a series of biophysical mechanistic studies utilizing HDPs and peptoids comprising orthogonal activity/selectivity profiles, displaying either high (or low) activity and high (or low) selectivity, providing a grid of mechanistic data that facilitates comparisons of selective and non-selective HDPs. Vesicle leakage, membrane depolarization, and TEM yielded surprising information concerning interrelationships between mechanisms by which peptide and peptoid variants exert damaging effects on pathogens vs. mammalian cells.

Unexpected insights into biology and physiology followed a realization that our most active antibiotic peptoids invariably had concomitant cytotoxicity; we discovered that this is also true of important classes of HDPs. And as many researchers have learned but few have reported in the literature given that it is a negative result from a drug development standpoint, high HDP selectivity in mammalian cell culture does not necessarily nor at all reliably predict low toxicity in living mammals.

An important result of our studies was that we have learned and will explain that bacteria, fungi, and mammalian cells have evolved powerful, complex mechanisms of resisting and recovering from different types of HDP-induced damage; some are internal to the cells, while others involve well-known extracellular constructions.

I will describe striking mechanistic similarities between antibacterial and cytotoxic activities of peptoid HDP mimics and of natural HDPs. The latter are phylogenetically ancient innate immune effectors. Particular peptides constitute one-of-a-kind weapons in the human proteomic arsenal, while also playing indispensible pleiotropic roles at many levels in mammalian physiology. Guided by these insights and the clinical literature, I hypothesize a central role for HDPs in the etiology of human degenerative diseases, including Alzheimer's and other amyloid diseases (atherosclerosis, diabetes, and other maladies) as a consequence of the HDPs' interrelated, highly potent mechanisms of action—both beneficial and damaging—which are profoundly concentration-dependent. Extrapolation of these insights predict that a major key to understanding human health and disease will be gaining a deep understanding of how human lifetime experiences dictate and control temporal and spatial dynamics and concentrations of HDPs and the nanoparticle complexes that they form within tissues and organs of the human body.

(242a) Simultaneous Adsorption and Heterogeneous Oxidation of Endocrine Disrupting Compounds in Wastewater Using Nano Metal Catalyst-Deposited Carbon Nanotubes

Cleveland, V., University of Hawaii at Manoa
Bingham, J. P.,
Kan, E., University of Hawaii at Manoa
Originally presented on: 11/5/2013 8:30:00 - 8:51:00

A process for the removal of bisphenol A (BPA), one of the endocrine disrupting compounds (EDC) from wastewater was developed using nano metal catalyst-deposited carbon nanotubes (CNT). For the past decade BPA have gained more attention as emerging contaminants, and have been linked to multiple cancers below the concentration of ppb. This process relies on simultaneous adsorption and heterogeneous *in-situ* Fenton oxidation for degradation of BPA. The present study investigated how acidic surface modifications of CNTs and the novel amendment of iron onto the CNTs affect the adsorption and oxidation of BPA. The virgin CNTs were treated with acid to increase acidic surface oxide groups while lowering the pH at the point of zero charge (pH_{PZC}) of the CNT, and improve electrostatic interactions with the iron catalyst for enhancing dispersion and deposition of iron onto the CNT. The novel iron amendment method encourages a uniform distribution of iron nanoparticles onto the surface of the CNTs. The SEM and TEM image analysis confirmed the uniform distribution of the iron over the CNT. Adsorption and desorption of BPA onto the CNTs was found to reach equilibrium in less than 5 min. The iron-amended acid-treated CNTs and the iron-amended untreated CNTs followed the Langmuir model of adsorption, indicating a single layer of adsorption. The iron-amended CNTs showed slightly lower maximum adsorption capacity for BPA in water compared to the virgin CNTs. The oxidation of BPA with the iron-amended acid-treated CNTs showed greater than 95% removal of BPA and a H_2O_2 half-life ranging from 36-1386 min at 4.5-45 mmol H_2O_2/mmol BPA at pH 3.5 over the repeated cycle of treatment of BPA. The heterogeneous Fenton oxidation of BPA in water revealed quick adsorption, effective degradation of BPA, and high regeneration of adsorptive capacity with no iron sludge production compared to homogeneous Fenton oxidation. Since the limitation of the process is the oxidation rate of BPA, further investigation will focus on enhancing the reaction kinetics of BPA oxidation by optimizing the ratio of BPA:H2O2:Fe and operating conditions such as temperature, pH and mixing.

(242b) Study of Carbon-Hydrogen Bond On the Microstructure of Hydrogenated Graphite

Zhang, Y., University of Birmingham; Waseda University

Originally presented on: 11/5/2013 8:51:00 - 9:12:00

There is a need to understand the hydrogen sorption mechanism(s). In this work, hydrogenated graphite samples were characterized by TGA-Mass Spectrometry, Raman spectros-copy and high resolution TEM. It was found that graphite milled for various times (0-40 h) had different d-spacings and CHx ($x \leq 3$) ratios. Graphite milled up to 10 h that was subsequently heated (under argon to 990 °C) desorbed only hydrogen, while above 10 h the amount of desorbed hydrogen decreased and an increasing amount of methane was also evolved. The 10 h sample had no measurable sp3 bonding, and desorbed 5.5 wt% hydrogen when heated. While the 40 h sample had a d-spacing of 0.48 nm, both sp2 and sp3 bonding, and desorbed 5.1 wt% pure hydrogen. The mechanism of graphite hydrogenation is discussed in terms of hydrogen atoms being predominately trapped at the graphite edges, rather than between graphene layers.

References

[1]. S. Orimo, G. Majer, T. Fukunaga, A. Zuttel, L. Schlapbach and H. Fujii, *Applied Physics Letters*, 1999, **75**, 3093-3095.

[2]. T. Ichikawa, H. Fujii, S. Isobe and K. Nabeta, *Applied Physics Letters*, 2005, **86**.

[3]. H. Miyaoka, T. Ichikawa, T. Fujii, W. Ishida, S. Isobe, H. Fuji and Y. Kojima, *J Alloy Compd*, 2010, **507**, 547-550.

(242c) Efficient Removal of Both Arsenic and Fluoride By Lanthanum Modified Carbon Nanoparticles

Li, G., National University of Singapore
Yu, Y., National University of Singapore
Chen, J. P., National University of Singapore
Ma, Y., National University of Singapore
Originally presented on: 11/5/2013 9:12:00 - 9:33:00

In this study, a novel lanthanum modified carbon nanoparticle was prepared and applied in the removal of both arsenic and fluoride. Electron microscopic investigation showed that the prepared carbon particles were around 50-70 nm in diameter. The adsorbent showed good adsorption efficiency of both arsenic and fluoride in a wide range of solution pH. In the kinetic study, it was observed that the adsorption process was fast and around 90% of its final adsorption was achieved within the first 10 min. The Langmuir equation well described the adsorption isotherm. At the optimal pH, adsorption capacities of arsenic and fluoride were around 102mg/g and 122mg/g, respectively. The ionic strength had a slight influence on the arsenic and fluoride removal. The effect of common existing anions such as PO_4^{3-} HCO_3^- SO_4^{2-} and SiO_3^{2-} and humic acid on the adsorption process was investigated in detail. From this investigation it was concluded that the La-immobilized carbon nanoparticle showed the fast removal of toxic arsenate and fluoride from water with very high adsorption efficiency which is beneficial to its application in the water treatment.

(242d) Efficient Removal of Fluoride By Lanthanum Modified Seaweeds Sargassum S.P

Li, G., National University of Singapore
Yu, Y., National University of Singapore
Chen, J. P., National University of Singapore
Originally presented on: 11/5/2013 9:33:00 - 9:54:00

In this study, a novel lanthanum modified seaweeds (Sargassum S.P.) was prepared and applied in the fluoride removal from the water. The adsorbent showed good adsorption efficiency on the fluoride removal in a wide range of solution pH and the maximum adsorption capacity reached 45mg/g at pH 4. In the kinetic study, it was observed that the adsorption process was very fast and around 75% of its final adsorption was achieved within the first 10 min. The Frenudlich isotherm equation well described the adsorption process. At the optimal pH, adsorption capacities of fluoride were around 125mg/g. The fluoride removal efficiency increased with the increase of the ionic strength in the solution. The effect of common existing anions such as PO_4^{3-}, HCO_3^-, SO_4^{2-}, $SiO_3^{2-}$and AsO_4^{3-} and humic acid on the adsorption process was also investigated in detail. The presence of SO_4^{2-} did not show a significant impact on the fluoride removal while the adsorption efficiency was decreased with the increase of the concentration of other ions and humic acid. From this investigation it was concluded that the La-immobilized seaweed showed the fast fluoride removal from water with very high adsorption efficiency which is beneficial to its application in the water treatment.

(242e) Size-Dependent Removal of Arsenite From Aqueous Solution By Akaganeite (β-FeOOH)

Wu, S. H., School of Chemical Engineering and Technology
Sun, Y. L. Sr., Department of Chemical Engineering, School of Chemical Engineering and Technology, Tianjin University,
Jia, S. Y. Sr., Department of Chemical Engineering, School of Chemical Engineering and Technology, Tianjin University,
Zan, T. T. Jr., Department of Chemical Engineering, School of Chemical Engineering and Technology, Tianjin University,
Wang, J. Sr.,
Originally presented on: 11/5/2013 9:54:00 - 10:15:00

Crystal size and morphology had a significant effect on its adsorption behaviors. Therefore, in the present study, we have synthesized three different sizes (from <10 nm="" to="">100 nm) and morphologies of akaganeite nanometer materials to investigate their different performance of the adsorption of the arsenite. The kinetics data of these three adsorbents showed little difference and were best described by the pseudo-first-order and pseudo-second-order equation which assumed that the adsorption rate depended on both physical and chemical adsorption. The adsorption equilibrium isotherm fitted a Langmuir isotherm model well and the maximum loading capacity could reach 790 mg of arsenite/g of adsorbent. The value was comparatively higher than previously reported. This was due to the high specific surface and the unique combination between akaganeite and arsenite. The influence of the pH on the adsorption rate was also discussed. The results showed that the adsorption rate of the size smaller than 100 nm had little change with the increasing of pH. While the adsorption rate of the size larger than 100 nm presented apparent difference as pH shifted from 7.3 to 10. A mechanism that chloride ion was replaced by arsenite had been proposed. The phosphate and silicate showed significant competitiveness with arsenite, while sulfate showed little influence. Besides, there was no change on the crystalline structure of the adsorbents after adsorption. This work offered a novel understanding on the adsorption mechanism of the arsenite by the akaganeite.

(242f) Removal of Ammonia From Water Using Nanoscale Zero Valent Irons Modified Fly Ash-Derived Zeolite Adsorbents

Li'an, H.,
Minmin, L.,
Originally presented on: 11/5/2013 10:15:00 - 10:36:00

Fly ash-derived zeolite A was impregnated with different weigh ratios of nanoscale zero valent irons to produce new adsorbents (x) Fe-Z. Adsorbents were tested to remove ammonia from water solutions. Simultaneous removals of ammonia were examined via batch adsorption experiments. The new adsorbents were characterized by the X-ray diffraction (XRD), Brunauer-Emmett-Teller-N2 surface areas, pore size distributions, Transmission Electron Microscope (TEM) and Hysteresis loops in order to probe effects of the weigh percentages of NZVI impregnated. Virgin and exhausted adsorbents were examined via X-ray photoelectron spectroscopy (XPS) analyses to elucidate adsorption mechanisms. Maximum adsorption capacity of the ammonia for (50) Fe-Z adsorbent was 50.2mg/g. The maximum adsorption capacity of (50) Fe-Z is more than that of other adsorbents for ammonia removal from water. The performance was affected by the pH of aqueous solution, the initial concentration and reaction temperature. The adsorption mechanisms of ammonia on (x) Fe-Z are attributed to the ion exchange and electrostatic adsorption. Some parts of NZVI were oxidized and substituted the aluminum in the framework of zeolite A in the synthesis process of (x) Fe-Z. It made more sodium ions be exchanged with positive ions after NZVI impregnated into zeolite A. In addition, the negative surface charge of (x) Fe-Z attracted the positive ammonia ions. Furthermore, the dispersion of NZVI on the surface of (x) Fe-Z was benefit for (x) Fe-Z to be separated magnetically. Moreover, the presence of iron hydroxide promoted adsorption of ammonia on the surface of (x) Fe-Z.

(242g) Drinking Water Disinfection Using Silver Nanoparticle Impregnated Activated Carbon Hybrid

Biswas, P., Indian Institute of Technology Bombay
Bandyopadhyaya, R., IIT Bombay
Originally presented on: 11/5/2013 10:36:00 - 10:57:00

In most household drinking water purification systems, activated carbon (AC) is extensively used as a filtration medium. However, prolonged use results in biofouling, which can be eliminated by utilizing the biocidal effect of silver (Ag). In the present work, Ag nanoparticle has been impregnated in AC, for disinfection and purification of drinking water.

Ag nanoparticles were synthesized by three different methods, namely, UV reduction (A), UV reduction in presence of ammonia (B) and thermal reduction (C). Silver nitrate was used as the precursor, while tri-sodium citrate acts as both a reducing as also a stabilizing agent, to synthesize a stable Ag nanoparticle dispersion in water. Methods A, B and C showed mean primary nanoparticle diameter of 28.1 nm, 44.8 nm and 26.7 nm respectively, with standard deviation of 12.6 nm, 19.4 nm and 11.7 nm respectively. So, method B was not desirable due to the larger particle size. Between the other two, method C does not work at higher precursor concentration and results in aggregation and sedimentation of nanoparticles. So method A was selected, as usage of higher initial precursor concentration was possible, due to the slow rate of reduction, resulting in higher quantity of Ag particles being formed. The AC used in the present work is hydrophobic in nature, with a contact angel with water of 112°. To improve wettability of an aqueous Ag dispersion during the subsequent wet-impregnation step, the AC granules were therefore first plasma treated under a pure oxygen atmosphere. Consequent to this, FTIR spectroscopy showed a definitive increase in polar functional groups like, carboxyl, ketones on the outer surface of AC, which hence resulted in Ag impregnation mostly on the outer surface of AC, instead of the pore-interiors of AC, during the wet-impregnation step. FEG-SEM images confirmed the presence of Ag nanoparticles on the outer surface of the Ag-AC hybrid, with a loading of 0.78 wt. % of Ag in the Ag-AC hybrid.

Disinfection studies were performed using these hybrid granules, in both batch and continuous mode, using a wild type model microorganism, namely, *E. coli* K12 (MTCC 1302), starting with an initial cell concentration of 10^4 CFU/ml, typical of contaminated drinking water. Cells were counted using the plate-count method and the antibacterial activity of Ag-AC was confirmed by the zone of inhibition method. For a typical loading of 8 mg of Ag-AC hybrid per ml of cell suspension, batch experiments showed a first order rate constant of 0.21 min^{-1} with respect to *E. coli* concentration. Subsequently, residence time required for continuous column-mode disinfection experiments were calculated based on batch kinetic data. To this end, a small column with a diameter of 1 cm and a height of 25 cm was used. Inlet water with a flow rate of 0.4 ml/min and having 10^4 CFU/ml *E. coli* was successfully treated by Ag-AC, so that no cells in the outlet water stream was detected.

In general from such hybrids, there is a slow loss of Ag due to formation of Ag-ions in presence of water. Our batch study however showed only 0.31 wt% of Ag release out of the total Ag loaded in the Ag-AC hybrid, even after a 108 h of contact period with water. The overall release rate was thus significantly low and confirms the potential of a long period of usage for antibacterial activities of Ag-AC. This was possible as Ag was impregnated with a good adhesion on the plasma treated functionalized surface of the hybrid material. Thus, the high biocidal activity coupled with a low rate of Ag release establishes the Ag-AC hybrid of the present study as a good material for drinking water disinfection. Hence, water purification, disinfection and biofouling prevention can be potentially achieved in a single water-filter column, suitable for potable quality water. Further ongoing work is aimed towards this goal.

(262a) Effects of Composition and Compositional Distribution On the Optoelectronic Properties and Function of Semiconductor Ternary Quantum Dots

Han, X., University of Massachusetts Amherst
Pandey, S. C., University of Massachusetts
Maroudas, D., University of Massachusetts Amherst
Originally presented on: 11/5/2013 8:30:00 - 8:49:00

The function of nanometer-size quantum dots (QDs) of ternary compound semiconductors, such as $In_xGa_{1-x}As$, $ZnSe_{1-x}S_x$, and $ZnSe_{1-x}Te_x$, used in the fabrication of optoelectronic and photovoltaic devices can be optimized by precise tuning of their electronic band gap through control of the QD composition (x) and diameter. In this presentation, we report results on compositional distributions in ternary QDs and how they affect the QDs' electronic band gap. We follow a hierarchical modeling approach that combines first-principles density functional theory (DFT) calculations and classical Monte Carlo simulations with a continuum model of species transport in spherical nanocrystals. In certain cases, the model predicts the formation of a concentration boundary layer near the nanocrystal surface, i.e., the formation of core/shell-like structures with shell regions rich in the surface segregating species. A systematic parametric analysis generates a database of transport properties, which can be used to design post-growth thermal annealing processes that enable the development of thermodynamically stable QDs.

Based on the continuum transport model, we have analyzed species interdiffusion kinetics in ternary semiconductor nanocrystals during their thermal annealing and used the modeling results to interpret the evolution of near-surface species concentration according to X-ray photoelectron spectroscopy (XPS) measurements. We have also explored the impact of composition and compositional distribution on the ternary QDs' electronic band gaps based on first-principles DFT calculations and found that ternary QDs with thermodynamically stable compositional distributions allow for precise band-gap tuning. Our findings lead to a proposal for an efficient one-step ternary QD synthesis method followed by annealing to promote self assembly of the thermodynamically stable configuration for optimal optoelectronic function in devices.

(262b) Molecular Modeling of Nanoparticles and Conjugated Polymers During Synthesis of Photoactive Layers of Organic Photovoltaic Solar Cells

Mortuza, S.,
Cisneros, C.,
Dela Cruz Bartolo, M.,
Banerjee, S.,
Originally presented on: 11/5/2013 8:49:00 - 9:08:00

Molecular modeling of nanoparticles and conjugated polymers during synthesis of photoactive layers of organic photovoltaic solar cells

S.M. Mortuza (1), Corinna Cisneros (1), Mark Dela Cruz Bartolo (2) and Soumik Banerjee (1)

1. School of Mechanical and Materials Engineering, Washington State University, Pullman, Washington, 99164, USA

2. California State University San Marcos, San Marcos, CA 92096

Solar cells are widely considered as potential alternative energy devices that can meet the increasing global energy demand. In particular, organic photovoltaic (OPV) solar cells, which employ organic nanoparticles in conjugated polymers as photoactive layers for energy conversion, have generated significant scientific interest due to their ease of manufacturing, relative low costs and flexibility. However, the power conversion efficiency (PCE) of OPV solar cells is still limited to 10-12% in part due to recombination of excitons before they can reach a nanoparticle-polymer interface. Such losses can be minimized by synthesizing photoactive layers with specific morphologies that result in large contact surface area between polymer and nanoparticles in the photoactive layer achieved through solution-based synthesis process. The scientific literature lacks detailed study of the effect of synthesis parameters, such as choice of solvent, processing temperature and relative concentration of nanoparticles and polymers, on the morphology of the photoactive layer. In order to understand the effects of these synthesis parameters on the agglomeration of nanoparticle, we performed molecular dynamics (MD) simulations of binary systems that comprise [6,6]-phenyl-C61-butyric acid methyl ester (PCBM), a commonly used acceptor, and various solvents in our previous study. The study suggested that indane is a better choice of solvent than other commonly used solvents for fabricating the photoactive layer of OPV solar cells. A systematic study of the effect of temperature and concentration of PCBM on agglomerate formation of PCBM showed that the extent of agglomerate formation increases with increasing temperature and concentration.
In an effort to further determine the agglomeration of PCBM in the conjugated polymer, we simulated systems comprising PCBM in poly-hexylthiophene (P3HT) that mimic the structure of a realistic

photoactive layer after spin coating. We evaluated the agglomeration behavior of PCBM by determining sizes of the largest clusters of PCBM in P3HT. To obtain further insight into the agglomerate structure of PCBMs, we evaluated the radial distribution functions (RDFs), which describe the coordination of one group of atoms or molecules, such as PCBMs or P3HTs, with the same or other group of atoms/molecules in the agglomerate. Our results indicate that the solubility of PCBM is greater in P3HT than in indane. Our simulations also demonstrate that PCBMs have a tendency to agglomerate in P3HT and the extent of agglomerate formation increases with increasing concentration of PCBM. We also analyzed the orientation of P3HTs to determine the extent of crystallinity. The results show that P3HTs have greater likelihood of parallel stacking in the system comprising PCBM and P3HT in weight ratio 1:2 than in the system that comprise 1:1 weight ratio of PCBM and P3HT. Figures 1 (a) and (b) show a magnified snapshot of a subsection of the simulation domain that illustrates the molecular arrangement of PCBM and P3HT in systems comprising PCBM and P3HT in weight ratios 1:1 and 1:2 respectively at 310 K. The aggregation pattern of PCBM and P3HT in Figure 1(a) demonstrates a greater contact area of PCBM and polymer compared to that in Figure 1 (b). Overall, our results show that relatively low concentration of PCBM in P3HT leads to sparse networks for charge transfer. In contrast, greater concentration of PCBM in P3HT leads to formation of suitable networks for transporting holes to photoanode and electrons to photocathode respectively. Results presented in this study provide fundamental insight that can help in selecting favorable processing parameters during synthesis of photoactive layer for OPV applications.

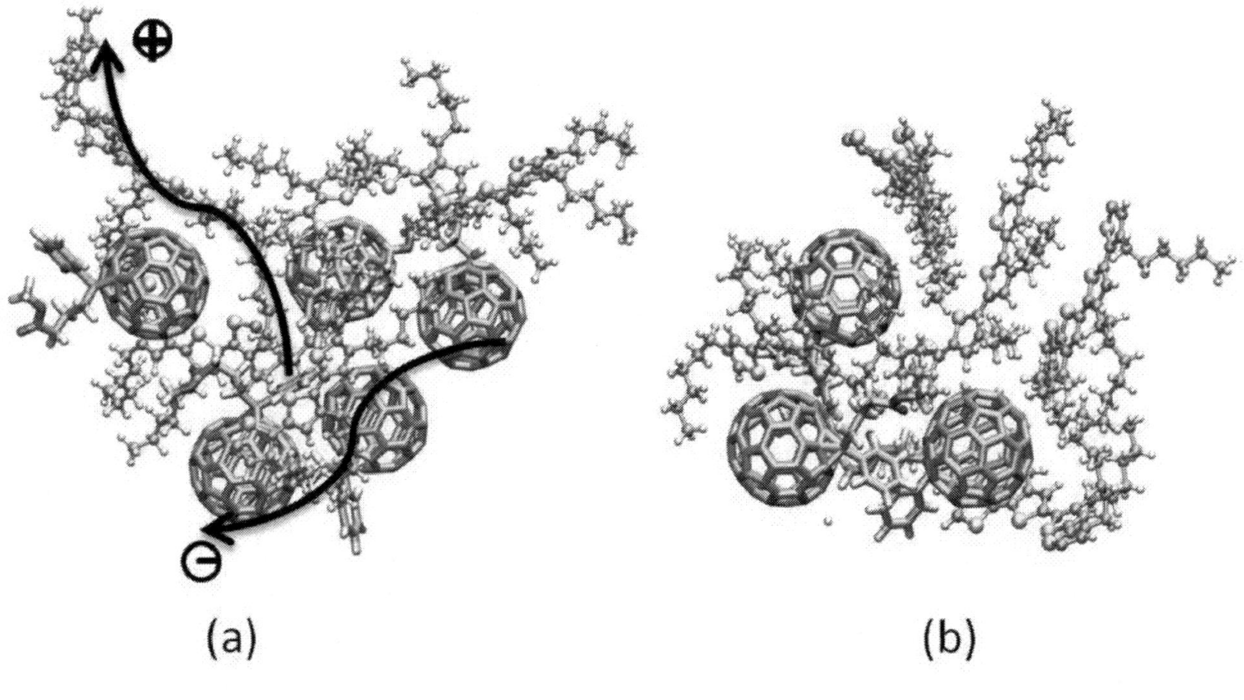

Figure 1: The molecular arrangement of PCBM and P3HT in systems of (a) 1:1 PCBM-P3HT and (b) 1:2 PCBM-P3HT at 300 K.

(262c) Integrating Photosystem I Proteins With Advanced Materials for Biologically Inspired Solar Energy Conversion

Jennings, G. K., Vanderbilt University
LeBlanc, G., Vanderbilt University
Gunther, D.,
Yang, S., Vanderbilt University
Cliffel, D., Vanderbilt University
Originally presented on: 11/5/2013 9:08:00 - 9:27:00

Photosystem I (PSI) is a ~10 nm protein complex that drives photosynthesis. Because of its nanoscale size and photodiode-like properties, PSI has drawn a surging interest for use in biologically inspired energy conversion devices upon extracting it from plants and assembling it as films on electrode surfaces. The research described in this presentation will focus on the molecular and electronic integration of spinach-extracted PSI at electrode surfaces, including metals, rectifying semiconductors, and graphene, to produce wet photoelectrochemical cells. Specifically, I will describe how monolayers of PSI can be deposited onto electrodes using self-assembly and Langmuir-Blodgett/Schaefer methods as well as how multilayers of PSI are deposited through a vacuum-assisted drop-casting approach. Thicker films of PSI absorb more light and yield much greater photocurrents as compared to monolayer films. The electrode that these films are deposited on is vitally important for the performance of the film. We have observed that micron-thick PSI films on p-doped silicon produce 200-times greater photocurrents than those obtained by deposition on gold. The position of the Fermi level combined with the band gap of silicon enables unidirectional electron transfer to PSI so that we avoid current cancellation from non-optimally aligned proteins. Adsorption of PSI on graphene provides flexibility in cell design due to the conductive, transparent nature of the atomically thin electrode. Furthermore, integrating a PSI film with reduced graphene oxide improves film conductivity and enhances observed photocurrents. These discoveries are leading to vastly improved performances for PSI-biohybrid solar cells.

(262d) Exciton Diffusion in Quantum-Dot Thin Films

Tisdale, W. A., MIT

Originally presented on: 11/5/2013 9:27:00 - 9:46:00

Colloidal quantum dots (QD), also known as semiconductor nanocrystals, are an intriguing material platform for solution-processable optoelectronic devices, such as solar cells, light-emitting diodes, thermoelectric modules, and flexible electronics. In QD thin films, the efficient movement of charge, heat, and excitonic energy is intimately linked to the QD surface chemistry and the amount of structural or energetic disorder in the material. In excitonic solar cells, the efficient diffusion of excitons to charge-separating interfaces is central to device operation. On the other hand, exciton diffusion to quenching interfaces in light-emitting diodes is a process that can limit luminescence efficiency. While the diffusion of singlet and triplet excitons in organic semiconductors has been studied extensively, exciton diffusion in colloidal QD thin films remains largely unexplored. In this talk, I will detail my group's combined experimental and computational efforts to obtain a deeper understanding of excitonic energy transport phenomena in colloidal QD materials. These efforts include spectrally-resolved transient photoluminescence spectroscopy, photoluminescence quenching, time-resolved optical imaging, and kinetic Monte Carlo simulation.

(262e) Photonic Curing of Nanocrystals for Photovoltaics

Stolle, C. J., The University of Texas at Austin
Harvey, T. B., The University of Texas at Austin
Pernik, D. R., The University of Texas at Austin
Du, J., The University of Texas at Austin
Rhee, D.,
Korgel, B. A., The University of Texas at Austin
Originally presented on: 11/5/2013 9:46:00 - 10:05:00

Nanocrystal inks are being explored as a route to low cost high efficiency photovoltaic (PV) devices. The efficiency of nanocrystal devices is limited by charge transport through the nanocrystal film and higher efficiencies can be achieved by sintering the nanocrystals into polycrystalline films. For copper-indium-selenide (CIS), this is done by thermal annealing under a selenium rich atmosphere above 500°C (called selenization). This processing step is relatively expensive and difficult to control.

Therefore, we have been exploring nanocrystal film processing strategies that might alleviate this step. Here, we present results on the use of rapid photonic curing using a rapid broadband pulsed light source to transiently heat the sample, which allows for sintering of CIS films on metallic contacts and helps prevent the outgassing of selenium from the film. We demonstrate a working solar cell fabricated from a photonically sintered CIS film. The choice of back contact material is critical for achieving working sintered devices, as the thermal conductivity and optical properties of the contact have a large impact on the film heating. We also demonstrate the removal of organic capping ligand from the film without inducing grain growth. In this case, nanocrystals come into direct contact with each other and charge transport is greatly increased. Devices made from nanocrystal films without capping ligands demonstrate high short circuit current and external quantum efficiency, although the overall device efficiency is limited due to a low open circuit voltage.

(262f) Chloride Surface Modified Cadmium Telluride Nanocrystals for Photovoltaics

Hellebusch, D. J., University of California, Berkeley
Alivisatos, A. P., Lawrence Berkeley National Laboratory
Originally presented on: 11/5/2013 10:05:00 - 10:24:00

Semiconductor nanocrystals (NCs) are of great scientific and commercial interest due to their optoelectronic properties. However, improvements in nanocrystals for solar cell applications, specifically cadmium telluride, require the ability to controllably manipulate their surface. Halides are ideal for nanocrystal surface functionalization as their small size provides advantages in passivation, coupling, and processing compared to the ubiquitous long-chained aliphatic surfactants. Here, we present the controlled surface modification of cadmium telluride nanocrystals with chloride while maintaining colloidal stability in organic solvents with a secondary labile ligand. After deposition of NCs from solution, the secondary ligand is easily removed with vacuum leaving behind an entirely organic-free NC film, offering a processing advantage for solution-based semiconductor thin films. X-ray Photoelectron Spectroscopy, Rutherford Backscatter Spectroscopy, and H-NMR supports successful substitution of the native ligand with the halide species while UV-Vis Absorption Spectroscopy and Transmission Electron Microscopy confirm maintenance of particle size, shape and solubility in organic solvents. Furthermore, we investigate the effect of the chloride on thermally treated, or sintered, NCs films. X-Ray Diffraction and Small Angle X-Ray Scattering reveal recrystallization and grain growth while Capacitance-Voltage, Field Effect Transistors, and Time-Resolved Photoluminescence probe enhancements in electrical and optical properties of sintered NC films. Our material system offers a unique method to deposit films from solution with easily removable secondary ligands that tend to contaminate films with electrically insulating materials. This study introduces a nanocrystal system with a quantitative handle on surface manipulations for a broad range of investigations and applications for photovoltaics.

(262g) Multistep Selenization of Copper Indium Gallium Selenide (CIGS) Nanocrystal Photovoltaics

Harvey, T. B., The University of Texas at Austin
Bogart, T., The University of Texas at Austin
Stolle, C. J., The University of Texas at Austin
Du, J., The University of Texas at Austin
Pernik, D. R., The University of Texas at Austin
Korgel, B. A., The University of Texas at Austin
Originally presented on: 11/5/2013 10:24:00 - 10:43:00

Copper Indium Gallium Selenide (CIGS) nanocrystal inks have been used to fabricate photovoltaic devices without post-deposition thermal treatment, but have been limited to relatively low efficiency (~3%) by poor electron transport. The insulating organic nanocrystal capping ligands are largely responsible for this. Annealing the nanocrystal films at high temperature under selenium atmosphere—i.e., selenization—can be used to improve device performance by removing ligands and increasing grain size in the film. The selenization process, however, is difficult to control due to non-uniform film sintering, the formation of insulating layers from residual carbon in the film, and film delamination with complete conversion of the Mo back contact to $MoSe_2$.

We have found that a one hour thermal anneal under argon at 525 °C before selenization of the nanocrystal film can significantly improve device efficiency. This pre-selenization anneal removes organic ligands, which appears to be important for uniform nanocrystal sintering. Lowering the anneal temperature from 525 °C to 425 °C drastically lowers device efficiency although the capping ligands are still removed. Sodium has been shown to improve CIGS grain growth in co-evaporated films and a similar connection is found in selenized nanocrystal film performance. X-ray photoelectron spectroscopy (XPS) measurements taken after the pre-selenization anneal show increased Na presence at the Mo surface when the annealing temperature is increased due to greater diffusion from the soda-lime glass substrate.

Additionally, this pre-selenization anneal accelerates grain growth in the film, shortening the necessary selenization time. Shortened selenization length reduces $MoSe_2$ formation at the back contact, enabling the use of repeated selenization treatments without film delamination. Devices with over 7% efficiency have been fabricated by depositing and selenizing successive CIGS nanocrystal layers.

(262h) Exciton Dissociation and Charge Carrier Generation At Core/Shell Heterojunction in Quantum Dots

Chakraborty, A., Syracuse University
Originally presented on: 11/5/2013 10:43:00 - 11:02:00

Electron-hole pairs (excitons) are generated by photoexcitation of electrons to excited electronic states. Exciton dissociation and generation of free charge carriers is central for light-harvesting applications of photoactive nanoparticles. This talk will focus on computational investigation of effect of heterojunctions on exciton dissociation in core/shell quantum dots. It has been shown in various experimental studies that presence of heterojunctions in nanocrystals can greatly facilitate exciton dissociation. This study will present the following four computed properties: (1) exciton binding energy, (2). electron-hole recombination probability (3). electron-hole separation distance and (4).electron-hole pair density as metrics for analyzing effect of core/shell structures in nanocrystals. Results for CdSe/ZnS core/shell quantum dots will be presented for a series of dots sizes. Effect of dot size and thickness of the shell of exciton dissociation will be presented and comparison with experimental results will be discussed. The electron-hole interaction energy is quantified by calculating the exciton binding energy and the spatial distribution of the charge carrier in the quantum dot is analyzed by electron-hole pair density functions. The local environment of the electron and hole quasiparticles in the quantum dot is probed using the average electron-hole separation distance and the electron-hole recombination probability. Exciton dissociation is a complex process and the four metrics mentioned above provide a multi-faceted description of this phenomenon.

(275a) Supercritical Carbon Dioxide (scCO2) Processing of Polystyrene/Clay Nano-Composites: Structures and Properties

Yang, F.,
Kriegel, R., Coca Cola
Kannan, R., Johns Hopkins School of Medicine
Originally presented on: 11/5/2013 8:30:00 - 8:48:00

Improving clay dispersion and clay-polymer interfacial interactions are keys to enhancing barrier property of nanocomposites,. A supercritical CO_2 ($scCO_2$) processing method was utilized to pre-disperse commercialized organic clays, and then those pre-dispersed clays were compounded with polystyrene (PS) to form nanocomposites with significant dispersion and interfacial enhancement. The effects of $scCO_2$ processing on clays pre-dispersion and polymer-clay dispersion and interaction in nanocomposites were investigated. SEM and WAXD of the clays indicated that after $scCO_2$ process the clays lost their long region ordered layer structure, the size of particles and the tightness of packing are reduced, among different clays, those with lower inter-modifiers interaction tend to be dispersed. WAXD and TEM of the PS/clay nanocomposites show that PS have penetrated into the inter galleries and the pre-dispersed clays lead to a disordered intercalated/ exfoliated structure with improved interfacial interaction rather than a disordered intercalated structure as "as-received" clay did. The effects of clay modifier, clay weight fraction and $scCO_2$ processing on clay dispersion and polymer-clay interactions in nanocomposites were investigated. Structure-barrier property relationship was set up based on several phenomenological models study. Results indicated that $scCO_2$ processing lead to a more homogenous dispersion and achieved a nearly 83% reduction of oxygen permeation in PS/clay system with only 3.3 vol% clay, which corresponding to an effective aspect ratio of 109.09±4.35. Moreover, among those phenomenological models, Cussler random array model shown the best fit and given a more realistic prediction of effective aspect ratio in nanocomposites.

(275b) Mechanical Reinforcement and Nanoparticle Dispersion of Hydrogen Bonded Supramolecular Polymer-Silica Nanocomposites

Neikirk, C. C., Princeton University
Priestley, R., Princeton University
Originally presented on: 11/5/2013 8:48:00 - 9:06:00

We have prepared supramolecular polymer nanocomposites using both poly(butyl acrylate) hydrogen bonded sidechain supramolecular polymers and poly(ε-caprolactone) linear supramolecular polymers blended with complementary hydrogen bond functionalized silica nanofillers. The use of complementary hydrogen bonding groups prevents hydrogen bonding between nanoparticles while allowing the nanoparticles to intimately participate in the supramolecular polymer network. Mechanical properties, including stress-strain behavior, dynamic modulus, and creep were investigated over a range of temperatures for various nanoparticle loadings and nanoparticle surface functionality levels. The effect on crystallinity for poly(caprolactone) supramolecular polymers was used to explain trends in reinforcement behavior for the semi-crystalline supramolecular polymer nanocomposites. Additional comparison of nanoparticle dispersion and nanocomposite reinforcement in materials made with self-complementary and complementary nanofillers provides a more complete understanding of these hydrogen bonded nanocomposite materials. In particular, significant improvement in the mechanical properties and nanoparticle dispersion was found for nanoparticles with complementary, rather than self-complementary, hydrogen bonding interactions. Finally, these nanocomposites' self-healing behavior was also investigated.

(275c) Improving the Electrical Conductivity of Polycarbonate Carbon Nanotubes Composites

Herrington, K., Virginia Tech
Quigley, J., Virginia Tech
Qian, C., Virginia Tech
Woohyun, C.,
Baird, D. G., Virginia Tech
Originally presented on: 11/5/2013 9:06:00 - 9:24:00

Carbon Nanotubes' (CNT) high aspect ratio and electrical conductivity (EC) make them ideal conductive fillers for polymers. However, in order to take full advantage of these properties, CNT need to be separated and dispersed evenly throughout polymer matrices. Current methods are either not commercially viable, inefficient, environmentally unfavorable, or severely damage the CNT. The primary goal of this project was to show that by using a super critical carbon dioxide (scCO2) expansion process, CNT could be deagglomerated before being melt blended in to the polymer matrix, leading to improved dispersion, EC, and percolation threshold.

It was found that for 3wt % multi-wall carbon nanotubes (Baytube C150 P) in polycarbonate samples treated with scCO2 showed a sixteen fold increase in surface conductivity (SC) from $3.1*10^{-6}$ S to $5.1*10^{-5}$ S. For 1wt % samples treated with scCO2, SC went from $<10^{-10}$ S to $3.1*10^{-8}$ S. In addition to the conductivity tests, transmission electron microscopy was employed to examine the impact on the morphology of the samples.

(275d) Graphene-Polymer Composite Materials

Chakraborty, I., University of Rhode Island
Heeder, N., University of Rhode Island
Shukla, A., University of Rhode Island
Bose, A., University of Rhode Island
Originally presented on: 11/5/2013 9:24:00 - 9:42:00

We present two novel techniques for generating graphene - polymer composite materials that serve to minimize the loading of graphene while retaining its electrical conductivity. In the first we add silica microspheres to the polymer (polystyrene) along with the graphene. The dispersed silica spheres reduce the volume available to the graphene, thereby reducing the graphene concentration required for getting the target electrical conductivity. Our experimental results and simple models corroborate this results from this technique. In the second technique we use electrostatics to coat polystyrene pellets with graphene, and use compression molding to create composites. We show electrical percolation at near theoretical limits. We optimize loading and processing conditions for both techniques, to provide an optimal balance between electrical and mechanical properties.

(275e) Nanostructure of a Novel Fluoroblock Copolymer Using Atom Transfer Polymerization: Poly(styrene)-b-Poly(2,3,4,5,6-Pentafluorostyrene)-b-Poly(2,2,3,4,4,4-Hexafluorobutyl methacrylate)

Guerrero-Gutierrez, E. M. A., University of Puerto Rico, Mayaguez
Perez, M., University of Puerto Rico
Suleiman, D., University of Puerto Rico, Mayaguez Campus
Originally presented on: 11/5/2013 9:42:00 - 10:00:00

This investigation studied the synthesis and characterization of ionic membranes composed of sulfonated copolymers with a novel fluoroblock copolymer. Poly(styrene)-b-poly(2,3,4,5,6-pentafluorostyrene)-b-poly(2,2,3,4,4,4-hexafluorobutyl methacrylate) [PS-P5FS-PHFBM] was synthesized using Atom Transfer and Polymerization (ATRP). The block copolymer's physical and thermal properties were measured using different characterization techniques such as gel permeation chromatography (GPC), thermogravimetric analysis (TGA) and differential scanning calorimetry (DSC). The chemical composition was monitored with Fourier Transform Infrared spectroscopy (FTIR) and Nuclear Magnetic Resonance (NMR) spectroscopy. The resulting block polymer was incorporated to an elastomeric matrix; therefore, physical blends of sulfonated poly(styrene-isobutylene-styrene) and unsulfonated PS-P5FS-PHFBM were casted and analyzed. The newly developed fluoro-membrane was characterized using several techniques including: elemental analysis (EA), thermogravimetric analysis (TGA), differential scanning calorimetry (DSC), Fourier Transform Infrared spectroscopy (FTIR) and small angle X-Ray scattering (SAXS). In addition, methanol, ethanol and isopropanol liquid permeabilities were measured to understand their transport properties and mechanism throughout the membranes and their connection to this unique nanostructure where chemistry and morphology play a critical role. The synthesis methodology was proposed as an alternative to obtain an advanced fluorinated, elastomeric and glassy polymeric membranes capable of being functionalized for applications such as specialty separations (*e.g.*, alcohols), or direct methanol fuel cells (DMFC).

(275f) Biorenewable Chitin Reinforced Polyethylene Oxide(PEO) Light Weight Composites

Wu, J.,
Meredith, J. C., Georgia Institute of Technology
Originally presented on: 11/5/2013 10:00:00 - 10:18:00

There is currently tremendous interest in the development of new bio-renewable sources of high-performance materials. A chief aim is to substitute portions of mineral, or petroleum-based materials, with high-strength sustainable materials sources. For example, recent research has focused on cellulose nanocrystals (CNCs) and nanofibrillated cellulose for high-strength, lightweight composites. Less well-explored, but equally promising, are chitin nanocrystals and nanofibers, which also show high-strength like CNCs. Chitin is also a poly(saccharide), structurally similar to cellulose, but is obtainable from the shells of shellfish and insect exoskeletons, as opposed to plant sources for cellulose. In this work, we explore the potential to use chitin nanocrystalline fibers as a strengthening and stiffening filler in polymers. Chitin nanofibers were extracted from crab shells and used to create composite films with polyethylene oxide (PEO). The films were tested for mechanical properties, including tensile strength, modulus, strain at break, as well as thermal properties via DMA, DSC and TGA. Interactions between chitin and PEO were also investigated using FTIR spectroscopy. This talk will report the key results and summarize the future prospects for utilization of chitin nanofibers as polymer fillers.

(275g) A Hydrated Salt/SiO2 Shape-Stabilized Phase Change Material Prepared Via Sol-Gel Process

Wu, Y., Tsinghua University

Originally presented on: 11/5/2013 10:18:00 - 10:36:00

The utilization of phase change materials (PCM), which can absorb, and release the latent heat over a defined temperature range while undergoing phase change, for thermal energy storage has received considerable attention. A novel shape-stabilized phase change material with a mixture of hydrated salts, $Na_2SO_4 \cdot 10H_2O$-$Na_2HPO_4 \cdot 12H_2O$ as phase change material and SiO_2 as supporting material was prepared via sol-gel process in this work. In the composite, $Na_2SO_4 \cdot 10H_2O$-$Na_2HPO_4 \cdot 12H_2O$ was embedded into the porous structure of SiO_2 and encapsulated by polyvinylpyrrolidone (PVP), so that there is no leakage of it during the solid-liquid phase change even when its maximum mass fraction reached 70wt%. The surface morphology was observed by scanning electron microscopy (SEM). Fourier transformation infrared (FT-IR) analysis was conducted to investigate the chemical compatibility between $Na_2SO_4 \cdot 10H_2O$-$Na_2HPO_4 \cdot 12H_2O$ and SiO_2. Thermal properties, such as latent heats and phase change temperatures, and thermal stability of the composite PCMs were determined using differential scanning calorimetry (DSC) and thermal gravimetric analysis (TGA). The DSC results showed that the melting temperatures and latent heat values of the composite PCMs are in range of 29-38°C and 41-89kJ/kg, respectively. The thermal cycling tests revealed that the composite PCMs had good thermal reliability and chemical stability.

(275h) Structure-Property Relationships for PDMS-Silica Nanocomposites

Luo, M., Clarkson University
Sankarasubramanian, M.,
Krishnan, S., Clarkson University
McLaughlin, J. B., Clarkson University
Originally presented on: 11/5/2013 10:36:00 - 10:54:00

This talk presents results for the nanostructure and mechanical properties of cross-linked polydimethylsiloxane networks containing silica particle fillers. The results were obtained using the Materials Studio FORCITE module. After equilibration, the nanocomposites were subjected to mechanical strain and the elastic modulus and strength were determined. The influences of cross-link density and filler volume fraction on microstructure and mechanical properties were evaluated. Simulation results will be compared with experimental data obtained using dynamic mechanical analysis.

(290a) Invited: Peptide Amphiphiles: Modular Construction Of Biomaterials

Tirrell, M., University of Chicago

Originally presented on: 11/5/2013 12:30:00 - 13:20:00

Peptide amphiphiles combine the functionality of of peptides and
proteins
with the ability create synthetic, modular bioactive materials. These
engineered molecules can be designed to have self-assembly
characteristics
driven by either hydrophobic interactions or electrostatic
complexation.
Examples will be given of tissue engineering matrices as well as
self-assembled structures with activities ranging from tumor targeting
to
atherosclerotic plaque identification to immuno-modulation.

(290b) Invited: Protein Nanomaterials From Bioconjugate Block Copolymer Self-Assembly

Olsen, B. D., Massachusetts Institute of Technology
Lam, C. N., Massachusetts Institute of Technology
Chang, D.,
Thomas, C. S., Massachusetts Institute of Technology
Kim, M.,
Xu, L., Massachusetts Institute of Technology
Sanoja, G.,
Originally presented on: 11/5/2013 13:20:00 - 14:10:00

Protein-based materials show a great deal of potential as biomaterials, catalysts, sensors, and optoelectronics, where the unique efficiency, selectivity, or activity of enzymes can be captured to improve the performance of these devices. Control over the structure and orientation of the protein in three dimensions is required to improve transport through the devices, increase the density of active sites, and optimize the stability of the protein. We demonstrate self-assembly of globular protein-polymer conjugates into nanostructured phases as an elegant and simple method for structural control in biomaterials. These conjugates may be conceptualized as diblock copolymers, where the first block is the globular protein and the second block is the synthetic polymer. In order to fundamentally investigate self-assembly in these complex block copolymer systems, proteins are site-specifically conjugated to synthetic polymer blocks to form well-defined model globular protein-polymer diblock copolymers.

Functional protein materials are obtained by solvent evaporation in order to access different pathways toward self-assembly using polymer-selective, non-selective, and protein-selective solvents. Similarly, solvent annealing using these different conditions is exploited as a means to both improve ordering and explore the thermodynamic stability of the as-cast nanostructures. Small angle scattering and transmission electron microscopy are used to explore the dependence of nanostructure formation on processing conditions and the molecular weight of the PNIPAM block. Wide angle X-ray scattering demonstrates that diblock copolymer self-assembly results in a noncrystalline structure within the protein nanodomains. Circular dichroism, UV/Vis spectroscopy, and Fourier transform infra-red (FTIR) spectroscopy show that a large fraction of the protein remains in its folded and active state after conjugation. The effect of coil fraction and hydrogen bonding additives on maintaining protein activity within nanostructured phases is also explored, demonstrating methods for fabricating structures with both a high protein density and a high fraction of active protein. The effect of plasticizing additives on thermal and chemical stability was also explored, illustrating the ability of these materials to dramatically enhance the stability of proteins in polymeric materials.

Phase diagrams for these materials have been prepared as a function of coil fraction and water content in the materials, providing insight into the type of self-assembled nanostructures that may be formed. Small-angle light scattering allows quantitative measurement of solvent-mediated interactions between the different components of the diblock copolymers, enabling a fundamental understanding of the relationship between molecular interactions and self-assembly. Comparison of copolymers with variable protein and polymer blocks enables the effects of protein shape and protein-protein interactions in these systems to be understood. Together, these results begin to lay a foundation for understanding the general principles of self-assembly in block copolymers containing globular proteins.

(290c) Invited: Micro and Nanoscale Fluid Shear Stimulation To Direct Stem Cell Differentiation

Murthy, S., Northeastern University

Originally presented on: 11/5/2013 14:10:00 - 15:00:00

The role of fluid shear forces in stimulating changes in cells has been well documented in several contexts. However, these studies are typically carried out with cells that are plated and spread on two-dimensional surfaces. Comparatively little is known about the impact of fluid shear forces on cells that are in suspension, particularly those cell types that are not normally present in circulating blood in vivo. This presentation will describe how simple microfluidic shear stimulation platforms along with low fluid flow rates can cause observable changes such as gene and surface marker expression. In human mesenchymal stem cells, such stimulation may accelerate soluble factor-mediated differentiation toward the adipogenic lineage. In cells derived from mouse embryos, such shear stimulation may accelerate differentiation toward hematopoietic lineages. Indeed such shear stimulation is known to occur in vivo in the embryo as a result of precise fluid flows generated upon initiation of the heartbeat. Our in vitro observations will be discussed in the context of what is currently known about such in vivo phenomena.

(319a) Core-Shell Structured Composite Microparticles With Ability to Store a Chemical Payload and Release a Defined Quantity "On Demand"

Kovacik, P., Prague Institute of Chemical Technology
Stepanek, F., Institute of Chemical Technology, Prague
Singh, M.,
Originally presented on: 11/5/2013 12:30:00 - 12:55:00

Core-shell structured composite microparticles with ability to store a chemical payload and release a defined quantity "on demand"

P. Kovacik, M. Singh, F.Stepanek

ICT Prague, Laboratory of chemical robotics, Technicka 5, 166 28 Prague,
Tel.: 2 2044 3046, e-mail: kovacikp@vscht.cz

The aim of this work was the preparation of biocompatible microparticles with a hollow interrior. The microparticles must be able to encapsulate and store a chemical payload for a certain time, followed by the release on demand of this payload. In our previous work [1] we have prepared silica microparticles with a hollow core and found that the diffusion across the mesoporous silica shell was strongly dependent on temperature. In this work, we used this dependence and attached iron oxide nanoparticles on the surface of the silica shell to create composite iron oxide/silica microparticles. The iron oxide nanoparticles were able to heat up in the presence of an alternating magnetic field. This property allowed us to use magnetic field as a tool for remote control of diffusion across the microparticle shell. To avoid spontaneous leakage of encapsulated payload in time, we have modified the surface of the composite microparticles with a layer of palm oil. Palm oil is a phase change material which is solid under 37 Â°C. We showed that the resultant composite microparticles are able to store a payload for several months and release a defined quantity on demand by the application of a magnetic field. The particles were characterised in shape, size, heating ability and their mass transport properties.

[1] Kovacik P., Kremlackova Z., Stepanek F., Investigation of radiofrequency induced release kinetics from magnetic hollow silica microspheres?, Micropor. Mesopor. Mater. 159, 119-125 (2012)

(319b) ZnO Nanorods Hermetically Encapsulated By a Nanothin Amorphous SiO2 Coating: Toxicological Profile and Optical Properties

Sotiriou, G. A., Harvard University
Murdaugh, K.,
Cohen, J., Harvard University, School of Public Health
Watson, C.,
Elder, A.,
Demokritou, P., Harvard School of Public Health
Originally presented on: 11/5/2013 12:55:00 - 13:20:00

Zinc oxide (ZnO) is a wide-bandgap metal oxide semiconductor that can be excited at room temperature by UV irradiation, and finds applications in a variety of fields and products including paints, pigments, batteries, photocatalysis, foods, to name a few. When in nanometer size range, it may be used in polymer nanocomposites and sunscreens as an efficient UV-filter with high transparency in the visible wavelength range. However, the photocatalytic activity of ZnO causes the degradation of the surrounding polymer rendering it unsuitable for long-term employment. Furthermore, ZnO nanoparticles are highly toxic and may pose risks to the public and the environment. Here, ZnO nanorods are made by scalable flame synthesis [1,2] and are in-situ encapsulated by an amorphous nanothin SiO_2 layer [3,4]. The as-prepared nanoparticles are characterized by electron microscopy (EM), X-ray diffraction (XRD), X-ray photoelectron spectroscopy (XPS) and N_2 adsorption. The hermetic nature of the SiO_2 coating is evaluated by a detailed dissolution study and the photocatalytic activity is monitored by the decomposition of methylene blue dye. The presence of SiO_2 facilitates the dispersion of these nanorods in relevant solutions. The core ZnO nanorods exhibit the characteristic optical properties as determined by UV/vis and luminescent spectrometry, while the SiO_2 coating eliminates the photocatalytic activity and minimizes the toxicity and genotoxicity to human cells, as illustrated in cellular studies using multiple cell lines. Therefore, these flame-made ZnO nanorods may be safely used as fillers in polymer UV-filter nanocomposites and in sunscreens exhibiting superior performance while mitigating their impact on environmental health.

References

[1] Sotiriou, G. A., Sannomiya, T., Teleki, A., Krumeich, F., Vörös, J. and Pratsinis, S. E. Non-toxic dry-coated nanosilver for plasmonic biosensors. *Adv. Funct. Mater.* **20**, 4250-4257, (2010).

[2] Sotiriou, G. A., Hirt, A. M., Lozach, P. Y., Teleki, A., Krumeich, F. and Pratsinis, S. E. Hybrid, silica-coated, Janus-like plasmonic-magnetic nanoparticles. *Chem. Mater.* **23**, 1985-1992, (2011).

[3] Sotiriou, G. A., Franco, D., Poulikakos, D. and Ferrari, A. Optically Stable Biocompatible Flame-Made SiO_2-Coated Y_2O_3:Tb^{3+} Nanophosphors for Cell Imaging. *ACS Nano* **6**, 3888-3897, (2012).

[4] Gass, S., Cohen, J. M., Pyrgiotakis, G., Sotiriou, G. A., Pratsinis, S. E. and Demokritou, P. Safer Formulation Concept for Flame-Generated Engineered Nanomaterials. *ACS Sustainable Chem. Eng.* **in press**, DOI: 10.1021/sc300152f, (2013).

(319c) Single-Pot Synthesis of Uniform Glucan Multilayers On Oxide Particles

Jankolovits, J.,
Gazit, O., University of California Berkeley
Nigra, M., University of California, Berkeley
Katz, A., University of California, Berkeley
Originally presented on: 11/5/2013 13:20:00 - 13:45:00

Title: Single-Pot Synthesis of Uniform Glucan Multilayers on Oxide Particles

Authors: Joseph Jankolovits, Oz M. Gazit, Michael M. Nigra, Alexander Katz

Department of Chemical and Biomolecular Engineering

University of California, Berkeley

Polymer coatings are used widely to tune the surface chemistry of solids in biomaterials, paints, and other materials. Covalently grafted polymer coatings provide advantageous stability, though their prevailing syntheses involve multiple step procedures, low solid yields and surface coverage, and/or non-uniformity on the nanoscale. A novel method for uniformly grafting polymer multilayers on inorganic oxide particles in a single-pot process will be demonstrated using a wide variety of characterization techniques. Implications of the grafted polymer multilayer coatings on inorganic oxide particles for the stabilization of paint pigment dispersions will be discussed. Efforts to prepare grafted polymer multilayers on nanoparticles will be described. The promise of these grafted polymer multilayer coatings on inorganic oxide particles in biomaterials and nanomedicine will be highlighted.

(319d) Continuous Polymer Coating of Nanoparticles: A Novel Method

Sirkar, K. K., New Jersey Institute of Technology
Chen, D., New Jersey Institute of Technology
Singh, D., New Jersey Institute of Technology
Pfeffer, R., Arizona State University
Originally presented on: 11/5/2013 13:45:00 - 14:10:00

Nanoparticle-based drug delivery systems are of significant interest in controlled release of drugs, delivery of anticancer drugs and imaging agents to tumors, tuberculosis treatment and as non-viral gene delivery vehicles. To develop controlled release of the drug as well as to protect the drug from demanding environments, appropriate coatings are employed on the drug nanoparticles. Conventional methods for coating or encapsulating micron–sized and nanoparticles utilize dry or wet approaches. Dry methods include physical vapor deposition, plasma treatment, chemical vapor deposition and pyrolysis of polymeric organic materials. Wet methods cover sol-gel processes, emulsification and solvent evaporation techniques. Supercritical fluid (SCF) processes such as Rapid Expansion of Supercritical Solutions (RESS), Supercritical Anti-Solvent (SAS), and Gas Anti-Solvent (GAS) processes employing supercritical CO_2 are alternative methods for nanoparticle coating or encapsulation of ultrafine particles. These SCF-based processes have many shortcomings. Some processes require demanding operating conditions, suffer from low polymer solubility etc. Most of these techniques are also batch processes. Fluidized bed-based processes, which can be continuous, face difficulties with nanoparticles caused by van der Waals and other interparticle forces. In all such processes, scale-up is also quite demanding. We will illustrate a technique for continuous polymer coating of nanoparticles that is easily scalable. The nanoparticles used were of silica. Polymers used for coating were Eudragit and PLGA.

(319e) Coating FINE Particles With ULTRA Thin FILMS Using Atomic Layer Deposition

Meesters, G., DSM
Valdesueiro, D., Delft University of Technology
van Ommen, J. R., Delft University of Technology
Originally presented on: 11/5/2013 14:10:00 - 14:35:00

Atomic Layer Deposition (ALD) is a gas-phase coating technique that can provide fine particles with ultrathin films, typically 1-50 nm. Due to the large value of the specific surface area, fine particles –diameter ranging from 1 nm to 50 µm– are used in a wide range of industrial fields such as electronics, pharmaceutics, energy storage and catalysis. The aim of the coating is to improve the performance of these materials, either by the activation of the surface of the particles, or protecting them against the environment.

ALD can deposit uniform inorganic layers [1] by a set of two self-limiting reactions (Fig. 1). The precursors, which are fed into the reactor in subsequent steps, react on the surface of the particles to form a very thin layer of coating material. The final thickness of the coating is controlled by repeating the cycle a desired number of times. This layer-by-layer growth mechanism can also be used to deposit polymeric compounds (Molecular Layer Deposition) [2] or inorganic-organic hybrid compounds [3]. The coating process is carried out in a fluidized bed reactor, which offers high mass and energy transfer, and the possibility to process large amounts of solids.

The production of core-shell particles via ALD has been already applied in fields such as the Li-ion batteries [1] (Fig. 2) or photo-catalysis [4]. Our current work aims at the production of non-permanent coatings to be applied in the controlled release of compounds for food and pharmaceutical applications.

Fig. 1. Schematic representation of the two ALD half-reactions.

Fig. 2. Li-ion cathode material ($LiMn_2O_4$) coated with 5 cycles of Al_2O_3 using ALD [1]

1. Van Ommen, J.R., et al., *Scalable gas-phase processes to create nanostructured particles.* Particuology, 2010. **8**(6): p. 572-577.

2. Adamczyk, N.M., A.A. Dameron, and S.M. George, *Molecular layer deposition of poly(p-phenylene terephthalamide) films using terephthaloyl chloride and p-phenylenediamine.* Langmuir, 2008. **24**(5): p. 2081-2089.

3. Yoon, B., et al., *Molecular layer deposition of hybrid organic - Inorganic alucone polymer films using a three-step abc reaction sequence.* Chemistry of Materials, 2009. **21**(22): p. 5365-5374.

4. Zhou, Y., et al., *Optimal preparation of Pt/TiO2 photocatalysts using atomic layer deposition.* Applied Catalysis B: Environmental, 2010. **101**(1-2): p. 54-60.

(319f) Surface Engineered Quantum Dots for Light Selective Polymer Films

Mumin, M. A., University of Western Ontario
Allan, J. M., Western University
Charpentier, P. A., Western University
Originally presented on: 11/5/2013 14:35:00 - 15:00:00

We have investigated the optimization of the optical properties of polymer films integrating engineered quantum dot (QD) nanocrystals to manipulate and control light transmission for applications in light selective polymer films. Highly luminescent CdS QDs and CdS/ZnS core/shell QDs with 5 nm size were synthesized using a facile approach based on the pyrolysis of single molecule precursors. After capping the CdS QDs with a thin layer of ZnS to reduce toxicity, the photoluminescence and photostability of the core-shell QDs was significantly enhanced. The ZnS shell was found to passivate the surface defects of the CdS dots while confining the electron-hole pair to the CdS core region.

To make both the bare and core/shell structure QDs more resistant against photochemical reactions, a mesoporous silica layer was grown on the QDs through a reverse microemulsion technique based on hydrophobic interaction and surface silanization. Mesoporous silica can protect completely both type of QDs from photobleaching and also improves the quantum yield by providing much stronger resistance to oxidation and Oswald ripening of QDs. The encapsulated QDs were then functionalized using silane coupling agent and integrated to poly ethylene vinyl acetate (pEVA) network both by melt mixing (using twin screw extruder) and grafting from polymerization in supercritical CO_2. Morphology, UV absorption, visible light transmission, and IR retention were observed and compared to that obtained from commercial pEVA greenhouse plastics. Mesoporous silica encapsulated QDs into EVA polymer films were found to provide simultaneous UV protection, infrared retention, and light selectivity of solar-polymer films.

(321a) Surface Hydrogen Stabilized Semiconductor Nanowire Synthesis

Sivaram, S., Georgia Institute of Technology
Shin, N. C., Georgia Institute of Technology
Chou, L. W., Georgia Institute of Technology
Filler, M. A., Georgia Institute of Technology
Originally presented on: 11/5/2013 12:30:00 - 12:45:00

The 3-D nature of semiconductor nanowires, and nanoscale materials in general, requires new chemical strategies with which to control growth processes and engineer structure. Unfortunately, there is little knowledge of the chemical bonding that underlies nanowire synthesis and, as a result, the *ab initio* manipulation of nanowire structure and function remains challenging. To this end, we report on a real-time *in situ* infrared spectroscopic investigation of sidewall hydrogen atoms during germanium nanowire growth via the vapor-liquid-solid (VLS) technique. We find that stable <111> oriented growth is only possible in the presence of adsorbed hydrogen and that these surface species modify taper as well as faceting. Arrays of epitaxially aligned nanowires were synthesized with a Au catalyst at a range of process conditions (210 – 280 °C, $5.0 \times 10^{-5} - 1.5 \times 10^{-4}$ Torr). Two absorption bands, present near 1980 cm^{-1} and 1965 cm^{-1}, exhibit intensities that are a function of substrate temperature, precursor pressure, and nanowire elongation time. A series of adsorption studies on substrates with different orientations reveal that these bands result from hydrogen covalently bonded at Ge(100)-like and Ge(111)-like surface sites. Experiments where atomic hydrogen atoms are delivered *in situ* unambiguously confirm that hydrogen is responsible for these effects and suggest new avenues to rationally manipulate nanowire structure.

(321b) Geometry-Dependent Formation of Multiple Twins in Si Nanowires

Shin, N. C., Georgia Institute of Technology
Chi, M., Oak Ridge National Laboratory
Filler, M. A., Georgia Institute of Technology
Originally presented on: 11/5/2013 12:45:00 - 13:00:00

The rational and atomically precise insertion of twin boundaries and stacking faults in semiconductor nanowires would create myriad new opportunities to engineer their electrical, optical, and thermal properties. While these structural motifs are commonplace in group III-V nanowires synthesized via the vapor-liquid-solid (VLS) technique, their presence in group IV nanowires is much less frequent. Here, we report on the appearance of consecutive, geometry-dependent twin boundaries in Si nanowires for the first time. Si nanowires with multiple twins were grown with a Au catalyst via a two-step process in a custom-built ultrahigh vacuum chamber. The first 10 min of nanowire elongation occurred with a substrate temperature of 490 °C and Si_2H_6 pressure of 2×10^{-4} Torr. The substrate temperature and Si_2H_6 pressure were then modulated to 410 °C and 5×10^{-4} Torr, respectively, for another 10 min. The first twin boundary (TB_1) appears at an axial position corresponding to the point where growth conditions were changed and was the same for all nanowires. Importantly, the location of the second twin boundary (TB_2) shows a clear diameter-dependence. High resolution electron microscopy reveals that thin {111} sidewall facets, which extend following TB_1 and distort the triple-phase line, likely underlie the formation of TB_2. We hypothesize that an increased amount of surface hydrogen, present as a result of lowering the substrate temperature and raising the Si_2H_6 pressure, favors the thin {111} facets. These results indicate that surface chemistry can influence defect formation in semiconductor nanowires and constitute an important step toward defect superstructure engineering in this ubiquitous material. [Research supported by ORNL's Shared Research Equipment (ShaRE) User Program, which is sponsored by the Office of Basic Energy Sciences, the U.S. Department of Energy.]

(321c) Phase Transformation of Silicon Nanowires As a Route for the Mass Production of Mg2Si Nanowires

Kang, Y., Texas A&M University
Brockway, L., Texas A&M University
Vaddiraju, S., Texas A&M University
Originally presented on: 11/5/2013 13:00:00 - 13:15:00

Solid state thermoelectric modules can convert unused waste heat energy directly into electricity without emitting any additional greenhouse gases. The performance of thermoelectric materials that serve as building blocks in the fabrication of thermoelectric modules can be gauged by examining their figure of merit, zT, which is given by $zT = S^2 \sigma T / (\kappa_e + \kappa_l)$. Here, S is the Seebeck coefficient of the material, σ is the electrical conductivity of the material, and κ_e and κ_l are the thermal conductivities of the material from electronic and lattice contributions, respectively. Enhancing the zT values of materials beyond the current state-of-the-art requires materials that exhibit lower thermal conductivities and higher electrical conductivities. Theoretical predictions indicate that accomplishing this task, while navigating the constraints imposed by the Wiedemann-Franz law, is possible by selectively lowering the κ_l of materials. Selective reduction of the κ_l of materials is possible by phonon confinement and phonon scattering when they are synthesized in ultra-thin, single-crystalline nanowire format. One of the chemical compositions of materials useful for the fabrication of thermoelectrics is magnesium silicide (Mg_2Si). Previous studies have indicated that bulk and nanocrystalline Mg_2Si is a promising thermoelectric material in the 500 to 800 K temperature range. It is also non-toxic, comprised of earth-abundant elements, and inexpensive. However, it is not yet clear whether Mg_2Si in nanowire format performs better than Mg_2Si in bulk or other nanocrystalline formats. In this context, there is a need to develop a method for the mass production of Mg_2Si nanowire powders and measure their thermoelectric performance.

A possible route for the mass production of Mg_2Si nanowire powders is the phase transformation of pre-synthesized silicon nanowires. To accomplish this, it is imperative to understand whether the phase transformation of silicon nanowires leads to the formation of Mg_2Si nanowires. Secondly, it is important to deduce the conditions under which the transformation of single-crystalline silicon nanowires leads to the formation of single-crystalline Mg_2Si nanowires. Hence, the aim of this presentation is to discuss phase transformation of silicon nanowires in detail. The silicon nanowires necessary for the phase transformation studies have been obtained by electroless etching of wafers, and the phase transformation of so-obtained silicon nanowires into Mg_2Si was accomplished by supplying magnesium vapor over their surfaces at elevated temperatures. This process resulted in the formation of polycrystalline Mg_2Si nanowires. The nucleation and growth steps involved in this process, along with the modification made to the experiments to attain single-crystalline silicon nanowire to single-crystalline Mg_2Si nanowire transformation, will be discussed in this presentation.

(321d) Growth and Transfer of High Aspect-Ratio Nanowires Onto Flexible Substrate

Xu, C., University of Florida at Gainesville
Zhao, Y., University of Florida at Gainesville
Ziegler, K. J., University of Florida
Originally presented on: 11/5/2013 13:15:00 - 13:30:00

Growth and Transfer of High-Aspect Nanowires onto Flexible Substrates

Cheng Xu[1,2*], Yang Zhao[1], Jie Liu[1], and Kirk J. Ziegler[1,2]

[1] *Department of Chemical Engineering*

[2] *Department of Materials Science and Engineering*

University of Florida, Gainesville, FL 32611

*Presenting Author (Ph: 352-392-3122; Email: chengotopia@ufl.edu)

Abstract:

Nanowire-based devices have gained tremendous interest because of their potential in fabricating devices with ultra-high energy conversion and storage, such as photovoltaics, thermoelectrics, capacitors, batteries, spintronics, and piezoelectrics. Since the device performance is often directly proportional to the active surface area, fabricating vertically aligned nanowire arrays with high density and aspect ratio is important to maximize their efficiencies beyond the current limit. More importantly, many of these devices require substrates with specific properties of flexibility, transparency, and light weight. Thus, the key challenge lying ahead is not only to improve methods to maximize surface area of nanowires but also fabricate them onto a desired substrate. Currently, researchers have not been able to produce nanowire devices with maximum density. Many methods struggle with difficulties in obtaining catalytic seeds with small diameter and maintaining nanowires with small diameter during growth due to the high surface energy associated with substrates. While templated approaches provide an excellent way of demonstrating nanowire density control, it is difficult to form and remove the back electrode and integrate these nanowires into useful devices. In this study, we focused on using the anodized alumina oxide (AAO) as a template for nanowire deposition and then transferring the vertically aligned nanostructure onto a transparent flexible substrate. AAO provides maximum density with its closely-packed hexagonal pore arrangement and, by carefully tuning the experimental conditions, the pore diameter and interpore distance can be precisely controlled. To demonstrate the transfer process, nanowires with various dimensions and materials, including Au, Ag, Cu, Pt, and ZnO, were fabricated using electrodeposition or chemical vapor deposition (CVD). These embedded nanowires were then lifted off and transferred onto a polydimethylsiloxane (PDMS) substrate. During the AAO etching and drying process, an electrostatic repulsion (ESR) technique was applied to prevent aggregation and keep the nanowires vertically aligned. The current-voltage (IV) response before and after nanowire transfer were measured and compared to demonstrate the efficiency of the transfer process. As a result, nanowires were fabricated on a flexible substrate with densities and aspect-ratios of $10^{11}/cm^3$ and 1000:1, respectively. This technique provides a new route of fabricating large-scale, cost-effective flexible optoelectronic devices.

Keywords: Dye sensitized solar cell (DSSC), ZnO nanowires, solar cell, alumina template, electrodeposition, flexible electronics

(321e) A Facile Route for the Synthesis of Nickel Nanoneedle

Gupta, V. K.,
Mehra, A., Indian Institute of Technology, Bombay (IITB)
Thaokar, R., Indian Institute of Technology, Bombay
Originally presented on: 11/5/2013 13:30:00 - 13:45:00

Down to nanoscale, properties of the material depend strongly on both the size and the morphology of the nanoparticles. Nickel nanoparticles find their applications in the areas that demand high specific surface area like catalysis, media for data recording, etc. In this work we propose a novel protocol for synthesis of nickel nanoneedles without the use of organic solvent at moderate temperature of 60 °C. Our method utilizes ability of hydrazine hydrate (HH) to simultaneously act as a reducing agent as well as a chelating agent. This enables us to eliminate the use of bulky organic molecules like surfactants or polymers that are commonly used as morphology modifiers in synthesis of anisotropic nanoparticles. This also implies that our protocol has potential for scale up as reuse and separation issues of the reagents are readily taken care of. These nanoneedles were synthesized by using Ni- acetate as metal precursor. High Resolution Transmission Electron Microscopy (HRTEM), High Resolution Scanning Electron Microscopy (HRSEM), Energy Dispersive Spectroscopy (EDS) and X-ray Diffraction (XRD) confirm material of the nanoparticles as nickel. Length of these nanoneedle ranges between 50 – 400 nm while diameter ranges between 8 – 14 nm. Effect of temperature (60 – 80 °C), Precursor concentration (35 – 105 mM), molar ratio of HH to precursor (2 – 10) and aging time (24 – 48 h) were investigated for understanding mechanism of formation of these nanoneedles. We have observed that concentration and temperature does not have significant effect on the size and shape of nanoneedles. Aging time however causes the increase in size of nanoparticles. Presence of amino group in HH seems to acts as shape inducing agent. A possible mechanism is consolidation of spherical particles to produce needle shaped particles. The exact mechanism is still under the investigation.

(321f) Surface Mineralization and Controlled Deposition of Biotemplated Palladium Nanorods Onto Gold Substrates

Freer, A., Purdue University
Smith, J., Purdue University
Macino, A., Purdue University
Harris, M. T., Purdue University
Originally presented on: 11/5/2013 13:55:00 - 14:10:00

Certain innate properties of the tobacco mosaic virus (TMV) make it ideal as a biotemplate for nanoparticle formation. Genetic modifications of the TMV coat proteins allow for an increase in the selective deposition and controlled growth of different metals onto the surface of the virus. In the current process, TMV is coated with uniform layers of palladium metal particles in aqueous solution under very mild conditions producing nanowires of 300nm in length and 30-40nm in diameter. In order to control the higher order assembly of these nanowires, the TMV is first deposited onto a substrate prior to coating. This helps to limit the uncontrolled aggregation of individual nanorods and arrange the TMV into nanostructures to be used in multiple applications ranging from catalysts to sensors to battery electrodes and other nanoelectronic devices. Varying the deposition pH, virus concentration, and substrate type can alter the aggregation and final orientation of the virus on the substrate surface. After TMV deposition, the virus is coated with palladium in an aqueous solution without the addition of an external reducing agent to form uniform palladium nanowires. Multiple characterization techniques are used to determine the consistency of the controlled orientation as well as the monodispersity of the nanowires.

(321g) Spectrally and Spatially Tailored Gold/Silica/Ytterbium, Erbium Doped Yttria Core/Shell Nanorods For Theranostic Applications

Jankovic, V., University of California, Los Angeles

Chang, J. P., University of California, Los Angeles

Originally presented on: 11/5/2013 14:10:00 - 14:25:00

Spectrally and Spatially Tailored Gold/Silica/Ytterbium, Erbium Doped Yttria Core/Shell Nanorods For Theranostic Applications

Besides their reduced size, the technologically disruptive potential of nanoparticles rests with their ability to gather several complementary properties in the same object. For example, fluorescent/magnetic nanoparticles combine the high sensitivity of fluorescence imaging with the high spatial resolution of magnetic resonance imaging, while plasmonic/fluorescent nanoparticles combine the photo-thermal properties of plasmonic materials with fluorescence imaging, enabling simultaneous bio-imaging and photo-thermal therapy.

Spectrally and spatially tailored, $Au/SiO_2/Yb{:}Er{:}Y_2O_3$ plasmonic/fluorescent core/shell nanorods were synthesized. The photonic modes of the fluorescent component were spectrally and spatially matched to the gold nanorod plasmonic modes by controlling the gold nanorod aspect ratio (spectral mode match) and the silica spacer layer thickness (spatial mode match). The spectrally-tailored hybrid plasmonic/fluorescent nanoparticles exhibited a more intense emission signal, a shorter spontaneous emission lifetime and a non-linear optical power dependence compared to the fluorescent material. The material also showed similar cyto-toxic and photo-thermal properties compared to polystyrene-coated gold nanorods. In photo-thermal therapy treatment tests, we were able to induce selective hyperthermia of breast cancer tissue both *in-vitro* and *in-vivo* using these hybrid nanomaterials as photosensitizers.

(a)

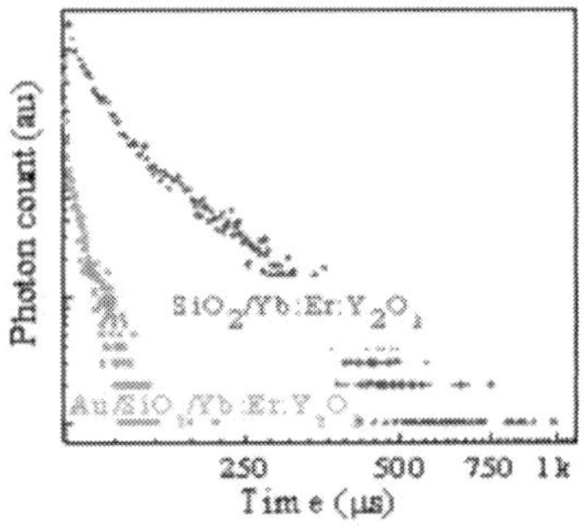

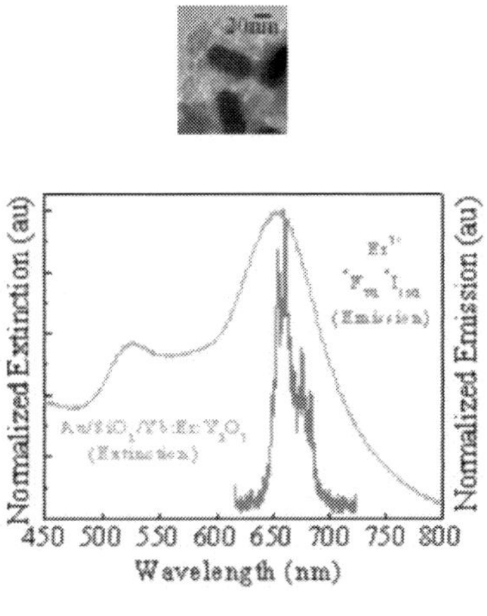

(b)

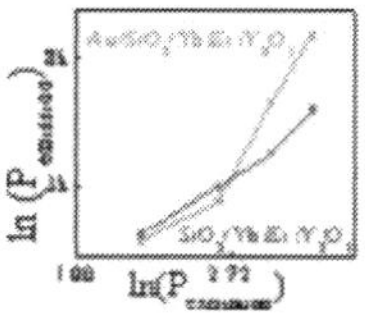

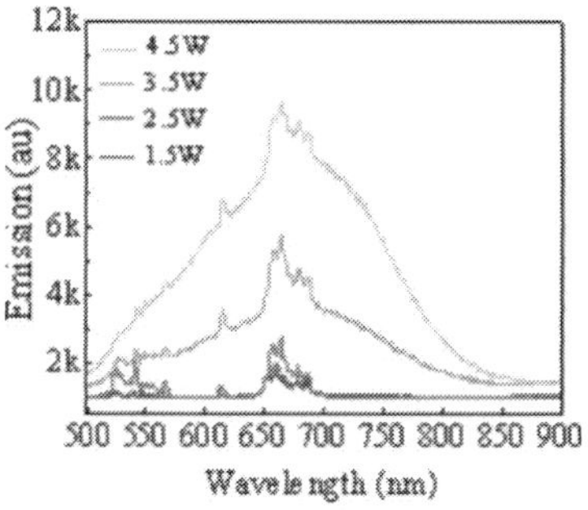

(c)

(321h) Synthesis of High Aspect-Ratio ZnO Nanowires By CVD Method

Zhao, Y., University of Florida at Gainesville
Xu, C., University of Florida at Gainesville
Zou, Y., University of Florida
Ziegler, K. J., University of Florida
Originally presented on: 11/5/2013 14:25:00 - 14:40:00

Synthesis of High Aspect-Ratio ZnO Nanowires by CVD Method

Yang Zhao[1,*], Cheng Xu[1,2], Jie Liu[1], Yongjie Zou[2], and Kirk J. Ziegler[1,2]

[1] *Department of Chemical Engineering*

[2] *Department of Materials Science and Engineering*

*University of Florida, Gainesville, FL 32611**Presenting Author (Ph: 352-392-3122; Email: yangzhao@ufl.edu)

Abstract:

Dye-sensitized solar cells (DSSCs) have been extensively studied for the past few years because of their low-cost and high-efficiency. The surface area of the photoanode, which determines the amount of photons absorbed, is one of the most critical factors to improve the efficiency of DSSCs. Although conventional TiO_2 nanoparticle-based DSSCs have shown promising results, their performance has been limited to ~11% because of diffusion and back reaction problems. Due to its excellent band gap (3.37eV) and high excitation bonding energy (60 meV) properties, one-dimensional (1D) ZnO nanowires are capable of overcoming the limitation of nanoparticle-based DSSCs by improving electron injection and axial electron diffusion. Based on models proposed by our group, the efficiency of DSSCs based on TiO_2 core-shell ZnO nanowires (40 μm in lengths, 40 nm in diameter and ~10^{11} /cm^2 of density) can significantly surpass current efficiencies. To reach our objective, we focused on using two-step anodized aluminum oxide (AAO) as a template to control the dimension of nanowires. The synthesis of ZnO nanowires was carried out using thermal decomposition of zinc acetate dehydrate in the AAO pores by low temperature, catalyst-free vapor-solid (VS) growth. After transferring embedded ZnO nanowires to a PDMS substrate, AAO was etched and an electrostatic repulsion technique was applied to provide vertically-aligned ZnO nanowires. After that, SEM, TEM, and XRD were applied to characterize these nanowires. By using this technique, it provides a new approach to obtain vertically-aligned ZnO nanowires with both high density and aspect-ratio through a large-scale process.

Keywords: Dye sensitized solar cell (DSSC), ZnO Nanowires, Alumina Template, Thermal Decomposition.

(321i) Fabrication of Tin-Doped Indium Oxide Nanowires for Dye-Sensitized Solar Cells

Li, L., University of Florida
Chen, S., University of Florida
Kim, J., University of Florida
Ziegler, K. J., University of Florida
Originally presented on: 11/5/2013 14:40:00 - 14:55:00

alternative energy sources have been the focus of research for the scientific community for decades. Compared with other sources, solar energy offers advantages such as being completely renewable, low operating cost, and environmental friendliness. The solar cell is an important class of devices that harvest solar energy and produce electricity. Among different types of solar cells, dye sensitized solar cells (DSSC) are promising and conventionally the photo-sensitized anode and the cathode are made of thin films. Nanowires (NWs) are widely investigated because of their potential applications in nanoscale devices and sensors. Nanowire-based DSSCs are more advantageous than their thin-film or nanoparticle-based counterparts due to the former's enhanced charge collection efficiency in the radial direction and faster charge transport in the axial direction. Tin-doped indium oxide (ITO) nanowires boast high conductivity and high transmittance in the visible spectra of sunlight and plays critical roles in DCCSs. In this work, free-standing ITO nanowires were synthesized via a vapor transport method on ITO-coated glass substrate. These nanowires have uniform diameters of ~ 200 nm and their length can be tailored in the range of 5-50 μm by controlling the growth time. A liquid phase deposition technique is used to deposit ~ 20nm-thick TiO_2 layers on NW surfaces. The coating crystallizes in anatase phase upon calcination, as evidenced by Raman spectroscopy and transmission electron microscopy. The electrical and optical properties of these core-shell nanowires are presented, which make them promising materials as the photoanode in DSSCs.

(321j) Titanium Dioxide Nanowires With Controllable Overpotential for Oxygen Evolution Reaction
Liu, B.,
Originally presented on: 11/5/2013 14:55:00 - 15:10:00

Titanium dioxide (TiO_2) is one of the most widely studied semiconductors in photocatalysis because it is nontoxic, abundant, stable and photoactive. However, the wide bandgap, low electron mobility and short minority carrier diffusion length of TiO_2 limit its quantum efficiency in these applications. In this work, we present a large scale flux method for growing transition-metal doped TiO_2 nanowires. In this method, commercially available TiO_2 nanoparticles act as both nucleation seeds and reactant supply for nanowire growth within a matrix of common salt mixtures. Furthermore, in situ dopant incorporation of various transition metals allows for the tuning of optical, electrical, and catalytic properties. With this combination of control, robustness, and scalability, the molten-salt flux scheme can provide high quality TiO_2 nanowires to satisfy a broad range of application needs ranging from photovoltaics to photocatalysis. The transition-metal doped TiO_2 nanowires show improved visible light absorption, better transport properties and enhanced catalytic activity for oxygen evolution reaction (OER).

(349a) Award Submission: Tunable Assembly of Biomaterials for Bone Tissue Engineering

Shah, N., David H. Koch Institute for Integrative Cancer Research at MIT
Hyder, M. N., Massachusetts Institute of Technology
Seeherman, H., Restituo LLC.
Padera, R., Brigham and Women's Hospital
Spector, M., Massachusetts Institute of Technology
Hammond, P. T., Massachusetts Institute of Technology
Originally presented on: 11/5/2013 15:15:00 - 15:30:00

Biomedical devices such as implants for total joint replacements and grafts to fill large bone defects replace native tissue, restore mobility, and alleviate morbidity. However, in some cases, these devices may fail by post-surgical loosening or stress shielding due to a modulus mismatch between the bone and implant. To attenuate these incidences, endogenous progenitor cells can be recruited to regenerate tissue to (i) promote the direct integration of the implant with the host bone tissue and (ii) repair critical size defects. We have developed modular nanostructured coatings containing osteogenic materials, including multiple biological growth factors and inorganic bone minerals, to mimic aspects of the natural bone wound healing cascade. This enabled the formation of new, functional bone tissue which bonded a permanent prosthesis to the native tissue with superior strength and rapidly filled large defects and restored tissue strength.

We used a layer-by-layer approach to coat implant surfaces with individual layers of materials with nanoscale control over their deposition. Angiogenic growth factor VEGF and osteoinductive growth factor BMP-2 were incorporated in gradient degradable coatings, such that VEGF released rapidly during the early acute phase of wound healing, along with slower releasing BMP-2 for longer term osteogenic effects. In vivo, coated polycaprolactone/β-tricalcium phosphate scaffolds were implanted in the rat quadriceps muscle pocket. Dual growth factor from the scaffolds surface resulted in superior de novo ectopic bone formation compared to single growth factor release as measured by μCT and histology. New bone formation was restricted to the scaffold. In a complementary study, surrogate orthopedic implants were biologically fixed to the bone tissue by recapitulating the bone matrix environment with films containing a combination of hydroxyapatite (HAP) and BMP-2. The films contained a permanent osteoconductive HAP base layer and a degradable BMP-2 releasing top layer. In vivo release of BMP-2 was monitored using fluorescence and ELISA of the marrow aspirates. In combination, HAP and BMP-2 accelerated the differentiation of human mesenchymal stem cells in vitro with concurrent upregulation of osteogenic markers. We observed that the mode and rate of growth factor release significantly influenced bone regeneration. When implanted in the rat tibia, implants bonded to the parent bone via maturing trabecular bone, with a temporal increase in the pull-out force when coated with a combination of these bone matrix materials. Long term studies indicated that the implant was stable over at least 18 months with minimal foreign body response. The interfacial tesile strength was 2-3 higher than clinically used bioactive bone cements.

These results demonstrate that surface modification alone can be used to induce tissue regeneration. The local availability of precise amounts of regulators is as a potent tool in tissue engineering and controlling the amount and duration is critical. The layer-by-layer technique is scalable, adaptable and highly tunable which makes it promising for clinical use. Precisely assembling materials using this technique in modular nanoscale assemblies can result in a synergistic response which accelerates bone regeneration and enables implants to last their natural lifetime and scaffolds to introduce rapid bone healing.

This work was funded by the NIH (R01 AG029601) and supported in part by the NCI (P30 CA014051) and the U.S. Army Research Office (under contract W911NF-07-D-0004).

(349b) Award Submission: Histidine-Mediated Hydrogen Bonding Enhances Stability and Silencing Activity of Peptide siRNA Polyplexes

Chou, S. T., University of Maryland
Hom, K., University of Maryland Baltimore
Zhang, D., University of Maryland
Tricoli, L., University of Maryland
Hustedt, J., University of Maryland
Lee, A., University of Maryland
Leng, Q., University of Maryland Baltimore
Shapiro, M. J., Pifzer
Seog, J., Univeresity of Maryland
Kahn, J., University of Maryland
Mixson, A. J., University of Maryland Baltimore
Originally presented on: 11/5/2013 15:30:00 - 15:45:00

Branched peptides containing histidines and lysines (HK) have been determined to be effective carriers for DNA or siRNA. We anticipate that elucidation of the binding mechanism of HK with siRNA will provide greater insight into the self-assembly and delivery of the polyplex. In siRNA silencing experiments with mammalian cells, a four-branched HK peptide siRNA polyplex maintained silencing activity even with prolonged pre-incubation with serum. In contrast, siRNA in complex with four-branched N3K4b peptide, in which histidines were substituted with asparagines, showed a marked decreased in silencing activity with pre-exposure to serum. Consequently, we explored the hypothesis that histidines might form non-covalent bonds with nucleic acids to enhance the stability of siRNA polyplexes. To accomplish this, we initially compared the biophysical properties of H3K(+H)4b with N3K4b and polylysine. Consistent with siRNA silencing experiments, gel electrophoresis analysis demonstrated that the HK siRNA polyplex maintained its integrity for more than 24 h incubation in 50% serum, whereas siRNA in complex to N3K4b or polylysine was degraded in a time-dependent manner. We next studied the thermodynamic profiles of various peptides binding to siRNA at pH 7.3 with isothermal titration calorimetry. While four-branched polylysine and linear lysine-alanine peptide (K4b and A2K) interacted with siRNA resulting in an endothermic reaction, branched and linear lysine-histidine peptides (H3K(+H)4b and H2K) both exhibited an exothermic reaction at the initial binding step. This indicates that an important part of the exothermic interaction was based on non-ionic bond formation of histidines with siRNA. To investigate the type of non-ionic bond, we studied the protonation state of the imidazole ring of a selectively N15 labeled H3K(+H)4b upon siRNA binding with heteronuclear single quantum coherence (HSQC) NMR. The peak of $N^{\delta 1}$-protonated tautomers of imidazole (type α, neutral form) shifted downfield (in the direction of deprotonation) by 0.5 to 1.0 ppm with addition of siRNA at pH 7.3, providing direct evidence that uncharged histidines formed hydrogen bonds with siRNA at physiological pH. Moreover, unshifted $N^{\varepsilon 1}$ tautomers described the directionality of hydrogen bonds. Taking together, these results indicate that histidine-rich peptides form non-ionic bonds, and in particular hydrogen bonds, with siRNA, thereby enhancing the stability and biological activity of the polyplex *in vitro* and *in vivo*.

(349c) Award Submission:Multi-Enzyme Co-Localization On Nanoparticles Facilitated By DNA Hybridization

Jia, F., Iowa State University
Narasimhan, B., Iowa State University
Mallapragada, S. K., Iowa State University
Originally presented on: 11/5/2013 15:45:00 - 16:00:00

Multi-enzyme complexes (MECs) in Nature exhibit highly efficient catalytic mechanism(s) in reaction cascades. Researchers have developed several strategies to co-localize multiple enzymes on nano-carriers to improve multi-enzyme catalytic efficiency by mimicking the structure and function of MECs. Numerous studies have indicated that the spatial arrangement and orientation of multiple enzymes in confined space are of paramount importance in facilitating cooperative enzymatic activity in multi-enzyme co-localization. Biomolecule scaffolds based on DNA hybridization are very attractive for multi-enzyme co-localization because of their ability to enable unique effective control of the relative positions of different enzymes. In this work, glucose oxidase (GOX) and horseradish peroxidase (HRP) were co-localized onto polystyrene nanoparticles via DNA hybridization. The rationale behind selecting the particulate geometry in contrast to a planar surface is that the particulate geometry is more suitable for co-localization of multiple layers of enzymes due to lower steric hindrance. Free DNA hybridization and co-localization efficiency were studied using Förster resonance energy transfer (FRET) techniques. Various capture DNA concentrations were compared and optimized. The immobilization of one enzyme was studied initially to investigate the steric hindrance effect. Evidence of co-localization of GOX and HRP was obtained by FRET studies of dye-labeled tag DNAs. Finally, it was demonstrated that co-localizing GOX and HRP via DNA hybridization significantly improved the overall reaction efficiency in comparison to a mixture of immobilized single enzymes. In summary, DNA-directed co-localization of multiple enzymes provides excellent control of the relative positions of the enzymes, thereby mimicking MECs, and leading to high reaction efficiency.

(349d) "Award Submission" Nanoscale Patterning of Membrane-Bound Proteins Formed Through Curvature-Induced Partitioning of Phase-Specific Receptor Lipids

Ogunyankin, M. O., University of California, Davis
Longo, M. L., University of California, Davis
Sasaki, D., Sandia National Laboratories
Originally presented on: 11/5/2013 16:00:00 - 16:15:00

Lipid mixtures self-assemble to form planar bilayers containing coexisting lipid phases in many compositional regimes. Coexisting lipid phases each have distinct compositions that can selectively partition optically active dyes, ligands, and membrane proteins. Therefore, precise arrangements of these phases in continuous lipid bilayers could be used to create displays of proteins, DNA strands, optically, chemically, or electrically active molecules, or nanoparticles. Curvature-based sorting of functionalized lipids that comprise or partition into specific lipid phases could serve as a way of templating active molecules. A variety of methods have been developed to attach functional molecules to lipid headgroups. These include covalent bonding and linkers such as single-stranded DNA, biotin, and glycan-phosphatidyl inositol. Lipid headgroups can also be functionalized with a metal chelate group such as iminodiacetic acid (IDA) and nitrilotriacetic acid (NTA). These chelate groups coordinate to divalent transition-metal ions such as Cu^{2+}, Ni^{2+}, and Zn^{2+} and sequester proteins rich with histidine moieties. This work describes a technique for forming high-density arrays and patterns of membrane-bound proteins through binding to a curvature-organized compositional pattern of metal-chelating lipids (Cu^{2+}-DOIDA or Cu^{2+}-DSIDA). In this bottom-up approach, the underlying support is a poly(methyl methacrilate) (PMMA) e-beam formed, square lattice pattern of hemispheres. This curvature pattern sorts Cu^{2+}-DOIDA to the 200 nm hemispherical lattice sites of a 600 nm ′ 600 nm unit cell in L_d - L_o phase separated lipid multibilayers. Binding of histidine-tagged green fluorescent protein (His-GFP) creates a high density array of His-GFP-bound pixels localized to the square lattice sites. In comparison, the negative pixel pattern is created by sorting Cu^{2+}-DSIDA in L_d - L_b phase separated lipid multibilayers to the flat grid between the lattice sites followed by binding to His-GFP. Lattice defects in the His-GFP pattern lead to interesting features such as pattern circularity. We also observe defect-free arrays of His-GFP that demonstrate perfect arrays can be formed by this method suggesting the possibility of using this approach for the localization of various active molecules to form protein, DNA, or optically active molecular arrays.

(349e) AWARD Session: Emergent Properties of Nanosensor Arrays: Applications to Monitoring IgG Affinity Distributions, Weakly-Affined Hypermannosylation, and Colony Selection for Biomanufacturing

Reuel, N. F., Massachusetts Institute of Technology
Strano, M. S., Massachusetts Institute of Technology
Originally presented on: 11/5/2013 16:15:00 - 16:30:00

It is widely recognized that an array of addressable sensors can be multiplexed for the label-free detection of a library of analytes. However, such arrays have useful properties that emerge from the ensemble, even when monofunctionalized. As examples, we show that an array of nanosensors can estimate the mean and variance of the observed dissociation constant (K_D), using three different examples of binding IgG with Protein-A as the recognition site, including polyclonal human IgG (K_D $\mu = 19\ \mu M$, $\sigma^2 = 1000\ mM^2$), murine IgG (K_D $\mu = 4.3\ nM$, $\sigma^2 = 3\ \mu M^2$), and human IgG from CHO cells (K_D $\mu = 2.5\ nM$, $\sigma^2 = 0.01\ \mu M^2$). Second, we show that an array of nanosensors can uniquely monitor weakly-affined analyte interactions *via* the increased number of observed interactions. One application involves monitoring the metabolically-induced hypermannosylation of human IgG from CHO using PSA-lectin conjugated sensor arrays where temporal glycosylation patterns are measured and compared. Lastly, the array of sensors can also spatially map the local production of an analyte from cellular biosynthesis, thus providing a powerful approach to clonal selection. As an example, we rank productivity of IgG-producing HEK colonies cultured directly on the array of nanosensors itself. This study opens new avenues for the use of nanosensor arrays for biomanufacturing applications.

(349f) Award Submission: Nanofunctionalized Organic Field-Effect Transistors for Selective, In Situ Biodetection

Hammock, M. L., Stanford University
Knopfmacher, O., Stanford University
Bao, Z., Stanford University
Originally presented on: 11/5/2013 16:30:00 - 16:45:00

Organic field-effect transistors (OFETs) are unique platforms for the detection of chemical and biological species. Such sensors are able to directly transduce an analyte-binding event into an electrical signal, making them highly compatible with sensing platforms requiring a digital readout. In particular, the detection of biologically relevant molecules lends itself to this detection platform because of the inherent charge associated with many biomolecules, which can be detected by the OFET. Additionally, OFETs are more biocompatible than their inorganic counterparts, facilitating their use as biosensors.

OFET sensor applications have historically been limited to the detection of small molecules in the vapor phase, due to the incompatibility of many organic semiconductors with water. Our group first demonstrated the real-time detection of several chemical species in aqueous media using an inherently water-stable organic semiconductor. These sensors were capable of low-voltage operation due to the incorporation of an ultrathin polymer dielectric layer, and were shown to be highly stable operating in aqueous conditions. However, while the limit of detection of these sensors has been demonstrated down to the part-per-billion (ppb) level, the response of the sensors was nonspecific, as the active layer itself displayed a serendipitous sensitivity to a number of small-molecule analytes.

In order to define sensitivity for a particular molecule of interest, a receptor group must be integrated into the device's architecture, preferably by a method that does not damage the OFET itself. We recently developed a novel OFET sensor platform that is capable of stable operation in an aqueous environment while also allowing for the selective detection of a user-defined analyte. Sensitivity for the targeted analyte is engineered by virtue of the ordered array of AuNPs that decorates the OFET's surface. This highly versatile platform is compatible with a large number of available, thiolated receptor groups that can be used to functionalize the AuNPs through the well-known gold-thiol (Au-S) linkage.

We have previously used this platform to demonstrate the highly selective detection of mercury(II) in solution. We have now expanded the sensing capabilities of this platform by demonstrating the selective detection of thrombin to 100 pM. Additionally, using this model system, we systematically investigated the effect of varying the ionic strength and pH of the buffer, as well as the average center-to-center distance between the receptor sites in order to form a more comprehensive picture of biodetection with OFETs. Using these parameters, we have devised a method to tune the dynamic range of our sensors in order to match that necessary for the particular application for which they will be used.

(349g) Award Submission: Effects of Acid Treatment On Structure, Properties and Biocompatibility of Carbon Nanotubes

Dong, C., West Virginia University
Campbell, A., West Virginia University
Eldawud, R., West Virginia University
Dinu, C. Z., West Virginia University

Originally presented on: 11/5/2013 16:45:00 - 17:00:00

Carbon nanotubes (CNTs) applications in fields as diverse as electronics, optics, nanotechnology, materials science of biomedical industry rely on improved understanding of CNTs physical structures and chemical properties as well as control over their user and environment compatibility. User-controlled strategies that allow changes in physical and chemical properties of CNTs can lead to improved performance of CNTs-based technologies in any of the fields of applications listed above. Herein we carried out a systematic investigation on how acid liquid oxidation treatment changes CNTs physical and chemical properties to lead to nanomaterials with increased biocompatibility. Specifically, by exposing CNTs to strong acid mixtures (sulfuric and nitric acid), we generated user-controlled libraries of CNTs with different characteristics, as recorded and analyzed using Raman energy dispersive x-ray spectroscopy, scanning electron microscopy, energy dispersive spectroscopy, atomic force microscopy, and solubility tests. Systematically characterized CNTs were subsequently tested for their biocompatibility in relation to human epithelial cells (BEAS-2B) or enzymes. Such selected examples are building pertinent relationships between CNTs biocompatibility and their intrinsic properties and suggest that acid oxidation treatment reduces CNTs toxicity. Further, such user-tailored libraries of CNTs with controlled physico-chemical properties can serve as the next generation of nanosupports to provide feasible platforms to be used for biomedical applications or the next generation of biosensors and diagnose tools.

(349h) Award Submission: Minimal â€œSelfâ€ Peptides That Inhibit Phagocytic Clearance and Enhance Delivery of Nanoparticles

Rodriguez, P. L., University of Pennsylvania
Harada, T., University of Pennsylvania
Christian, D. A., University of Pennsylvania
Pantano, D. A., University of Pennsylvania
Tsai, R., University of Pennsylvania
Discher, D. E., University of Pennsylvania
Originally presented on: 11/5/2013 17:00:00 - 17:15:00

Foreign particles and cells are rapidly cleared from the body by professional phagocytes that more often encounter and recognize "self" cells. From studies of mice, CD47 appears to be a factor in multiple glycoprotein complexes that impede phagocytosis of self cells, with signaling of 'self' occurring through a highly polymorphic, species-specific receptor, SIRPα. Here, short peptides were computationally designed from human-CD47, synthesized with anchoring groups, and attached to virus-size nanoparticles for injection into NOD.SCID (NSG) mice that are known to exhibit uniquely high compatibility with human cells. The hCD47-peptides delay splenic clearance of particles by macrophages with an exponential advantage in persistent circulation, which we exploit in enhancing by >10-fold the near-infrared imaging of human tumor xenografts. The affinity of hCD47 for NSG-SIRPα proves weak but within the broad range (0.1~5 mM) measured for ten constructed variants of human-SIRPα; several versions of hCD47-peptide are likewise shown to bind and potently inhibit nanoparticle uptake by a surprising cytoskeletal mechanism. The reductionist approach reveals the importance and utility as well as some limits of a human 'Marker of Self'. (Rodriguez et al., Science 2013)

(349i) Award Submission: Dynamic Microstructures for Controlling Spatial Organization of Biological Entities and Materials Within Defined Geometries

Tekin, H.,
Khademhosseini, A., Massachusetts Institute of Technology
Langer, R. S., Massachusetts Institute of Technology
Originally presented on: 11/5/2013 17:15:00 - 17:30:00

Introduction: The architectural complexity and intricate interactions between multiple cell types in their native extracellular microenvironment start in the early developmental stages and continue through the life cycle of a living system. These complex interactions play a role in regulating tissue morphogenesis, organ functions, cellular decisions, and cancer.[1] Recreating these microenvironments could improve our understanding of the early developmental stages and allow us to fabricate disease models for drug discovery and tissues for regenerative medicine. Replicating native biological complexities has remained a challenge. Previous methods, such as the use of static microstructures and patterning techniques,[2,3] have not allowed for targeted spatial control. Herein, we introduce dynamic microstructures to control spatial organization of living materials and biomaterials within defined microenvironments.[4]

Materials and Methods: Prepolymer solutions of poly(N-isopropylacrylamide) (PNIPAAm), a thermoresponsive polymer, were used to generate dynamic microstructures.[4,5] Dynamic microstructures with different patterns, such as microgrooves, circular microwells and square microwells, were fabricated using soft lithographic techniques. The dynamic microstructures responded to temperature by changing their patterned areas. Precursors of an agarose mixture containing different cell types were sequentially molded within dynamic microstructures, which allowed for controlled spatial arrangements of different cell types in different compartments of microgels. In addition, fluorescent beads of different colors were used to show proof of concept for spatially immobilizing chemical compounds within different compartments of a microscale hydrogel. Human hepatoblastoma (HepG2) cells, NIH-3T3 fibroblasts, and Human Umbilical Vein Endothelial cells (HUVECs) were used in the study.[4]

Results: The dynamic microstructures were subjected to different temperatures to test their responsiveness. Their patterned areas gradually increased with temperature increase. The change in patterned areas allowed for patterning biomaterials (containing either biological entities or chemical compounds) at different temperatures. The first layers of microgels were patterned at room temperature and the second layers were fabricated at physiological temperature. Multicompartment striped microgels containing fibroblasts in one compartment and HUVECs in another compartment were fabricated as a model for tissues having striped geometries. Cylindrical and cubic multi-layered hydrogels encapsulated HepG2s in the first layer and HUVECs in the second layer, served as models for tissue engineering and drug discovery. Different fluorescent beads were successfully immobilized into different compartments of the multi-layered microgels, suggesting potential applications for drug delivery or for delivery of soluble factors to microscale complex tissues. With the presented method, controlled spatial distribution of multiple cells and microbeads within a geometrically controlled biological material was achieved.[4]

Conclusion: We have introduced dynamic microstructures for a wide range of applications in life sciences. The shape changing properties of these microstructures could be used to generate complex biological architectures. One could substitute different cells and gels with the ones used in this study to create different microscale complex tissue constructs for regenerative medicine or to study cancer dynamics. Dynamic microstructures may also be useful for studying developmental biology by recapitulating the intricate extracellular organizations of embryonic stages. Thus, the presented methods with dynamic microstructures could be potentially useful for applications in tissue engineering, cancer biology, developmental biology, and drug discovery.[4]

References: **(1)** C. M. Nelson et al., In *An. Rev. of Cell and Dev. Bio.* 2006; 22, 287 **(2)** E. E. Hui et al, PNAS 2007, *104*, 5722. **(3)** Y.-s. Torisawa et al., *Int. Bio.* 2009, *1*, 649. **(4)** H. Tekin et al., JACS 2011, *133*, 12944. **(5)** H. Tekin et al., *Lab Chip* 2010, *10*, 2411.

(349j) Award Submission: Highly Efficient Polymer-Based Microrockets and Their Biomedical Applications

Gao, W., University of California, San Diego

Originally presented on: 11/5/2013 17:30:00 - 17:45:00

The motion of synthetic micro/nano objects is of great fundamental and practical interest, and has thus stimulated major research efforts over the past decade in connection to diverse biomedical applications. Here we demonstrated a new polymer-based catalytic tubular microrocket synthesized using a template based electrodeposition method. The effects of different electropolymerized outer layers, including polypyrrole (PPy), poly(3,4-ethylenedioxythiophene) (PEDOT), polyaniline (PANI) and of various inner metal surfaces (Ag, Pt, Au-catalase), upon the movement of such bilayer microtubes are evaluated and compared. These oxygen bubble propelled Pt based microrockets are only 8 μm long, can operate in very low levels of the hydrogen peroxide fuel (down to 0.2%) and achieve a consistently remarkable high speed of 1400 body lengths/s in physiological temperature (the speed record of all artificial nanomotors). These microrockets can also move efficiently in various fuel enhanced biological media such as human serum and can thus serve as an ideal platform for diverse biomedical and environmental applications. For example, lectin modified PANI/Pt microrockets are used for selective bacteria (*E. Coli*) isolation from food, clinical and environmental samples. Poly(3-aminophenylboronic acid) (PAPBA)/Ni/Pt microrockets coupling the selective monosaccharide recognition of the boronic-acid-based outer polymeric layer with the catalytic function of the inner platinum layer. The resulting boronic-acid based microrocket itself provides the 'built in' target recognition capability. 'On-the-fly' capture, transport and release of yeast cells (containing glucose on the cell wall) are illustrated. Molecularly imprinted polymers (MIPs) based microrockets can concentrates the fluorescent-tagged protein target onto the polymer outer layer without the need for additional external functionalization, allowing motion based extraction and isolation of Av-FITC from raw serum and saliva samples along with real-time visualization of the protein loading and transport. Besides the catalytic Pt based microrockets which rely on the hydrogen peroxide fuel, we also demonstrate a new hydrogen bubble propelled micromotor which can be driven by the natural environments (such as human stomach), without any additional chemical fuel. The tubular polyaniline/zinc microrockets display effective autonomous motion in extreme acidic environments (with a speed of over 100 body lengths per second). The ejection of hydrogen bubbles from the exposed zinc inner layer, upon its contact with acid, provides a powerful directional propulsion thrust. The observed speed-pH dependence holds promise for sensitive pH measurements in extreme acidic environments. The attractive guided cargo transport capabilities of these acid driven microrockets hold great promise for practical biomedical applications such as targeted drug delivery.

(374a) Simulation Tools for Nanoparticle-Based Composite Processing and Property Prediction

Bolintineanu, D. S., University of Minnesota
Lechman, J. B., Sandia National Laboratories
Schunk, P. R., Sandia National Laboratories
Originally presented on: 11/5/2013 15:15:00 - 15:40:00

We present a set of tools for the modeling of various aspects of solution-based nanomanufacturing processes. First, we discuss several discrete element modeling (DEM) techniques for the simulation of nanoparticle/colloidal suspensions, including efficient algorithms for the inclusion of hydrodynamic interactions, capillary forces, long-range colloidal interparticle forces, granular contact forces, solvent evaporation and highly aspherical particles. We apply these techniques to the simulation of a battery electrode manufacturing process, and show that the predicted microstructures provide relevant insights for the processing parameters. In the second half of the talk, we discuss methods for the prediction of the bulk properties of disordered, particle-based composites. We present several approaches that rely on a finite element model (FEM) representation of a sample of the material, which can be extracted from a DEM simulation or taken from a tomographic image of the real composite. Overall, this work presents an extensive, multiscale simulation toolset for connecting particle-based nanomanufacturing processing conditions to resulting microstructures and ultimately to material properties.

(374b) Tin Oxide Nanowires and Their Hybrid Architectures for Kinetically Fast Redox Couples in Dye-Sensitized Solar Cells

Vendra, V. K., University of Louisville
Nguyen, T., University of Houston
Druffel, T., Conn Center for Renewable Energy Research
Sunkara, M., University of Louisville
Amos, D. A., University of Louisville
Originally presented on: 11/5/2013 15:40:00 - 16:05:00

The redox electrolyte is an important component of the dye-sensitized solar cells. Iodide/tri-iodide electrolyte used in high efficiency cell is not suitable for practical application due to its corrosive nature, toxicity, competition between dye and tri-iodide for light aborption. One of the major drawback of using alternate redox couples is their fast recombination kinetics and mass transport limitations.

In this study, we show that tin oxide nanowires show a substantial improvement in the photovoltaic performance when compared to titania nanoparticles, when electrolytes having fast recombination kinetics are employed in dye-sensitized solar cells (DSCs). State of the art methods in literature employ titania nanoparticles use passivating layers of Al_2O_3 or poly(methyl siloxane) to lower the electron recombination rate. There have been reports investigating the performance of nanowire in conjunction with alternate redox couples. The single crystalline nature allows for fast electron transport in tin oxide nanowires. Further, prior work from our group has also shown that tin oxide nanowires exhibit two orders of magnitude higher electron lifetimes when compared to titania nanoparticles when iodide/tri-iodide is used as an electrolyte. In this study we employed tin oxide nanowires as photoanodes for DSCs and investigated their performance when kinetically fast electrolytes such as ferrocene/ferrocenium,tempo/tempo,ferricyanide /ferrocyanide are used as redox electrolytes.

Surface modifications such as $TiCl_4$ treatment and atomic layer deposition of titania and alumina were performed to further lower electron recombination in tin oxide nanowires. Fundamental properties such as electron transport time scales and electron recombination time scales were determined for tin oxide nanowires and hybrid architectures where the surface of tin oxide nanowires is coated with titania and alumina.

(374c) Multiscale Sustainability Assessment in Nanocoating Material Design and Manufacturing

Song, H., Wayne State University
Uttarwar, R., Wayne State University
Huang, Y., Wayne State University
Originally presented on: 11/5/2013 16:05:00 - 16:30:00

Nanopaint has drawn great attention in the coatings industry, as it may offer surface coatings with significantly improved performance or new functionalities and properties, such as surface texture alteration control, self-healing, toxic chemical/acid/corrosive agent repelling, glass opaqueness control, building exterior's infrared reflecting/absorbing properties control, and even pathogen growth control. Nevertheless, it is under scrutiny for the environmental and health risks which may pose in nanopaint manufacturing and nanocoating realization. Therefore, it is crucial to analyze the sustainability performance of nanocoating materials and prepare a roadmap for their successful development and application in the future.

In this paper, we introduce a systematic sustainability assessment methodology by resorting to multiscale science and sustainability science. Different from known methodologies focusing on the assessment of manufacturing process and manufactured product, this methodology initiates the assessment starting at the nanomaterial formulation design stage. A variety of material design parameters, such as those about polymer system, nanomaterial type, shape, size, and orientation, solvent and additives, are analyzed using a number of sustainability indicators, which will be incorporated into the analysis of coating manufacturing stages. Multiscale models are developed to predict nanoparticle emission severity in nanopaint spray and curing, nanocoating properties, product realization process performance on quality assurance, waste reduction, and operational safety. The efficacy of the multiscale sustainability assessment methodology will be demonstrated by the case studies on nanocoatings with super-scratch resistance performance and barrier property. The introduced methodology is generic and can be applied to study the sustainability performance of a wide range of nanocoating materials in their early design stage.

(374d) Continuous Nanoparticle Sizing and Characterization Via Microfluidic Interfacial Fluorescent Complexation

Meng, F., Texas A&M University
Ugaz, V. M., Texas A&M University
Originally presented on: 11/5/2013 16:30:00 - 16:55:00

A variety of characterization tools are available to sensitively analyze collected nanoparticles (e.g., condensation particle counters, differential mobility analyzers, scanning mobility particle sizers (SMPS), ion chromatography, mass spectrometry, and electron microscopy). But these methods often require a dedicated laboratory infrastructure to operate optimally, and are therefore not straightforward to adapt for online use. We have developed a new microfluidic-based approach that overcomes many of these limitations by harnessing the inherently steep chemical gradient established at the interface between co-flowing streams containing a nanoparticle-laden suspension and a fluorescent dye. This sharp mismatch acts to localize adsorptive dye-nanoparticle complexation interactions within a narrow interfacial zone, instantaneously producing an intense and easily detectable fluorescence signature. These interactions are inherently dominated by phenomena occurring at the nanoparticle surface, introducing the exciting possibility to extract information about particle size and morphology from the fluorescence profile. We demonstrate this capability experimentally using ZnO and TiO_2 nanoparticles, and introduce a reaction model that enables interfacial fluorescence to be predicted so that kinetic parameters associated with the underlying surface complexation can be quantified. Our approach offers potential for continuous operation that is particularly attractive in manufacturing settings where robust on-line characterization tools are needed.

(374e) Safer Formulation Concept for Flame-Generated Engineered Nanomaterials

Sotiriou, G. A., ETH Zurich
Gass, S.,
Cohen, J., Harvard University, School of Public Health
Pyrgiotakis, G., University of Florida
Pratsinis, S. E., ETH Zurich
Demokritou, P., Harvard School of Public Health
Originally presented on: 11/5/2013 16:55:00 - 17:20:00

The likely success or failure of the nanotechnology industry depends on the environmental health and safety of engineered nanomaterials (ENMs). While efforts toward engineering safer ENMs are sparse, such efforts are considered crucial to the sustainability of the nanotech industry. A promising approach in this regard is to coat potentially toxic nanomaterials with a biologically inert layer of amorphous SiO_2. Core-shell particles exhibit the surface properties of their amorphous SiO_2 shell while maintaining specific functional properties of their core material. A major challenge in the development of functional core-shell particles is the design of scalable high-yield processes that can meet large-scale industrial demand. Here, we present a safer formulation concept for flame-generated ENMs based on a one-step, in flight SiO_2 encapsulation process, which was recently introduced by the authors as a means for a scalable manufacturing of SiO_2 coated ENMs. Firstly, the versatility of the SiO_2-coating process is demonstrated by applying it to four ENMs (CeO_2, ZnO, Fe_2O_3, Ag) marked by their prevalence in consumer products as well as their range in toxicity. The ENM-dependent coating fundamentals are assessed and process parameters are optimized for each ENM investigated. The effects of the SiO_2-coating on core material structure, composition and morphology, as well as the coating efficiency on each nanostructured material, are evaluated using state-of-the-art analytical methods (XRD, N_2 adsorption, TEM, XPS, isopropanol chemisorption). Finally, the biological interactions of SiO_2-coated vs. uncoated ENMs are evaluated using cellular and animal bioassays, providing valuable evidence for reduced toxicity for the SiO_2-coated ENMs. Results indicate that the proposed 'safer by design' concept bears great promise for scaled-up application in industry in order to reduce the toxicological profile of ENMs for certain applications.

References

[1] Sotiriou G.A., Sannomiya T., Teleki A., Krumeich F., Vörös J. and Pratsinis S.E. Non-toxic dry-coated nanosilver for plasmonic biosensors. *Adv. Funct. Mater.* **20**, 4250-4257, (2010).

[2] Sotiriou G.A., Hirt A.M., Lozach P.Y., Teleki A., Krumeich,F. and Pratsinis S.E. Hybrid, silica-coated,

Janus-like plasmonic-magnetic nanoparticles. *Chem. Mater.* **23**, 1985-1992, (2011).

[3] Gass S., Cohen J.M., Pyrgiotakis G., Sotiriou G.A., Pratsinis S.E. and Demokritou P. Safer Formulation Concept for Flame-Generated Engineered Nanomaterials. *ACS Sustainable Chem. Eng.* **in press**, DOI: 10.1021/sc300152f, (2013).

[4] Demokritou P., Gass S., Pyrgiotakis G., Cohen J.M., Goldsmith W., McKinney W., Frazer D, Ma J, Schwegler-Berry D, Brain J, Castranova V., An in vivo and in vitro toxicological characterisation of realistic nanoscale CeO_2 inhalation exposures *Nanotoxicology* **in press**, DOI: 10.3109/17435390.2012.739665 (2013).

(374f) Green Pathways for Development of Nanostructured Aerogel Photocatalysts Effective At Both UV and Visible Range

Li, H., University of South Florida
Sunol, S. G., University of South Florida
Sunol, A. K., University of South Florida
Originally presented on: 11/5/2013 17:20:00 - 17:45:00

Nano-structured highly porous $TiO_2/WO_3/Fe^{3+}$ aerogel photocatalyst are prepared, characterized, and tested for model photo-catalytic reactions. The catalyst structure is tailored to capture environmental pollutants and destruct them under both u-v and visible light through oxidation.

A novel and green method is utilized to prepare the unique surfactant templated aerogel photo-catalyst that has highly porous nanostructure with high surface area and tailored pore size distribution. Sol-gel process is combined with supercritical extraction, deposition and drying. Supercritical drying with heat treatment results in titanium dioxide anatase crystal form. Templates used further enable retention and tuning of the nano-pore structure and surface properties.

The catalyst was characterized using AFM, SEM, FIB, XRD, XPS, BET and porosimetry prior to using in reactions. The band gap of the catalyst particles were determined using diffuse reflectance.

Resulting aerogel $TiO_2/WO_3/Fe^{3+}$ has similar photo-catalytic capability compared to highly optimized commercial Degussa P25 under UV exposure and offer much superior photo-catalytic capability under visible light exposure. The model reaction utilized employed Methylene Blue photo oxidation under Visible and UV light.

(397bp) Green-One Step Synthesis of Silver Nanoparticles in PVA Polymer Solution

Gonzalez-Campos, J. B., Universidad Michoacana de San Nicolas de Hidalgo
Hernandez-Vargas, J., UNIVERSIDAD MICHOACANA DE SAN NICOLAS DE HIDALGO
Perez-Nava, A.,
Lopez-Castro, Y.,
del Rao, R. E., Universidad Michoacana de San Nicolas de Hidalgo

In recent years the synthesis of silver nanoparticles (AgnP's) has been subject of focusing research due to their unique optical, electrical, and thermal properties coupled with their wide potential applications including molecular diagnostics and photonic devices. Different routes for AgnP's synthesis have been reported including ultraviolet irradiation, lithography, laser ablation and thermal decomposition to name just a few. However, the development of efficient green chemistry methods for synthesis of metal nanoparticles has become a very important issue to find an eco-friendly technique for their production. Some metal nanoparticles green syntheses have been reported. The green synthesis techniques are generally synthetic routes that utilize relatively non-toxic chemicals to synthesize nanomaterials, and use non-toxic solvents such as water or plant extracts for reduction. Even though biological processes are eco-friendly, they usually are elaborate processes that involve multiple steps from the collection of the plant to silver nanoparticles synthesis which in turns translates into time and money. The chemical reduction of metal ions is still the most widely applied technique. Different reducing agents could be used; however, sodium borohydride ($NABH_4$) and sodium citrate ($Na_3C_6H_5O_7$) are the most commonly used. This method also could differ in the use of different sources to drive the reduction (i.e. microwave, ultraviolet, gamma, thermal decomposition or thermal radiation). Although chemical reduction is the most commonly used method for metal nanoparticles synthesis, it is considered hazardous to environment because it employs many chemical agents.

On the other hand, the chemical reduction synthesis of metal nanoparticles in polymer solutions is a practical technique since polymers can act as stabilizing agents, and polymer/metal nanoparticles composites can be simultaneously produced. Polyvinyl alcohol (PVA)/ AgnP's is a very attractive combination for composites production, since PVA is a biocompatible, biodegradable, non-toxic and water soluble polymer for potential biomedical applications and combined with AgnP's could produce an organic/inorganic hybrid material with antibacterial activity, which is highly desired in applications where hygiene is a key factor; especially in biomedicine. There are several works on PVA/AgnP's composite synthesis via chemical reduction, and two green synthesis without the use of any reducing agent and using different stirring times at high temperature (90°C and 70°C) have also been reported. However, as it can be seen, artificial energy to rise temperature and uphold stirring to drive the reduction is still used. This paper reports a green-one step silver nanoparticles synthesis method in PVA polymer solution without using any reducing agent, using water as the solvent and with no artificial energy to drive chemical reduction, providing a facile, economic, quick, efficient and more eco-friendly method for the synthesis of metal nanoparticles that results in the production of a PVA/AgnP's composite were PVA acts simultaneously as reducing and stabilizing agent.

(397bl) Molecular Understanding of Electrospinning: Polymer Solutions in An External Field

Nezbeda, I., J. E. Purkinje University
Jirsak, J.,
Moucka, F., J. E. Purkinje University

Nanospider technology™is a method to produce nanofibers by electrospinning from the free surface of polymer solutions. The process runs only in a narrow window of conditions (strength of the electric field, composition of the solution, temperature, etc.) and the lack of understanding of the process at a molecular level means that its determination has been based so far on the trial-and-error approach only.

It will be shown how molecular simulations carried out on aqueous solutions of electrolytes in an external electric field can mimic the entire process. Specifically, development of the Taylor cones and the onset of spraying and jetting. Experimentally detected regular rippling of the electromagnetic field before the onset of spraying is explained and the crucial role of the presence of ions at the surface of the solution for the transition from spraying to jetting has been identified. Simulations make it possible to define molecular layers and determine the average molecular properties in individual layers.

(397aj) Development of Self-Aligned Contact Using Negative Tone Lift-Off Process for Electron Beam Lithography On Phost/Pmgi Bilayer Resists

Liang, Y. C., University of Houston
Litvinov, D.,

In this article, a simple, low cost self-aligned contact fabrication technique for metal line patterns is presented. The use of a new electron beam bilayer resists comprised of poly-(4-hydroxystyrene) (PHOST) as the top layer and poly-(dimethylglutarimide) (PMGI) as the bottom layer is demonstrated. The characterization of electron beam exposure on PHOST/PMGI has first been investigated, showing that deploying PMGI as underlayer has no significant influence on the exposure of PHOST. The well defined undercut configuration with 100nm undercut length in the PMGI layer was generated. An Ar ion milling etching on PHOST/PMGI was performed to remove unprotected Ta layer, showing that PHOST is an effective mask layer. A lift-off method using 2% TMAH aqueous solution was conducted after sputter deposition of 70nm thick Cu layer on the substrate, demonstrating good contact connection at both transverse sides of 400nm wide Ta line patterns. As a result, self-aligned contacts have been successfully obtained.

(397d) Organized Self-Assembly of Janus Catalytic Nanomotors

Gao, W., University of California, San Diego
Wang, J., University of California, San Diego

Locomotion of synthetic nano-/microscale objects through fluid environments is one of the most exciting and challenging areas of nanotechnology. Inspired by animal interactions, the ability of synthetic nanoscale motors to produce self-organized structures is of considerable interest, owing to their future implications in nanomedicine, nanomachinery, transport systems, and chemical sensing. Organized self-assemblies of Janus catalytic motors, induced by hydrophobic surface interactions involving multiple motor/motor and motor/nonmotor particles, display controlled coordinated self-propulsion. These assemblies were prepared by octadecyltrichlorosilane (OTS) modification of the surface of a silica microparticle and addition of a catalytic Pt hemispheric coating. The influence of the self-assembled structures upon the motion behavior is investigated. Different bonding orientations between these hydrophobic Janus motors induce different forms of motion. The relative orientation of each motor in the assembly changes its contribution to the net propulsion force and rotational moment. The hydrophobic interactions between individual micromotors and micromotor assemblies can promote a continuous growth of the assembly during its movement and lead to dynamic changes in the motion behavior. Organized assemblies of multiple motor/nonmotor particles are also illustrated toward optimal cargo transport and delivery. Such controlled structures and motion of chemically powered Janus nanomotor assemblies hold considerable promise for the creation of intelligent nanomachines that perform collective tasks.

(397e) Enhanced DNA Segmental Dynamics in Nanofluidic Channels

Chen, Y. L., Academia Sinica
Luzhbin, D.,

Investigations of DNA molecules in strongly confined environments depend on the increased DNA bending stiffness due to spatial restrictions. We examine DNA molecules in nanofluidic devices using multi-scale computational models and how confinement causes non-monotonic change in DNA conformation, correlation lengths, and the segmental relaxation. Classical theories of deGennes and Odijk predict increased polymer relaxation time and polymer size in confinement with dimensions smaller than the radius of gyration. When the confinement dimensions become smaller than the polymer persistence length, the segmental correlation length conforms to the confinement geometry, leading to enhanced correlation lengths in the unconfined dimension but shortened correlation length in the confined dimension. We report the segmental relaxation times and correlation lengths in nano-confinements of height H, from H larger than the polymer radius of gyration Rg to smaller than the polymer persistence length P. A non-monotonic transition in the DNA relaxation time is observed in nanochannels, as found in previous experimental observations [1]. In contrast, no qualitative change of the segmental dynamics is observed in nanoslits [2,3]. The change in DNA correlation length is also characterized in semi-dilute and dense solutions, with implications for the mechanical properties for glass formation in thin films.

[1] W. Reisner et al., Physical Review Letters 94, 196101 (2005).

[2] PK Lin et al., Macromolecules 45, 2920 (2012).

[3] J. Tang et al., Macromolecules 43, 4368 (2010).

(397ak) Synergistic Photocatalytic Activity of Vox-TiO2 Nanocomposite Under UV and Visible Light Conditions

Kim, S., Korea Electronics Technology Institute
Suh, M. S.,
Lee, C. S., Korea Electronics Technology Institute

VOx-TiO2 nanocomposite made by one-pot synthesis showed the different catalytic activity from the mixture of VOx and TiO2 nanoparticles. Photocatalytic activity of VOx-TiO_2 nanocomposite with various compositions was investigated by following photodegradation of methylene blue in aqueous solutions. Nanocomposites show significant enhancement of decomposition of methylene blue and the reason is considered by the synergistic effect between two different materials. However, this synergy was not obtained from the mixture of VOx and TiO2 nanoparticles. Morphologies of synthesized catalysts were characterized by scanning electron microscopy, transmission electron microscopy, and UV-visible diffuse-reflectance spectroscopy (UV-vis DRS). Composition of nanocomposite catalysts wase varied by vanadium precursor amount and the photocatalytic activity was analyzed based on pseudo first-order reaction kinetics under 250 W Xe-lamp radiation with various longpass filters for UV and visible light conditions. Obtained results show that the as-prepared VOx-TiO2 nanocomposites have different degradation abilities of methylene blue dependent on the amount of vanadium oxide and the wavelength of exposed light source.

(397al) Effect of Nanoparticles On the Viscoelastic Properties of Poly(acrylamide) Hydrogels

Zaragoza, J., Santa Clara University
Truong, W.,
Thomin, J., Northwest State College
Asuri, P., Santa Clara University

It is well established that the addition of nanoscopic filler particles to polymer systems can have a large impact on mechanical properties, even for very low weight fractions of nanoparticles. However, despite over two decades of research using experimental studies and modeling, the mechanisms behind this reinforcement are still under debate. Our work aims to shed light on the molecular underpinnings of reinforcement in polymer nanocomposites so that specific mechanical properties can be engineered in a wide range of polymer nanocomposite systems. In our current study, we investigated the viscoelastic response of poly(acrylamide) hydrogels containing silica nanoparticles of various sizes and concentrations. The polymer density and crosslink density of the hydrogels were controlled by manipulating the relative concentrations of acrylamide monomer and N,N-methylenebis(acrylamide) crosslinker. Initial results suggest that the elastic modulus of the hydrogels depends strongly on particle size and concentration. In addition, the relative increase in the elastic modulus of the nanocomposite compared to the pure polymer appears to depend strongly on the overall concentration of the gel. Furthermore, we find that incorporating nanoparticles can result in gel stiffnesses greater than the maximum that is obtainable through purely chemical crosslinking. Future work will explore the fundamental nature of this reinforcement.

(397am) Cooperative Catalytic Activity of Ã¡-Cyclodextrin and Ag Nanoparticles in Spherical Polyelectrolyte Brushes

Liu, J.,
Wang, J.,
Li, L., East China University of Science and Technology
Guo, X., East China University of Science and Technology
Lincoln, S., University of Adelaide
Prud'homme, R. K., Princeton University

Spherical poly(acrylic acid) brushes (SPB) were prepared by photo-emulsion polymerization. Positive silver ions were absorbed and concentrated in SPB due to the Donnan effect. Silver nanoparticles (Ag-NP) were synthesized in-situ by reduction of Ag^+ upon addition of $NaBH_4$. The nano-composite particles were characterized using TEM, DLS, ICP-AES and UV-vis spectroscopy. While the size distribution of the Ag-NP synthesized at a given temperature is quite uniform this distribution varies with change in synthesis temperature. The lower temperature results in smaller silver nanoparticles. The Ag-NP loaded in SPB have a strong catalytic effect on the reduction of 4-nitrophenol (4-NP) to 4-aminophenol (4-AP) by $NaBH_4$. The addition of α-cyclodextrin (α-CD) further accelerates this 4-nitrophenol reduction and modifies the associated activation parameters. This is attributed to α-CD complexing 4-nitrophenol in the vicinity of the Ag-NP surface and thereby aiding the catalysis. This work opens a new way to prepare metal nano-particles and to improve their catalysis by addition of CD.

(397f) Synthesis and Characterization of Magnetic Biodegradable Poly(D,L-lactide-co-glycolide) Nanocapsules

Xu, P.,

The objective of this study was to develop biodegradable/biocompatible polymer-coated magnetic nanocapsules for potential biomedical applications. Air-filled magnetic nanocapsules were prepared by an ultrasound emulsification technique and multiple emulsion-solvent evaporation method utilizing highly concentrated magnetic Fe_3O_4 nanoparticles and poly(D,L lactide-co-glycolide) (PLGA). Hydrophilic and hydrophobic Fe_3O_4 nanoparticles separately modified with monolayer and bilayer of oleic acid, exhibiting good miscibility in inter water solution and biopolymer solution, and were induced to the core and shell area of PLGA nanocapsules during the procedure of emulsification. Air-filled nanocapsules with magnetic Fe_3O_4 nanoparticles embedded were obtained with a mean diameter of about 700 nm. We characterized these nanocapsules in terms of external morphology, microstructure, size and magnetite content and distribution within the nanostructures. Both the core-embedded and shell-embedded magnetic nanocapsules showed efficient contrast in vitro for ultrasound imaging, and suitable as a potential platform for the design of magnetically guided drug delivery and other in vivo biomagnetic applications.

(397ao) Solar Hydrogen Production From Metal Sulfide (ZnS-CuS-CdS) Photocatalysts

Hong, E., University of Seoul

Kim, J. H., University of Seoul

SOLAR HYDRPGEN PRODUCTION FROM METAL SULFIDE (ZnS-CuS-CdS) PHOTOCATALYSTS

Eunpyo Hong, and Jung Hyeun Kim, University of Seoul, Seoul (Korea)

1. Introduction

Hydrogen has long been considered as a clean and eco-friendly energy source. However, conventional methods such as steam reforming and electrolysis of the water require high-energy consumption for hydrogen production. In that sense, hydrogen production from water splitting reaction using photocatalysts is one of the ideal ways because it uses only water and solar light, as shown below for the complete consumption and regeneration cycle.

Generation: H_2O + energy(solar) °æ H_2 + $1/2O_2$ (1)

Consumption: H_2 + $1/2O_2$ °æ H_2O + energy (2)

Such a photocatalytic process was first developed by Fujishima and Honda [1], and later many researchers have devoted to improve photocatalytic performance in various ways. In this study, the ZnS photocatalyst is used as a base material. However, the ZnS photocatalyst solely is not so effective to use because of its relatively large band gap energy. Therefore, we applied two primary concepts for improving the electron usage efficiency on hydrogen production.

First, CuS nanoparticles were introduced on the ZnS photocatalyst surface to improve interfacial electron transfer. The main role of the CuS nanoparticles is providing a driving force for electron movement. The CuS materials accept electrons from ZnS surface, and these electrons reduce the protons (H^+). The similar phenomenon is observed from the CuS/ZnO (nanowire) and CuS/ZnS (nanosheet) [2, 3]. The major electron transfer mechanisms on photocatalysts under light illumination are shown below.

CuS/ZnS + $h\nu$ °æ CuS(e^-)/ZnS(h^+) (3)

2CuS + $3e^-$ °æ Cu_2S + S^{2-} (4)

Cu_2S + $2H^+$ + S^{2-} °æ 2CuS + H_2 (5)

Second, the third material is used for composite semiconductor fabrication for expanding the light harvesting region. Various composite semiconductor materials such as CdS, CdSe, CdTe, and CuO were reported. Amongst them, the CdS material is a promising light sensitizer considering the band gap energy and band position (valence and conduction bands).

Therefore, the ZnS-CuS-CdS semiconductor compounds are utilized for hydrogen production under visible light illumination condition in order to achieve high electron usage efficiency.

2. Results and Discussion

The ZnS nanoparticles were formed in ionic solutions (Zn^{2+} + S^{2-}), and a cation exchange reaction was used for the ZnS-CuS compound fabrication. Finally, the CdS quantum dots were deposited on the as-prepared particle surface by a chemical bath deposition method. Schematic view of the ZnS-CuS-CdS nanoparticle compounds and the ideal electron paths are demonstrated in Fig. 1.

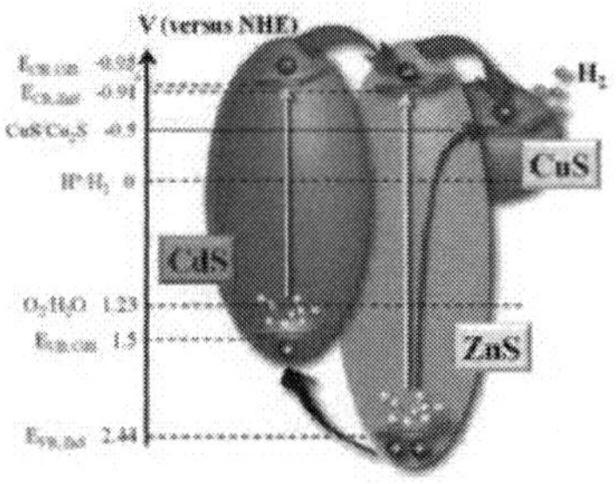

Fig. 1. Schematic energy diagram of ZnS-CuS-CdS composite photocatalysts.

Three major experimental variables were used for fabrication of semiconductor nanoparticle materials: 1) thermal sintering of the ZnS nanoparticles, 2) varying Cu contents for the ZnS-CuS compound, and 3) varying Cd content for the ZnS-CdS. Fig. 2 shows hydrogen production results from only ZnS case and ZnS-CuS composites. The ZnS-CuS composite evolved much more hydrogen than from the only ZnS nanoparticle case. Thermal treatment of the ZnS nanoparticles improves hydrogen production activity for both the only ZnS case and ZnS-CuS composites.

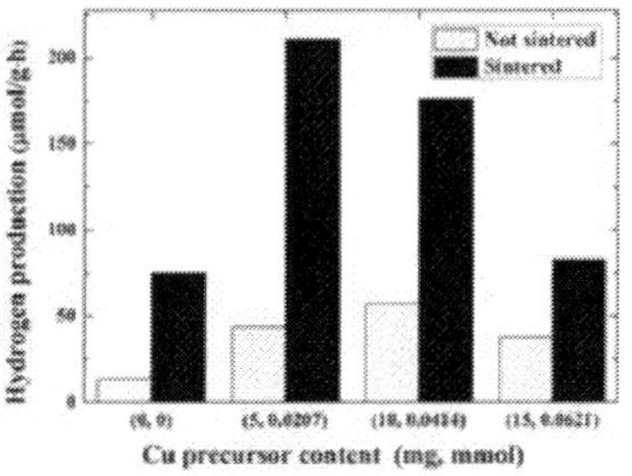

Fig. 2. Amount of evolved hydrogen of the ZnS-CuS particles as a function of Cu precursor content.

Fig. 3 shows amounts of evolved hydrogen from the photocatalyst samples during the experimental time period. As a result, photocatalytic performance is greatly increased by introducing two concepts (CuS and CdS junction), in our experimental system. This result demonstrates that the electron transfer through the semiconductor composite material (CdS °æ ZnS °æ CuS) is very efficient and its usage for hydrogen production is remarkably improved.

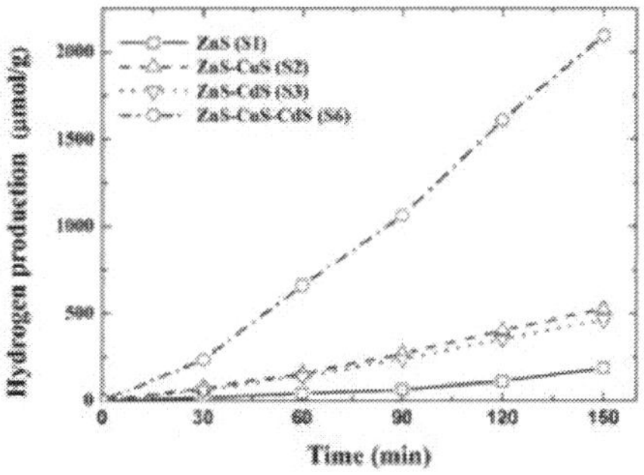

Fig. 3. Amounts of evolved hydrogen from four different types of photocatalysts.

Other material properties characterized by HR-TEM, XRD, UV-vis, and XPS analyses are also helpful in understanding this improved hydrogen production phenomenon with semiconductor composite photocatalysts under the solar light illumination.

References

1. A. Fujishima, K. Honda (1972), Nature, 238, 37.

2. J. Zhang, J. Yu, Y. Zhang, Q. Li, and J.R. Gong (2011), Nano Letters, 11, 4774.

3. M. Lee and K. Yong (2012), Nanotechnology, 23, 194014.

(397ap) Design and Synthesis of N-Doped Mesoporous Carbon Materials Using Modified-Pyrrole Monomer and Block Copolymer Template

To, J., Stanford University
He, J., Stanford University
Gary, B., Stanford University
Stack, D., Stanford University
Wilcox, J., Stanford University
Bao, Z., Stanford University

Porous materials such as activated carbons, zeolites, silica gels, and inorganic oxides are of great interest because of their potential applications in adsorption, catalysis, separation, purification processes, optics, electronics, etc. Among these materials, ordered mesoporous carbons have attracted considerable attention owing to their high surface area, tunable pore structure and narrow pore size distribution. On the other hand, nitrogen doped carbons have shown enhanced mechanical, thermal and electrical properties; therefore, there are increasing effort in making of N-doped mesoporous carbon, which combine the high porosity of the mesoporous framework and unique chemical functionality. N-doped mesoporous carbons are often prepared by nanocasting method using a sacrificial template, usually porous silica. The template is impregnated with nitrogen containing organic molecules, followed by carbonization and removal of silica template or through post-synthesis treatment of mesoporous carbon using acetonitrile or ammonia CVD. These multiple-steps processes are costly and time consuming. Other methods include the co-assembly of nitrogen containing monomer (melamine) with structural directing agent; however, the resulting porous polymer is not rigid enough to withstand harsh carbonization condition.

In this work, we report successful synthesis of nitrogen-doped mesoporous carbon using modified-pyrrole monomer as nitrogen sources. Through molecular co-assembly with triblock copolymer and oxidative polymerization during sol-gel preparation, ordered mesoporous polymer was made. Synthesis of mesoporous polymer is often challenging due to macrophase separation and polymer demixing. Nitrogen rich mesoporous carbon was resulted from carbonization of the respective polymer without structure collapse. High surface area and nitrogen loading enable its application in carbon capture and electric double-layer capacitors.

(397an) Observation of Nano-Sized Spherical Polyelectrolyte Brushes By Small Angle X-Ray Scattering

Yu, X., East China University of Science and Technology
Wang, W., East China University of Science and Technology
Li, L., East China University of Science and Technology
Guo, X., East China University of Science and Technology

Small angle X-ray scattering (SAXS) is a well-established characterization method in materials science and colloidal physics. SAXS is highly valuable for a comprehensive analysis of polymer colloids, and can give information on the radial structure of nanoparticles. Spherical polyelectrolyte brush (SPB) is a nano-sized core-shell particle with a polymer core densely grafted by linear polyelectrolyte chains, which has been synthesized by photo-emulsion polymerization. SPB can be used to create a large well-defined surface in an aqueous environment, which is suitable for the in-situ preparation of metal nano-particles or immobilization of proteins. The SPBs were observed by SAXS, and the scattering data were analyzed by two new methods, which can reflect an apparent core-shell pattern as well as the interaction among SPB. The particle size change upon changing pH and salt concentration of the solution as observed by SAXS agreed well with the observation by dynamic light scattering (DLS). SAXS can also determine the change of SPB by immobilzation of metal nanoparticles or adsorption of proteins.

(397g) Simulation of Worm Like Micelle Assisted Assembly of Linear Nanostructures

Chhatre, A., Indian Institute of Technology Bombay
Thaokar, R., Indian Institute of Technology, Bombay
Mehra, A., Indian Institute of Technology, Bombay (IITB)

In surfactant assisted synthesis of nanoparticles, morphologies of the micelles play the role of guiding evolution of shape of nanoparticles. Depending upon the concentration of surfactant and counter-ions, micelles exhibit a range of morphologies such as spheres, cylinders, worm like micelles (WLM), etc. Formation of silver halide nanorods of high aspect ratio, assisted by worm like micellar phase at room temperature, have been demonstrated by Gupta et. al., (2012) (Colloids and Surfaces A 393 (2012) 73–80) recently. In the present work we propose a mechanism for formation of these nanorods.

The WLM phase is characterized by high degree of entanglement amongst the worm-like chains; resulting in an increase in the viscosity by over four orders of magnitude (as compared with an aqueous phase). This causes a substantial reduction of the particle diffusivity which in turn reduces frequency of collision between the (primary) spherical particles. The attractive Van der Waals interaction between the particles and the chains therefore leads to adsorption of the particles on backbone of WLM, resulting in decoration of the WLM chains with spherical nanoparticles. During their relaxation by reptation the micelles bring particles closer to each other and promote the formation of linear assemblies of the particles. This mechanism is simulated using a frame work of slithering snake dynamics for WLM. In this study the worm like micelles are represented by flexible polymers of fixed contour length. Further, a rule based intermicellar particle exchange protocol is formulated and simulated on a periodic lattice. Simulations reveal that the particles start accumulating slowly on few of the micellar backbones; towards the end the fraction of micelles carrying no particles increases from 15% to nearly 70% which is typical behavior observed in coagulation processes. The particulate masses accumulated on the WLMs are then converted to their respective lengths and diameters. Further simulations like effect of precursor concentration and surfactant volume fraction are being done to gain more understanding of the system.

(397as) Sonolytic Synthesis Of Single and Binary Metal-Based Magnetic Nanostructured Materials

Moore, G. L., North Carolina A&T State University
Wicker, J. P., North Carolina A&T State University
Roberts, K. L., North Carolina A&T State University

In this project iron oxide and iron oxide-based magnetic nanomaterials were manufactured using sonolytic & chemical synthesis techniques for biomedical applications. Technology challenges involved with dimensional consistency (or the control of size and morphology), biocompatibility, and the scale up of bench-scale syntheses to produce large batch sizes of nanomaterials. When specifically considering the application of bimetallic nanomaterials, additional design considerations will include the control of relative metal compositions and requirements of production consistency regarding the layering of active metal on the surfaces of the nanomaterials. Magnetic nanomaterials were produced using wet chemistry / reactive processing and sonolytic methods. Superparamagnetic iron oxide (SPIO) nanoparticles were synthesized with iron chlorides using sonochemical methods. All magnetic nanomaterials will be characterized using x-ray diffraction (XRD), elemental analysis, BET surface area analysis, scanning and transmission electron microscopies (SEM and TEM) and superconducting quantum interference device (SQUID) magnetometric analysis. The effects of binary metal oxide composition on the magnetic and structural properties will be discussed.

(397b) Polydispersity Control In the Liquid Phase Synthesis of Amphipathic, Self-Assembling Polypeptides

Kubilius, M., City College of New York
Tu, R., The City College of City University of New York

Synthesizing a useful, periodically-sequenced, amphipathic polypeptide is a challenge because the polydispersity index increases with increasing molecular weight. To circumvent this hurdle, we designed synthetic amino acid dimers that are both water-soluble and amphipathic. The resulting peptide backbone has alternating hydrophilic/hydrophobic side groups, the typical periodicity for beta-sheet forming polypeptides. Using this approach, we can influence the polydispersity of the growing polypeptide chains, controlling the kinetics of growth through transport-limited chain elongation. Our experiments show that in the absence of an interface, standard bulk-phase condensation polymerization occurs. With an interface present, the amphipathic character of the peptide chain increases with increasing molecular weight, resulting in a polypeptide that partitions to interface as a function of the degree of polymerization. This type of kinetically limited growth serves to narrow the polydispersity of our periodically sequenced polypeptide. We quantify the dynamics of chain elongation and interfacial assembly using multi-angle light scattering, and we define the evolving sheet-like secondary structure using circular dichroism for various peptides of differing amino acid pairs.

(397aq) Dynamic Adsorption of Bovine Serum Albumin Onto Spherical Polyelectrolyte Brushes Triggered By pH

Wang, S., East China University of Science and Technology
Chen, K., Shanghai Jiaotong University
Yu, X., East China University of Science and Technology
Wang, W., East China University of Science and Technology
Li, L., East China University of Science and Technology
Guo, X., East China University of Science and Technology

The adsorption of bovine serum albumin (BSA) in spherical poly(acrylic acid) brushes (SPB) was observed by a combination of turbidimetric titration, dynamic light scattering (DLS), zeta potential measurement, UV spectroscopy, and small angle X-ray scattering (SAXS) techniques. Adsorption of BSA onto SPB occurred in the pH range from 7.0 to 4.6 as observed by turbidimetric titration, DLS, and zeta potential measurement. Adsorption, aggregation, and desorption of BSA from SPB were triggered by changing pH. Ionic strength can modulate the extent of the pH regions for adsorption, aggregation, and desorption. The adsorption amount of BSA in SPB was determined by UV spectroscopy, which was dependent on pH and ionic strength. The dynamic adsorption of BSA onto SPB can be controlled through adjusting the pH and ionic strength of the system. The analysis of SPB loaded with BSA by SAXS demonstrated that the protein molecules enter deeply into the brush layer at low ionic strength. The positive charge patches of BSA become multivalent counterions of the poly(acrylic acid) chains grafted on SPB.

(397ar) Zeolite-Confined Sulfonated Graphene-Nafion Composite Membrane for Self-Humidifying PEMFC

Han, W., The Hong Kong University of Science and Technology
Poon, H. Y., The Hong Kong University of Science and Technology
Sim, V., The Hong Kong University of Science and Technology
Yeung, K. L., The Hong Kong University of Science and Technology

Perfluorosulfonic acid (PFSA) membrane is the most common electrolyte membrane used for PEMFC. Nafion (PFSA) membrane displays high proton conductivity (about 0.1 $S{\cdot}cm^{-1}$) and excellent long-term stability (about 6000 hours) under fully hydrated conditions. However, the proton conductivity of Nafion membrane decreases sharply under low humidity, whereas excessive water can result in electrode flooding. External humidification equipment and special-designed water drainage unit are often required for hydrogen PEMFC resulting in complicated design and operation of system, and a lower overall energy efficiency. Self-humidifying proton exchange membrane that can regulate better hydration in PEMFC is attractive.

Many self-humidifying composite membranes using Pt-containing inorganic fillers such as Pt/Carbon-Nafion, Pt/Oxide-Nafion and Pt/Zeolite to Nafion matrix generate and retain water by catalysis and adsorption. However, increased transport resistance, particle aggregation and phase separation are major concerns for these traditional composite membranes. There are also increasing evidences that platinum deposited in proton exchange membrane promotes the formation H_2O_2 and HO· free radicals that are responsible for membrane degradation. Moreover, electron-conducting carbon material can cause internal short-circuit. Recently, we explored the effects of narrow confinement of PFSA in subnanoliter volume of zeolite-coated pores. These zeolite-confined Nafion composite membranes displayed improved thermal and mechanical stabilities as well as much higher proton conductivity under the conditions of high temperature and low humidity. The hierarchical design could be used to uniformly distribute inorganic fillers within PFSA and isolate electron-conducting fillers in a discontinuous phase. Heteroatom (N, B, or P)-doped carbon nanotube, graphene and graphite are shown to be interesting oxygen reduction electrocatalysts. Graphene derivatives in PFSA could therefore in principle catalyze water generation. This work examines a series of zeolite-confined sulfonated graphene-Nafion composite membranes and their performances for self-humidifying PEMFCs.

A silicalite-1 (Sil-1) coating was hydrothermally grown on porous stainless steel mesh (SSM) as an electron-insulating and anticorrosive layer. Graphene was prepared via modified Hummers method from commercial graphite flakes, followed by sulfonation with sulfanilic acid. The sulfonated graphene was dispersed in Nafion PFSA and casted into Sil-1-coated SSM to prepare Sil-1-confined sulfonated graphene-Nafion composite membranes. For comparison, Sil-1-confined sulfonated graphite (Sgrap)-Nafion and sulfonated carbon nanotube (SCNT)-Nafion were prepared. The membrane-electrode assembly was fabricated by hot-press method using 0.5 mg/cm^2 Pt/Vulcan XC-72 catalyst.

The results of micro-Raman spectroscopy and transmission electron microscopy show that sulfonated graphene oxide (SGO) and sulfonated graphene (SG) have multi-layered structures of less than 10 layers. All sulfonated carbon materials were well dispersed in the Nafion PFSA, and the composite membranes are uniform black in color. Compared to Sil-1-confined Nafion composite membrane, Sil-1-confined SGO-Nafion and SG-Nafion composite membranes have weaker fluorescence due to the quenching effect of graphene materials, which further indicates membrane homogeneities. The fuel cell with Sil-1-confined SGO-Nafion composite membrane gives a power density of 728 mW/cm^2 at 70°C without humidification, which is 16.5 times higher than those of standard fuel cell with Nafion 117 membrane. It can also tolerate much higher temperature even compared to Sil-1-confined Sgrap-Nafion and SCNT-Nafion composite membranes. Improved glass transition temperatures (verified by differential scanning calorimetry curves) and oriented proton transport channels (suggested by EDX elemental mapping images) are contributing reasons for the excellent performance.

(397bm) Substrate-Imprinted Lipase Nanogel for Enzymatic Catalysis in Organic Solvents

Wang, R., Tsinghua University
Ge, J., Tsinghua University
Liu, Z., Tsinghua University

In recent years, the employment of enzymes in industrial catalysis has become an increasingly promising alternative to conventional chemical catalysis due to their high specifity, selectivity and activity. Enzymatic catalysis in organic solvents has a lot of advantages, such as the increased solubility of non-polar substrates, enantioselectivity, and prohibition of water-dependent side reactions. However, enzymes quite often display very low activity and low stability in organic solvents, restricting wide applications of enzymes in organic solvents. We introduced a new method of preparing a 'substrate-imprinted' lipase nanogel that displays high activity and selectivity in organic solvents [1].

The enzyme nanogel was synthesized using lipase from *Thermomyces lanuginosus* (Lipozyme TL 100L) as the model enzyme. A two-step procedure including surface acryloylation and *in-situ* aqueous polymerization was applied to encapsulate lipase into a nanogel. The lipase nanogel was then lyophilized in the presence of a substrate of lipase, palmitic acid. Then palmitic acid was removed from the lyophilized lipase nanogel by solvent extraction.

The imprinting treatment brought about 2-fold and 1.4-fold increase in the catalytic activity of transesterification reaction between para-nitrophenyl palmitate and ethanol, ascorbic acid and vinyl palmitate, respectively. The effects of temperature, solvent and molar ratio of reactants on the yield and selectivity of the enzymatic synthesis of chloramphenicol palmitate were examined. One-step synthesis of chloramphenicol palmitate catalyzed by the imprinted lipase nanogel gave a yield of ~99% and a purity of ~99% within 12 hours at 20 °C, whereas the imprinted free lipase gave a yield below 60% in 20 hours. The increased catalytic activity could be attributed to the imprinted molecular 'cavities' in polymer shells which enhanced substrate transport and recognition. The hydrophilic micro-environment of the polyacrylamide network also facilitated the conformational transition of the encapsulated lipase.

To further elaborate the mechanism underlying the increased activity of the imprinted lipase nanogel, lipase nanogel imprinted with polyethylene glycol of different molecular weights were prepared to compare their catalytic activities with palmitic acid-imprinted lipase nanogel. By this study, it will be revealed that whether the porous structures or the increased affinity of molecular cavities in polymer shells after imprinting resulted in the increased enzymatic activity. *Candida antarctica* lipase type B (CAL-B) was chosen as the model enzyme in the study, which also demonstrated the effectiveness of the imprinting treatment for different enzymes.

1. Wang, R., Zhang, Y. F., Huang, J. H., Lu, D. N., Ge, J., Liu, Z. Substrate imprinted lipase nanogel for one-step synthesis of chloramphenicol palmitate. Green Chemistry, 2013. 15: p.1155-1158

(397at) Preparation, Characterization and Application of Microfluidic-SERS Sensors

Parisi, J. Jr., University of Connecticut
Su, L., University of Connecticut
Liu, Y., University of Connecticut
Lei, Y., University of Connecticut

Preparation, Characterization and Application of Microfluidic-SERS Sensors

Joseph Parisi, Liang Su, Yixin Liu, Yu Lei

Department of Chemical, Materials, & Biomolecular Engineering, University of Connecticut, 191 Auditorium Road, Unit 3222, Storrs, Connecticut 06269

Research for the detection of chemical and biological species is consistently moving forward and creating more accurate and sensitive devices. However, many of these devices require multiple steps for not only fabrication of the sensor material but also for preparation of the sample for testing. In many of these examples there is no simple approach to economically fabricate novel 3-D sensing materials and package them within a microfluidic channel for future applications. Our work focuses on *in-situ* preparation of 3-D nanostructured electrodes within a microfluidic channel using electrodeposition and galvanic replacement reactions. These nanostructures are fabricated in a fraction of the time of traditional lithography, CVD and hydrothermal techniques, require less material and limit any complicated fabrication steps. Once the nanostructures are fabricated they are immediately ready for species detection using surface enhanced Raman scattering (SERS). The detection of various species is demonstrated using the developed device.

(397au) Investigation of Black Silicon As SERS Substrate for High Sensitivity Detection

Deng, Y. L.,
Juang, Y. J., National Cheng Kung University

Surface enhanced Raman scattering (SERS) has been utilized for bio-detection, environmental monitoring, single molecule detection and toxic material analysis. Various SERS substrates such as silver coated nano sphere array, gold or silver nano particles cluster and electrochemically roughened metallic surface have been fabricated and the enhancement effect has been examined. In this study, we have investigated the effect of the surface morphology of the black silicon substrate on SERS. The ratio of O_2/SF_6 flow rate was adjusted to produce the black silicon with different tip densities (tip numbers per unit area). It was found that the higher the tip density, the greater the ehancement effect. Moreover, by increasing the thickness of the deposited gold layer on the black silicon substrate, the SERS effect can be increased as well. This could be owing to generation of more hot spots by bridging individual silicon tips through the thick gold deposit. The detection limit of 10fM R6G solution with uniform SERS effect across the SERS substrate was achieved. The single algal cell will be analyzed and the SERS spectrum will be acquired by using the black silicon substrate as fabricated.

(397av) Rate-Limiting Nutrient Delivery System for Microbial Enhanced Oil Recovery
Li, W.,

Rate-limiting nutrient delivery system for microbial enhanced oil recovery

Weiwei Li, Stephen Johnson, Jenn-Tai Liang

Tertiary Oil Recovery Project, University of Kansas, 66045

Abstract

Microbial enhanced oil recovery (MEOR) techniques stimulate naturally occurring reservoir microbes or inject specially selected consortia of natural bacteria into the reservoir. The stimulated microbes produce specific metabolites such as surfactants, polymers, acid or gases that lead to the extraction of oil trapped in capillary pores of the formation rock or in areas not swept by the classical or modern enhanced oil recovery (EOR) methods. Although the potential benefits of MEOR applications are considerable, improvement of oil recovery via the manipulation of microbial metabolism in the reservoir still remains an unproven concept. One of the major issues of MEOR application is that biomass and extracellular polymeric substances produced in response to high injected nutrient concentrations can cause plugging close to the injection point (near wellbore plugging).

In this study, a novel approach to MEOR is discussed. Polyelectrolyte complex nanoparticles (PECs) were used to encapsulate and propagate nutrients, and release the nutrient substrate slowly, to enable stimulation of the microbes to occur over an extended distance and prevent near wellbore plugging.

In the present study, polyethylenimine-dextran sulfate (PEI-DS) polyelectrolyte complexes were used to entrap nutrients commonly used in microbial stimulation. Stability and reproducibility of PEC nanoparticles was assured over time. The toxicity of PEI and, DS to a bacterial consortium isolated from Wellington oil field, Wellington, KS was also monitored by aerobic batch culture. The microbes survived at a DS concentration up to 6000 ppm and a PEI concentration up to 3000 ppm.

Bacteria growth, as revealed by plate counting, was delayed, compared to equivalent systems where the nutrient mixture was not entrapped (positive control group). This is consistent with the hypothesis that PEC nanoparticles delay the growth of microbes by entrapping their nutrient scource. Compared to the positive control group, the entrapped nutrient group has a 10^2 magnitude decrease of microbial number for the first 72 hours; also the entrapped nutrient group requires at least 96 hours more to reach the stationary phase.

Future work will include propagation of the PEC-entrapped nutrients and microbial growth in porous media.

(397aw) Pollutant Removal From Water By Reduced Graphene Oxide Membranes and Their Electrochemical Regeneration

Sheath, P., Monash University
Majumder, M., Rice University

In current water purification practice, spent adsorbent materials (usually activated carbon) are subject to thermal or chemical reactivation which contributes significantly to the cost of the overall purification process. Reduced graphene oxide (RGO), as derived from the exfoliation of graphite by oxidation followed by its subsequent chemical or thermal reduction, has recently been reported to have exceptional capacity to absorb pollutants from water. In particular, we utilise the conductivity of RGO to attempt regeneration of spent RGO adsorbents by electrochemical means. Films of conductive RGO, synthesised by vacuum filtration methods, are adsorbed with model pollutants after which, by applying an electrical potential to the spent film, adsorbed pollutants can be removed and the film subsequently re-used. By analysis of the adsorption capacity and by electrochemical impedance spectroscopy, we correlate the electrochemical behaviour of the RGO films with their regenerative capability in order to demonstrate the feasibility of their potential as a unique addition to the suite of water purification methodologies.

(397ax) Electrostatic Coating Poly-L-Lysine and Chitosan On the Perfluorocarbon Emulsion and the Conformation Analysis of Coating Materials

Wu, C. J., University of Utah
Ostafin, A. E., University of Utah

Stability and performance control of surfactant-stabilized nanoemulsions in drug delivery, imaging and other applications is enhanced when they are coated with other molecules that can provide chemical responsiveness, diffusional gating, or target recognition function. Little is known about the characteristics of these coatings at the molecular level in part due to measurement difficulties. Here we present a method to characterize the construction and properties of nanoemulsion coatings using two distinct molecules poly-L-lysine (PLL) and chitosan (CS) as representative coatings. Dynamic light scattering and zeta potential are used to follow and quantify the coating construction and used to detect subtle changes in coating molecule conformation at the surface of the nanoemulsion consistent with fluorescent quenching experiments and thermodynamic analyses. Opposing results for PLL and CS coated emulsions were obtained: PLL molecules compress and CS molecules expand on the nanoemulsion surface at higher coating concentration, and PLL molecules expand or CS molecules compress at lower coating concentration. Such phase behavior is also observed in thin films or Langmuir-Blodgett films and reflects effects of local confinement of molecules at interfaces. The results show that the properties of molecules used as nanoemulsion coatings can be quite complex.

(397ay) Nanocapsulation of Ascorbyl Dipalmitate in Chitosan Via Rapid Expansion of Subcritical Solutions Followed By Ionic Crosslinking

Noiklam, T., Kasetsart University
Inta, O.,
Yoksan, R., Kasetsart University
Sane, A., Kasetsart University

Ascorbyl dipalmitate (ADP) nanoparticles encapsulated in chitosan were successfully produced by rapid expansion of subcritical solution (RESOLV) process followed by ionic crosslinking with chitosan-tripolyphosphate. The rapid expansion of supercritical solution process provided ADP nanoparticles, while ionic crosslinking at the particle surface with chitosan-tripolyphosphate imparted coating and/or encapsulation of the ADP nanoparticles with chitosan. In general, ADP nanoparticles obtained from RESOLV were spherical with a narrow size distribution between 30 and 70 nm, independent of ADP concentration, pre-expansion temperature and pressure, as well as tripolyphosphate concentration. Encapsulating ADP nanoparticles in chitosan slightly increased the particle sizes to ~40-90 nm. The size of ADP nanoparticles encapsulated in chitosan were unaffected by changes in concentration and content of chitosan solution. However, increasing the concentration and content of chitosan solution led to increased amount of chitosan in the coating layer and decreased the fraction of ADP in the nanocapsules. Increasing ADP concentration and pre-expansion temperature and pressure did not significantly influence on the size of nanoparticles but increased the degree of particle agglomeration instead.

To obtain ADP nanoparticles coated by chitosan with a narrow size distribution and high loading content of ADP, suitable RESOLV processing conditions, i.e. low pre-expansion temperature and pressure as well as ADP concentration, are required to consistently produce ADP nanoparticles with less degree of particle agglomeration. In addition, suitable encapsulation/coating conditions, i.e. low concentration and amount of chitosan solutions, are also required to obtain high loading capacity of ADP. It is possible to control the loading content of ADP in nanoparticles by adjusting the concentration of tripolyphosphate, with the loading capacity being decreased as the tripolyphosphate concentration increased. With suitable processing conditions, we were able to produce ADP nanoparticles encapsulated in chitosan with an average size range of 55–60 nm and relatively high loading contents of ADP ranging between 43 and 74 %. The release profiles of ADP in phosphate buffer solution exhibited a rapid burst release within 4 hr followed by a lag phase (~5-24 hr) and a sustained release (during 1-10 day). In addition, the stability of encapsulated ADP increased as compared with its non-encapsulated counterpart.

(397az) Application of a Smart Antimicrobial Coating in Facilities to Combat Indirect Infection

Leung, H. H.,
Sim, V., The Hong Kong University of Science and Technology
Yeung, K. L., The Hong Kong University of Science and Technology

The World Health Organization (WHO) has stated that indirect transmission is one of the major means of disease transmission. Indirect transmission refers to the transfer of an infectious agent via a contaminated object or person. Inanimate surface is considered as a common medium that acts as a reservoir for microorganisms to thrive in until the next user touch it.

An antimicrobial coating containing antimicrobial agents has been developed to inhibit the survivability and growth of microorganisms on surfaces. In this work, chlorine dioxide was encapsulated with polymer. The water in oil in water encapsulation provides a slow and controlled release of chlorine dioxide that is responsive to touch. The release of antimicrobial agent will be faster upon an increase in temperature and moisture. Therefore, the designed coating can provide both contact killing and release killing properties.
In laboratory testing, the antimicrobial coating can obtain a reduction of 4log10 bacteria within 1 min of contact time for gram negative and gram positive bacteria.
The bactericidal property of the coating can last for at least one month. The antimicrobial coating is also able to kill several Multidrug Resistance Organisms(MDRO), including Vancomycin, Methicillin-resistant Staphylococcus Aureus (MRSA).

The antimicrobial coating has undergone field testing in hospital settings, including an Intensive Care Unit and an orthopedic ward in a local hospital in Hong Kong. The result showed that the microbial contamination on inanimate surfaces has been significantly reduced compared surfaces routinely cleaned by bleach solution.
Our results show that the antimicrobial coating is a potent agent to reduce surface contamination, and therefore, a possible solution to reduce disease transmission.

(397ba) Catalysis of Gold Nanoparticles Within Cross-Linked Lysozyme Crystals

Liang, M., Tianjin University

Qi, W., Tianjin University

Huang, R., Tianjin University

Su, R., Tianjin University

He, Z., Tianjin University

Preparation of bio¨Cnano hybrid materials has been the focus of numerous studies because of their broad range of potential applications in catalysis, drug delivery, medical imaging, and chemical sensing [Wei, Wang et al., 2011]. Among various biological components that could be used as templates for the fabrication of bio-nanocomposites, protein assemblies with different nanostructures, such as fibrils, tubes, and cages, have received special attention due to diverse structures and promising functionalities [Abe, Tsujimoto et al., 2012]. Nevertheless, bridging from the nanoscale to the macroscale and to create macroscopic hybrid materials for practical application remains a challenge. Therefore, it is desirable to explore macroscopic protein assemblies as templates for the fabrication of novel composite materials with functional properties [Liang, Wang et al., 2013].

Here we show that lysozyme molecules in their crystalline state can be treated as macroscopic templates for the construction of novel composite catalysts with good recyclability (Scheme 1). Firstly, tetragonal lysozyme crystals were prepared according to our previous crystallization conditions [Liang, Jin et al., 2013]. The stabilization of native crystals with glutaraldehyde resulted in cross-linked lysozyme crystals (CLLC) that were pale yellow in color and robust physically. Then, soaking the CLLC in $HAuCl_4$ solution allowed the accumulation of Au^{3+} in the interior pores of crystal. Subsequently, the coordinated Au^{3+} underwent progressive reduction to form Au nanoparticles (AuNPs) *in situ* after the addition of NaOH. Furthermore, we demonstrated that these prepared AuNPs-at-CLLC (Au@CLLC) hybrid materials exhibited good catalytic activity toward nitrophenol reduction with excellent recyclability. The main ideas and conclusions are summarized as follows.

(1) Tetragonal lysozyme crystals that comprised four relatively large (110) side faces capped by four inclined (101) faces can be readily crystallized using our previous conditions. It is known that protein crystals present highly ordered three-dimensional arrangements of protein molecules with the properties of high porosity (0.5 to 0.8), large surface area (800 m^2/g to 2000 m^2/g), and a wide range of pore sizes (2 nm to 10 nm). In addition, various functional groups that responsible for coordinating metal ions are exposed on the surface of solvent channels of crystal lattices. Given these advantageous properties, protein crystals can be treated as promising matrices for nanoparticle immobilization. Light microscope of CLLC did not show obvious structural distortion originated from the stabilization treatment. Meanwhile, SAXRD pattern of the CLLC shows characteristic diffraction at 2¦È = 2.88°, 3.47°, 4.51°, and 6.91°, indicating the crystalline form of CLLC has remained largely unchanged.

(2) AuNPs were synthesized within the solvent channels of CLLC using a reductant-free and green synthetic route. The CLLCs were immersed in $HAuCl_4$ solution for the sequestration of metallic precursors. Reduction of the coordinated Au^{3+} by tyrosine and tryptophan residues of lysozyme under alkaline condition resulted in intact red metallized CLLC. Meanwhile, no characteristic SPR of AuNPs was observed in the supernatant solution as shown in the UV-vis spectrum. HRTEM measurements of the AuNPs formed within CLLC demonstrated that the particle sizes were almost 1.5 nm to 3.0 nm with narrow size distribution. These results indicated the crystalline porous structures of CLLC can effectively limit the migration and aggregation of AuNPs. Moreover, the small sized crystalline AuNPs was demonstrated by the relative weak characteristic diffraction peaks as shown in XRD patterns. FTIR analysis confirmed the involvement of tyrosine and tryptophan residues in the reduction of Au ions.

(3) we choose the reduction of 4-nitrophenol as a model system to evaluate the catalytic properties of the as-prepared Au@CLLC hybrid catalysts. The reaction over Au@CLLC catalyst was almost completed within 360 s for the first round. The pseudo-first-order kinetics with respect to 4-nitrophenol was used to evaluate the kinetic reaction rate of the current reaction, and the apparent rate constant was estimated from the slop to be 14.92 °Á 10^{-3} s^{-1}. The recovered hybrid catalysts exhibited similar catalytic activity in the first three rounds of the reaction, and can be successfully recycled and reused in twenty successive reactions with a conversion of >96%. These results clearly indicate that the CLLCs are a good supporting material for AuNPs because of their physical robustness, high specific surface area, and three-dimensional nanostructure property.

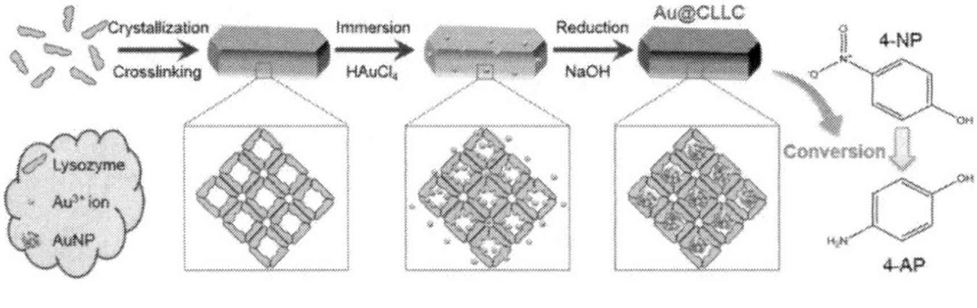

Scheme 1. Schematic illustration of the lysozyme crystallization, crosslinking and preparation of Au@CLLC composite materials for the catalytic conversion of 4-nitrophenol.

This work was supported by the Natural Science Foundation of China (Nos 21276192, 51173128 and 31071509), the Ministry of Science and Technology of China (Nos 2012YQ090194, 2012BAD29B05 and 2012AA06A303), and the Ministry of Education (No. NCET-11-0372).

References

1. Wei H., Wang Z. D., Zhang J., House S., Gao Y. G., Yang L. M., Robinson H., Tan L. H., Xing H., Hou C. J., Robertson I. M., Zuo J. M. and Lu Y. Time-dependent, protein-directed growth of gold nanoparticles within a single crystal of lysozyme, Nature Nanotechnology, 6, 93-97, 2011

2. Abe S., Tsujimoto M., Yoneda K., Ohba M., Hikage T., Takano M., Kitagawa S. and Ueno T. Porous protein crystals as reaction vessels for controlling magnetic properties of nanoparticles, Small, 8, 1314-1319, 2012

3. Liang M., Wang L. B., Su R. X., Qi W., Wang M. F., Yu Y. J. and He Z. M. Synthesis of silver nanoparticles within cross-linked lysozyme crystals as recyclable catalysts for 4-nitrophenol reduction, Catalysis Science & Technology, DOI: 10.1039/c3cy00157a, 2013

4. Liang M., Jin F. M., Liu R., Yu Y. J., Su R. X., Wang L. B., Qi W. and He Z. M. Shape evolution and thermal stability of lysozyme crystals: effect of pH and temperature, Bioprocess and Biosystems Engineering, 36, 91-99, 2013

(397bb) Protein-Mimetic Inhibition of Enzyme Activity By Nanoparticles

Cha, S. H., University of Michigan
Bahng, J. H., University of Michigan
Yeom, B., University of Michigan
Tong, W., University of Michigan
Kotov, N., University of Michigan

Enzyme inhibitors are as ubiquitous as enzymes and are central to wide spectrum of clinical and technological problems. Enzyme inhibitors are typically viewed as organic molecules with inter-molecular lock-and-key matches to sites of target substrates. Unconventional approaches to new enzyme inhibitor designs are required to circumvent biodegradation as well as to develop new momentum and impart new findings to the field. Based on the general nanoparticle-protein analogy, we hypothesized inorganic nanoparticles with specific shapes can be efficient enzyme inhibitor surrogates. We verified the hypothesis through inhibition activity and kinetics of ß-galactosidase (GAL) by ZnO nanoparticles of different shapes. Amongst the ZnO nanoparticles, which include nanospheres, nanoplates and nanopyramids, the inhibition was the strongest for ZnO nanopyramids. Such originated from the geometrical and chemical match between their apexes and grooves on GAL's surface thereby mimicking biomacromolecular inhibitors. These findings open the road for nanoscale engineering of NPs to fit specific enzymes with biotechnological or biomedical relevance.

(397bc) Assessing the Potential Permeability and Salt Rejection of Membranes Incorporating Carbon Nanotubes

Corry, B., The Australian National University, Canberra, ACT 0200,

The incorporation of carbon nanotubes (CNTs) into water filtration membranes has been suggested as a potential route to improve the efficiency of water filtration and desalination processes. Experimental studies have shown that water and gases are able to move through CNTs at rates considerably higher than expectations derived from continuum hydrodynamic theories. Simulations have also suggested that CNTs may be able to prevent the passage of salts and large solutes, which when coupled with rapid transport could provide an ideal basis for new filtration membranes.

While considerable progress has already been made in fabricating these materials, a number of questions have to be answered if this technology is to be realised. Using molecular dynamics simulations we have addressed the following issues:

1. Why does water pass so quickly through CNTs?
2. What is the theoretical maximum membrane permeability that could be achieved?
3. Why are salts rejected by some CNTs and how can this rejection be improved?
4. How does CNT diameter and length influence water transport and salt rejection.
5. How does chemical functionalisation influence transport and rejection?

Rapid water flux is seen to result from the near frictionless walls of the CNT which in turn is due to their physical smoothness and lack of localised electrostatic interactions. Salt rejection arises in narrow CNTs as a consequence of the energetic costs of ions having to dehydrate in order to fit inside the pores. An upshot of this is that salt rejection is strongly related to the CNT diameter. Appropriate chemical modification of the CNTs can improve salt rejection by simple charge repulsion or attractive binding interactions. However any chemical modification is likely to reduce the water transport rate by increasing the friction of the walls. Together these results indicate some of the design principles that should be considered when fabricating membranes incorporating carbon nanotubes for water filtration applications.

(397bd) Kinetically Controlled Self-Assembly of Redox-Active Ferrocene-Diphenylalanine: From Nanospheres to Nanofibers

Wang, Y., Tianjin University

Huang, R., Tianjin University

Qi, W., Tianjin University

Su, R., Tianjin University

He, Z., Tianjin University

Self-assembly is a common process in biological and synthetic systems, which has been emerging as a powerful, bottom-up approach in the construction of well-ordered nanostructures. The inherent biocompatibility, ease of synthesis and facile chemical modification of peptide molecules render them excellent building blocks in the construction of well-defined supramolecular architectures. A well-known example is that of the diphenylalanine peptide (L-Phe-L-Phe, FF), which is the core recognition motif of Alzheimer's Ab polypeptide.[1] Through a combination of specific molecular interactions such as van der vaals, hydrogen bonding, hydrophobic and ¦Ð-¦Ð stacking under different conditions (e. g. solvent or surface), Phe-Phe peptide can self-assemble into various nanostructures such as nanotubes,[1] nanowires,[2] microcrystals,[3] and vertically aligned nanoforest.[4] Due to the specific molecular interactions, Phe-Phe can be employed as a recognition module to design novel molecular building blocks. It is suggested that the modified motifs play a determinant role in the creation of novel functional peptide nanomaterials.

Putting metals into self-assembling peptides may provide us an opportunity to design molecular building blocks with new desirable properties. Ferrocene (Fc),[5] one of the most stable and useful organometallic compounds, has been used as metal-containing component for the construction of various hybrid molecules from small molecules to polymers with various applications in biosensor, magnetic and stimuli-responsive materials, and asymmetric catalysis. Here, we showed the kinetically controlled self-assembly of a novel bioorganometallic molecule via the conjugation of ferrocene with diphenylalanine (ferrocene-diphenylalanine, Fc-FF, Figure 1a), and directly observed a morphological transition from metastable nanospheres to well-defined nanofibers by introducing an external mechanical force (i.e. shaking), which finally led to the formation of a stable self-supporting hydrogel (Figure 1b and c). Moreover, as the Fc moiety in this hybrid molecule serves as a redox-active site, its self-assembly process can be reversibly controlled by altering the redox state of Fc group (Figure 1d and e). The main ideas and conclusions are summarized as follows:

(1) Strong hydrophobic interaction among molecular units may lead to the formation of nanomaterials that depend largely on the pathway or kinetic control. For Fc-FF assemblies, after diluting the concentrated Fc-FF methanol solution in water, the solvent environment around Fc-FF molecules changed intensively, giving rise to strong hydrophobic interactions which were adequate for achieving efficient kinetic trapping, thus leading to formation of nanospheres. After exerting slight shaking, the metastable Fc-FF nanospheres can transform into thermodynamically stable nanofibers with simultaneously molecular rearrangement from unordered structure to ¦Á-helix. It is obvious that the self-assembly of Fc-FF is a kinetically controlled process. High-uniform nanospheres were formed by the fast self-assembly, whereas the nanofibers with helical superstructures were obtained from the slow self-assembly. External mechanical force can be used to overcome the energy barrier and thus to initiate the 'slow' self-assembly process. The strong hydrophobic interactions introduced by Fc group in water may play an essential role in this transition process.

(2) For Fc-FF molecule, the Fc group in its reduced state is hydrophobic, while its oxidized form (Fc^{+}) is hydrophilic (Fig. 1d). This can give rise to a remarkable change in the hydrophobic-hydrophilic balance of Fc-FF, thus altering its self-assembly behavior. To demonstrate this, 1 mg ml^{-1} Fc-FF nanofibers was dispersed in 10% CH_3OH/90% H_2O. After introducing an appropriate amount of $CeSO_4$ as oxidant, the yellow turbid Fc-FF nanofibers suspension turned into a blue transparent solution, and continuous addition of GSH to the solution transformed it back to yield a yellow turbid suspension again (Figure 1e, insets). SEM analysis revealed that the Fc-FF units could reversibly self-assemble into nanofibers after the oxidation-reduction cycle (Figure 1e). After the oxidation of Fc-FF into Fc^{+}-FF, apart from an amorphous film deposited on the surface of the microscope glass coverslip, no characteristic morphology was observed. This indicated that the Fc-FF nanofibers may disassemble or disrupt into monomers or much smaller aggregates as the oxidation of Fc-FF into Fc^{+}-FF. Moreover, continuous reduction of the Fc^{+}-FF molecule into Fc-FF recovered its self-assembling ability, thus leading to the reassemble of well-defined Fc-FF nanofibers.

(3) Due to the stimulus (i.e. mechanical force, redox) responsive properties, the Fc-FF nanostructures may have great potential applications in areas such as biosensor, tissue engineering, controlled drug release. Moreover, considering the structural diversity of FF assemblies, the Fc-FF molecule may have the ability to form other novel nanosturcutres by controlling the self-assembly conditions.

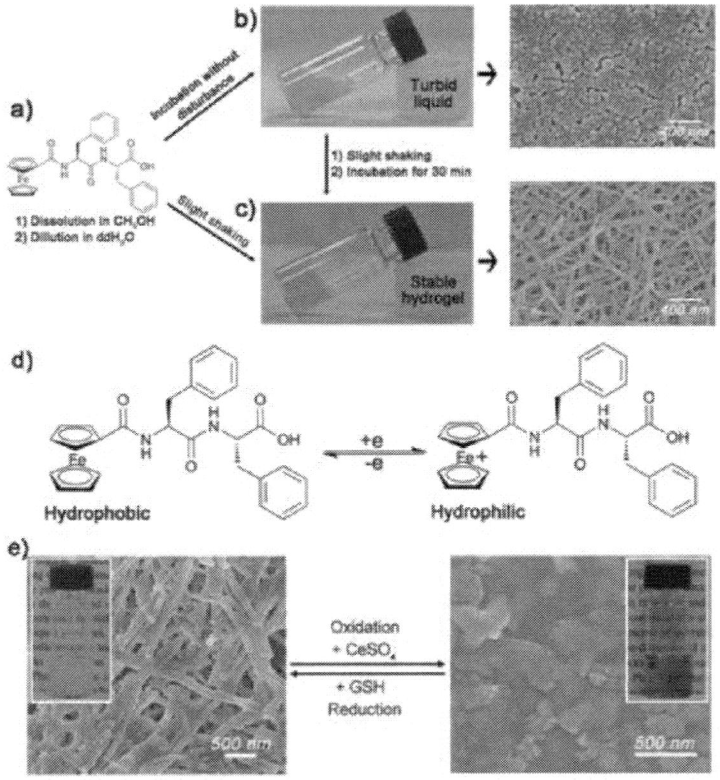

Figure 1. Self-assembly behavior of ferrocene-diphenylalanine. a) Molecular structure of ferrocene-diphenylalanine (Fc-FF). b) Photograph of the mixture incubated without any external disturbance. c) Photograph of the mixture after incubation by introducing slight shaking. d) Redox-control of the ferrocenyl group. e) SEM analysis showed the self-assembling behavior of Fc-FF molecules in their reduced and

oxidized state. The insets showed the photographic images of 1mg ml^{-1} Fc-FF nanofibers dispersed in 10% CH_3OH/90% H_2O before and after oxidation.

This work was supported by the NSF of China (51173128, 31071509), the Ministry of Science and Technology of China (Nos. 2012YQ090194, 2013AA102204, 2012BAD29B05), the Program for New Century Excellent Talents in Chinese University (NCET-08-0386), and Beiyang Young Scholar Program (2012).

References

1. Reches, M., Casting Metal Nanowires Within Discrete Self-Assembled Peptide Nanotubes. *Science* 2003, *300* (5619), 625-627.

2. (a) Kim, J.; Han, T. H.; Kim, Y.-I.; Park, J. S.; Choi, J.; Churchill, D. G.; Kim, S. O.; Ihee, H., Role of Water in Directing Diphenylalanine Assembly into Nanotubes and Nanowires. *Adv. Mater.* 2010, *22* (5), 583-587; (b) Huang, R. L.; Qi, W.; Su, R. X.; Zhao, J.; He, Z. M., Solvent and surface controlled self-assembly of diphenylalanine peptide: from microtubes to nanofibers. *Soft Matter* 2011, *7* (14), 6418-6421.

3. Zhu, P.; Yan, X.; Su, Y.; Yang, Y.; Li, J., Solvent-Induced Structural Transition of Self-Assembled Dipeptide: From Organogels to Microcrystals. *Chem. Eur. J.* 2010, *16* (10), 3176-3183.

4. Reches, M.; Gazit, E., Controlled patterning of aligned self-assembled peptide nanotubes. *Nat. Nanotechnol.* 2006, *1* (3), 195-200.

5. Togni, A., *Ferrocenes: homogeneous catalysis, organic synthesis, materials science*. Wiley-VCH: 1995.

(397c) Effect of Molecular Architecture On the Morphology and Properties of Bio-Nanostructured Soft Materials

Koufos, E., Rutgers University
Dutt, M., Rutgers University

Our objective is to design effective controlled-release vehicles for lab-on-chip devices for applications in the areas of bionanotherapeutics, sensing and catalysis. In this presentation we present our results on the role of molecular architecture on the morphology and mechanical properties of bio-nanostructured soft materials. We focus on four lipid architectures with variations in the head group shape and the hydrocarbon tail length. Each lipid species is composed of a hydrophilic head group and two hydrophobic tails. We characterize the soft material via the thickness, interfacial tension, molecular diffusion, area compression modulus, bending modulus, and total energy of the lipid bilayer. We demonstrate the equilibrium morphology of these bilayer membranes to be determined by factors such as the average area of the lipid, lipid head group orientation, and lipid tail length (Koufos et al., 2013.) We use a Molecular Dynamics-based mesoscopic simulation technique called Dissipative Particle Dynamics that captures the molecular details of the components through soft-sphere coarse-grained models and reproduces the hydrodynamic behavior of the system over extended time scales. The DPD method is highly effective in resolving the dynamics of complex fluids over extended spatiotemporal scales due to the soft-repulsive interaction forces.

(397bn) A Lipase-Cyclodextran Conjugate With Improved Enantioselectivity

Zhang, Y., Tsinghua University
Lu, D., Tsinghua University
Ge, J., Tsinghua University
Liu, Z., Tsinghua University

Enzymatic catalysis has been widely applied in the preparation of chiral compounds with high activity and selectivity. In general, asymmetric synthesis and chiral resolution are the two main approaches to produce chiral chemicals. Lipase, a class of ester hydrolase, has been employed in the kinetic resolution of racemic esters. However, the enantioseletivity of native lipase sometimes still cannot meet the criteria of practical use because of the unsatisfied selectivity, leaving the rational design and engineering of lipase a tough challenge. Here, we present a new method to alter the enantioselectivity of lipase by conjugating the enzyme with a chirality 'recognizer' such as β-cyclodextrans (β-CD). *Candida antarctic* lipase B (CALB) was selected as a model enzyme to conjugate with β-CD *via* a click reaction. (R,S)-methyl mandelate and (R,S)-methyl cyclohexylphenylglycolate were used as substrates to examine the stereoselectivity of the β-CD modified CALB. Since the β-CD has different binding abilities towards R- and S-substrates, the lipase-CD conjugate showed an altered enantioselectivity. For instance, the conjugation of β-CD facilitated the enzymatic hydrolysis of (R)-methyl mandelate but had no distinct influence on the (S)-isomer. A typical result showed that the enantiomeric ratio (E ratio) was improved from 5 to 30 (at a conversion of 40%) and the enantiomeric excess (ee) of reached higher than 85%. The mechanism of the improvement of enzymatic enantioselectivity will be also discussed in this presentation. This type of enzyme-β-cyclodextran conjugate offers a novel chemical method for improving, controlling or even inversely altering the enzymatic enantioseletivity, which would have much promise in the applications of enzymatic resolution.

(397be) Hierarchical FeOx@SiO2-ZnO Koosh Ball Nanostructure With Tunable Magnetic Core, Fluorescent Nanowire Shell and Enhanced Photocatalytic Property

Ren, Z., University of Connecticut
Guo, Y., University of Connecticut
Gao, P., University of Connecticut

A mild, wet chemical strategy has been developed to synthesize three-dimensional (3D) multicomponent koosh ball nanoarchitectures comprised of tunable magnetic iron oxide core and 3D fluorescent ZnO nanowire shell, which demonstrates enhanced photocatalytic property towards dye degradation under UV irradiation. The synthesis involves a hydrothermal preparation of magnetite particles followed by the coating of an amorphous silica layer using Stober sol-gel process. ZnO nanowires are successfully assembled on the seed mediated surface by low temperature solution synthesis. The phase transition of magnetic iron oxide core and the native defects induced fluorescence of the nanowire shell can be simultaneously manipulated by post-hydrogen annealing to produce various koosh balls with retained morphology in Figure 3 but with different magnetic and photocatalytic properties. Specifically, ambient annealing at 350 °C will lead to the formation of nano koosh balls with maghemite cores, while 600 °C ambient annealing will produce koosh balls with hematite cores. The 5% hydrogen treatment at 400 C transforms maghemite cores into bcc-iron and the 1% hydrogen treatment at the same temperature can make hematite into magnetite. During the hydrogen annealing, the surface defect of ZnO nanowires can be tuned. The photoluminescence in the visible light region is quenched . The reduced population of surface defects contributes to the worse photocatalytic performance of Rhodamine B degradation. This unique koosh ball architecture initiates the three-dimensional nanowire growth on micro-scale spherical substrate and enables the rational combination of multiple desired functionalities originated from dissimilar constituents. The obtained nano koosh ball will open up new opportunities towards environmental application of nanomaterials such as photocatalytic water treatment.

Reference

[1] **Z. Ren**, Y. Guo, G. Wrobel, D. A. Knecht, Z. Zhang, H. Gao and P. X. Gao, *J. Mater. Chem.*, **2012**, 22, 6862-6868.

(397bg) Changing the Mesostructure of MOF-Derived Carbons By Doping MOF-5 With Ni2+ for Bulk Synthesis of Carbon Nanofibers

Lobo, D. E., Monash University
Balakrishnan, S., Monash University
Banerjee, P. C., Monash University
Remadevi, R. C. N., Monash University
Williams, T.,
Sheath, P., Monash University
Hill, M. R., CSIRO
Majumder, M., Rice University

High quality production of carbon nanostructures has been an extensive area or research with a great emphasis placed on Chemical vapor deposition (CVD). Here we illustrate a novel concept in tailoring the architecture of Metal-Organic Framework-derived carbons (MDCs) by doping Metal-Organic Framework (MOF) crystals with transition metal ions prior to its use as a template in CVD growth of carbon nanofibers. MOF-5 was isomorphously doped with Ni^{2+} during its synthesis. The Ni-doped MOF-5 was then used as a template in the CVD process (at ~550 – 650 °C) and yielded a 3D hierarchy composed of a labyrinth of carbon nanofibers with electrical conductivity (~ 550 S/m) in sharp contrast to a 'flake-like' morphology and lack of electrical conductivity in MOF-5 derived carbons. The nanofibers were short, having variable diameters, (~ 10-40 nm) but possess a conical arrangement of graphene sheets as observed by HRTEM, while electrochemical measurements show good charge storage capability, suggesting their potential in double layer capacitor devices. This approach can be used for production of bulk quantities of carbon nanofibers. The uniformly distributed metals joined by aromatic carbon groups in the MOFs could provide a new method to develop carbons with tunable architecture.

(397bk) Robust SERS Substrate With Hot Spots Derived From Massive Nanogaps Between Silver Nanoparticles and Nanowires

Tsai, P. R.,
Cheng, S. C., National Cheng Kung University
Wen, T. C., National Cheng Kung University

Robust SERS substrates with hotspots on a large scale from massive nanogaps can be fabricated by assembling Ag nanocubes around nanowires on the ITO glass. Noble metal nanomaterials are commonly used as SERS-active substrates and referred to hotspots because the localized surface plasmon resonance (LSPR) occurs on the surface of metal nanoparticles, especially when nanoparticles are gathered together. Herein, Ag nanowires and nanocubes are synthesized by polyol method and then nanowires and nanocubes were put on the ITO glass step by step. Massive nanogaps form strong LSPR would lead to link Ag surfaces in a nano-scale gap for incurring the substantial increase with several orders of magnitude in SERS signals due to the plasmonic interaction.

The plasmonic interaction of substrate was corroborated by UV-Vis spectra because the frequency change of plasmon resonance indicates the coupling of plasmon. Since Ag nanowire can form large area porous network on the ITO glass, Ag nanocubes could be dispersed homogeneously on the substrate. The adhesion of two metallic surfaces would bring Ag nanocubes and nanowires closely to form lots of nanogaps, which can lead to strong LSPR. From the spectra of UV-Vis, the strength and frequency change of plasmon resonance indicate the strong electrical field coupling occurred on our substrate. Since the intensity of Raman is proportional to the fourth power of the electric field strength, the huge changes in the electromagnetic field of our substrate would result in the substantial increase in SERS signals. In order to clarify the performance of our substrate, the rhodamine 6G (R6G), tryptophan, Lysozyme (LYZ), and Bovine Serum Albumin (BSA) are used as model compounds to confirm SERS results.

R6G	10^{-9}M,EF=5×10^{8}
Tryptophan	10^{-6}M in alkaline solution
Lysozyme (LYZ)	10^{-10}M in Di water/10^{-9}M in Eq point
Bovine Serum Albumin (BSA)	10^{-10}M in Di water/10^{-11}M in Eq point

Table 1 shows the Raman results for different analytes. With the hot spot on a large scale, and strong LSPR, the SERS results prove the supreme performance of the robust substrate by detecting rhodamine 6G solution with high sensitivity (enhancement factor 5×10^8) on the basis of 1 cm × 2 cm area. Furthermore, several biomolecules were detected in low concentration (as list in Table1), indicating highly potential in analysis by using our substrate.

SERS is a useful and rapid tool for low concentration detection. Our robust substrate, prepared by assembling Ag nanocubes around Ag nanowires represents the homogeneous distribution of nanogaps, possessing the high enhancement factor on a large scale, indicating highly potential application in biomolecules measurements.

(397bf) Facile Method to Synthesize Graphene-ZnS Nanocomposites and Their Application in Bioelectrochemistry of Hemoglobin

Wang, L.,

Qi, W., Tianjin University

Su, R., Tianjin University

He, Z., Tianjin University

Electrochemical technology is a useful way to probe the protein conformation and protein electron transfer. Thus, the direct electron transfer (DET) between biomolecule and electrode has aroused growing interest due to its significance in both theoretical and practical applications[1,2]. In recent years, many researches have focused on the application of nanomaterials to modify electrode. For example, graphene (GR), a two-dimensional sheet of carbon atoms bonded through sp^2 hybridization with the properties of high electrical conductivity, exceptional thermal and mechanical stability, is showing promising applications in electrode modification[3]. Due to the excellent optical and electronic properties, semiconductor quantum dots (QDs) has also been used in the fields of electrochemical as well as photoelectrochemical biosensors[4]. Among the QDs materials, CdS nanostructures were most generally used. However, considering that CdS might be harmful and carcinogenic to the human beings[5], ZnS nanomaterial, a semiconductor with large band-gap energy and good conductivity, has been a good substitute of CdS owing to its low cost and nontoxicity[6].

In this work, we explored a new method to fabricate graphene-ZnS (GR-ZnS) nanocomposites for electrochemical applications based on noncovalent functionalization of graphene (Figure 1). Compared with the harshly oxidative approaches, herein, the surface of pristine graphene was firstly modified with 1-aminopyrene via a strong ¦Ð-¦Ð stacking mechanism between the pyrenyl groups and the carbon rings of the graphene. The introduction of active amine group on the graphene could not only improve the dispersibility of graphene but also provide nucleation centers on the surface of graphene for the uniform growth of ZnS nanoparticles. The as-prepared GR-ZnS nanocomposites were characterized and identified by means of TEM and EDS analysis, indicating the uniform formation of ZnS nanoparticles on the surface of graphene. FT-IR spectroscopic results confirmed that hemoglobin (Hb) remained its native structure in the nanocomposite material. The GR-ZnS nanocomposites could efficiently promote the direct electron transfer between Hb and electrode with the electron transfer rate constant of 3.42 s^{-1}. The modified electrode was then used for the determination of H_2O_2 based on the electrocatalytic activity of Hb towards H_2O_2, which exhibited a linear range from 10 to 250 μM with a detection limit of 1.12 μM. The proposed method to fabricate graphene-based hybrid nanomaterials would have a great potential for a wide range of applications.

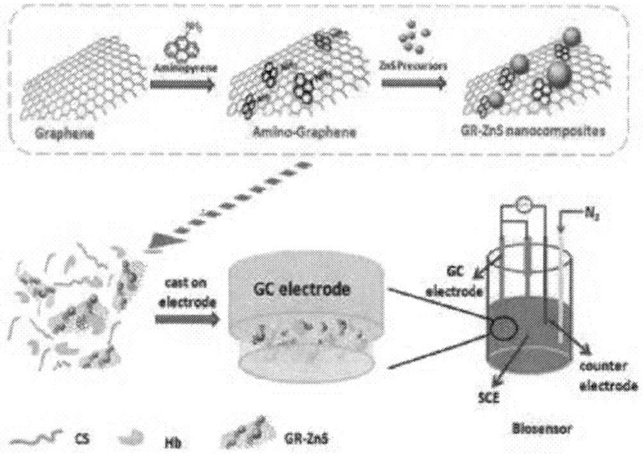

Figure 1. Fabrication of a H_2O_2 biosensor.

This work was supported by the NSF of China (51173128, 31071509), the Ministry of Science and Technology of China (Nos. 2012YQ090194, 2013AA102204, 2012BAD29B05), the Program for New Century Excellent Talents in Chinese University (NCET-08-0386), and Beiyang Young Scholar Program (2012).

References

[1] Baby TT, Aravind S, Arockiadoss T, Rakhi R, Ramaprabhu S. Metal decorated graphene nanosheets as immobilization matrix for amperometric glucose biosensor. Sensors and Actuators B: Chemical, 2010, 145: 71-77.

[2] Yılmaz Ö, Demirkol DO, G¨¹lcemal S, Kılınç A, Timur S, Çetinkaya B. Chitosan-Ferrocene Film as a Platform for Flow Injection Analysis Applications of glucose oxidase and gluconobacter oxydans. Biosensors. Colloids and Surfaces B: Biointerfaces, 2012, 100: 62-68

[3] Rao CNR, Sood AK, Subrahmanyam KS, Govindaraj A. Graphene: The New Two-Dimensional Nanomaterial. Angewandte Chemie International Edition, 2009, 48: 7752-7777.

[4] Sun W, Wang D, Li G, Zhai Z, Zhao R, Jiao K. Direct electron transfer of hemoglobin in a CdS nanorods and Nafion composite film on carbon ionic liquid electrode. Electrochimica Acta, 2008, 53: 8217-8221.

[5] Miyawaki T, Ichimura M. Fabrication of ZnS thin films by an improved photochemical deposition method and application to ZnS/SnS heterojunction cells. Materials Letters, 2007, 61: 4683-4686.

[6] Nakaoka Y, Nosaka Y. Electron spin resonance study of radicals produced by photoirradiation on quantized and bulk ZnS particles. Langmuir, 1997, 13: 708-713.

(397bh) Robust SERS Substrates With Hotspots On a Large Scale Derived From Massive Nanogaps Between Nano-Structured and Massed Silver Surfaces

Wen, T. C., National Cheng Kung University

Cheng, S. C., National Cheng Kung University

Robust SERS substrates with hotspots on a large scale from massive nanogaps can be fabricated by assembling Ag nanocubes on the massed Ag surface via 1, 2-ethanedithiol monolayer as an ultrathin spacer. Noble metal nanoparticles such as gold nanoparticles and silver nanoparticles are used as SERS-active substrates and referred to as hotspots because the localized surface plasmon resonance (LSPR) occurs on the surface of metal nanoparticles. The hotspot effect is commonly observed from a magnified perspective during SERS measurement when nanoparticles are gathered together. Ag surfaces were linked in a nano-scale gap for incurring the substantial increase with several orders of magnitude in SERS signals due to the plasmonic interaction. Plasmonic resonance creates highly enhanced local electric fields in confined nanoscale locations. In order to take the advantage of strong LSPR around nanogaps, we envisage that it might generate hotspots on a large scale by assembling Ag nanocubes on the massed Ag surface due to its massive surface. Ag nanocubes are synthesized by polyol method and be assembled homogeneously on the massed Ag surface via 1, 2-ethanedithiol.

X-ray spectroscopyis used to confirm the existence of nanogaps. With the binding energy shift of Ag3d core level and feature peaks of the sulfur atoms on the substrate treated by 1, 2-ethanedithiol, we can know that Ag nanocubes self-assembled on the massed Ag surface by Ag-S chemical bonds and the nanogaps were formed between Ag nanocubes and the massed Ag surface.

The plasmonic interaction between Ag nanocubes and the massed Ag surface was corroborated by UV-Vis spectra because the frequency change of plasmon resonance indicates the coupling of Ag nanocubes and the massed Ag surface. The strength and frequency change of plasmon resonance indicates the strong electrical field coupling between Ag nanocubes and the massed Ag surface. Since the intensity of Raman is proportional to the fourth power of the electric field strength, the huge changes in the electromagnetic field of Ag nanocubes self-assembled on Ag surface result in the substantial increase with several orders in magnitude in SERS signals. The rhodamine 6G (R6G) is used as a model compound to confirm SERS results.

Three subatrates were used in Raman measurments: (a)$10^{|1}$ M R6G solution on a microscope slide, (b) $10^{|7}$ M R6G solution on SA, and (c) for $10^{|9}$ M R6G solution on SA. With the homogeneous distribution of Ag nanocubes, and strong LSPR, the SERS results prove the supreme performance of the robust substrate by detecting $10^{|9}$ Mrhodamine 6G solution with high sensitivity (enhancement factor 2.8×10^{8}), high reliability (6.6% standard deviation from 20-sites measurements), and high precision (calibration line with 99.9% correlation coefficient). on the basis of 1 cm × 2 cm area. Furthermore, the straight calibration line with 99.9% correlation coefficient is obtained by using the concentration of R6G solution from $10^{|6}$ M to $10^{|9}$ M, confirming its practical application in quantitative analysis.

It is very difficult to arrange nanoparticles regularly on a substrate in the general case. The substrate with high degree of regularity used for Raman quantitative analysis is usually manufactured by FIB or E-beam etching which is heavy in cost and time. Our robust substrate, prepared by assembling Ag nanocubes on the massed Ag surface represents the homogeneous distribution of nanocubes, possessing the high EF and the low standard deviation for the potential application in quantitative analysis.

(397bi) Efficiency Analysis of Nanoemulsions Systems Containing Silicone Polyether As Demulsifying and Antifoaming Agents for Petroleum

Fraga, A. K., Federal University of Rio de Janeiro
Mansur, C. E., Federal University of Rio de Janeiro

In the petroleum industry gravity vessels separators are used to segregate the oily, water and gas phases. Within these vessels occurs the formation of a large amount of foam due to constant agitation. The objective of this work is to obtain and evaluate oil in water (o/w) nanoemulsion systems as demulsifying and anti-foaming agents for oil.

For the composition of the nanoemulsion oily phase polypropylene glycol and Solbrax ECO were selected at concentrations of 5, 7 and 10% and compounds of silicone polyether of different polarities were used as surfactants (concentrations of 12 and 20%). The nanoemulsion systems were obtained by employing the high energy method with the aid of a high-pressure homogenizer and further characterized by Particle size analyzer Zetasizer Nano ZS in order to obtain the size range of the droplets dispersed in the nanoemulsion as well as their stability. To evaluate the efficiency of nanoemulsions as demulsifies the bottle test methodology is being applied and the evaluation of them as anti-foaming agents will be made through a methodology developed in the laboratory [1,2].

Results obtained show that, using polypropylene glycol as the oily phase, systems with 5% of oily phase and 12% of different types of surfactants presented a higher stability. However, the efficiency as demulsifing agents was not observed. A higher efficiency of the nanoemulsion as a demulsifying agent (about 60%) was observed when the surfactant concentration was increased to 20%.

References

1- V.F. Pacheco, L. Spinelli, E.F. Lucas, C.R.E. Mansur, *Energy Fuels*, 2011, 25, 1659.

2 – A. K. Fraga; R. F. Santos; C. R. E Mansur. *Journal of Applied Polymer Science*, 2012, 124, 4149-4156.

(397bj) Comparative Localized Surface Plasmon Resonance Between Silver Nanocubes and Nanospheres On the Massed Silver Surfaces

Cheng, S. C., National Cheng Kung University

Wen, T. C., National Cheng Kung University

Strong localized surface plasmon resonance (LSPR) can be derived from silver nanocubes and nanospheres self-assembling on massed silver surfaces. Ag surfaces were linked in a nano-scale gap to become hot spots for incurring the substantial increase in plasmonic interaction and lead to high performance with several orders of magnitude in SERS. Since the LSPR is sensitive to the geometry of nanoparticles, we synthesized Ag nanocubes and nanospheres by polyol method and self-assembled them homogeneously on the massed Ag surface via 1, 2-ethanedithiol to used for robust SERS substrates. With a 1, 2-ethanedithiol monolayer as the dielectric spacer, nanoparticles are separated from a continuous metal surface. The existence of the dielectric spacer offered space between two metallic surfaces that provided more opportunity for the generation of localized surface plasmon to prevail the strong plasmonic interaction at the nanogaps. In order to elucidate the plasmon coupling in our substrates, the electric field simulations were performed using finite-difference time-domain (FDTD) method.

The induced electric field of the substrate without dielectric spacer occurs near the vertex on the top of Ag nanocube and at the bottom with the decay strength because of the complete contact. Figure 2c and 2d show the EM fields for the substrate with spacer. The strong induced electric field appears in nanogaps between the Ag nanocubes and the massed Ag surface rather than the sharp corner, indicating strong plasmon coupling happened in nanogaps between two Ag surfaces. Otherwise, the hot region of the substrate made by self-assembling Ag nanocubes on the massed Ag surface is much larger than the substrate with nanosphere, showing the importance of the dielectric spacer.

Due to the low surface energy, nanosphere is the commonly used nanoparticle. In order to creat a roubst SERS substrate, we use nanocubes to replace the commonly used nanosphere as our nanoparticle. Figure 3 shows the electric field simulations around an Ag nanocube and nanosphere placed on Ag surface with 2 nm spacer. The strength of induced electric field in the nanogap between two Ag surfaces are huge different. The substrate with nanocubes shows stronger induced electric field than the former the substrate with nanospheres and larger hot spot area, indicating the powerful interaction happened when Ag nanocubes were assembled on the massed Ag surface with the ultrathin spacer, being a suitable material to form strong plasmon coupling. Since the intensity of Raman is proportional to the fourth power of the electric field strength, the huge changes in the electromagnetic field of Ag nanocubes and nanospheres on massed Ag surface would result in the substantial increase with several orders in magnitude in SERS signals. The rhodamine 6G (R6G) is used as a model compound to confirm SERS results.

Three substrates were used for SERS measurments: (a)$10^{|1}$ M R6G solution on a microscope slide, (b) $10^{|6}$ M R6G solution on the substrate by assembling Ag nanospheres on massive Ag surface, and (c) for $10^{|9}$ M R6G solution on the substrate by assembling Ag nanocubes on massed Ag surface. The substrate by assembling Ag nanocubes on massed Ag surface shows about 100 times EF of nanospheres (2.9×10^{6}), being attributable to the geometry effect of nanoparticles. The SERS results corresponded to the calculation results from finite-difference time-domain method, indicating strong LSPR can be derived from silver nanoparticles self-assembling on massed silver surfaces, especially using nanocubes as nanoparticles.

It is very difficult to arrange hot spots in a large scale on a substrate in the general case. For efficiently and promptly analysis, nanoparticles are regarded as SERS substrates owing to the rapidly, easily, and low-cost production. Our robust substrate, prepared by assembling Ag nanocubes on the massed Ag surface represents the homogeneous distribution of nanocubes, possessing the high EF and the low standard deviation, generating strong LSPR, indicating highly potential application in SERS measurements and quantitative analysis.

(397k) Augmenting the Size-Selective Fractionation of Metal and Metal Oxide Nanoparticles Using a Modified Gas-Expanded Liquid Process

Vengsarkar, P. S., Auburn University
Roberts, C. B., Auburn University

The term nanotechnology is employed to describe the creation and manipulation of materials with structural features in between those of atoms and bulk materials (two and three dimensional assemblies of molecular scale), with at least one dimension in the nanometer range. It is an important topic in modern research due to its multidisciplinary applicability and future potential in many underdeveloped applications. Nanoparticles have unique size-contingent physicochemical properties which arise due to quantum effects and properties such as chemical reactivity and melting point which are attributed to the nanoparticle size. The synthesis and post-synthesis processing procedures used to generate monodisperse nanoparticles usually necessitate the use of expensive reagents & equipment, large quantities of solvents and high temperatures. A post-synthesis processing technique to size-selectively precipitate and fractionate ligand-stabilized gold and silver nanoparticle dispersions in organic solvents has been developed in our laboratory which utilizes the pressure tunable physicochemical properties of CO_2 Gas-eXpanded Liquids (GXLs). This size-selective fractionation technique is based on the controlled reduction of the solvent strength through increases in the concentration of CO_2 (a known nonsolvent for aliphatic stabilizing ligands) via pressurization. These changes in solvent strength affect the subtle balance between the osmotic repulsive forces and the van der Waals forces of attraction between differently sized nanoparticles necessary to maintain a stable dispersion. Through modest changes in CO_2 pressure, increasingly smaller sized nanoparticles can be controllably precipitated from the dispersion.

Previous studies have shown that using metal (gold) nanoparticles with branched ligands, it is possible to manipulate the process pressure and hence the size-selective fractionation of metal nanoparticles using the GXL system. However, employing the same apparatus that was used in the previous studies, various anomalies were observed when size-selective fractionation of metal oxide (iron oxide) nanoparticles was attempted. These anomalies included non-uniform, location-specific precipitation of nanoparticles and poor fractionation results when using iron-oxide nanoparticles coated with fatty acid ligands. This particular study aims to increase the efficiency and effectiveness of the GXL process through the development of several process modifications. These modifications include improvements in the cascaded vessel apparatus, such as conversion into a packed column and using a metal jacket to cause particles with lower Hamaker constant values to adhere more effectively. The precipitation characteristics of ligand-stabilized gold nanoparticles were quantified by measuring the absorbance of the nanoparticle dispersion using UV-vis spectroscopy at various levels of CO_2 pressurization. The characterization of these nanoparticles was performed using transmission electron microscopy (TEM) to analyze their size distribution and hence judge the efficacy of the size-selective fractionation. The pressure range over which the nanoparticles precipitate and the fractionation efficiency from each of these process modifications provides new insights into the GXL fractionation process that help further our fundamental understanding of the GXL precipitation phenomenon.

(397l) Sulphur-Infiltrated 3D Porous Carbon Microsphere Nanoarchitecture for High Energy Lithium-Sulphur Batteries

Zhao, C., University of Wisconsin-Milwaukee
Liu, L., University of Wisconsin-Milwaukee
Zhao, H., University of Wisconsin-Milwaukee
Krall, A.,
Li, Y., University of Wisconsin-Milwaukee

Sulfur has received increasing attention as a cathode material for lithium-sulfur batteries due to its high theoretical specific capacity of 1675 mAh g^{-1}, which is 5 times higher than that of conventional lithium-ion batteries. However, poor electronic conductivity of sulfur, dissolution of the lithium polysulfide intermediates and large volume expansion of sulfur during cycling limit the commercialization of Li-S batteries. Herein, we demonstrate the design of a smart-structured sulfur-infiltrated 3D multi-modal porous carbon microsphere to overcome these obstacles. This 3D multi-modal porous microsphere carbon (PMC) have large surface area, uniform pore size, high pore volume, and interconnected pore structure, that will ensure the access connection of the electrolyte through the PMC and uniformly loading for sulfur particles in the spaces within the 3D structure, thus limiting the volume expansion and dissolution of lithium polysulfide. Hence, a higher capacity, better cycling performance and stable Coulombic efficiency are expected. An initial capacity of 1289 mAh g^{-1} at 1C and Coulombic efficiency of 90% over 200 cycles are achieved.

(397n) "Smart" Nanocomposites for Enhanced Oral Drug Delivery

Fan, X., Auburn University
Milton, A., Auburn University
Soundarrajan, D., Auburn University
David, A. E., The University of Michigan

Oral delivery is the most desirable route for drug administration because of its convenience and improved patient compliance. Delivery of drugs, particularly macromolecular drugs, through the oral route, however, faces several problems. Poor stability, solubility and permeability of these drugs often limit their bioavailability to sub-therapeutic concentrations. Nanocomposites have attracted increasing attention due to their ability to protect drugs from degradation in the GI tract and to improve penetration across the mucus barriers. Utilization of the mucoadhesive properties of biopolymers or the "stealth" properties of poly(ethylene glycol) (PEG) has been shown to further improve oral drug delivery. It is possible that a balance between the stealth and mucoadhesive properties of the nanocomposites would enhance oral drug bioavailability. The goal of our research is to elucidate the nature of the interaction between nanocomposite surfaces and biological membranes, through rheology measurements, and to use the understanding gained to optimize the oral delivery of drugs. Here we present our *in vitro* and *in vivo* analysis on the effects of surface properties, particle size, and composition on the mucoadhesive properties of nanocomposite materials.

(397o) Carbon Nanopatterns and Nanoribbons From Directly Nanoimprinted Polyacrylonitrile: Correlation Between Crystallite Orientation and Nanoimprint Process

Zhang, Z., University of Colorado at Boulder
Molina Piper, D., University of Colorado at Boulder
Son, S. B.,
Kim, S. C.,
Oh, K. H., Seoul National University
Lee, S. H., University of Colorado at Boulder
Ding, Y., University of Colorado at Boulder

The ability to create carbon nanostructures such as carbon nanofibers (CNF) and patterns with controlled dimensions, position and ordering, is crucial for engineering all-carbon devices. We present the fabrication of lithographically defined carbon patterns and nanoribbons using a common carbon precursor, polyacrylonitrile (PAN). Thermal-embossing NIL (TE-NIL) is used to directly fabricate PAN films that can be subsequently cyclized and carbonized to yield highly uniform carbon patterns and nanoribbons over a large area. We demonstrate that this technique is reliable and capable of nanofabrication over a large surface area at low cost, compared with current carbon-patterning techniques. The evolution of the nanostructures during heat treatment was studied. Most interestingly, the deformation profile of the PAN during the imprinting process resulted in a distribution of aligned PAN chains within the patterns, which led to a similar anisotropic correlation of the carbon crystallites in the carbonized structures.

(397p) Using Photoluminescence and Proton NMR As Probes to Understand the Interaction of Surfactants and Other Adsorbents With Single Walled Carbon Nanotubes

Xu, J., University of Florida
Clar, J. G., University of Florida
Bonzongo, J. C. J., University of Florida
Ziegler, K. J., University of Florida

The ability to understand and control the surface structure of nano materials is a difficult barrier that must be overcome for their practical application in biology and medicine. Single walled carbon nanotubes (SWCNTs) have a large specific surface area and, thus, a high adsorption capacity. The high adsorption capacity can be utilized for practical applications in drug loading but also has important implications on their fate and toxicity as the materials enter natural systems during production and disposal. The changes that occur to the localized region surrounding individual SWCNT can be tracked by photoluminescence (PL), which has been proven to be a quick and reliable probe of SWCNT surfaces. In this approach, the SWCNTs act like a chemical sensor that qualitatively detects the adsorption of molecules around itself, such as drugs, chemicals, or environmental species. While PL provides information about direct contact with the SWCNT, Proton NMR is another probe that we have started to use to track the specific interaction of surfactants with SWCNTs and other molecules adsorbed onto their surface. Changes to the NMR chemical shift of the surfactant molecules provides information about their interaction with either the SWCNTs or other molecules. These changes are shown to correspond to molecular structural changes. By combining PL and Proton NMR analysis, we aim to understand the formation and morphology of both surfactant and solvent structures around SWCNTs. We find characteristic trends in the PL and Proton NMR changes that help elucidate the different interactions occurring between SWCNTs, surfactants, and adsorbents. The ability to understand and control the structure around SWCNTs while interacting with different adsorbents has novel applications in drug delivery and will also improve our understanding of the fate and transport of SWCNTs in natural systems.

(397bo) Evaluation Of A New Method For Encapsulating Two Drugs In PLGA Nanoparticles: Physical-Chemical Charaterization

Espanol, L., Sao Paulo University
Carranza, M. V., Polytechnic School of the University of Sao Paulo
Giudici, R., Universidade de Sao Paulo
Prado, M. A.,

Currently one of the rheumatic diseases that most affects the world population is rheumatoid arthritis (RA), systemic disease, whose main clinical manifestation is chronic inflammation of the synovial membrane and deformation. Approximately 1-2% of the adult population has rheumatoid arthritis, focusing mainly in the elderly. Brazil has approximately 14.1 million of elderly people, and for 2025 they will grown 16 times. In this sense, many researches are developed in different parts of the world to find a more efficient alternative for the treatment this disease.

This research shows the development of a method for encapsulation of two anti-inflammatory drugs, in order to increase their efficiency in the treatment of that disease. The method involves the preparation of nanoparticles (NPs) of the spherical core-shell type morphology; the core is formed by one Non Steroidal Anti-Inflammatory Drug (NSAID) and a glucocorticoid, and the shell casing is a biocompatible polymer PLGA. The physical chemical characterization of the nanoparticles was performed considering the morphological analysis by using a scanning electron microscopy; particle size by dynamic light scattering using a photon correlation espectrometer Coulter NP4 Plus; the load surface using a device for measuring the zeta potential and the thermal analysis by using a calorimeter differential scanning (DSC).

This system loader is an optimal alternative because it has advantages such as: allows modulation of the release of molecules with different physicochemical characteristics, the drug release is more specific and controlled, reduces local and systemic adverse effects, improves treatment effectiveness and reduces costs for the treatment of rheumatoid arthritis.

(397q) Processing of Uniform Metal and Alloy Nanocrystals for Catalytic Applications

Yang, H., University of Rochester
Pan, Y. T., University of Illinois at Urbana-Champaign
Wu, J., University of Rochester
Shi, M.,
Yin, X., University of Illinois at Urbana and Champaign
Zhou, W.,
Kim, J.,

With the recent development of non-aqueous synthetic approaches, new understandings and synthetic systems for making well-defined nanoparticles are emerging. In non-aqueous system, surfactants are often used and play important roles in the synthesis of nanocrystals. We recently used carbon monoxide as a reducing and capping agent in the synthesis of a range of metal and metal alloy nanoparticles. In this presentation, we will show non-aqueous synthetic media can be used to study the formation of nanocrystals through a highly controllable manner, including the uncommon Pt icosahedral nanocrystals, which is hard to achieve in aqueous solution due to the complex, fast reaction kinetics. This poster will also discuss the newly identified functions that organic ligands play in such reaction systems and the applications of such nanoparticles in electrocatalysis.

(397j) Molten Droplet Synthesis of CdSe Hollow Nanoparticles

Gullapalli, S., Rice University
Grider, J. M., Rice University
Bagaria, H. G., Rice University
Lee, K. S., Electronics and Telecommunication Research Institute
Cho, M., Rice University
Colvin, V., Rice University
Jabbour, G., Arizona State University
Wong, M. S., Rice University

Many colloidal synthesis routes are not scalable to high production rates, especially for nanoparticles of complex shape or composition, due to precursor expense and hazards, low yields, and the large number of processing steps. The recognition that optical, electronic, catalytic, and other material properties of nanoparticles can be manipulated by shape, in addition to size, drives the current interest in devising new synthesis methods to non-spherical nanoparticles. A new strategy has been devised to readily synthesize hollow nanoparticles (HNPs) out of metal chalcogenides, based on the slow heating of a low-melting-point metal salt, an elemental chalcogen, and an alkylammonium surfactant in octadecene solvent. Colloidally stable and monodisperse CdSe HNPs with an outer diameter of 15.6±3.5 nm and a shell thickness of 5.4±0.9 nm are characterized extensively using UV-visible spectroscopy, high resolution TEM, X-Ray diffraction, dynamic light scattering and electron energy loss spectroscopy. The HNP synthesis is proposed to proceed with the formation of alkylammonium-stabilized nano-sized droplets of molten cadmium salt, which then come into contact with dissolved selenium species to form a CdSe shell at the droplet surface. In a reaction–diffusion mechanism similar to the nanoscale Kirkendall effect it is speculated that the cadmium migrates outwardly through this shell to react with more selenium, causing the CdSe shell to thicken. The proposed CdSe HNP structure comprises a polycrystalline CdSe shell coated with a thin layer of amorphous selenium. Photovoltaic device characterization indicates that HNPs have improved electron transport characteristics compared to standard CdSe quantum dots, possibly due to this selenium layer. The HNPs are colloidally stable in organic solvents even though carboxylate, phosphine, and amine ligands are absent; stability is attributed to octadecene-selenide species bound to the particle surface. This scalable synthesis method presents opportunities to generate hollow nanoparticles with increased structural and compositional variety.

(397r) Comparison of Size Distributions of Hollow Shell Nanoparticles Determined By Particle Tracking and Direct Imaging

Shin, J., University of Minnesota
Ogunyankin, M. O., University of California, Davis
Zasadzinski, J. A., University of Minnesota

Particle tracking methods determine nanoparticle size distributions by measuring the independent diffusive Brownian motions of a large number of individual nanoparticles for fixed time intervals, and converting the diffusive motion into an effective "spherical" particle diameter using the Stokes – Einstein equation. Little information is available about the validity of these methods for non-spherical particles or porous hollow nanoshells. We are interested in hollow gold nanoshells (HGNs) as tunable near infra-red light absorbers; the absorption maximum depends on the ratio of the nanoshell thickness to the nanoshell diameter. Hence, a reliable method of measuring the nanoshell diameter is essential to optimizing HGN synthesis. Currently, the HGN synthesis, which involves the galvanic exchange of gold on a silver nanoparticle template, provides a relatively polydisperse population, which broadens the absorption maximum (1, 2). Direct imaging of the nanoparticles by depositing the particles on electron microscope grids and imaging with transmission electron microscopy (TEM), provides a precise measure of the size distribution, although the sample size is small and sizing and counting individual particles from the images is cumbersome and time consuming.

The alternative to both sizing and estimating the particle concentration in solution is particle tracking using a commercial Nanosight particle analysis system. However, it is unclear how accurate particle tracking and the Stokes-Einstein equation might be for hollow, porous, spheroidal nanoparticles. A direct comparison of the mean sizes and size distributions obtained via TEM and particle tracking will provide the necessary calibration and show if particle tracking is a viable method of obtaining size distribution for hollow, porous nanoparticles. We can then isolate the steps in the synthesis that lead to the greatest polydispersity; likely choices include the synthesis of the silver templates, the variation in gold nanoshells thickness caused during the reduction step, or the choice of capping ligands such as citrate or thiols to promote HGN stability in salt solutions. The effects of decreasing the polydispersity on the near infra-red light absorption will also be discussed.

1. Prevo BG, Esakoff SA, Mikhailovsky A, & Zasadzinski JA (2008) Scalable routes to gold nanoshells with tunable sizes and response to near-infrared pulsed laser irradiation. *Small* 4:1183-1195.
2. Wu G, *et al.* (2008) Remotely triggered liposome release by near-infrared light absorption via hollow gold nanoshells. *J Am. Chem. Soc.* 130:8175-8177.

(397s) Complexation of Various Polyelectrolytes With Gold Nanoparticles, Colloidal Properties

Rabbani Esfahani, M., The Center for the Management, Utilization and Protection of Water Resources,Department of Chemical Engineering , Tennessee Technological Universsity
Pallem, V. L., Tennessee Technological University
Stretz, H. A., Tennessee Technological University

Complexation process between polyelectrolytes and nanoparticles plays important roles in many branches of nanotechnology and nano-science such as nanoparticles stabilization and dispersion in aquatic environment. The presence of a charged nanoparticles can significantly modify the acid/base and colloidal properties of a polyelectrolyte in water. Global structural parameters such as the polyelectrolyte size and zeta potential, nanoparticle size, nanoparticles zeta potential, solution pH, were investigated. Dynamic light scattering (DLS) and zeta potential measurements for at least two polyelectrolyte (Humic acid (HA) and poly (Styrene sulfonic acid malic anhydride (PSMA)) are reported as titrations and compared to complex of mixture of them and gold nano- particles(Aqueous dispersions of citrate-coated gold nanoparticles 50 nm (GNP-50) . The isoelectric behavior was shifted in a manner that is synergistic and this is explained by comparison to literature based Monte Carlo simulations.

(397u) Development of a Novel in-Vitro Tumor Cell Extravasation Model

Fischer, J., University of Kentucky
Trinkle, C., University of Kentucky
Richards, C.,
Anderson, K. W., University of Kentucky

Cancer is the second leading cause of death in the United States, with metastatic lesions resulting from circulating tumor cells (CTCs) most often causing the mortality. The basic steps of the metastatic cascade, from initial tumor formation to growth at a secondary site, are a linear sequence of steps composed of a network of mechanisms that have not been fully elucidated. The goal of this work was to develop a novel nanoarray microfluidic platform allowing for the examination of the steps in the metastatic cascade, particularly extravasations, at the single-molecule level. It is hypothesized that this platform may also be used as both a screening tool for drug discovery efforts as well as to further basic scientific research in identifying inhibitors of metastasis. Single-molecule assessments offer several distinct advantages to conventional bulk scale analyses. In addition to being highly sensitive and requiring small amounts of reagents, single-molecule evaluations allow for interactions to be measured one a time, eliminating the potential for lost distribution information in ensemble averaging. Additionally, structural and functional information can be directly probed and these molecules can be evaluated at native concentrations. In this work, the optical and spatial resolution attainable using a nanoarray was exploited to study receptor-ligand interactions between cell surface adhesion molecules which are significant in each step of metastatic progression. It has recently been demonstrated that single molecules on the surface of living cells can be examined singularly and with high parallelism using these arrays since filipodia-like extensions on the plasma membrane are readily drawn into the array holes. In these nanoarrays, single-molecule events such as nanoscale binding, conformation and stoichiometry may be both observed and quantified. The focus of this work was to develop the nanoarray microfluidic device and complete proof of concept binding studies examining vascular endothelial growth factor (VEGF) receptors (VEGFRs) to observe dimerization of the receptors upon binding on VEGF which has not yet been examined on the single molecule scale. VEGF was selected for study since it has been demonstrated to modulate extravasations and VEGFR antagonists have already been approved for cancer treatment. The extension of this nanoarray platform to include a microfluidic channel beneath the array allowed for selective stimulation of the basal side to determine the effects of various stimulants on binding properties. A discussion of device fabrication and verification as well as preliminary results from extravasation and binding studies will be presented.

(397t) Magnetic Nanoparticles As Multifunctional Carriers for the Diagnosis of Prostate Cancer

Anani, T., Auburn University

Prostate cancer is currently the most common site for cancer in men in the developed world. Despite decades of effort in the fight against cancer, current diagnostic techniques are inaccurate and invasive. Nanoparticles, due to their large surface area and the several functional groups on their surfaces, offer new and exciting opportunities to overcome the shortcomings of conventional diagnostic techniques. We use magnetic iron oxide nanoparticles (MNP) as our multifunctional system due to their intrinsic diagnostic capabilities as MRI contrast agents and the ability to add functionalities, through appropriate surface modifications, to include targeting moieties, fluorescence dyes and radioactive agents. We developed a synthesis method of long-circulating MNPs for sustained tumor exposure that allows for precise control over the physical characteristics of magnetic nanoparticles. Several active targeting strategies will be explored to optimize cancer cell specificity and enhance delivery efficiency based on prostate cancer-specific markers while avoiding nonspecific toxicity to non-tumor sites. Peptide conjugated MNPs will be used to detect proteolytic activity in vivo. Peptides are attractive targeting moieties due to their small size, low immunogenicity and low cost. Phage display techniques will be used to identify peptides capable of detecting Matrix Metalloproteinase (MMP) activity in the tumor site and its microenvironment. The identified peptides will then be optimized through modification of their sequences, and the optimal method of conjugating the peptide to the surface of the MNPs as well as the optimal targeting ligand density will be investigated. The effect of nanoparticle size on biodistribution will also be investigated, using a mouse model, utilizing particles with diameters of 10, 50, 100, and 200 nm.

(397v) Increased Osteoblast Adhesion and Proliferation On Electrophoretic Deposition Coated Nano Hydroxyapatite On Titanium-6,4

Bhardwaj, G.,

Mathew, D.,

Sun, L., Northeastern University

Webster, T. J., Northeastern University

Manivasagam, G.,

INCREASED OSTEOBLAST ADHESION AND PROLIFERATION ON ELECTROPHORETIC DEPOSITION COATED NANO HYDROXYAPATITE ON TITANIUM-6,4

Garima Bhardwaj[1], Dennis Mathew[1], Linlin Sun[2], Thomas J.Webster[2] and Geetha Manivasagam[3].

[1]Department of Biomedical Engineering, VIT University,Vellore-632014,Tamil Nadu,India.

[2]Department of Chemical Engineering, Northeastern University, Boston, MA-02115, USA.

[3]Department of Mechanical Engineering, VIT University,Vellore-632014,Tamil Nadu,India.

Introduction: Improving the performance of titanium implants for orthopedic applications has been a concern for many years.This can be completed by designing surface properties to promote osseointegration. One manner to improve orthopedic implant performance is to coat metals with bioactive ceramics, such as nanocrystalline hydroxyapatite, which have been shown to improve bone formation compared to micron crystalline hydroxyapatite and other ceramics. However, current hydroxyapatite coating methods are time consuming, may be expensive, and involve high heat which transforms nanocrystalline hydroxyapatite into micron crystalline hydroxyapatite decreasing bioactivity and decreasing coating adhesion strength.The goal of this present research was to develop a novel, quick, and inexpensive implant coating process to improve bone cell responses.

Materials and Methods: All of the tests in this study were performed on a titanium alloy, Ti-6Al-4V, which was cut into small pieces with dimensions of 1cm X 1 cm. Nanophase hydroxyapatite (HAP) was prepared by wet chemical synthesis using Ca $(NO_3)_2 4H_2O$, KH_2PO_4,distilled water, ammonia and acetone with a

stirring time of 1 hour and an aging time of 24 hours followed by filtering then sintering at 900°C. The HAP

was then coated onto the alloy using electrophoretic deposition (EPD). The electrolyte was prepared by dissolving HAP in isopropyl alcohol and stirring for 10 minutes followed by overnight settling. The mixture was again stirred for 2 hours before starting the coating and Ti-6,4 was used as a working electrode(-ve) and 316 stainless steel was used as a counter electrode(+ve).Voltage varying from 40 to 80 V was used for 2 to 5 minutes and an optimal coating was achieved at 60 V for 4 minutes. Samples were characterized using X-ray Diffraction (XRD), SEM, and contact angle analysis to confirm crystallinity, surface roughness, and chemistry of the coatings. Osteoblasts (ATCC-C12720) were cultured in growth medium supplemented with 1% penicillin.A 4 hour MTT adhesion assay was performed to determine cell viability.A MTT proliferation assay was performed after 1,3 and 5 days of culture.All experiments were conducted in triplicate and repeated three times each.

Results and Discussion XRD results of the nanophase HAP (Figure 1) and SEM images of the samples taken after the coating process (Figure 2) confirmed the uniformity and nanoscale character of the coatings. Contact angle results of the coated samples showed an increased hydrophilic nature of the coated samples. Results from the MTT assay showed increased osteoblast adhesion (Figure 3) and proliferation (Figure 4) on the coated samples compared to controls (tissue culture polystyrene and uncoated Ti-6Al-4V).

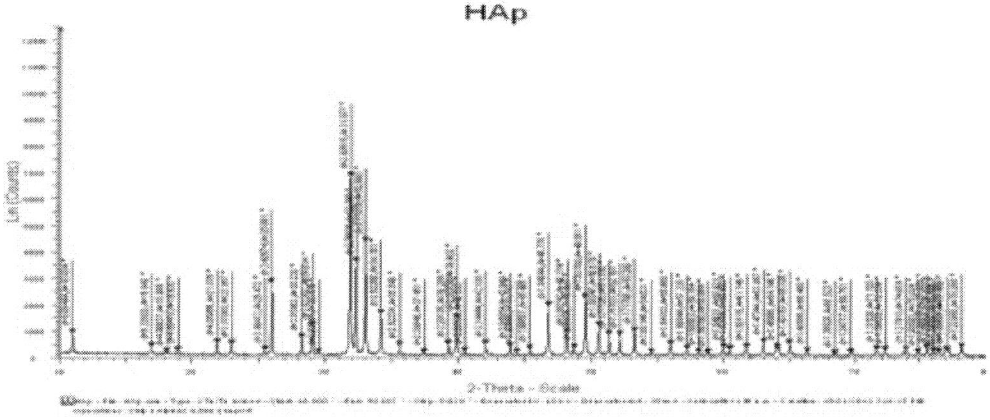

Figure 1: XRD of nanocrystalline HA used in the present study

Figure 2. SEM of the EPD nanocrystalline HA on titanium-6,4

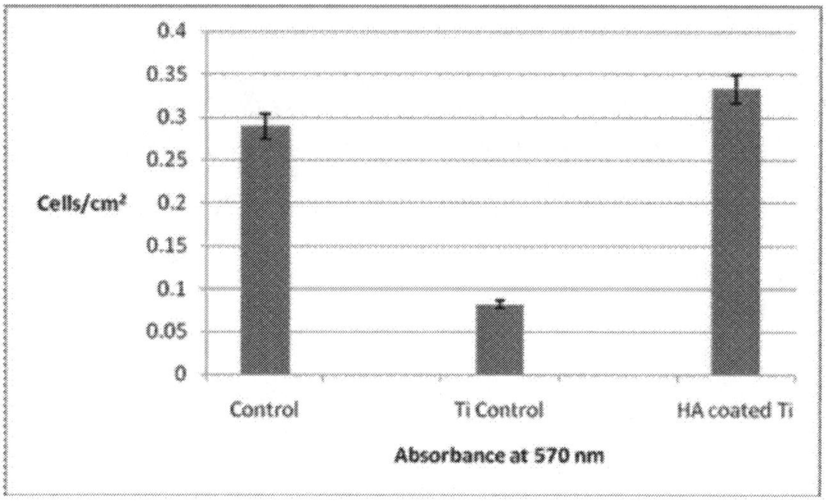

Figure 3. Increased osteoblast adhesion on HA (EPD) coated Ti compared to Ti control. Data = mean +/- SEM; n = 3 * $p < 0.01$ compared to Ti control,F-test value=0.000356.

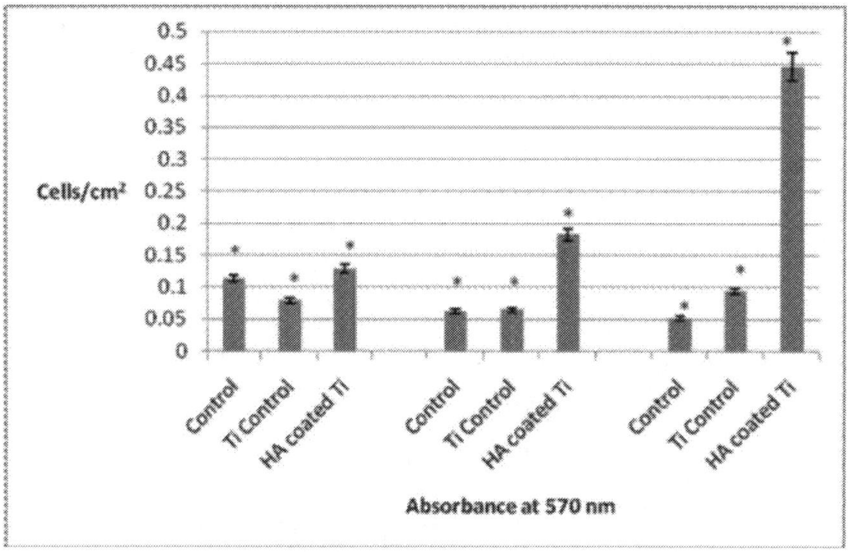

Figure 4: Osteoblasts proliferation results, day 1, 3 and 5 respectively.Data = mean +/- SEM; n = 3; * $p < 0.01$ compared to Ti control,F value for Day 1=0.001872,F value for Day 3=0.0017608,F value for Day 5=0.007834

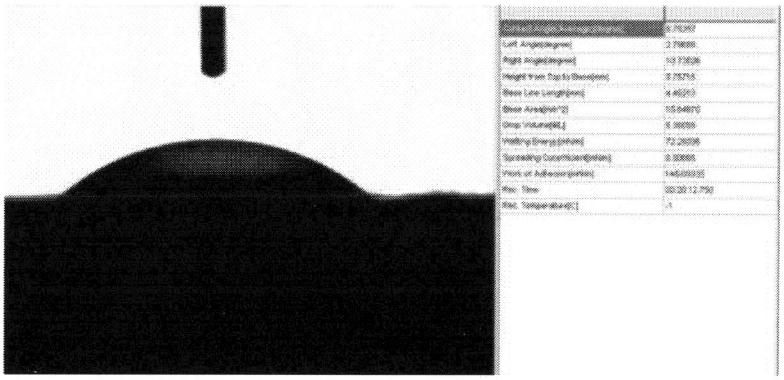

Figure 5:Contact angle analysis results(actual picture and values of drop and surface contact) for uncoated Ti-6,4 obtained using distilled water as a fluid.

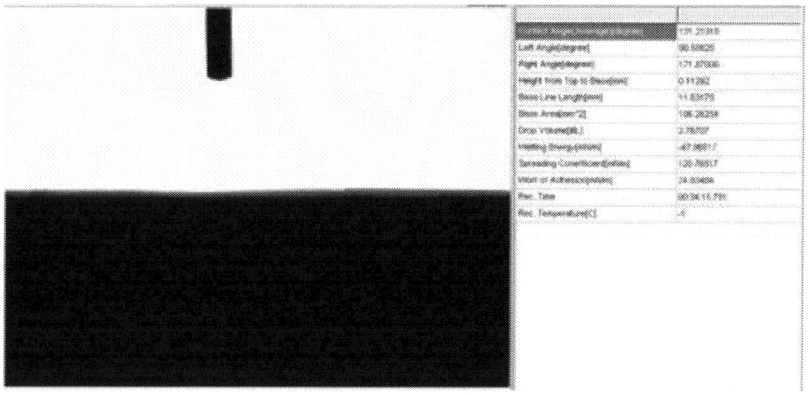

Figure 6:Contact angle analysis results(actual picture and values of drop and surface contact) for Ti-6,4 coated with nano-HA by EPD obtained using distilled water as a fluid.

Conclusions: Electrophoretic deposition is a novel, inexpensive and quick coating process that leads to a strong and optimal coating to increase bone cell compatibility. Hence, this coating process should be further explored for enhancing orthopedic implant efficacy.

Acknowledgements: The authors thank Northeastern University and VIT University for funding.

(397i) Effect of Salts With Different Cations On the Stability of Self-Assembled Two Component Nanoparticle System

Gao, Y., University of Kansas
Dhar, P., University of Kansas
Liang, J. T., University of Kansas
Chowdhury, M.,

Effect of Salts with Different Cations on the Stability of Self-assembled Two Component Nanoparticle System

Yan Gao, Roy Chowdhury, Monica, Jenn-Tai Liang and Prajnaparamita Dhar

Department of Chemical and Petroleum Engineering, The University of Kansas, Lawrence, Kansas, 66045

Abstract

Novel surfactant/polyelectrolyte complexes are of great interest for consumer products, biotechnology and enhanced oil recovery due to simplicity of preparation and well-controlled properties. In a previous study, self-assembled polyelectrolyte nanoparticles (PEC) containing entrapped surfactant molecules were formed by mixing polyethylenimine (PEI) and sodium dodecyl sulfate (SDS) in nonstoichiometric amounts. These positively charged nanoparticles can form stable colloidal solution in deionized water without any change in particle size and zeta potential over several months.

However, in several applications, the nanoparticles may be exposed to saline environments. Understanding the stability of PEC nanoparticle in different salt solutions is therefore important from an application perspective.

In this study, we investigated the effects of salts, NaCl, $MgCl_2$, $CaCl_2$, on the structure parameter change of PEI/SDS nanoparticles, focusing on the influence of valence of the cations and entrapped surfactant concentrations on colloidal stability. Our particle size and zeta potential results show that, contrary to our expectations (Schulz-Hardy rule), positively charged PEI/SDS nanoparticles were more stabilized against flocculation in the presence of divalent salts compared with sodium chloride. At low SDS concentrations (below 3 mM), PECs aggregate at a critical ionic strength of about 0.2 mM in the presence of NaCl. However, within the same range of ionic strength for divalent salts, the particle size changes less than 10 percent. In fact, the critical ionic strength for aggregation increases to 0.6 mM in case of $MgCl_2$. Furthermore, the critical ionic strength for aggregation decreases with increasing SDS concentration, suggesting higher aggregation potential for a more concentrated solution.

(397w) Antibacterial Studies On Titania Polyurethane Nanocomposite Coatings

Azhie, K., Western University
Charpentier, P. A., Western University

While temporarily disinfection of a surface is possible with the help of strong cleaners, tremendous interest exists for the control of microorganisms on surfaces by effective, durable antimicrobial coatings to prevent multidrug resistant pathogens. There is a wide spectrum of potential applications for antibacterial coatings, spanning from industrial surface coatings to biomedical applications, where sterile conditions are crucial. This work examined the synthesis of nanotitanium dioxide/polyurethane ($nTiO_2$/PU) composite coatings prepared by using various bifunctional monomers. The antibacterial behavior of virgin PU, $nTiO_2$/PU and silver doped $nTiO_2$/PU composites were investigated qualitatively and quantitatively against both gram-negative (*Escherichia coli*) and gram-positive (*Micrococcus luteus*) bacteria. The effect of exposure time was investigated using a UV lamp and solar simulator by monitoring the growth of bacterial populations in the presence and absence of the above-mentioned nanocomposites. The distribution of $nTiO_2$ in the polymer matrix was enhanced by monomer functionalization in which $nTiO_2$ was chemically attached to the backbone of the polyurethane polymer matrix with a bifunctional monomer. Since silver is one of the most effective antibacterial agents with low toxicity used for its high degree of biocompatibility and its long-term antibacterial effectiveness against many different bacterial strains, it was also examined as a TiO_2 dopant. The quantitative examination of bacterial activity was determined by the survival ratio as calculated from the number of viable cells, which form colonies on the petri dishes with nutrient agar. Good inhibition results were observed and demonstrated visually with close to 97% of bacteria killed after 2 hours. In summary the functionalized $nTiO_2$/PU, and silver doped $nTiO_2$/PU composite coatings displayed considerable antibacterial activity against both gram-positive and gram-negative bacteria under UV irradiation while silver doped $nTiO_2$/PU coatings displayed considerable antibacterial activity even under irradiation of visible light.

(397x) Fabricating Ultra-High Density Nanowires for Dye-Sensitized Solar Cells

Xu, C., University of Florida at Gainesville
Zhao, Y., University of Florida at Gainesville
Ziegler, K. J., University of Florida

Fabricating Ultra-High Density Nanowires for Dye-Sensitized Solar Cells

Cheng Xu[1,*], Yang Zhao[1], Jie Liu[1], and Kirk J. Ziegler[1]

[1]Department of Chemical Engineering, University of Florida at Gainesville, 1006 center drive, Gainesville, FL 32611, US

*Presenting Author (Ph: 352-392-3122; Email: chengotopia@ufl.edu)

Abstract

Dye-sensitized solar cells (DSSCs) use titania nanoparticles as the photoanode and have been extensively studied for the past decade as a promising technology for low-cost, high-performance solar cells. However, ever since the efficiency of DSSCs reached 11% in 2001, little progress has been made in efficiency due to energy loss from the random walk of electrons and the recombination with defects and surface trap sites that result in slow trap-limited electron diffusion in the titania nanoparticles. A ZnO core-shell nanowire network offers unique advantages over the traditional nanoparticle-based DSSCs. By eliminating the random walk of electrons and the loss mechanism from interfacial charge recombination and back reactions, improvements are observed in the charge injection at the interface and electron diffusion within ZnO nanowires. However, it is still challenging to fabricate these nanowires with similar surface area as the nanoparticle system. In this study, anodized alumina (AAO) is used as a template to control the density and dimension of nanowires. Multiple growth methods are used to grow highly crystalline ZnO nanowires, including hydrothermal, chemical vapor deposition (CVD), electrodeposition, and vapor-liquid-solid (VLS) methods. These vertically aligned nanowires can also be transferred onto a foreign flexible transparent substrate. The impact of nanowire dimensions on DSSC performance is investigated to obtain a better understanding of the impact of photoanode morphology on charge transport dynamics. Comparisons of a theoretical model and experimental observations can help understand how to control interfacial charge transfer, leading to significant improvements to DSSC efficiency due to the ultra-high photoactive surface area in these nanowire systems.

(397y) Electronic Platform to Quantify Cellular Mechanisms Associated With Carbon Nanotubes

Eldawud, R., West Virginia University
Dong, C., West Virginia University
Sargent, L. M., National Institute for Occupation Safety and Health
Rojanasakul, Y., West Virginia University
Dinu, C. Z., West Virginia University

Carbon nanotubes (CNTs) exhibit unique properties that make them attractive candidates for the delivery of therapeutic molecules, genes and drugs. However, advances in the biomedical applications of CNTs are being hindered by uncertainties associated with their cellular uptake and delivery mechanisms as well as their fate inside the biological systems. Recent reports have shown that CNTs toxicity can be attributed to metal impurities, length, size, coating, uptake, or internalization. However, there are no reports **that provide fundamental understanding of the toxicological and pharmacological profiles of cellular systems exposed to CNTs.** In this research, a new approach is used to assess the toxicity of a wide library of surface modified CNTs incubated with human lung epithelial cells (BEAS-2B) in real time. This non-invasive technique relies on an electrical cell impedance sensing system (ECIS) used as a proxy to measure morphological changes and cellular interactions upon exposure to different concentrations and functionalities of CNTs. Such measurements are correlated to the cellular behavior under the influence of different parameters like endocytotic inhibitors or electroporation and are further complemented with cellular and optical assays to derive mechanisms associated with cytotoxic, genotoxic and apoptotic events. Our real time results provide an insight to the underlying cellular mechanisms responsible for the uptake and delivery of CNTs to the biological systems and promises to extend our understanding of such cellular-based interfaces to advance the biomedical applications of CNTs.

(397z) Interactive Forces Between SDS-Suspended Single-Wall Carbon Nanotubes and Agarose Gels

Clar, J. G., University of Florida
Silvera-Batista, C., National Institute Of Standards And Technology
Youn, S.,
Bonzongo, J. C. J., University of Florida
Ziegler, K. J., University of Florida

The separation of metallic and semiconducting fractions of as-prepared single-walled carbon nanotubes (SWCNT) is of great interest for their inclusion in a wide variety of devices. A commonly used method to produce large-scale separations of the two fractions is based on the flow of SWCNTs through a column packed with agarose gel. Prior studies have attributed the mechanism of separation to the selective adsorption of semiconducting SWCNTs onto the agarose gels. However, few details of the forces driving this selective adsorption are understood. Here we probe the driving forces that occur during the interaction of sodium dodecyl sulfate (SDS) suspended SWCNT with agarose gels using a combination of non-equilibrium and equilibrium studies. By altering the agarose backbone, the role of ionic, purely hydrophobic interaction, and predominantly $\pi - \pi$ interactions on the selective adsorption could be examined. The results indicate that the surfactant orientation on the SWCNT surface is an important parameter during separation and that the primary mode of attraction is ion-dipole interactions between SDS-SWCNTs and hydroxyl groups on the agarose gel backbone. The selectivity of adsorption is then driven by the inherent differences in polarizability between the metallic and semiconducting species. Furthermore, we report that simple Langmuir models for the adsorption behavior of SWCNTs onto agarose gels is not appropriate for all systems despite persistent use in the literature. We recommend that further attempts to understand the entropic and enthalpic contributions to the free energy of these systems be measured with a series of calorimetric studies.

(397aa) Microfabricated Non-Linear Architectures for Propagation and Differentiation of Neural Progenitors

Moustafa, M.,
Rao, R. R., Virginia Commonwealth University
Yadavalli, V. K., Virginia Commonwealth University

Challenges in the development of successful cell therapies involve engineering and control of cues to regulate the balance between differentiation and self-renewal. However, the complexity of architecture and function make this an intriguing problem in the context of forming functional connections. Here we present the design and fabrication of microstructured scaffolds that present a *biomimetic* framework along which neural progenitors can organize into oriented constructs. Specifically, we show microfabricated non-linear architectures that promote cellular fate related to propagation of human neural progenitors and directed differentiation towards neurons. By mimicking biological networks that allow for spreading of the cells instead of confining them in a groove or a well, a nonlinear configuration can promote a relaxed, self-supportive stem cell niche. The tailoring of non-homogeneous adhesion sites via the geometry and the compliance and roughness of the substrate allows a versatile microenvironment that promotes propagation and neuronal differentiation.

(397ab) Layer-By-Layer Synthesis of Polymeric Carriers for Drug Delivery

Anselmo, A. C., University of California Santa Barbara
Mitragotri, S., University of California Santa Barbara

Polymeric drug delivery carriers are the most investigated approach for development of therapies for disease treatment. While promising, polymeric carriers have yet to make a lasting impact in the clinic mainly due to the rapid clearance from blood and limited targeting to diseased sites. The future of polymeric carriers for drug delivery depends on the synthesis and application of biomaterials that can address these challenges. Layer-by-layer synthesis of drug carriers allows for better control of biomaterial composition than that of typically used synthesis methods. Specifically, layer-by-layer synthesis allows for the precise control and modification over important parameters for targeted therapies such as size, shape, biocompatibility, surface chemistry, and drug encapsulation. Recently, layer-by-layer synthesis has been used to create: (i) synthetic cells able to mimic their real life counterparts in terms of size, shape, and functionality [1, 2], (ii) biocompatible rod-shaped nanoparticles which provide distinct advantages over their spherical counterparts for targeted drug delivery [3], and (iii) drug-containing patches capable of adhering to the surface of immune cells for the targeted delivery to diseased tissues [4]. This poster will highlight the exploitation of layer-by-layer synthesis to create polymeric carriers (as described above) and their applications towards addressing the main challenges in the field of drug delivery.

1. Doshi, N., et al., *Platelet mimetic particles for targeting thrombi in flowing blood.* Adv Mater, 2012. **24**(28): p. 3864-9.

2. Doshi, N., et al., *Red blood cell-mimicking synthetic biomaterial particles.* Proc Natl Acad Sci U S A, 2009. **106**(51): p. 21495-9.

3. Zhou, Z., A.C. Anselmo, and S. Mitragotri, *Synthesis of Protein-Based, Rod-Shaped Particles from Spherical Templates using Layer-by-Layer Assembly.* Adv Mater, 2013.

4. Doshi, N., et al., *Cell-based drug delivery devices using phagocytosis-resistant backpacks.* Adv Mater, 2011. **23**(12): p. H105-9.

(397ac) Understanding Surfactant/Graphene Interactions Using a Graphene Field Effect Transistor: Relating Molecular Structure to Hysteresis and Carrier Mobility

Shih, C. J., Massachusetts Institute of Technology
Paulus, G. L. C., Massachusetts Institute of Technology
Wang, Q. H., Massachusetts Institute of Technology
Jin, Z., Massachusetts Institute of Technology
Blankschtein, D., Massachusetts Institute of Technology
Strano, M. S., Massachusetts Institute of Technology

Manipulation of transport hysteresis on graphene transistors and understanding electron transfer between graphene and polar/ionic adsorbates are important for the development of graphene-based sensor devices and nonvolatile memory electronics. We have investigated the effects of commonly used surfactants for graphene dispersion in aqueous solution on transport characteristics of graphene transistors. The adsorbates are found to transfer electrons to graphene, scatter carrier transport, and induce additional electron−hole puddles when the graphene is on an SiO_2 substrate. We relate the change in transport characteristics to specific chemical properties of a series of anionic, cationic, and neutral surfactants using a modification of a self-consistent transport theory developed for graphene. To understand the effects of surfactant adsorbates trapped on either side of the graphene, suspended devices were fabricated. Strong hysteresis is observed only when both sides of the graphene were exposed to the surfactants, attributable to their function as charge traps. This work is the first to demonstrate the control of hysteresis, allowing us to eliminate it for sensor and device applications or to enhance it to potentially enable nonvolatile memory applications.

(397ae) Application of Core-Shell Titania-ITO Nanowires in DSSCs

Li, L., University of Florida
Chen, S., University of Florida
Kim, J., University of Florida
Ziegler, K. J., University of Florida

With increasing demand for more energy, fossil fuel depletion, and global warming, finding alternative energy sources have been the focus of research for the scientific community for decades. Compared with other sources, solar energy offers advantages such as being completely renewable, low operating cost, and environmental friendliness. The solar cell is an important class of devices that harvest solar energy and produce electricity. Among different types of solar cells, dye sensitized solar cells (DSSCs) are promising and conventionally the photo-sensitized anode and the cathode are made of thin films. Nanowires (NWs) are widely investigated because of their potential applications in nanoscale devices and sensors. Nanowire-based DSSCs are more advantageous than their thin-film or nanoparticle-based counterparts due to the former's enhanced charge collection efficiency in the radial direction and faster charge transport in the axial direction. Tin-doped indium oxide (ITO) nanowires boast high conductivity and high transmittance in the visible spectra of sunlight and plays critical roles in DCCSs. In this work, free-standing ITO nanowires were synthesized via a vapor transport method on ITO-coated glass substrate. These nanowires have uniform diameters of ~ 200 nm and their length can be tailored in the range of 5-50 μm by controlling the growth time. A liquid phase deposition technique is used to deposit ~ 20nm-thick TiO_2 layers on NW surfaces. The coating crystallizes in anatase phase upon calcination, as evidenced by Raman spectroscopy and transmission electron microscopy. The electrical and optical properties of these core-shell nanowires are presented, which make them promising materials as the photoanode in DSSCs.

(397ad) Understanding the pH-Dependent Behavior of Graphene Oxide Aqueous Solutions: A Comparative Experimental and Molecular Dynamics Simulation Study

Shih, C. J., Massachusetts Institute of Technology
Lin, S., Massachusetts Institute of Technology
Sharma, R., Lawrence Berkeley National Laboratory
Strano, M. S., Massachusetts Institute of Technology
Blankschtein, D., Massachusetts Institute of Technology

Understanding the pH-dependent behavior of graphene oxide (GO) aqueous solutions is important to the production of assembled GO or reduced GO films for electronic, optical, and biological applications. We have carried out a comparative experimental and molecular dynamics (MD) simulation study to uncover the mechanisms behind the aggregation and the surface activity of GO at different pH values. At low pH, the carboxyl groups are protonated such that the GO sheets become less hydrophilic and form aggregates. MD simulations further suggest that the aggregates exhibit a GO-water-GO sandwichlike structure and as a result are stable in water instead of precipitating. However, at high pH, the deprotonated carboxyl groups are very hydrophilic such that individual GO sheets prefer to dissolve in bulk water like a regular salt. The GO aggregates formed at low pH are found to be surface active and do not exhibit characteristic features of surfactant micelles. Our findings suggest that GO does not behave like conventional surfactants in pH 1 and 14 aqueous solutions. The molecular-level understanding of the solution behavior of GO presented here can facilitate and improve the experimental techniques used to synthesize and sort large, uniform GO dispersions in a solution phase.

(397af) Synthesis and Characterization of Maghemite Nanobricks

Duttagupta, S., Indian Institute of Technology Bombay
Thaokar, R., Indian Institute of Technology, Bombay
Mehra, A., Indian Institute of Technology, Bombay (IITB)

Magnetic iron oxide nano-bricks of size range 100-700 nm were synthesized using Ferric Chloride ($FeCl_3$) and Ammonium Hydroxide (NH_4OH) as the precursor material in a water-in-oil microemulsion system. The as prepared ironoxide nanoplates were characterised by X-ray di raction (XRD), high resolution transmission electron microscope (HRTEM), field emission scanning electron microscope (FESEM) and energy dispersive spectroscopy (EDS).

Here in we discuss a simple one pot facile protocol which allows us to prepare a large number of nano-brick structures.

Normal reverse micelles yield spherical nanoparticles, whereas one-dimensional nanotubes have been prepared using cylindrical reverse micelles, synthesized in
the presence of high concentration of $FeCl_3$. When aqueous $FeCl_3$ solution is added to the
AOT/heptane mixture, the surfactants generate cylindrical micelles due to the solvation and hydrogen bonding of polar groups. $FeCl_3$ acts as a promoter that increases the ionic strength of a solvent and also decreases the second critical
micelle concentration of AOT. The anionic polar head groups of AOTs also have the capability to extract Fe^{3+}ions from the aqueous solutions to the reverse micellar phase. As a result, iron cations concentrated in the reverse micelles are capped by the anionic groups of AOT. We decided on using this as a template to create maghemite nanorods, but serendipitously synthesized nano-bricks.

(397ag) Tin Selenide Nanocrystals for Near-Infrared Applications

Qi, Y., University of Massachusetts
Mountziaris, T. J., University of Massachusetts

Near-infrared (NIR) absorbing semiconductor nanocrystals have exciting applications in photothermal therapy, photovoltaics, and infrared absorbing glass coatings or interlayers. However, the most common NIR materials contain heavy metals, such as Cd, Hg or Pb, that are not environmentally friendly. Tin Selenide (SnSe) has emerged as a heavy-metal-free alternative NIR material. The typical synthesis of SnSe colloidal nanocrystals or quantum dots (QDs) involves the use of toxic and pyrophoric organometallic precursors (e.g., Baumgardner et al., *J. Am. Chem. Soc.*, 2010, 132 (28), pp 9519–9521). We developed a new synthesis method for SnSe QDs that employs an air-stable tin(II) chloride-oleylamine (OLA) complex and selenium powder dissolved in trioctylphosphine (TOP) as precursors. SnSe QDs with size tunable bandgap and average diameters between 1.7 and 2.5 nm were synthesized by injecting the precursors into a hot coordinating solvent consisting of OLA and TOP. The growth rate and final morphology of the nanocrystals were studied by TEM as functions of the processing conditions. Operating conditions that enable precise control over the final particle size and prevent particle aggregation were identified. The ultimate objective of our work is the development of a commercially viable method for manufacturing SnSe nanocrystals for NIR applications using readily available, air-stable precursors.

(397ah) Inhibiting Bacterial Growth --- a Novel Biomedical Application of Ceria Nanoparticles

Wang, Q., Brown University
Webster, T. J., Northeastern University
Perez, J. M.,

Inhibiting Bacterial Growth --- a Novel Biomedical Application of Ceria Nanoparticles

Qi Wang[1], J. Manuel Perez[2], Thomas J. Webster[1,3]

[1]Bioengineering Program, [3]Department of Chemical Engineering, College of Engineering, Northeastern University, Boston, MA 02115

[2]Nanoscience Technology Center, University of Central Florida, Orlando, FL 32826

As a highly abundant material, ceria (or cerium oxide, CeO_2) is technologically important due to its wide-range of applications, for example in catalysts for the elimination of toxic auto-exhaust gases,[1-2] oxygen sensors,[3] electrochromic thin-films[4] and so on. Upon the transition from a macro-sized particle into a nanocrystalline particle, ceria significantly changes its physicochemical properties to usually possess a high density of nanocrystalline ceria interfaces.[5] Thus, nanostructured ceria has attracted extensive attention in such applications due to improvements in redox properties, transport properties, and high surface to volume ratios.

However, although ceria nanoparticles have been studied for numerous applications in traditional science and engineering applications, there have been almost no studies regarding the potential biomedical applications of ceria until recently when researchers demonstrated that ceria nanoparticles possess antioxidant properties at physiological pH values. The ability of these nanoparticles to act as an antioxidant lies in their ability to reversibly switch from Ce^{3+} to Ce^{4+}.[6] In recent studies, this antioxidative ability enables the application of ceria nanoparticles to protect against radiation damage, oxidative stress and inflammation.[7]

To build off of these exciting applications of ceria nanoparticles, the present study investigated for the first time the influence of ceria nanoparticles on bacteria growth. Specifically, dextran or polyacrylic acid coated ceria nanoparticles were studied in the bacterial experiments involving *Pseudomonas aeruginosa.* Results showed that the growth of *P. aeruginosa* was significantly inhibited in the presence of dextran or polyacrylic acid coated ceria nanoparticles. Ceria nanoparticles with either coating inhibited bacteria growth by about 20% compared with blank controls after treatment for 1 hour. After 6 hours of treatment, the inhibition of bacterial growth reached 99.94% and 98% for dextran coated ceria nanoparticles and polyacrylic acid coated ceria nanoparticles, respectively, compared with blank controls. After 24 hours of treatment, there were at least 98.69% bacteria inhibition for polyacrylic acid coated ceria nanoparticles and almost complete (~100%) inhibition for dextran coated ceria nanoparticles. This study revealed a promising biomedical application of ceria nanoparticles to prevent bacterial infections that is worthwhile to be further explored.

Reference:

[1] Trovarelli, Catalytic properties of ceria and CeO_2-containing maerials, *Catal. Rev. Sci. Eng.*, 1996, 38, 439-520.

[2] J. Kaspar, P. Fornasiero and M. Graziani, Use of CeO_2-based oxides in the three-way catalysis, *Catal. Today*, 1999, 2, 285-298.

[3] P. Jasinski, T. Suzuki and H. U. Anderson, Nanocrystalline undoped ceria oxygen sensor, *Sens. Actuators, B*, 2003, 95, 73-77.

[4] N. Ozer, Optical properties and electrochromic characterization of sol-gel deposited ceria films, *Sol. Energy Mater. Sol. Cells*, 2011, 68, 391-400.

[5] V. K. Ivanov, A. B. Shcherbakov, and A. V. Usatenko, *Usp. Khimii*, 2009, 78, 924-928.

[6] R. W. Tarnuzzer, J. Colon, S. Patil and S. Seal, Vacancy engineered ceria nanostructures for protection from radiation-induced cellular damage, *Nano Lett.*, 2005, 5, 2573-2577.

[7] A. Asati, S. Santra, C. Kaittanis, S. Nath and J. M. Perez, Oxidase-like activity of polymer-coated cerium oxide nanoparticles, *Angew. Chem., Int. Ed.*, 2009, 48, 2308-2312.

(397ai) Synthesis of High Aspect-Ratio ZnO Nanowires By Both 2-Step Cbd and CVD Methods

Zhao, Y., University of Florida at Gainesville
Xu, C., University of Florida at Gainesville
Zou, Y., University of Florida
Ziegler, K. J., University of Florida

Synthesis of High Aspect-Ratio ZnO Nanowires by Both 2-step CBD and CVD Methods

Yang Zhao[1,*], Cheng Xu[1,2], Jie Liu[1], Yongjie Zou[2], and Kirk J. Ziegler[1,2]

[1] *Department of Chemical Engineering*

[2] *Department of Materials Science and Engineering*

*University of Florida, Gainesville, FL 32611**Presenting Author (Ph: 352-392-3122; Email: yangzhao@ufl.edu)

Abstract:

Dye-sensitized solar cells (DSSCs) have been extensively studied for the past few years because of their low-cost and high-efficiency. The surface area of the photoanode, which determines the amount of photons absorbed, is one of the most critical factors to improve the efficiency of DSSCs. Although conventional TiO_2 nanoparticle-based DSSCs have shown promising results, their performance has been limited to ~11% because of diffusion and back reaction problems. Due to its excellent band gap (3.37eV) and high excitation bonding energy (60 meV) properties, one-dimensional (1D) ZnO nanowires are capable of overcoming the limitation of nanoparticle-based DSSCs by improving electron injection and axial electron diffusion. Based on models proposed by our group, the efficiency of DSSCs based on TiO_2 core-shell ZnO nanowires (40 μm in lengths, 40 nm in diameter and ~10^{11} /cm^2 of density) can significantly surpass current efficiencies. To reach our objective, we focused on using two-step anodized aluminum oxide (AAO) as a template to control the dimension of nanowires. The synthesis of ZnO nanowires was carried out both in 2-step chemical bath deposition method and using thermal decomposition of zinc acetate dehydrate in the AAO pores by low temperature, catalyst-free vapor-solid (VS) growth. After transferring embedded ZnO nanowires to a PDMS substrate, AAO was etched and an electrostatic repulsion technique was applied to provide vertically-aligned ZnO nanowires. After that, SEM, TEM, and XRD were applied to characterize these nanowires. By using this technique, it provides a new approach to obtain vertically-aligned ZnO nanowires with both high density and aspect-ratio through a large-scale process.

Keywords: Dye sensitized solar cell (DSSC), ZnO Nanowires, Alumina Template, Thermal Decomposition.

(397a) Fundamental Studies on the Origin of Reduced Graphene Oxide Enhancements in Energy Storage Applications

Radich, J. G.,
Kamat, P. V., University of Notre Dame

Graphene oxide (GO) offers unique opportunities to generate conductive composites consisting of electrochemically-active nanomaterials useful in energy storage applications. The electronegative oxygen moieties present on the GO surface provide negatively-charged species to which positive cationic precursors common to energy storage materials such as metal oxides can adsorb. This mechanism for producing electrode composites yields high surface coverage of the 2-D GO sheets. After reduction of GO to reduced graphene oxide (RGO), which can be accommodated in a multitude of ways, conductivity is restored and the active materials are in direct contact with an electron shuttle for effective delivery of electrons to otherwise poorly-conducting metal oxide intercalation materials. At the same time the drying of RGO sheets during solid electrode preparation enhances the natural wrinkling of the graphene sheets and produces a 3-D porous structure through which Li^+ ions are able to more rapidly diffuse. Here we present a systematic study of the electrochemical enhancements RGO offers to electrode architectures in energy storage applications. Specifically, we build a composite electrode consisting of MnO_2 nanowires and RGO and test the electrodes in lithium half-cell construct using an in-house cell design allowing for post-characterization of the electrode. Coulometric techniques were utilized to study the kinetics of the electron transfer reactions, Li^+ ion diffusion and adsorption, and the electron storage capabilities of the RGO sheets. Cyclic voltammetry in conjunction with XRD also shows a self-healing mechanism for oxygen deficient sites on the MnO_2 nanowires when composited with RGO. Finally, Transmission and scanning electron microscopy were used to characterize the surface morphology of the MnO_2 nanowires and the RGO composite.

(422a) Engineering Protein Delivery Systems to Protect Pyrethroids (λ-Cyhalothrin) Against Photodegradation

Lee, H., University of Illinois at Urbana-Champaign
Thirumalai, P.,
Feng, H., University of Illinois at Urbana-Champaign

Originally presented on: 11/6/2013 8:30:00 - 8:55:00

Engineering Protein Delivery Systems to Protect Pyrethroids (λ-Cyhalothrin) against Photodegradation

Hyoungill Lee[a], Pranav P. Thirumalai[a], Hao Feng[a*]

[a]- Department of Food Science and Human Nutrition, University of Illinois at Urbana-Champaign, Urbana, IL, USA-61801

[*] Corresponding author and Associate Professor, Department of Food Science and Human Nutrition, University of Illinois at Urbana-Champaign, 382F-AESB, 1304 West Pennsylvania Avenue, Urbana, IL 61801, USA.

Phone: 01-217-244-2571; Fax: 01-217-333-9329; Email: haofeng@uiuc.edu

Abstract:

In recent years, Liu et al. have shown that pyrethroid pesticides have been widely used due to their long lasting stability and high insecticidal activity. Several studies have shown the adverse effects of chemical and/ or photodegradation of pesticides in food and water systems. A review by Legrini et al. shows Advanced Oxidation Processes (AOPs) as a potential in this area. In this work we have used soy protein stabilized nanoemulsions and nanocomplexes with pyrethroids, such as λ-Cyhalothrin, to protect the pesticides against UV exposure. Soluble Soy Protein Isolate (SPI) was prepared using pH shifting (pH 8, 9, 10, 11 and 12) and a combination of pH shifting and ultrasound (100% amplitude for 1, 2, 3, 4 and 5 minutes). The combination of pH shifting and ultrasound showed significantly greater recovery and surface hydrophobicity of soluble SPI, irrespective of the pH and ultrasound treatment time. The turbidity and the particle size (nm) of nanoemulsions increased from 0.6783 to 3.9445 and 85.1 nm to 241.9 nm respectively, with an increase in oil:protein (w:w) ratio from 0.25:1 to 2:1. Soy protein stabilized (pH 12 + sonication) nanocomplexes with λ-Cyhalothrin and nanoemulsions showed at least 40% more protection than λ-Cyhalothrin in distilled water, when exposed to UV radiation for 6 hours. Thus, both the nanocomplexes and nanoemulsions prepared from 90% soluble SPI with a combination of pH shifting and ultrasound showed greater protection of pesticide against photodegradation.

Keywords: Photodegradation, pyrethroids, ultrasound, nanotechnology, nanoemulsion, nanocomplex

(422b) Toxicity Effects Associated With Exposure of Lung Epithelial Cells to Polymer Nanocomposites and Nano-Platelets of Silicates

Agarwal, S., West Virginia University
Eldawud, R., West Virginia University
Dinu, C. Z., West Virginia University
Gupta, R. K., West Virginia University
Originally presented on: 11/6/2013 8:55:00 - 9:20:00

Montmorillonite, a layered silicate in the form of platelets having a thickness of about 1 nm, an aspect ratio exceeding 200 and a surface area of about 650 m^2/g, is commercially compounded into polymers such as nylons and polyolefins. Even at low filler loading levels (about 5 wt%), the resulting polymer nanocomposites (PNCs) have improved properties, including reduced flammability when compared to the matrix polymer. Such enhanced property is observed when the nanofiller is deagglomerated and uniformly distributed throughout the plastic, with the nanofillers contributing to the formation of char that acts as a heat and mass transfer barrier. However, when such polymer nanocomposites are burnt, inhalation exposure to the nanoparticles contained within the PNCs can occur. Here we started to investigate the toxicological effects of associated with exposure of human epithelial cells to silicate nanoplatelets to determine the any harmful effects. For this, rectangular strips of PNCs were first ignited in air using a methane flame, and a Combustion™ DMS500 Differential Mobility Spectrometer was used to measure the number concentration of particles emanating from the burning sample. Measurements indicated the presence of nanoparticles in the 10 to 100 nm range and a concentration that was as much as $2.7 \times 10^9/cm^3/s$; these numbers are comparable to those encountered in the exhaust from heavy-duty diesel engines. Next, human lung epithelial cells were exposed to various concentrations of nanoplatelets and the effects of the nanoplatelets on the cells morphology, structure and functions were analyzed. Our studies allowed the characterization of the cytotoxicity associated with the combustion products of nanoplatelets and promises to help develop a tool for the prediction of large-scale burning behavior of polymer nanocomposites.

(422c) The Age of Carbon Nanotubes and Biomolecular Convergence for Cancer Therapeutics

Dinu, C. Z., West Virginia University

Originally presented on: 11/6/2013 9:20:00 - 9:45:00

Carbon nanotube (CNT) is the new "it" in nanotechnology revolution. CNT's ease of production as well as their exceptional chemical, mechanical and physical properties, and the ability to be functionalized with molecules of interest based on their high stability and high aspect-ratios, have recently prompted CNTs use both for delivery of drugs as well as for effective killing of tumor cells. In particular, interest was given to improving the engineering technologies that allow CNT synthesis and functionalization for safe and highly efficient delivery of the therapeutic agents. Nevertheless, given the nascent state of nanotechnology, much remains to be learned about the properties, characteristics and effects of carbon nanomaterials when exposed to biological systems, and the complexity of nano-bio-interface-based reactions when considering developing carbon nanomaterial-based therapeutics. In particular, the assessment of CNTs exposure risks and complementary, the benefits of using carbon nanomaterials in the biomedical fields, requires information and vision. This presentation reviews the current status in toxicological and pharmachological profiles of CNTs, the current efforts in my group in understanding nanotube toxicity in relationship to nanotube structure, properties and aspect ratio and proposes a paradigm shift for controlling immunomodulatory activity of nanotubes for providing the next generation of targeted therapeutics that require convergence to minimize host toxicity, while maximizing cancer cell genotoxicity.

(422d) Monitoring Ligand Exchange On the Surface of Gold Nanoparticles Using Isothermal Titration Calorimetry

Kitchens, C. L., Clemson University
Hart, A. E., Clemson University
Mefford, O. T., Clemson University
Powell, B. A., Clemson University
D'Unger, D., Clemson University
Originally presented on: 11/6/2013 9:45:00 - 10:10:00

Ligand exchange is an important factor in the function and end fate of nanomaterials. A better understanding of binding strengths of various functional groups to the surface of gold nanoparticles, GNPs, is needed in order to better predict how they will behave in natural and biological environments. Fluorescent Resonance Energy Transfer (FRET) was used to develop a hierarchy of competitive binding strengths of 5 fluorescently tagged functional groups using fluorescence spectroscopy. The functional groups investigated were; COOH, NH_2, OH, SH, and SO_3.GNPs bound with fluorescently tagged ligands were titrated into a known concentration of non-fluorescently tagged unbound ligands and the fluorescence intensity was monitored until ligand exchanged reached equilibrium. The study resulted in the following hierarchy with the lowest observed binding strength listed first; OH $<$ NH_2 $<$ COOH $<$ SO3 $<$ SH. Kinetics of the FRET data was used to calculate competitive binding constants. Isothermal Titration Calorimetry, ITC, was performed with the same ligands used in the FRET experiments in order to confirm the binding energies obtained from the FRET results.

(422e) Regulatory Association of Cell Responses Induced By Metal and Metal Oxide Nanoparticles

Liu, R., University of California, Los Angeles
France, B., University of California, Los Angeles
George, S., University of California, Los Angeles
Zhang, H., University of California, Los Angeles
Xia, T., University of California, Los Angeles
Nel, A. E., University of California, Los Angeles
Bradley, K., University of California, Los Angeles
Rallo, R., Universitat Rovira Virgili
Cohen, Y., University of California, Los Angeles
Originally presented on: 11/6/2013 10:10:00 - 10:35:00

A data mining approach was developed and applied to discover relationships among fourteen different types of cell responses (including ten signaling pathway activities and four cytotoxicity effects) for two different cell lines in response to exposure to six different metal and metal oxide nanoparticles. Association rules were identified for a High-Throughput Screening (HTS) data developed for murine macrophage (RAW264.7) and transformed bronchial epithelial (BEAS-2B) cells. The dataset consisted of a total of 55,296 data readings for nanoparticle-regulated pathway activities over an exposure period up to 24 h and exposure concentration range of 0.39-200 μg/mL. HTS data were processed for outlier removal and standardized via inter/intra-plate normalization with subsequent hit-identification (for significantly regulated pathways), which provided a reliable multivariate (multi-pathways) toxicity profiles for the nanoparticles. The association rules identified from the multivariate toxicity profiles were then pruned to remove redundant rules. The final set of non-redundant association rules revealed pathways that are co-regulated such that "significant regulation" of one or more pathways implied the existence of other (associated) specific pathways that are also significantly regulated. In order to explore the validity of data-driven hypothesis generation, specific pathway-activity studies were carried out with ZnO nanoparticles confirming the association rule identified for the p53 and Mitosox pathways. This pathway relationship revealed that blocking of the transcriptional activity of p53 lowered the Mitosox signal (i.e. p53 leads to mitochondrial dysfunction). The present approach to data-driven hypothesis generation (i.e., association rules discovery) has important implications for streamlining multi-parameter HTS assays, support for improved understanding of toxicity mechanisms, and selection of endpoints for the development of nano-Structure Activity Relationships (nano-SAR).

(422f) Investigations of the Oral Uptake of Titanium Dioxide Nanoparticles Via the Buccal Mucosa

Teubl, B.,
Leitinger, G., Medical University of Graz
Frohlich, E.,
Schneider, M.,
Lehr, C. M.,
Zimmer, A., University of Graz
Roblegg, E., University of Graz
Originally presented on: 11/6/2013 10:35:00 - 11:00:00

The development of engineered nanomaterials (ENM) and their commercialization for application in (consumer) products, medical and diagnostic devices and/or pharmaceutical drug delivery vehicles presents an enormous challenge for the scientific, regulatory, industrial and public field. Titanium dioxide (TiO_2) nanoparticles, for example, are manufactured worldwide in large quantities and TiO_2 pigments (< 2.5 µm) account for 70% of the total production volume [1]. However, decrease of particle size (from the microscale to the nanoscale) has been identified as a key parameter for an increased toxicity behavior. Regarding the respiratory tract it has been clearly demonstrated that TiO_2, which is one of the priority materials (concerning risk assessment) of the Organisation for Economic Co-operation, causes respiratory tract cancer [2]. However, studies dealing with the oro-gastrointestinal route are still lacking and the basic understanding of how nanoparticles interact with oral biological tissues is still missing. Previously, we demonstrated that nanoparticles are small enough to overcome the barriers of the oral cavity, likely entering systemic circulation and hence, translocate in the human body [3,4]. Thus, in this study we aimed at investigating interactions of three distinct TiO_2 nanoparticles (NM 100, pigment, anatase; NM 101, 7 nm, anatase; NM 105, 22 nm, anatase/rutile) with the oral cavity, in particular the buccal mucosa (lining the cheek area). To correlate particles´ properties to their biological interactions, characterization was carried out with photon correlation spectroscopy (PCS) in physiological media and with electron microscopy. Furthermore, the surface hydrophobicity was quantified with the Rose Bengal adsorption method. Ex-vivo penetration/permeation studies were investigated with static diffusion cells using porcine buccal mucosa. The penetration depth of TiO_2 nanoparticles was visualized by multiphoton microscopy (MPM). Furthermore, intracellular localization was examined via transmission electron microscopy (TEM) and the detected particles were verified with energy filtered TEM (EFTEM) combined with electron energy loss spectroscopy (EELS). Additionally, potential cytotoxicity effects were evaluated in-vitro with human oral TR146 cells and production of reactive oxygen species (ROS) was studied.

The data revealed that the investigated TiO_2 nanoparticles were hydrophilic and exhibited a high aggregation tendency in aqueous and physiological media. The particles penetrated/permeated the buccal epithelium in a size-dependent manner: NM 100 (two primary size distributions: 148 nm and 34 nm) and NM 105 (primary particle size: 22 nm) permeated the upper parts and penetrated into the deepest parts of the mucosa (i.e., basal lamina and connective tissue). However, NM 101 (7 nm particles) only penetrated the upper epithelium. Regarding intracellular localization, most of the particles were found freely distributed in the cytoplasm and not in endocytotic vesicles. These findings suggest that further entering mechanisms, apart from endocytosis, exist in the buccal mucosa. The cell viability/integrity was not compromised and no acute toxicity was observed after 4 and 24 h exposure time. However, NM 105 caused a significant increased metabolic activity ($P<0.001$) after 4 h, indicating a stress situation of cells. This was due to the production of reactive oxygen species (ROS) after short-time incubation.

It can be concluded that all investigated TiO_2 nanoparticles are able to rapidly penetrate the buccal mucosa. However, the penetration depth is mainly dependent on the particle size. Larger particles (i.e., NM 100 and NM 105) penetrate the upper and the lower parts of the buccal mucosa, suggesting particles′ distribution in the systemic blood circulation and furthermore, translocation in different organs. Smaller particles (i.e., NM 101) fail to penetrate lower parts and are only detected in the superficial layer of the oral epithelium. However, additional studies are necessary to evaluate possible interactions of freely distributed particles (in the cytoplasm) with intracellular proteins, organelles and cell nuclei. Additionally, ROS production of NM 105 may result in inflammation, mitochondria-mediated cell death and/or DNA damage. These data strongly emphasize that the oral cavity should be considered in further risk assessment, since it acts as the first line in defense, not only for oral but also for inhaled nanoparticles.

[1] Baan R., Straif K., Grosse Y., Secretan B., Ghissassi F., Cogliano V., *Carcinogenicity of Carbon Black, Titanium Dioxide and Talc*, Lancet Oncol 2006; 7: 295-296

[2] Trouiller B., Reliene R., Westbrook A., Solaimani P., Schiestl R., *Titanium Dioxide Nanoparticles Induce DNA Damage and Genetic Instability in vivo in Mice*, Cancer Res 2009; 69: (22) 8784-878

[3] Roblegg E., Fröhlich E., Samberger C., Zaversky M., Teubl B., Zimmer A., *Evaluation of a Physiological In-Vitro System to Study the Transport of Nanoparticles through the Buccal Mucosa*, Nanotoxicology 2012; 6: 399-413.

[4] Teubl B., Meindl C., Eitzelmayr A., Zimmer A., Fröhlich E., Roblegg E., *In-vitro Permeability Studies of Neutral Polystyrene Particles Through the Buccal Mucosa*, Small, 2013; 9(3): 457-66. doi: 10.1002/smll.201201789

(435a) Models for Scale Up of UV Roll-to-Roll Imprint Lithography

Jain, A., The University of Texas at Austin
Bonnecaze, R. T., The University of Texas at Austin
Originally presented on: 11/6/2013 8:30:00 - 8:55:00

Roll-to-roll nanoimprint lithography (R2RNIL) is a process based on batch nanoimprint lithography to replicate nano- and micro-scale patterns from the surface of a patterned mold wrapped on a roller to a large area, flexible or rigid substrate. One of the most appealing features of the R2RNIL is the ability to manufacture continuous rolls of patterned substrate. The throughput of the process and the quality of pattern replication is governed by the UV-exposure time, the peeling mechanism, and the forces on the substrate and the roller. However, there is a lack of understanding of the quantitative effects of the associated process parameters and material properties on the dynamics of the imprinting process.

Here we model the process and identify key parameters for scale up from an analysis of the fluid, curing and peeling dynamics. This work focuses on the roll-to-roll nanoimprint lithography process with UV curing and inkjet dispensing on a rigid substrate. The process includes merging of droplets of imprint material, curing of the imprint material from a viscous liquid to elastic solid resist, and pattern replication and detachment of the resist from template. The time and distances on the web or rigid substrate over which these processes occur are determined as function of the physical properties of the uncured liquid, the cured solid and the roller configuration. The upper convected Maxwell equation is used to model the curing viscoelastic liquid and to calculate the force on the substrate and the torque on the roller. The available exposure time is found to be the rate limiting parameter and is a function of the radius of the roller, the minimum gap between the roller and web, and the velocity of the web. It is found that the residual layer thickness of the resist must be larger than the gap between the roller and the substrate to ensure complete feature filling and optimal pattern replication. For lower residual layer thickness, the droplets may not merge to form a continuous film for pattern transfer. The results of this work help determine the optimal operating conditions for this roll-to-roll nanoimprint lithography process.

(435b) Templated Evaporative Lithography for High Throughput Fabrication of Nanopatterned Films

Arshad, T. A., The University of Texas at Austin
Bonnecaze, R. T., The University of Texas at Austin
Originally presented on: 11/6/2013 8:55:00 - 9:20:00

A new method for the fabrication of well-defined nanostructured deposits by evaporation-driven directed self-assembly of nanoparticles is proposed and studied theoretically. The technique comprises a film of suspended nanoparticles drying with its surface in contact with a topographically patterned membrane which promotes spatially varying evaporation, resulting in a patterned deposit. Membrane thickness and topography (in conjunction with the initial film height and concentration) allow the feature and residual layer dimensions to be controlled independently. Numerical solutions of equations governing the dynamics of the process show how the concentration profile evolves as a result of flow driven by heterogeneous evaporation. Analysis yields bounds on the dimensions of the dried deposit, and provides processing parameters to achieve specific patterns. It is estimated that films with 10 nm to 100 μm features can be fabricated with a drying time of 0.1–10 seconds per 10 μm of feature height above the residual layer (depending on membrane thickness), making this a promising method for high throughput pattern deposition.

(435c) Liquid Coating Processes for Nanoparticle Films With Defined Microstructure

Kraus, T., ETH Zurich / IBM Research GmbH

Originally presented on: 11/6/2013 9:20:00 - 9:45:00

Particles with diameters below a micrometer are versatile building blocks for thin films applicable in electronics, photovoltaics, sensors, packaging, batteries, and many other fields. The arrangement of particles in the films affects its properties: percolating networks of metallic particles facilitate electric conductivity, alignment of nanorods yields polarizing films, for example. Conventional wet coating methods provide little control over the microstructure.

In this contribution, I will discuss how doctor blading and spray-coating of nanoparticles alone and in combination with polymers can be used to obtain layers with defined microstructures. Tuning liquid flows and colloidal interactions provides structural control without templates [1]. Doctor blading can be combined with template surfaces to create fully defined particle arrangements [2]. Emulsions direct the structure of spray-coated layers and enable aqueous processing on unpolar particles [3]. I will discuss which deposition rates can be achieved using different methods, the requirements on process control, and the degree of structural control achievable.

In-situ observation of the deposition provides direct feedback and simplifies the screening of the very large parameter spaces. I will present data from setups for wet deposition with microscopic online analysis of the process. Models are used to fit this data and make predictions on suitable ranges of process parameters to yield desirable film structures.

In summary, this contribution will present new processing routes for functional thin films based on nanoparticles. Different strategies that yield films with controllable microstructures using standard industrial coaters will be introduced and discussed.

[1] T. Geyer, P. Born and T. Kraus: Switching between crystallization and amorphous agglomeration of alkyl thiol-coated gold nanoparticles. *Physical Review Letters* 109 (2012) 128302

[2] T. Kraus, L. Malaquin, H. Schmid, W. Riess, N. D. Spencer and H. Wolf: Nanoparticle printing with single particle resolution. *Nature Nanotechnology* 2 (2007) 570–576

[3] J. Lacava, P. Born and T. Kraus: Nanoparticle clusters with Lennard-Jones geometries. *Nano Letters* 12 (2012) 3279–3282

(435d) Continuous Production of Magnetic and Metal-Organic Nanoparticles With Netmix Reactor

Leblebici, M. E., LSRE - Laboratory of Separation and Reaction Engineering - Faculty of Engineering - University of Porto

Fonte, C. M., Faculdade de Engenharia da Universidade do Porto

Costa, M. F., LSRE - Laboratory of Separation and Reaction Engineering - Faculty of Engineering - University of Porto

Ataíde, F.,

Garcia, M. P.,

Silva, V. T., LSRE - Laboratory of Separation and Reaction Engineering - Faculty of Engineering - University of Porto

Devic, T., Institut Lavoisier

Horcajada, P., Institut Lavoisier

Tavares, P.,

Araújo, J. P., IN-IFIMUP, Universidade do Porto

Oliveira, R., REQUIMTE/CQFB, Universidade Nova de Lisboa

Dias, M. M., LSRE - Laboratory of Separation and Reaction Engineering - Faculty of Engineering - University of Porto

Lopes, J. C. B., LSRE - Laboratory of Separation and Reaction Engineering - Faculty of Engineering - University of Porto

Faria, J. F., LCM – Laboratory of Catalysis and Materials - Faculty of Engineering - University of Porto

Originally presented on: 11/6/2013 9:45:00 - 10:10:00

Continuous Production of Magnetic and Metal-Organic Nanoparticles with NETmix Reactor

M. Enis Leblebici[a], Carlos M. Fonte[a], Marcelo F. Costa[a], Filipe Ataíde[b], M. Paz Garcia[c], Viviana Silva[a], Thomas Devic[d], Patricia Horcajada[d], Pedro Tavares[e], João Pedro Araújo[c], Rui Oliveira[b], Joaquim L. Faria[a], Madalena M. Dias[a], José Carlos B. Lopes*[a]

[a]Laboratory of Separation and Reaction Engineering, Faculty of Engineering, University of Porto, Porto, Portugal

[b]REQUIMTE/CQFB, Departamento de Química, Faculdade de Ciências e Tecnologia, Universidade Nova de Lisboa, Caparica, Portugal

[c]IFIMUP/Departamento de Física, Universidade do Porto, Porto, Portugal

[d]Institut Lavoisier (CNRS 8180) Institut universitaire de France, Université de Versailles, Versailles, France

[e]CQVR, Departamento de Química, Universidade de Trás-os-Montes e Alto Douro, Vila Real, Portugal

*e-mail: lopes@fe.up.pt

More than half a century passed over the legendary lecture of Professor Richard Feynman, "There's Plenty of Room at the Bottom" which is usually referred as the muse of nanotechnology. Today, nano-scale products find there way into a variety of promising services from more humane cancer treatments to increasing computer power/capacity, while their production scales are getting from laboratory towards industrial [1].

Chemical coprecipitation in batch stirred vessels has been a suitable, low-energy method to achieve nanometer scale crystals. However the focus now is not only to produce nanoparticles, but to control the particle size with a narrow particle size distribution (PSD). [2, 3]. Since coprecipitation in general results in wide PSD, high energy methods such as thermal decomposition, hydrothermal, and microwave synthesis were proposed and have been shown effective for size control of nanoparticles [2]. However, the productivity and scalability of these methods are lower than coprecipitation. Coprecipitation is done by vigorously mixing two or more reactant solutions, resulting in a product, which has low solubility in the medium. This supersaturated medium results in immediate nucleation [4] and particle growth. To produce uniform particles, it is crucial to keep a homogeneous concentration of the precursor molecules/clusters throughout the medium from the moment the reactant solutions contact until the end of the particle growth stage. Maintaining the homogeneity is the reason for the vigorous mixing in stirred vessels where batch coprecipitation takes place and it is the key factor on product quality, namely the PSD and on the energy input of the process.

This work revisits coprecipitation method focusing on continuous production of magnetic nanoparticles (MNP) and metal – organic frameworks (MOFs) with high productivity without compromising PSD. The aim is to produce MNP and MOFs nanoparticles on a continuous reactor, namely the NETmix Reactor. The NETmix Reactor is a new technology consisting of a network of mixing chambers interconnected by transport channels [5]. Networks are generated by the repetition of unit cells where each unit cell consists of one chamber and two inlet and two outlet channels oriented at a 45° angle from the main flow direction. Above a critical channel Reynolds number, the system evolves to a self-sustained oscillatory laminar flow regime inside the mixing chambers inducing local strong laminar mixing. This network system of identical units, results in a flexible static mixer, which can be scaled-up easily without changing the design of mixing units.

Magnetite synthesis with NETmix reactor was performed by mixing an aqueous solution of Fe^{2+} and Fe^{3+} salts with a base solution continuously to result in a pH value of 9.5 under inert atmosphere and 45°C. Trisodium citrate was used to stabilize particles. At this temperature and pressure, the solubility of magnetite is of order of 10^{-8} mol/kg the solubility of water [6]. This degree of supersaturation favors nucleation rather than particle growth therefore resulting in small crystals. Maturation time of 10 minutes was proven sufficient to ensure total conversion of iron salts to magnetite crystals. The washed crystals were characterized using dynamic light scattering, X-Ray diffraction, high-resolution transmission electron microscopy, Raman spectroscopy and supercooled quantum interference device. The pure magnetite crystals were found to be 4.8±0.25 nm, monocrystalline and superparamagnetic. The space-time yield of the process was 4.3 tons/$m^{3/}$day.

Two different iron-based metal-organic frameworks were also synthesized using the NETmix reactor namely iron fumarate MIL-88A [7] and iron trimesate MIL-100 [8]. An equilibrium model for iron (III) and the polycarboxylic acids was developed, in order to avoid formation of amourphous phases thus eliminating kinetic limitations. In this way, the nucleation of the MOFs is fast, which is further favored by the immediate homogenization provided by NETmix reactor which resulted in smaller and more uniform particles. The materials were characterized using dynamic light scattering, X-Ray diffraction, high-resolution transmission electron microscopy, Fourier transform infra-red spectroscopy, thermogravimetry and nitrogen adsorption (BET specific surface area). The product was shown to be of nanoscale with a narrow PSD. The space-time yield increased up to 2 tons/$m^{3/}$day since the amount of reaction time was drastically reduced.

In conclusion, continuous production of MNP and MOFs were performed with NETmix reactor. MNP with precise size control and high productivity was reached. Production of two iron-based crystalline nanoMOFs, using different organic ligands was successfully achieved with high productivity.

1. Wei, L., M. Hervé, and P. Edouard, *Use of different rapid mixing devices for controlling the properties of magnetite nanoparticles produced by precipitation.* Journal of Crystal Growth, 2012. **342**(1): p. 21-27.

2. Lu, A.-H., E.L. Salabas, and F. Schüth, *Magnetic Nanoparticles: Synthesis, Protection, Functionalization, and Application.* Angewandte Chemie International Edition, 2007. **46**(8): p. 1222-1244.

3. Park, J.A., Kwangjin Hwang, Yosun Park, Je-Geun Noh, Han-Jin AU - Kim, Jae-Young AU - Park, Jae-Hoon AU - Hwang, Nong-Moon Hyeon, Taeghwan, *Ultra-large-scale syntheses of monodisperse nanocrystals.* Nat Mater, 2004. **3**(12): p. 891-895.

4. Gribanov, N.M., et al., *Physico-chemical regularities of obtaining highly dispersed magnetite by the method of chemical condensation.* Journal of Magnetism and Magnetic Materials, 1990. **85**(1–3): p. 7-10.

5. Lopes, J.C.B., et al., *Network mixer and related mixing process. PCT/IB2005/000647, February 2005. European Patent EP172643 B1, October 2008.* 2005.

6. Zarembo, V.I., et al., *Solubility of magnetite in hot nuclear power station water under reducing conditions.* Soviet Atomic Energy, 1988. **64**(3): p. 283-286.

7. Ferey, G., C. Serre, C. Mellot-Draznieks, F. Millange, S. Surble, J. Dutour, and I. Margiolaki, *A hybrid solid with giant pores prepared by a combination of targeted chemistry, simulation, and powder diffraction.* Angew Chem Int Ed Engl. **43**, 6296-6301,2004.

8. Horcajada, P., S. Surble, C. Serre, D.Y. Hong, Y.K. Seo, J.S. Chang, J.M. Greneche, I. Margiolaki, and G. Ferey, *Synthesis and catalytic properties of MIL-100(Fe), an iron(III) carboxylate with large pores.* Chem Commun (Camb). 2820-2822,2007.

(435e) Chiral Plasmonic Nanoparticles From An Achiral Template

McPeak, K. M., ETH Zurich
Park, J. H., ETH Zurich
Blome, M.,
Burger, S.,
Govorov, A. O., Ohio University
Norris, D. J., Swiss Federal Institute of Technology (ETH ZÃ¼rich)
Originally presented on: 11/6/2013 10:10:00 - 10:35:00

Chiral plasmonic nanoparticles have been predicted to exhibit both strong circular dichroism in the visible spectral range and non-linear optical effects. These properties open up applications for chiral nanoparticles in sensing, enantiomer separations, and non-linear optics. Unfortunately, to date the fabrication of colloidal nanoparticles with chiral shape has posed a significant challenge due to a very limited ability to transfer chirality from chiral template molecules to nanoparticles. Here we report a novel route to enantiopure chiral gold nanoparticles. Importantly, we do not use in principle any chiral molecules or biomolecules and therefore create chiral shapes artificially. Template stripping is applied to achiral silicon wafers to prepare nanoparticles with chiral shape of only one handedness and with sizes between 100 nm and 1 micron. Both experimental and simulated circular dichroism spectra from suspensions of the template-stripped chiral nanoparticles will be discussed.

(435f) Catalytic Honeycomb Manufacturing By Three Dimensional Assembly of Hierarchical Co3O4 Nanoarray for High Performance NO Oxidation

Ren, Z., University of Connecticut
Guo, Y., University of Connecticut
Gao, P., University of Connecticut
Originally presented on: 11/6/2013 10:35:00 - 11:00:00

A novel noble metal free catalytic honeycomb has been successfully developed by incorporating hierarchically three dimensional Co_3O_4 nanoarrays into monolithic honeycomb with a facile one pot hydrothermal strategy followed by ambient annealing[1-3]. The morphology evolution with various precursors has led to Co_3O_4 nanostructures of different sizes and shapes. The Co_3O_4 nanoarrays-based honeycomb exhibits a remarkable efficiency as high as 80% for catalytic NO oxidation at high space velocity (~50,000 h^{-1}) but relatively low temperature (~275 °C). The nanoscale manufacturing of Co_3O_4 nanowire arrays has successfully improved the catalytic capability compared with the wash-coated Co_3O_4 powder. The three dimensional ordered geometric arrangement of nanoscale building blocks is thus believed to facilitate the gas solid interaction and efficient diffusion, which are important for catalytic reactions. Detailed investigation on grain size, surface area, pore size as well as surface chemistry comes to the conclusion that Co^{3+} on surface determines the catalytic activity. The nanocatalysts maintain high efficiency during cyclic tests and show little degradation in catalytic activity under both isothermal and temperature fluctuation condition. This work opens up new opportunity for fabricating catalytic device with well controlled nanostructure geometry and large scale spatial arrangement. In addition, the as-prepared ultraefficient but cost effective nanocatalytic honeycomb will serve as a good candidate in environmental applications. Specifically, the high performance NO conversion into NO_2 will potentially increase the efficiency of current nitrogen oxide removal technology.

Reference

[1] Y. Guo†, **Z. Ren**†, (†equally contributed) W. Xiao, C. Liu, H. Sharma, H. Gao, A. Mhadeshwar and P. X. Gao, *Nano Energy*, 2013, in press.

[2] **Z. Ren**, Y. Guo, Z. Zhang, C. Liu and P. X. Gao, submitted, 2013.

[3] P. X. Gao, Y. Guo, **Z. Ren** and Z. Zhang, US non-provisional patent filed, (2012)

(436a) Hydrogen Production From Water

McDaniel, M. D.,
Ekerdt, J. G., University of Texas at Austin
Originally presented on: 11/6/2013 8:30:00 - 9:00:00

Silicon-based photocathodes show promising potential for photoelectrochemical (PEC) water splitting for renewable production of hydrogen. Trade-offs between stability and efficiency have limited the success of Si-based photoelectrodes. To overcome this issue, we utilize a thin layer (≈ 2 nm) of strontium titanate, $SrTiO_3$ (STO), which is grown epitaxially on p-doped Si (001) by molecular beam epitaxy. The STO layer greatly enhances the stability of the Si-based photocathode. In addition, the high quality STO layer and clean, low defect interface between STO(001) and Si(001) provides minimal trap states for charge carriers. Moreover, there is no conduction band offset between STO and Si. This allows the photo-generated electrons to easily move from the Si substrate to the STO surface, where hydrogen ions are reduced to form gaseous hydrogen. Based on our electrochemical measurements, we find that the STO-Si photocathode is both stable and exhibits high conversion efficiency. Studies with continuous Ti/Pt films for H_2 evolution yield a current density above 15 mA/cm^2; patterned metal layers are expected to give higher current densities. The results indicate that Si-based photocathodes for PEC water splitting can be improved using a thin STO layer, providing a viable method for clean hydrogen production.

(436b) Core-Shell Photoelectrochemical Electrodes for Water Splitting

Peng, Q., Duke University
Kalanyan, B., North Carolina State University
Hoertz, P.,
Miller, A.,
Kim, D. H., North Carolina State University
Hanson, K., UNC-Chapel Hill
Alibabaei, L., UNC-Chapel Hill
Liu, J., Duke University
Meyer, T., UNC-Chapel Hill
Parsons, G. N., North Carolina State University
Glass, J. T., Duke University
Originally presented on: 11/6/2013 9:00:00 - 9:18:00

Photoelectrochemical (PEC) processes offer one of most promising ways to convert solar energy into fuels (e.g. H_2, and CH_3OH), especially H_2. In order to synthesize high efficient and stable PEC water splitting electrodes, one must optimize the light absorption, charge separation, and charge transfer processes of the electrode simultaneously. Core-shell nanostructures with transparent conductive cores and semiconductor shells show great promise for optimizing PEC performance of photoactive materials, such as TiO_2. However, limited research has been performed in this field.

We will present the result of using transparent conductive network consisting of low cost, solution-processed antimony doped tin oxide nanoparticles (nanoATO) and tin doped indium oxide nanoparticles (nanoITO) for enabling core-shell TiO_2 PEC electrodes. In detail, we will present the synthesis of TiO_2 PEC electrodes by coating solution-processed nanoATO on FTO glass with TiO_2 through atomic layer deposition (ALD). The conductive, porous nanoATO film-supported TiO_2 electrodes, yielded a highest photocurrent density of 0.58 mA/cm^2 under AM 1.5G simulated sunlight of 100 mW/cm^2. This is approximately 3× the maximum photocurrent density of planar TiO_2 PEC electrodes on FTO glass. Transient photocurrent measurements showed that nanoATO films reduce charge recombination by accelerating transport of photoelectrons through the less defined conductive porous nanoATO network. Further, we will show that the effect of interfacial reaction during the TiO_2 ($TiCl_4/H_2O$) ALD on transparent nanoATO and nanoITO on their PEC performance. The scalable solution-processed porous transparent nanoATO and nanoITO films are promising as a framework to replace other conductive scaffolds (Si nanowire forest, CNTs, and $TiSi_2$ nanonets) for PEC electrodes. The results are not only valuable for PEC, but also informative for the related DSSC research.

(436c) Design of Photocatalysts Active in Infrared and Visible Range

Lee, D. C., Korea Advanced Institute of Science and Technology (KAIST)

Originally presented on: 11/6/2013 9:18:00 - 9:36:00

Infrared- and visible-active semiconductors are promising candidates for light-harvesting materials. When the semiconductor nanocrystals are within quantum confinement regime, they exhibit size tunable bandgap, enabling light-harvesting even more controllable. Furthermore, the dynamics of excited states can also be controlled through a design of heterostructure semiconductor nanocrystals. We synthesized PbSe/CdSe/CdS core/shell/shell nanocrystals with varying structural design, e.g., spheres, pyramid and tetrapod. These nanocrystals show varying electron-hole exchange interactions and radiative recombination lifetimes depending on their dimensions. We also prepared semiconductor/metal nanojunction and used the nanomaterials as photocatalysts for reduction of various dyes by growing metal tips at the corners of semiconductor nanocrystals. Interestingly, the morphology change leads to the difference in the tip growth pattern. In addition, the growth of metal particles affects the Fermi level of the entire hybrid composites, leading to the change in their photocatalytic activity. In the presentation, the implications of the heterostructure quantum dots and their hybrid composites with metal nanoparticles will be discussed in the context of energy applications.

(436d) Efficient Photoelectrochemical Water Splitting With Si-Based Metal-Insulator-Semiconductor Photoelectrodes

Esposito, D. V., University of Delaware
Lee, Y., National Instittue of Standards and Technology
Talin, A. A., Sandia National Laboratory
Moffat, T. P., National Institute of Standards and Technology
Originally presented on: 11/6/2013 9:36:00 - 9:54:00

Photoelectrochemical (PEC) water splitting is an attractive pathway for renewable production of hydrogen, but the efficiency and stability of semiconducting photoelectrodes must be improved. One promising approach to achieve both high efficiency and good electrochemical stability is the metal-insulator-semiconductor (MIS) photoelectrode architecture.[1,2] The MIS design consists of catalytic metal structures, or collectors, deposited on a 1-3 nm thick oxide-covered semiconductor. Of great importance to this design is the thin insulating oxide layer, which must protect the semiconductor from the potentially corrosive electrolyte while simultaneously mediating minority carrier tunneling between the semiconductor and collectors. In recently published work,[2] we have demonstrated significant improvement in the performance of Si-based MIS photocathodes through interfacial engineering. These planar photocathodes exhibit excellent stability and appreciable photoelectrode conversion efficiencies, up to 2.9%, but further improvement in efficiency must be achieved in order to make this process economically viable.

In this work, nano- and micro-structured Si-based MIS photoelectrodes fabricated by reactive ion etching and vapor-liquid-solid chemical vapor deposition are explored as a means to further increase the efficiency of Si-based photoelectrodes. The use of 3D-structured electrodes offer several potential advantages for achieving higher efficiency, including increased surface area for catalysis, improved collection of photo-generated carriers, and decreased optical reflection losses. However, the use of 3D photoelectrodes also introduces detrimental effects that create trade-offs between optical, catalytic, and electronic properties. In this talk, design principles for optimizing the performance of 3D-structured photoelectrodes are described and a comparison of the performance of planar and 3D-structured MIS photoelectrodes is presented. The incorporation of MIS photoelectrodes into stand-alone PEC water splitting devices will also be discussed. Calculations based on the method of detailed balance show that PEC devices utilizing Si-based MIS photoelectrodes are capable of achieving solar-to-hydrogen conversion efficiencies greater than the 2015 efficiency target of 15% set by the Department of Energy.[3]

References

[1.] H.J. Lewerenz, et al., *Electrochem. Acta*, **2011**, 56, 10726.

[2.] D.V. Esposito, I. Levin, T.P. Moffat, A.A. Talin, *Nature Materials*, **2013**, (in press).

[3.] Fuel Cells Technologies Office Multi-Year Research, Development and Demonstration Plan. US Department of Energy Efficiency and Renewable Energy. Available from: http://www1.eere.energy.gov/hydrogenandfuelcells/mypp/ (2012)

(436e) Graphene Oxide Photocatalysts for Water Splitting and Its Upconverted Photoluminescence

Yeh, T. F.,

Teng, H.,

Originally presented on: 11/6/2013 9:54:00 - 10:12:00

Graphene-Oxide Photocatalysts for Water Splitting and Its Upconverted Photoluminescence

Te-Fu Yeh[1] and Hsisheng Teng[1,2,*],

[1]*Department of Chemical Engineering and Research Center for Energy Technology and Strategy, National Cheng Kung University, Tainan 70101, Taiwan*

[2]*Center for Micro/Nano Science and Technology, National Cheng Kung University, Tainan 70101, Taiwan*

**E-mail: hteng@mail.ncku.edu.tw*

Abstract

Graphene oxide (GO) is a p-type semiconductor with a suitable band positions for photocatalytic watersplitting. GO sheets have a hydrophilic 2D carbon structure that allows for extensive chemical modification. This study presents strategies of tuning the conductivity type of GO from p-type to n-type for simultaneous evolutions of H_2 and O_2 under visible light irradiation. Further converting the n-type graphene oxide sheets to quantum dots results in stoichiometric evolution of H_2 and O_2 at 2-to-1 ratio. The quantum dots also serve as an energy-converting medium with upconverted photoluminescence to promote the activity of metal-oxide photocatalyts.

Introduction

Polymeric semiconducting molecules, which have a large contact area with water, attract significant attention as alternatives to metal-containing photocatalysts. Graphene oxide (GO) derived from graphite oxidation is a molecule-like semiconductor [1]. GO can be extensively dispersed in water to molecular scale and has a large exposed area. Graphene is a zero gap semiconductor whereas the GO gap increases with the oxidation level [2,3]. The bonding with electron-withdrawing oxygen functionalities results in the p-type conductivity of GO. N-type conductivity appears when graphene covalently bonds to electron-donating nitrogen functionalities [4-6].

In addition to chemical doping, we exploit the quantum-confinement effect to modulate electronic properties. Unlike two-dimensional graphene, the zero-dimensional graphene quantum dots (GQDs) have a finite band gap depending on their size [7]. The confined p-electron in sp^2 domain gives the quantized discrete levels [8]. With these features, GQDs exhibit upconvered photoluminescence (PL) [9].

This study presents the potential of GO as a photocatalyst for water splitting. P-type GO can catalyze H_2 evolution with the presence of electron donors. Ammonia-treated GO (NGO) exhibits n-type conductivity and promising photocatalytic activity for generating H_2 and O_2 simultaneously, but with the ratio below the stoichiometric value (that is, 2). Miniaturized N-doped graphene oxide quantum dots (NGO QDs) function as photocatalysts for stoichiometric H_2 and O_2 generation from water cleavage under visible light irradiation.

We also demonstrate the upconversion ability of NGO QDs.

Experimental

GO was prepared using a natural graphite powder through a modified Hummers' method. NGO was obtained through nitridation of the as-prepared GO performed at room temperature using a flow of NH_3 gas. NGO QDs were prepared from oxidation of nitrogen-doped graphene. Photocatalytic reactions were conducted at approximately 25 ¢XC in a gas-enclosed inner irradiation system or a side irradiation system.

Result and Discussion

We demonstrated the photocatalytic activity of GO in the gas-enclosed circulation system with mercury-lamp irradiation. Continuous H_2 evolution over metal-free GO was observed for reactions conducted in pure water and an aqueous methanol solution [1]. The total evolution of H_2 from aqueous methanol solution far exceeds the amount of H_2 obtained from pure water. GO cannot catalyze O_2 generation even with the presence of the electron scavenger Ag^+ ion.

We modified the as-synthesized GO photocatalyst by ammonia treatment to form NGO, which exhibited n-type conductivity. The change of conductivity type was contributed by the electron-donating property of the introduced amino groups. NGO effectively catalyzes O_2 evolution under mercury-lamp irradiation [6]. The n-type conductivity of NGO attracts hole migration toward the interface region for O_2 evolution. NGO gave rise to simultaneous evolutions of H_2 and O_2 from the side irradiated system under visible light irradiation [6]. The molar ratio of the evolved H_2 and O_2 was smaller than the stoichiometric ratio 2-to-1.

Quantum confinement effect elevates the conduction band edge, promoting the electron injection for H_2 evolution from water. We found stoichiometric evolutions of H_2 and O_2 from pure water over NGO QDs with visible-light irradiation. Figure 1 shows the PL spectra of NGO QDs excited by long-wavelength light (660-900 nm), with the upconverted emission located in the range of 450-530 nm. The upconverted PL emission of NGO QDs excited $NaTaO_3$ to catalyze H_2 evolution from an aqueous methanol solution under visible light irradiation.

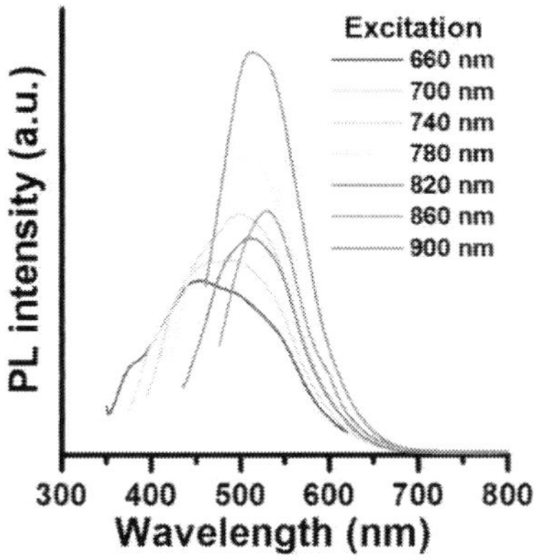

Figure 1. Upconverted PL from NGO QDs. The excitation wavelength ranges from 660 nm to 900 nm. The

upconverted emission located in the range of 450-530 nm.

Conclusions

The p-type GO catalyzed H_2 evolution from an aqueous methanol solution under mercury-lamp illumination. The introduction of nitrogen functionalities converted the p-type GO to an n-type NGO, which was capable of catalyzing simultaneous H_2 and O_2 evolutions from water under visible light. The NGO QD catalyst achieved overall water splitting ($H_2/O_2 = 2$) under visible light irradiation. The upconversion ability of NGO QDs made wide-gap photocatalysts active under visible-light illumination.

Acknowledgements

This research is supported by the National Science Council of Taiwan (101-2221-E-006-243-MY3 and 101-2221-E-006-225-MY3) and the "Aim for the Top-Tier University and Elite Research Center Development Plan" of National Cheng Kung University.

References

[1] T.F. Yeh, J.M. Syu, C. Cheng, T.H. Chang, H. Teng, *Adv. Funct. Mater.* **2010**, 20, 2255.

[2] T.F. Yeh, F.F. Chan, C.T. Hsieh, H. Teng, *J. Phys. Chem. C* **2011**, 115, 22587.

[3] T.F. Yeh, J. Cihlář, C.Y. Chang, C. Cheng, H. Teng, *Materials Today* **2013**, 16, 78.

[4] X. Wang, X. Li, L. Zhang, Y. Yoon, P. K. Weber, H. Wang, J. Guo, H. Dai, *Science* **2009**, 324, 768.

[5] C. C. Hu, H. Teng, *J. Phys. Chem. C* **2010**, 114, 20100.

[6] T.F. Yeh, S.J. Chen, C.S. Yeh, H. Teng, *J. Phys. Chem. C* **2013**, 117, 6516.

[7] L. A. Ponomarenko, F. Schedin, M. I. Katsnelson, R. Yang, E. W. Hill, K. S. Novoselov, A. K. Geim, *Science* **2008**, 320, 356.

[8] G. Eda, Y. Y. Lin, C. Mattevi, H. Yamaguchi, H. A. Chen, I. S. Chen, C. W. Chen, M. Chhowalla, *Adv. Mater.* **2010**, 22, 505.

[9] H. Li, X. He, Z. Kang, H. Huang, Y. Liu, J. Liu, S. Lian, C. H. A. Tsang, X. Yang, S. T. Lee, *Angew. Chem. Int. Ed.* **2010**, 49, 4430.

(436f) The Study of Cu2ZnSnS4 Nanocrystal/TiO2 Nanorod Arrays Heterojuction Photoelectrochemical Cell for Hydrogen Generation

Ho, T. Y.,
Chen, L. Y.,

Originally presented on: 11/6/2013 10:12:00 - 10:30:00

In this study, we reported TiO_2 nanorods arrays (TiO_2 NRAs) synthesized by hydrothermal method with controllable densities, diameters and lengths for photoelectrochemical hydrogen generation. The densities and diameters of the TiO_2 NRAs could be easily controlled by the concentration of titanium precursor (titanium isopropoxide, TTIP) and the amount of the HCl. With different reaction time, the lengths of TiO_2 NRAs could be increased from 2 μm to 9 μm. The photocurrent density and conversion efficient were increased with the lengths of TiO_2 NRAs. The maxima current density and conversion efficiency are $2.5mA/cm^2$ at 1.23V vs RNHE and 0.95% with 4-5 μm in length under $100mA/cm^2$ Xenon lamp illumination with AM1.5 filter, respectively. To improve the conversion efficiencies, various sensitizers, such as dye or quantum dot, are usually used to cover the absorption in the visible light range. Herein, Cu_2ZnSnS_4 (CZTS), which owns direct band gap semiconductor with 1.5eV and suitable band positions, was employed to extend the absorption range. In this study, CZTS nanoparticles were direct decorated on TiO_2 NRAs by solvothermal method with metal chloride and thiourea as precursor and ethanol as solvent to improve the conversion efficiency for hydrogen generation.

(436g) Development of New Generation of Sub-Nanometer Catalysts for Sustainable Energy Applications

Wu, Q., Stony Brook University/BNL
Shen, P., Stony Brook University / BNL
Zhao, S., Stony Brook University
Ramakrishnan, G.,
Su, D., Brookhaven National Laboratory
Li, Y.,
Orlov, A., Stony Brook University / BNL
Originally presented on: 11/6/2013 10:30:00 - 10:48:00

Developing novel heterostructured nanomaterials for energy applications as water splitting under visible light has enormous potential for energy production. Currently the existing materials have very low quantum efficiency, which indicate that new approaches are required in catalysts design. In this work we have developed and characterized sub nanometer Au and Pt nanoparticles, which greatly enhanced rate of hydrogen production of CdS. These heterostructured materials were characterized by TEM, STEM, XRD, UV-Vis spectroscopy and DRIFTS. We have also conducted comprehensive studies of these catalysts in photocatalytic reactors, where we observed a substantial hydrogen production under visible light.

(436h) ZrO2 Incorporated ZnO/TiO2 Mesoporous Photocatalyst for H2 Generation

Hussein, A., South Dakota Shool of Mines and Technology
Shende, R., South Dakota School of Mines and Technology
Originally presented on: 11/6/2013 10:48:00 - 11:06:00

The present research deals with the synthesis of ZrO_2 incorporated coupled ZnO/TiO_2 photocatalyst and its effectiveness on H_2 generation from aqueous methanol solution. Ordered mesoporous ZrO_2 supported coupled TiO_2/ZnO materials were synthesized using surfactant as structure directing agent, titanium isopropoxide, zinc acetate dihydrate and zirconium n-propoxide precursors. The ZrO_2 incorporated coupled ZnO/TiO_2composites were characterized by XRD, UV-Vis spectra, SEM/TEM, XRF, Raman and N_2 adsorption-desorption isotherm. Photocatalytic hydrogen evolution was carried out in a 7 mL quartz reactor at ambient temperature. The focused intensity and area on the quartz cuvette were 0.60 W and 2 cm^2, respectively. In a typical experiment, 2.0 mg of $ZrO_2/ZnO\text{-}TiO_2$ photocatalyst was dispersed in 3.5 mL of H_2O containing 0.5 mL methanol or 10 mg biomass as a sacrificial agent. A 300 W xenon lamp (Oriel) equipped with a UV-filter (240 nm - 400 nm) was employed as the light source. Prior to irradiation, the suspension of the catalyst was degassed with ultrapure nitrogen for 30 min to completely remove the dissolved oxygen and to ensure the reactor in an anaerobic condition. From 3 mL reactor overhead space, 0.1 mL of gaseous samples were withdrawn periodically and analyzed using the gas chromatograph (GC), from which photocatalytic effectiveness towards H_2 generation was evaluated and compared with coupled TiO_2/ZnO material.

(485a) Multimedia Environmental Distribution of Nanomaterials

Liu, H. H., University of California, Los Angeles
Cohen, Y., University of California, Los Angeles
Originally presented on: 11/6/2013 12:30:00 - 12:49:00

Nanotechnology has brought about many benefits and improvements to commercial and industrial applications and products, but its potential environmental impact must also be systematically assessed. In this regard, it is essential to assess the potential exposure levels (i.e. concentrations, exposure time, etc.) of ecological receptors to engineered nanomaterials (eNMs) in order to manage risks due to possible releases (chronic and/or accidental) of eNMs to the environment. Although a number of models have been developed to predict the multimedia environmental distribution of organics chemical contaminants, such models are not directly applicable to eNMs. For example, the concentration driving forces for transport of chemicals across environmental phase boundaries is typically constrained by equilibrium partitioning; in contrast, the transport of particulate matter is primarily driven by physical processes of driving forces that are unconstrained by equilibrium partitioning. In an effort to quantify the distribution of ENMs in the environment, "substance flow analysis" estimations have been employed in recent years as part of life cycle analysis (LCA) to assess the potential environmental impact of ENMs. Such approaches, however, are not a priori predictive and generally do not consider mechanistic descriptions of fundamental transport processes. Accordingly, in the present study, a generalized model of the multimedia environmental distribution of nanomaterials (MendNano) was developed to estimate the environmental distribution of nanomaterials subject to various release scenarios. In this model, the multimedia environment is considered as a collection of environmental compartments and subcompartments, with each compartment treated as either well-mixed (e.g. atmosphere, water), or non-uniform compartments (e.g. soil, sediment). Fundamental intermedia transport processes (e.g. dry deposition, rain scavenging, resuspension, runoff) link various compartments while considering the particle size distribution (PSD) of eNMs in both the atmospheric and aquatic compartments. In order to enable rapid "what-if?" scenario analyses, a web-based graphical user interface (GUI) was developed for the model as part of the MendNano cloud-based modeling system. The user defines the modeling scenario by specifying relevant properties of the eNM under consideration, geographical and meteorological parameters (i.e., size of regions, temperature, wind speed, rain rates, etc.), transport process parameters and source emissions (i.e., release rates, etc.). The time-evolution of eNM concentrations and mass in each environmental compartments are then calculated by solving the coupled series of differential equations and the results are displayed within the GUI. Simulations with MendNano for different geographical areas in the U.S. have shown that, when ENMs are released primarily to air the majority of their mass in the environment is accumulated in the soil and in certain cases in the soil compartment. Simulations confirmed that increased attachment efficiency of eNM to suspended solids resulted in significant increased concentration in the sediment compartment, and a modest concentration decreases for the water compartment. Additionally, for eNM with measurable solubility (e.g. Ag, ZnO), dissolution of the eNM had a significant impact in decreasing their accumulations in the water and sediment compartments. Lastly, risk indices were computed based on eNM concentrations and information from high throughput screening (HTS) of cellular toxicity studies, suggesting that ZnO, Fe_3O_4, and TiO_2 ranked high among 6 ENMs considered in the study. It is envisioned that the present multimedia analysis tool can assist regulators, industrial, and researchers in rapidly assessing the potential environmental implications of ENMs releases to the environment.

(485b) Origin, Cure and Control of Nanosilver Toxicity

Sotiriou, G. A., Harvard University
Fujiwara, K.,
Pratsinis, S. E., ETH Zurich
Originally presented on: 11/6/2013 12:49:00 - 13:08:00

Nanosilver is considered the most commonly used engineered nanomaterial, for example, in antibacterial textiles, polymer films for food packaging, paints and pigments, filters for water or air treatment to name a few applications (1). Concerns however about released Ag^+ ions during dispersion of nanosilver in liquids have limited its broad use. Here nanosilver supported on nanostructured silica is made with closely controlled Ag size both by dry (flame aerosol) and wet chemistry (impregnation) processes without, however, any surface functionalization that could interfere with its ion release (2). The dispersion method of nanosilver in solutions affects its dissolution rate but not the final Ag^+ ion concentration. By systematically comparing nanosilver size distributions to their equilibrium Ag^+ ion concentrations, it is revealed, that the latter correspond precisely to dissolution of one to two surface silver oxide monolayers, depending on particle diameter. When, however, nanosilver is selectively conditioned by either washing or H_2 reduction, the oxide layers are removed, drastically minimizing Ag^+ ion leaching and its antibacterial activity against *E. coli*. That way the bactericidal activity of nanosilver is confined to contact with its surface rather than to rampant ions. This leads to silver nanoparticles with antibacterial properties that are essential for medical tools and hospital applications. Complete cure of nano-Ag toxicity is achieved by its hermetic coating by nanothin inert silica films.

1. G.A. Sotiriou, S.E. Pratsinis, "Engineering nanosilver as an antibacterial, biosensor and bioimaging material", *Curr. Opinion Chem. Eng.* **1**, 3-10 (2011)

2. G.A. Sotiriou, A. Meyer, J.T.N. Knijnenburg S. Panke, S.E. Pratsinis, "Quantifying the origin of nanosilver ions and their antibacterial activity", *Langmuir*, **28**, 15929–15936 (2012).

(485c) Bacterial Colonization of Surfaces Displaying Adhered Silver Nanoparticles

Wirth, S. M., Carnegie Mellon University
Lowry, G. V., Carnegie Mellon University
Tilton, R. D., Carnegie Mellon University
Originally presented on: 11/6/2013 13:08:00 - 13:27:00

Silver nanoparticles (AgNPs) are among the most commonly employed engineered nanoparticles in consumer products, where the broad spectrum biocidal properties of silver are exploited to endow products with a measure of antimicrobial performance. Silver ions have well established microbial toxicity. In some cases AgNPs are embedded in a matrix; in others they are deposited on a surface. In either case, AgNPs provide a source of dissolved silver ions. The possibility exists as well for toxic consequences of direct microbe contact with the AgNPs, although this appears to be rather context-dependent. Often, materials that display AgNPs are envisioned as a means to inhibit biofouling or to enhance sterility. This can be particularly attractive in medical or food processing situations. AgNP leaching from consumer products under normal usage provides a pathway to environmental exposure, mainly via wastewater treatment plants. Concern over the possibly detrimental effects of engineered AgNPs in soil and aquatic ecosystems has motivated active research into the potential toxicity of these nanomaterials to microbes, small organisms and plants, as well as the chemical and physical transformation processes and transport processes that ultimately dictate the potential for broad environmental impact of the introduced nanoparticles.

While AgNPs can be highly toxic to planktonic bacterial cultures, recent work indicates that established bacterial biofilms, which represent the vast majority of naturally occurring populations, are considerably more tolerant to the introduction of AgNPs and the dissolved silver they release. Instead of AgNP effects on established biofilms, this presentation focuses on the ability of planktonic bacterial populations to colonize a surface that has been coated by adhesion of AgNPs from aqueous suspension before the introduction of bacteria to the system. Silver nanoparticles in this study were stabilized by poly(vinylpyrrolidone). Biofilm growth and viability were monitored on silica glass substrates, with or without adherent AgNPs, that were immersed in agitated suspensions of *Pseudomonas fluorescens*. The critical parameter that dictates whether or not the AgNP-decorated surface inhibits biofilm formation is the ratio of silver to planktonic biomass, or equivalently, the total cell population for a fixed amount of surface-adhered AgNPs. Biofilm formation is only prevented when the initial planktonic bacteria population is sufficiently low to suffer 100% lethality. Higher bacterial populations may initially suffer substantial loss of viability, but if this is less than 100%, the population recovers and establishes a thriving biofilm on the AgNP-decorated surface. An important aspect of this phenomenon is the binding of silver ions to suspended bacteria, which makes it unavailable to other bacteria. In this way, bacteria populations can overwhelm the release of toxic silver ions from AgNPs by sequestering silver ion in a non-bioavailable form. This enables the recovery from initial viability losses when the population grows at a faster rate than silver ion can be replenished in solution. Having sequestered most of the dissolved silver ions, the bacteria then readily colonize the AgNP-decorated surface. A particularly noteworthy effect is the preferential colonization of bacteria at early times on surface regions that have abundant AgNPs compared to AgNP-free regions. After several days, AgNP-rich and AgNP-free surface regions are indistinguishable in terms of biofilm biomass and viability. The AgNPs themselves provide no direct deterrence to biofilm formation. The system is dominated by dissolved silver toxicity, which can be readily overcome at high biomass:silver ratios. This work implies that even AgNPs that have managed to avoid any of the expected chemical transformations that are known to decrease silver toxicity, a "worst case scenario", may have little detrimental effect on bacterial biofilms in the environment.

(485d) Carbon Nanotubes Induced Cellular Biomeahanic Changes Is Depended On Treatment Time

Dong, C., West Virginia University
Eldawud, R., West Virginia University
Kashon, M. L., National Institute for Occupation Safety and Health
Lowry, D., National Institute for Occupation Safety and Health
Sargent, L. M., National Institute for Occupation Safety and Health
Dinu, C. Z., West Virginia University
Originally presented on: 11/6/2013 13:27:00 - 13:46:00

Carbon nanotubes (CNTs) are applied for a variety of applications from nanocircuits, to hydrogen storage devices, and from opticals fibers to conductive plastics. Recently, their functionalization with biomolecules offer opportunities for exciting biological and biomedical applications in drug delivery or bioimaging. However, because of their interactions with biological systems and their ability to translocate and persist into the circulatory and lymphatic systems and tissues, concerns about CNTs intrinsic toxicity have risen. It is thus necessary to develop significant and sensitive advances analysis technologies for real time investigation of CNTs toxicity. Our research aims at elucidating the mechanisms involved in CNTs internalization and association with cellular systems by evaluating the overall effects on of CNTs on cellular biophysical properties-biomechanics. To unravel the biomolecular mechanisms of such interactions, we used Fluorescence Activated Cell Sorting (FACS) and Atomic Force Microscopy (AFM). Our results suggest that cellular exposure to CNTs leads to higher stiffness and distinctive elastic modulus distribution for human lung epithelial cells, especially in cell nucleus region. Further, those changes are depended on CNTs uptake and cellular translocation time. Such findings provide necessary data for understanding the complex mechanotransduction processes within the cells interacted to nanomaterials which may be functionally linked to CNTs induced cytotoxicity and/or genotoxicity, as well as their potential for cancer progress.

(485e) A Simple Kinetic Model Describing Nanoparticle Interactions With a Tethered Lipid Bilayer

Liu, Y., Michigan State University
Worden, R. M., Michigan State University
Originally presented on: 11/6/2013 13:46:00 - 14:05:00

In this paper, we present a simple mechanistic kinetic model to describe interactions of engineered nanomaterials (ENM) with a tethered bilayer lipid membrane (tBLM). The model was developed to help elucidate molecular processes responsible for time-dependent changes in tBLM resistance (R_m) following ENM exposure. The model uses two kinetic constants to describe the rates of ENM binding to, and lipid removal from, the tBLM. The model was fit to R_m vs. time data for interaction of a tBLM formed using 1,2-dioleoyl-sn-glycero-3-phosphocholine (DOPC) with two types of ENM: functionalized polypropargyl glycolide nanoparticles and functionalized polystyrene nanoparticles. The model was able to reproduce diverse trends in the R_m vs. time data, including a continuous decrease in R_m and an initial increase in R_m followed by a decrease. The model also predicted that interaction of only a few nanoparticles could generate measurable changes in the tBLM's R_m value. Hierarchical clustering was applied to pairs of optimized kinetic constants for various ENM. The resulting dendrograms indicated that ENM having different properties (composition, size, surface charge) could be statistically distinguished using this approach. Results of this study provide insight into fundamental mechanisms by which ENM interact with biomembranes and may lead to improved methods to rapidly screen ENM libraries for desired functional and biosafety profiles.

Keywords: tBLM; engineered nanomaterials; model; kinetic constant; hierarchical clustering; lipid bilayer; resistance

(485f) Effect of Protein Corona On Nanoparticle Cellular Uptake

Nangia, S., Syracuse University
Jiang, W.,
Kamani, H.,
Originally presented on: 11/6/2013 14:05:00 - 14:24:00

Understanding the interface of nanomaterials and the environment is becoming increasingly important with the advances in biomedical applications of the nanoparticles. In the biological environment nanoparticles get exposed to serum proteins resulting in the formation of protein corona. Using chemically faithful coarse grained molecular dynamics simulations, the adsorption of specific serum proteins on a variety of nanoparticles have been studied to provide chemical insight into the into the dynamical equilibrium process. The role of physiochemical properties of nanoparticle such as size, shape, and surface charge on the corona formation has been investigated systematically. Furthermore, the dynamics of the nanoparticle-corona complex with the cell membrane is simulated to determine the mechanism of their cellular uptake. For comparison, we will provide estimates of the mechanistic and kinetic data for the cellular uptake of a various nanoparticle shapes both with and without protein corona. The impact of the corona on cell-adhesion and loss of cell specificity is evaluated.

(485g) Lessons From Nature: Analogies Between Protein and Nanoparticle Interactions With Lipid Membranes

Bothun, G., University of Rhode Island
Bobba, C., University of Rhode Island
Xi, A., University of Rhode Island
Kitchens, C. L., Clemson University
Originally presented on: 11/6/2013 14:24:00 - 14:43:00

Nanoparticle-cell membrane interactions play an important role in nanotoxicology and nanomedicine. Changes in membrane structure, phase behavior, and permeability due to nanoparticle binding are dependent upon nanoparticle size and surface functionality. Knowledge of nanoparticle-membrane interaction mechanisms can be used for the *a priori* design nanoparticles that are biocompatible or exhibit a desired membrane activity. Taking a lesson from nature, insight can be gained from our understanding of protein-membrane interactions. For example, theories describing lateral capillarity and changes in lipid phase behavior caused by hydrophobic proteins may be extended to hydrophobic nanoparticles embedded within lipid bilayers. Similarly, techniques used to evaluate interactions between hydrophilic (charged) proteins and membranes may be extended to hydrophilic nanoparticles. In this presentation we will describe the use of lipid bilayers and monolayers to determine nanoparticle interactions and draw analogies to proteins as a function of nanoparticle size and surface charge, and lipid composition and phase state. Our studies have been focused on engineered nanoparticles common to environmental and biomedical applications.

(485h) Reaction-Diffusion Model Describing Antioxidant Depletion in Polyethylene-Clay Nanocomposites

Ahmad, I.,
Hsuan, G.,
Li, C.,
Cairncross, R., Drexel University
Originally presented on: 11/6/2013 14:43:00 - 15:02:00

Polyethylene (PE) pipes are gradually replacing metal pipes in water and drainage transportations, but are inferior to metal pipes in mechanical properties. Many properties of PE are improved by adding clay to form nanocomposites. Long-term degradation of PE properties is primarily due to environmental oxidative degradation which is stabilized by antioxidants (AO). Although clay improves mechanical properties, in its presence AO fails to stabilize. Therefore a better understanding of underlying mechanisms of AO depletion in the nanocomposites can guide formulation of proper AO to ensure their long term durability. This paper presents research on modeling of AO (hindered-phenol) depletion in samples whose thicknesses are about 3 mm. The mathematical model considers reaction kinetic scheme for degradation and stabilization of neat PE and its nanocomposite. The model not only describes the AO depletion with time but also evolution of its profile throughout the depth of sample. Both chemical and physical losses of AO have been considered. Experimental 'oxidation induction time' (OIT) data showing variation in depletion with depth through the sample has guided development of the model. In nanocomposite samples there are two zones of AO depletion: a 'core zone' and 'depleted surface zone'. In the 'core zone' AO depletion is slower and uniform, and in the 'depleted surface zone', AO depletes more rapidly to produce a surface layer which is void of active AO.

In the model, alkyl free radical oxidizes rapidly compared to O_2 diffusion. Due to limited amount of alkyl free radicals present in the samples, all of them are oxidized in time less than 30 days. Once all alkyl free radicals are oxidized, O_2 diffuses into the sample and uniformly distributes throughout it. The role of AO is to stabilize all those oxidized free radicals, and thus stop their propagation. But continuous depletion of AO is brought about by generation of more free radicals. These additional free radicals are generated by decomposition of hydroperoxides due to catalytic effect of clay. The hydroperoxide decomposition keeps on supplying free radicals in a cyclic fashion resulting in continuous depletion of AO.

The surface depletion of AO can be brought about by evaporation of AO to the surrounding air combined with slower rate of diffusion in the nanocomposites. This can also be brought about by even lower diffusivity of AO in highly clay-orientation layer adjacent to sample surfaces.

In addition, the model predicts AO depletion profiles for a number of different experimental conditions: (1) thermal aging in stagnant air; (2) thermal aging in inert atmosphere; (3) initial AO concentration beyond saturation limit; and (4) thermal aging at a different temperature.

(498a) Sequential Infiltration Synthesis On Polymers for Nanofabrication

Peng, Q., Duke University
Tseng, Y.,
Darling, S. B., Argonne National Laboratory
Elam, J. W., Argonne National Laboratory
Originally presented on: 11/6/2013 12:30:00 - 12:50:00

Intensive research has focused on generating patterned nanomaterials by selective, localized synthesis in the microdomains of self-assembled block copolymers (BCPs) template. For sophisticated control of the final nanomaterials, it is critical to understand the behavior of the complex processes involved in the localized synthesis within BCP films. However, limited research has been done to understand these phenomena.

Sequential infiltration synthesis (SIS) method with BCPs template represents a hybrid soft and hard lithographic process and has been demonstrated as a promising solution to controllably synthesize patterned 2D and 3D nanofeatures of different materials with well-defined dimensions (down to sub-10 nm). Owing to its step-wise, substrate reactive site-limited heterogeneous reaction mode, SIS offers a great opportunity to investigate the complex diffusion and localized chemical reactions during the synthesis through the step-by-step mode.

In this work, we utilized sequential infiltration synthesis (SIS) of TiO_2 as the platform to investigate the behavior of reactant infiltration and localized reaction within polystyrene-*block*-poly(methylmethacrylate) (PS-*b*-PMMA). Similar studies were also performed onto two pure homopolymer (PS and PMMA) films. The *in-situ* quartz crystal microbalance (QCM) analysis and *ex-situ* SEM and XPS showed that the step-wise, substrate reactive site-limited heterogeneous growth mode in SIS is critical to eliminate undesired homogeneous reactions, and enable unprecedented control over the localized synthesis over previous methods. Furthermore, the experimental result illustrated that the unique segregated nanodomains in the self-assembled PS-*b*-PMMA thin film function as active channels to deliver reactants for the selective localized reaction in the PMMA domains during the SIS. Such mechanistic understandings shed lights on the mechanisms and potential problems in other localized synthesis methods involving BCPs as the substrate.

(498b) Advanced Nanomaterials Assembly and Nano-Joining Based On Lead-Free Nanosolders

Gao, F., University of Massachusetts Lowell
Gu, Z., University of Massachusetts Lowell
Originally presented on: 11/6/2013 12:50:00 - 13:10:00

Due to the requirements for high-performance interconnects in advanced packaging and nanoelectronics manufacturing, the development of nanotechnology has to meet the dramatic scaling down of the electronics packaging size from micron-scale to nanoscale. Both the micro and nano-joining are key enabling technologies in the construction of micro/nanoelectronic devices. In this study, advanced nanosolder materials are synthesized and developed for micro/nanoelectronics assembly and packaging applications. The micro and nano-joining are shown to be approached by utilizing nanosolders based on nanoparticles and nanowires. Both approaches are useful for the construction of different functional assemblies and nanodevices. The interactions between the nanosolders and substrates, including atomic diffusion, melting and wetting behavior, are studied from both the one-dimensional and two-dimensional perspectives. The nano-joining process is also developed to match the compatibility with the regular electronics assembly and packaging techniques.

(498c) Anisotropic Shrinkage of DNA Origami After a Wet-to-Dry Transition On Mica Surface

Verma, A., Indian Institute of Technology Kanpur
Schulman, R., Johns Hopkins University
Originally presented on: 11/6/2013 13:10:00 - 13:30:00

DNA nanotechnology is a fast emerging area with many potential applications such as nanofabrication, targeted drug delivery and self-assembly. DNA origami, for example, can serve as a scaffold for the assembly of functional materials. While most studies on DNA origami are done in aqueous solution, their application as templates for functional structures will inevitably occur in the dry state. Therefore, it is important to study the wet-to-dry transition of DNA origami and how this transition is affected by intrinsic factors such as origami design and external variables such as interfacial forces, buffer and drying conditions. Here we present a comparative study of the wet-to-dry transition in two different types of rectangular DNA origami on mica surface. The first type of origami is the tall rectangle (TR) with 10.67 bp/turn designed by Rothemund (2006).[1] This structure is distorted from its designed flat structure because of the deviation from a relaxed double stranded DNA configuration (i.e. 10.5 bp/turn).[2] We also considered a flatter tall rectangle (TRF) with 10.44 bp/turn designed by Woo and Rothemund (2011) in which twist in the direction perpendicular to helix stacking was reduced by introducing an irregular crossover pattern and leaving out parts of the scaffold unused in origami design.[2] While both structures shrink about 8-15% in transition from wet to dry state, TRF shrinks more anisotropically as compared to TR. Anisotropic shrinkage in TRF may be attributed to a more relaxed design where the inter-helix gap can be reduced more freely. Further, the unused parts of the DNA scaffold provide a cushion that facilitates shrinkage between two helices.[2] These differences in the wet-dry transition may be due to the unidirectional curvature of TRF, which allows preferential shrinkage along the length. In contrast, twist in both the directions perhaps puts more constraints in TR and the shrinkage is nearly isotropic. In the presence of weaker helical interaction, origami may be more likely to reduce interhelical spacing during drying.

The results suggest that it is important to consider the shrinkage during drying of origami while designing and assembling precision circuits as a change of few nanometers may compromise the functionality of the device.

References

1. Rothemund, P. W. K. Folding DNA to create nanoscale shapes and patterns. Nature 440, 297?302 (2006).

2. Woo, S. and Rothemund, P. W. K. Programmable molecular recognition based on the geometry of DNA nanostructures. Nat. Chem. 3, 620-627 (2011).

(498d) Building a 3D Graphene/Graphene Oxide Matrix for Cooperative Catalysis

Haag, D., Northwestern University
Kung, H. H., Northwestern University
Originally presented on: 11/6/2013 13:30:00 - 13:50:00

Reduced graphene oxide and graphene oxide (GO) based materials have many unique properties that make them intriguing candidates for catalyst supports. These materials have high surface areas, high adsorption capacities, can transfer π-electron density to supported catalytic metal particles, can have high electron mobility which can facilitate electron transfer during catalytic reactions, and also can be chemically transformed through the large number of oxygenate groups on the surfaces. One of the challenges encountered when utilizing graphene oxide materials for catalyst supports is the regraphitization and reduction in surface area when the catalyst is dried to a powder. The reduced graphene oxide layers restack via the strong π-π interactions between layers. This leads to a reduction in access to active sites on the interior basal planes of the catalyst. A number of methods have been explored to reduce this restacking interaction, for example, by creating 3D structures such as graphene-organic-frameworks (GOFs), graphene aerogels, and graphene hydrogels.

We have explored using 3-D graphene/graphene oxide based framework to create catalytic ?pockets? in which we can engineer a micro- or nano- environment that can influence catalytic performance. These ?pockets' are designed to create a partially nanoconfined or sterically hindered environment in which functional groups are placed near each other to work cooperatively and influence the catalytic active site or the reaction mechanism pathways. We made use of graphene oxides synthesized with nanosized in-plane defects or ?holes' that also have a high density of carboxylic acid groups and that allows for quicker diffusion of reactants and products to the interior areas of the 3-D structure. By reacting the GO support with compounds such as the diamine compound, 4-aminobenzylamine, the reaction of the two primary amine groups with the carboxylic acid or epoxy groups on the surface of the graphene oxide support would form a bridge between two separate graphene oxide sheets. The phenyl ring of this compound provides enough rigidity to increase the d-spacing of the material, forming pockets, and thus preventing re-graphitization. The introduction of catalytically relevant amine functional groups is accomplished by the utilization of another diamine complex, N-Boc-p-phenylenediamine or N-Boc-ethylenediamine, in which one of the amine functional groups is temporarily protected to prevent it from also becoming a bridging species. The protecting N-Boc group is then chemically or thermally removed which then result in a catalyst with a free primary amine can be located near residual surface carboxylic acid groups. The synthesis and characterization of 3-D structures prepared by this and similar methods will be described, including their structural, chemical and catalytic properties.

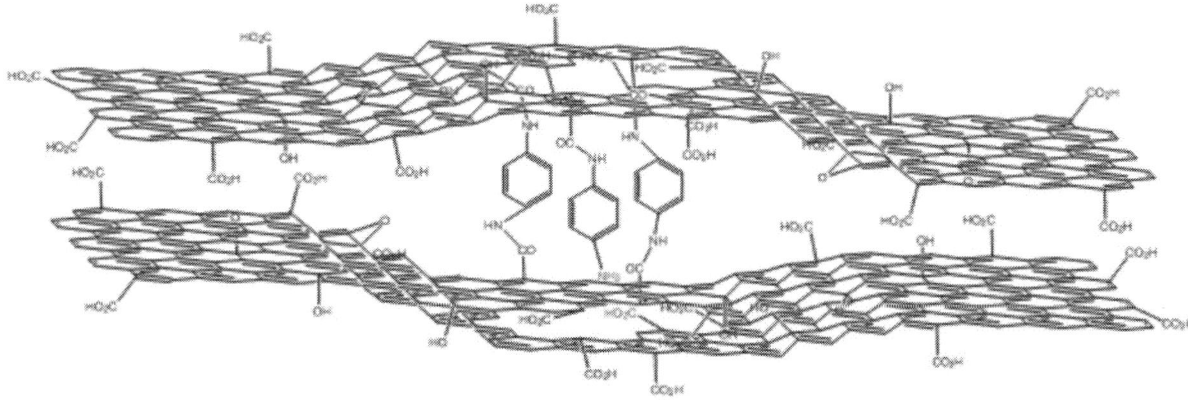

(498e) Molecular Dynamics Simulation-Based Study of MoS2 Solubilization: Predicting Solvent Performance in Liquid-Phase Exfoliation

Sresht, V., Massachusetts Institute of Technology
Blankschtein, D., Massachusetts Institute of Technology
Originally presented on: 11/6/2013 14:00:00 - 14:20:00

The intense research focus on graphene has ignited interest in the investigation of other two-dimensional 'wonder' materials, many of which exhibit superior chemical and opto-electronic properties. Experimental characterization of 2D nanomaterials, such as MoS2, hBN, and WS2, offers tantalizing glimpses of future applications in high-performance sensors and transistors. However, the large-scale production of these layered materials presents a key impediment to commercialization. Although liquid-phase exfoliation provides a scalable manufacturing route for high-quality layers, precise guidelines for the selection of a good exfoliating agent for a given nanomaterial remain elusive.

With the above in mind, we have used molecular dynamics (MD) simulations to garner several insights into the organization of molecules of commonly employed solvents, including N-methylpyrrolidone (NMP), dimethylformamide (DMF), N,N'-dimethylsulphoxide (DMSO), Cyclohexyl-pyrrolidinone (CHP), N-vinyl-Pyrrolidinone (NVP), and water (H2O), around and in-between the MoS2 sheets. We focus on the correlations between the structural features of these solvents and their effect on the energy barrier hindering the aggregation of the MoS2 sheets. The combination of MD-based computation of the potential of mean force (PMF) between pairs of MoS2 sheets and the application of theories of colloid aggregation offer a detailed picture of the mechanics underlying the stability of MoS2 sheets in these popularly-employed solvents.

In this talk, we will present a theoretical framework that harnesses MD simulations and colloidal theories to predict and rank the relative dispersive capabilities of the solvents considered with respect to MoS2. The excellent agreement between the predicted and the experimentally-observed solvent efficacies enables us to suggest design principles to assist in the selection of solvents which are suitable for the liquid-phase exfoliation of MoS2.

(498f) Heat Transfer Measurement of Oil-Based Copper Nanofluids; A Different Approach to Data Analysis Using Self-Organizing Feature Maps (SOFM)

Tatarko, J. L. Jr.,
Willing, G. A., University of Louisville
Originally presented on: 11/6/2013 14:20:00 - 14:40:00

Large amounts of experimental data can be correlated via regression analysis yet even when plotted the results can be confounding. Self-organizing feature maps (SOFM) provide a powerful method of classifying multidimensional data into 2d or 3d plots which might visually clarify meaning. This is one in a series of papers that describes the enhancement of heat transfer characteristics of various base oils when infused with nanoparticles. In this study, nanofluids were prepared by dispersing three different sizes and amounts of spherical Cu nanoparticles into poly-alpha-olefin, (PAO), formulated motor oil. Heat transfer coefficients of base/nanofluids were measured using a heat transfer test-rig built at the University of Louisville. Analysis of the results using dimensionless numbers was illuminating yet did not provide all of the answers. A seven-dimensional set of 1519 data points (217 x 7), was submitted as input to a mapping program. The results give a surprising visual prescription for optimization of heat transfer enhancement: small particle size, low operating temperatures and low to medium Reynolds, Prandtl, and Peclet numbers. SOFM provide yet another analytical tool for the discriminating scientist/engineer.

(498g) Influence of Dry Ball Milling and Wet Ball Milling Conditions On Reduction in Size of ZnO Particles and On Their Morphology

Middha, E., Indian Institute of Technology Madras, India
Nagarajan, R., Indian Institute of Technology, Madras
Originally presented on: 11/6/2013 14:40:00 - 15:00:00

Nanoparticles of Zinc Oxide have attracted the attention of lot of researchers with their unique chemical properties, low fabrication cost and variety of applications as a photo catalyst, in food packaging and in optoelctronics and biosensing. Reduction of Zinc Oxide from micron sized particles to nanoparticles has mainly been done using the technique of Planetary ball milling and Sono-fragmentation, however, as the size of ZnO reduces, its tendency to agglomerate increases due to high Van der waals forces of attraction. This agglomeration can be reduced in two ways: by maximizing the electrostatic repulsion force or by providing a barrier of steric hindrance. Wet ball milling with polymer of high molecular weight in water or in a solvent of high dielectric constant enjoys an advantage over dry ball milling with respect to agglomeration. During wet ball milling, high dispersibility, de-agglomeration and maximum reduction in size was noticed using the technique of Optical Microscopy in the case of Polymer PAA Poly Acrylic Acid (molecular weight ~ 100,000) in water and solvent Ethylene Glycol (dielectric constant : 38) as compared to surfactant SDS Sodium Dodecyl Sulphate (molecular Weight : 288.372) and solvent Toluene (dielectric constant : 2.38) respectively. The results were also examined by the technique of scanning electron microscopy (SEM) and a maximum change in morphology was noticed in the case of dry ball milling with a rugged, spherical and shortened structure and the least change in morphology was noticed in the case of wet ball milling with PAA in water and Ethylene Glycol as solvent.

(499a) SiC Photoanodes for Solar Water Splitting

Bohn, C., National Institute of Standards and Technology
Sharifi, F.,
Kang, M. G.,
Szalai, V., National Institute of Standards and Technology
Originally presented on: 11/6/2013 12:30:00 - 12:49:00

AIChE 2013 Session: Nanomaterials for Hydrogen Production and Fuel Cells

Title: SiC Photoanodes for Solar Water Splitting

Authors: C.D. Bohn, M.G. Kang, F. Sharifi, V.A. Szalai

SiC ceramics offer a rich field for exploration for solar water splitting.Â Polytypes include 4H-SiC with a bandgap of 3.23 eV, 6H-SiC with a bandgap of 3.05 eV and 3C-SiC with a bandgap of 2.36 eV.Â Single crystal substrates of some of these materials were obtained from commercial vendors.Â The effect of nanoscale morphology and the addition of Pt nanocrystals at the interface between the SiC and electrolyte were systematically investigated in a pH=7 potassium phosphate buffer under AM 1.5 illumination.Â Gas chromatography provided insight into the chemical products formed by the n-type SiC photoanodes.Â Additionally, linear sweep voltammetry, impedance spectroscopy and scanning electron microscopy were used to monitor the structural evolution of the semiconductor surface.Â Results are discussed in the broader context of photoelectrochemical hydrogen production and recommendations for improving the efficiency of water splitting with SiC are proposed.

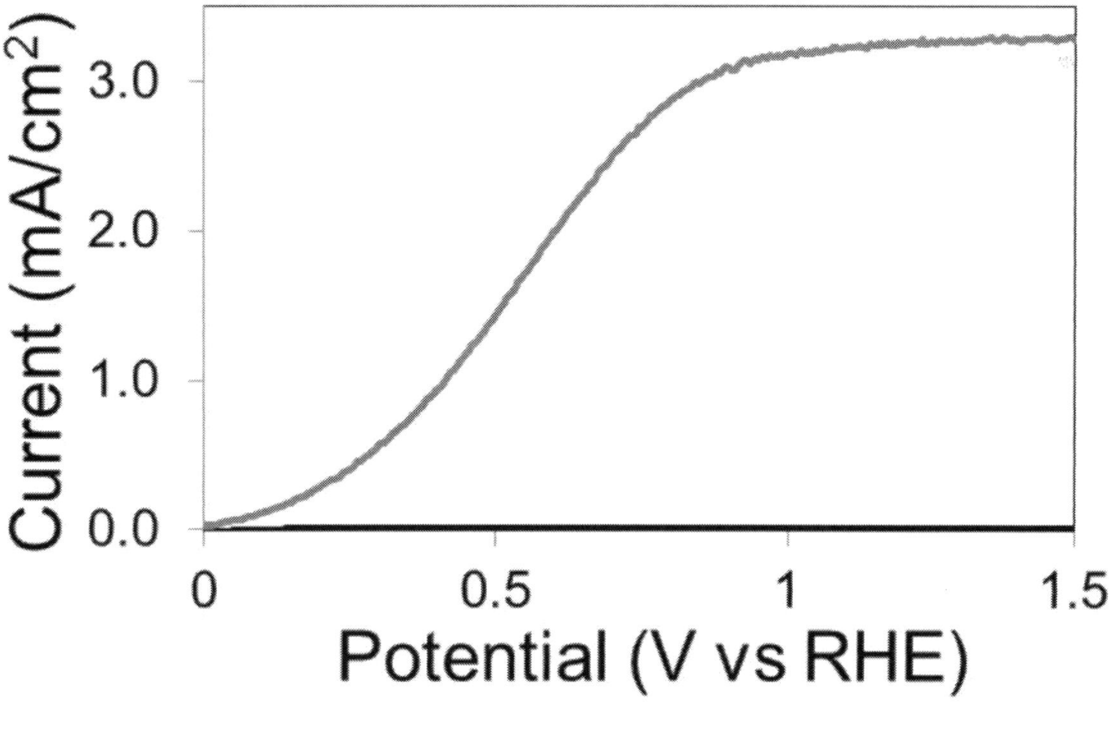

Keywords: solar water splitting, SiC, photoanode

(499b) Advanced Electrocatalysts From Organic Solution Synthesis

Wang, C., Johns Hopkins University

Originally presented on: 11/6/2013 12:49:00 - 13:08:00

Alloy nanoparticles (NPs) have attracted increasing interest due to their superior performance in various catalytic applications. For example, Pt alloys with transition metals (PtM with M = Fe, Co, Ni, etc.) have been found to be highly active for the oxygen reduction reaction in fuel cells. A lot of efforts have been dedicated to the synthesis of alloy catalysts, which are usually in the form of alloy NPs dispersed in high surface area carbon matrix. The approaches mostly include co-precipitation of metal salts in aqueous solution, impregnation, and electrodeposition. In spite of the successful preparation of various types of alloy catalysts, it yet remains challenging to synthesize monodisperse and homogeneous alloy catalysts with robust control over size and composition. Here we introduce our organic solvothermal approach towards high-quality alloy nanomaterials for electrocatalytic applications. Control over particle size, alloy composition and alloy homogeneity has been achieved by matching the different reduction and growth rates of the alloying elements, which has enabled systematic studies of the effects of these parameters on electrocatalysis. Our studies thus provide clues for the design and optimization of alloy nanomaterials for catalytic applications.

(499c) Nanostructured Ti –Fe2O3/Cu2O Bilayered Thin Films for Photoelectrochemical Water Splitting

Sharma, D., Dayalbagh Educational Institute,Dayalbagh

Upadhyay, S., Dayalbagh Educational Institute

Choudhary, S., Dayalbagh Educational Institute

Satsangi, V. R., Dayalbagh Educational Institute

Shrivastav, R., Dayalbagh Educational Institute

Dass, S., Dayalbagh Educational Institute

Originally presented on: 11/6/2013 13:08:00 - 13:27:00

Nanostructured Ti-Fe_2O_3/Cu_2O Bilayered Thin Films for Photoelectrochemical Splitting of Water

Dipika Sharma, Sumant Upadhyay, Surbhi Choudhary, Vibha R. Satsangi[?],

Rohit Shrivastav and Sahab Dass,

Department of Chemistry, [?]Department of Physics & Computer Sciences

Faculty of Science, Dayalbagh Educational Institute (Deemed University),

Agra-282110 (India)

Nanostructured thin films of pure Fe_2O_3, Ti ?Fe_2O_3, Cu_2O, and bilayered Fe_2O_3/Cu_2O, Ti-Fe_2O_3/Cu_2O were deposited on ITO glass substrate using spray pyrolysis method. Among various metal oxide semiconductors, cuprous oxide Cu_2O [Eg=2.4eV] and Iron oxide Fe_2O_3 [Eg=2.1eV] were
chosen for the present study on account of being abundant, non-toxic, environmentally safe, having low band gap and possess staggered type band edges which is suitable for efficient charge separation in water splitting process.

All samples were characterized using XRD, SEM and UV?Vis spectrometry. Photo electrochemical properties were also investigated in a three-electrode cell system. UV-Vis absorption spectrum for pure Fe_2O_3, Ti ?Fe_2O_3, Cu_2O, and Fe_2O_3/Cu_2O, Ti-Fe_2O_3/Cu_2O bilayered thin films shows absorption in visible light region. Nanostructured thin films for all the samples were used as photoelectrode in the Photoelectrochemical cell for water splitting reaction. Maximum photocurrent density of 2.22 mA/cm^2 at 0.95 V/SCE was exhibited by 454 nm thick Ti-Fe_2O_3/Cu_2O bilayered photoelectrode (Fig.1). Increased photocurrent density and enhanced photoconversion efficiency, exhibited by the bilayered thin films may be attributed to improved conductivity and efficient separation of the photogenerated charge carriers at the

Ti-Fe_2O_3/Cu_2O interface.

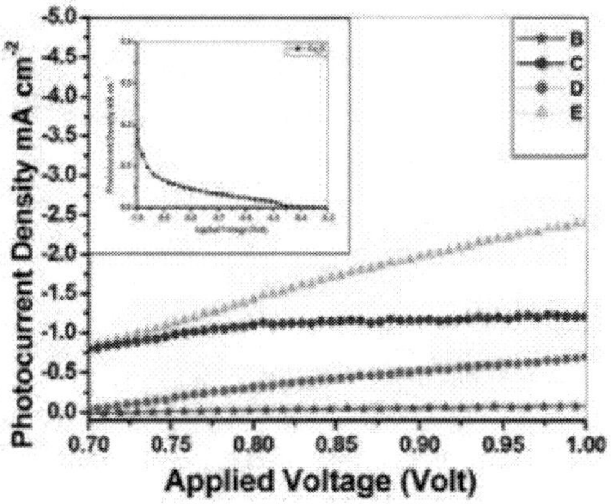

Fig. 1 Photocurrent density vs. Applied potential curve for pure Cu_2O(A), Fe_2O_3(B), Ti ?Fe_2O_3(C), and Fe_2O_3/Cu_2O(D), Ti-Fe_2O_3/Cu_2O(E) bilayered thin films respectively ,under visible light illumination in 0.1M NaOH electrolytic solution using 150W visible light source of irradiation 150 mW cm^{-2} at the position of sample.

Keywords: *Photoelectrochemical, Water splitting, Bilayered, Cu_2O, Fe_2O_3, Ti- Fe_2O_3, Spray-Pyrolysis*

(499d) Synthesis and Characterization of Mn2O3-ZnO Nanocomposites for Hydrogen Generation Via Photoelectrochemical and Photocatalytic Splitting of Water

Singh, N., Dayalbagh Educational Institute
Choudhary, S., Dayalbagh Educational Institute
Satsangi, V. R., Dayalbagh Educational Institute
Dass, S., Dayalbagh Educational Institute
Shrivastav, R., Dayalbagh Educational Institute
Originally presented on: 11/6/2013 13:27:00 - 13:46:00

Synthesis and Characterization of Mn_2O_3-ZnO Nanocomposites for Hydrogen Generation *via* Photoelectrochemical and Photocatalytic Splitting of Water

Nirupama Singh, Surbhi Choudhary, Vibha R. Satsangi[#], Sahab Dass and Rohit Shrivastav*

Department of Chemistry, [#]Department of Physics & Computer Science,

Dayalbagh Educational Institute, Dayalbagh, Agra-282 005, India

*E-mail: rohitshrivastav_dei@yahoo.co.in

The growing consumption of energy, which is now mainly produced from fossil fuels leads to a further depletion of the limited fossil reserves, increased dependency on fossil fuel imports from often politically unstable counties and furthermore gives rise to an increased concentration of green house gases mainly CO_2 in the atmosphere. The report deals with the synthesis of Mn_2O_3-ZnO nanocomposites, their characterization, and possible use in photoelectrochemical and photocatalytic splitting of water for hydrogen generation. Powdered samples of nanocomposites were prepared by co-precipitation at varying Mn/Zn molar ratios. XRD analysis revealed the evolution of wurtzite ZnO and Mn_2O_3 phases in samples, while optical absorption threshold was found shifted towards higher wavelength with increasing Mn amount. The Zn/Fe molar ratio in nanocomposites was optimized, yielding significantly enhanced photoelectrochemical current and photocatalytic hydrogen yield. Plausible explanations for observed change in photoresponse of samples have been offered.

(499e) High Performance Fluorine Doped Oxygen Evolution Reaction (OER) Electro-Catalysts for PEM Based Water Electrolysis

Kadakia, K., University of Pittsburgh
Datta, M. K., University of Pittsburgh
Velikokhatnyi, O.,
Kumta, P. N., University of Pittsburgh
Originally presented on: 11/6/2013 13:46:00 - 14:05:00

Identification of electro-catalysts containing non noble metal or significantly reduced amounts of expensive noble metals (e. g. RuO_2) is highly desirable. Development of such a catalyst with comparable electrochemical performance to the standard noble metal oxide counterpart for the proton exchange membrane (PEM) based water electrolysis would constitute a major advancement in generation of hydrogen by water electrolysis. We have accordingly developed a two-pronged theoretical first principles and experimental approach to generate a nano-structured solid solution of SnO_2:10wt.% F containing only 20at.% RuO_2 [*e.g.* $(Sn_{0.80}Ru_{0.20})O_2$:10F] that displays a remarkably similar electrochemical activity and moreover, comparable or even much improved electrochemical stability and durability compared to pure the noble metal counterpart, RuO_2. Results of these studies will be presented and discussed.

(499f) Multiwalled Carbon Nanotubes Drive the Activity of Metal@Oxide Core-Shell Catalysts in Modular Nanocomposites

Cargnello, M., University of Pennsylvania

Originally presented on: 11/6/2013 14:05:00 - 14:24:00

Multiwalled Carbon Nanotubes Drive the Activity of Metal@oxide Core-Shell Catalysts in Modular Nanocomposites

M. Cargnello,[1,2] M. Grzelczak,[1] B. Rodriguez-Gonzalez,[3] Z. Syrgiannis,[1] K. Bakhmutsky,[4]

V. La Parola,[5] L. M. Liz-Marzán,[3] R. J. Gorte,[4] M. Prato,[1] and P. Fornasiero[1]

[1] *Department of Chemical and Pharmaceutical Sciences, ICCOM-CNR, Consortium INSTM, University of Trieste, via L. Giorgieri 1, 34127 Trieste, Italy*

Â [2] *Department of Chemistry, University of Pennsylvania, 231 S. 34th Street, Philadelphia, PA 19104, USA*

[3] *Departamento de QuÃmica FÃsica, Universidade de Vigo, 36310 Vigo, Spain*

[4] *Department of Chemical and Biomolecular Engineering, University of Pennsylvania, 311A Towne Building, 220 S. 33rdÂ Street, Philadelphia, PA 19104, USA*

[5] *Istituto per lo Studio dei Materiali Nanostrutturati (ISMN-CNR), Via Ugo La Malfa 153, Palermo I-90146, Italy*

The ability to build hierarchical structures by arranging different building blocks with nanometer-scale precision is one of the most useful aspects of nanotechnology. This concept can produce novel materials with properties that are different from those expected from the simple sum of the individual blocks. Indeed, the interactions between the constituent parts in nanometer-scale ensembles can lead to novel electronic, optical, or catalytic properties not available in the initial building blocks. In the field of heterogeneous catalysis in particular, the interactions between the active phase, supports, and promoters is critical for obtaining high performance materials. These interactions can be both electronic and geometric. Furthermore, the demands in terms of geometry and binding energy of the active sites for different catalytic reactions can be very different.

In this contribution, rational nanostructure manipulation has been used to prepare nanocomposites in which multiwalled carbon nanotubes (MWCNTs) were embedded inside mesoporous layers of oxides (TiO_2, ZrO_2, or CeO_2), which in turn contained dispersed metal nanoparticles (Pd or Pt). We show that the MWCNTs induce the crystallization of the oxide layer at room temperature and that the mesoporous oxide shell allows the particles to be accessible for catalytic reactions. In contrast to samples prepared in the absence of MWCNTs, both the activity and the stability of core-shell catalysts is largely enhanced, resulting in nanocomposites with remarkable performance for the water-gas-shift reaction, photocatalytic reforming of methanol, and Suzuki coupling. The modular approach shown here demonstrates that high-performance catalytic materials can be obtained through the precise organization of nanoscale building blocks.

(499g) Separation of Sulphur Dioxide and Oxygen in Thermochemical Hydrogen Production

Elder, R. H., University of Sheffield
He, G., University of Sheffield
Cumming, D. J.,
Allen, R. W. K., University of Sheffield
Originally presented on: 11/6/2013 14:24:00 - 14:43:00

Dense Yttria-stabilised-zirconia (YSZ) membranes have been shown to have great potential for separating SO_2 and O_2 in the sulphur family of thermochemical cycles for hydrogen production. Use of a catalytic YSZ membrane reactor in the sulphuric acid decomposition stage of these cycles could drive the reaction forward and significantly increase the yield of SO_2 and O_2. Whilst SO_2/O_2 separation has been shown to be feasible at temperatures between 750°C and 950°C, significant performance degradation is seen over time. This performance degradation is shown to be due to sulphur poisoning (from both sulphur adsorption and sulphide formation) of the triple phase boundaries. A similar effect is seen in solid oxide fuel cells, where ppm concentrations of sulphur severely diminish cell performance.

Here we investigate the effect of voltage pulsing on membrane performance recovery. Significant performance recovery is seen after the application of a short voltage pulse. The optimum pulse duration and magnitude were found to be 0.08 s and 10 V respectively. The lower the magnitude of voltage pulse, the longer the duration required to achieve maximum recovery. Impedance spectroscopy and SEM results are presented, showing the contribution of different resistances and the effect of the voltage pulsing on the membrane cell.

Consideration of oxygen reduction mechanisms at the triple phase boundary shows that as voltage across the membrane increases, the rate controlling step changes from oxygen adsorption and surface diffusion, to oxygen availability in the feed. The higher voltage pulses are beneficial to performance recovery as polarisation resistance and charge transfer resistance decrease due to the removal of sulphur from the surface. Pulses that have too great a magnitude, or are too long, dramatically promote oxygen reduction and lead to concentration polarisation due to lack of oxygen in the feed.

Longer term testing (20 hours) showed that repeated, regular voltage pulsing can significantly improve membrane performance compared to those that receive no voltage pulsing.

(499h) Chemisorption, Physisorption, and Hysteresis of Hydrogen On Carbon Nanotubes

Barghi, S., University of Southern California

Tsotsis, T. T., University of Southern California

Sahimi, M., University of Southern California

Originally presented on: 11/6/2013 14:43:00 - 15:02:00

We used a magnetic suspension balance, coupled with a residual gas analyzer, to investigate hydrogen adsorption/desorption on carbon nanotubes (CNTs) for pressures up to 100 bars at 25 °C. Such a combination of gravimetric and mass analysis systems, used for the first time, is capable of overcoming the shortcomings of the common hydrogen adsorption measurement techniques. The most crucial drawbacks of the conventional measurement methods are as follows. (i) The inability to distinguish between hydrogen adsorption and chemisorptions; (ii) the relatively large errors caused by the leakage of hydrogen out of the system, and (iii) the difficulty of measuring hydrogen uptake capacity under operating conditions that differ from those of hydrogen storage systems used in practice.

In the experiments multi-walled carbon nanotubes (MWCNTs) were exposed to hydrogen, and were subsequently kept under vacuum for 2 consecutive cycles, in order to study hydrogen adsorption/desorption in the nanostructured materials. The difference between the first and the second hydrogenation cycles was in the initial preparation of the MWCNTs. For the first cycle, the MWCNTs were first degasified at an elevated temperatures (120 °C). For the second cycle the MWCNTs were not degasified at the elevated temperatures and, therefore, hydrogen residue from the first adsorption/desorption cycle was still adsorbed on the MWCNTs at the beginning of the second cycle.

The first hydrogenation cycle was started after pretreatment of the CNT at the elevated temperature. Hydrogen adsorption kinetics for the cycle indicates that the adsorption equilibrium occurs within 4 - 5 hours. Careful examination of the recorded adsorption kinetics revealed the fascinating fact that a considerable part of hydrogen adsorption in the MWCNTs takes place *instantaneously*. The measured equilibrium data revealed some hysteresis for hydrogen desorption in this cycle, which does not vanish, even when the system's pressure approaches zero. This indicates that hydrogen adsorption in the first hydrogenation cycle is a combination of physisorption and chemisorption.

After the first cycle was finished, the CNTs were exposed to hydrogen for the second time without any pretreatments at the elevated temperature. Measurements indicated no hysteresis for during the second cycle. Therefore, only physisorption took place during the second cycle. Comparison between the first and the second hydrogen isotherms, together with the measured hydrogen adsorption kinetics, made it possible to quantitatively distinguish between the physisorption and chemisorption portions of hydrogen adsorption in the MWCNTs. According to our data, for pressures less than 20 bars physisorption and chemisorption are both important, with the former becoming the dominant mechanism as the pressure increases. To the best of our knowledge, this is the first time that hydrogen physisorption and chemisorption on carbon nanotubes have been measured separately.

(540a) Biomimetic Properties of Carbon Nanotubes In Vivo

Villegas, J. C.,
Valiente, R.,
Rodriguez-Fernandez, L.,
Gonzalez, J.,
Fanarraga, M. L.,
Originally presented on: 11/6/2013 15:15:00 - 15:37:00

Multiwalled carbon nanotubes (MWCNTs) are well documented to invade the cytoplasm of eukaryotic cells and to interact with actin microfilaments and microtubules due to the many attributes they share with these cytoskeletal filaments. In particular, MWCNTs and microtubules have a similar length-to-with ratio and resiliency; they are both hollow cylinders; they both self-assemble, and form bundles to enhance their resistance. These features that provide MWCNTs with the intrinsic ability to interact with microtubules have also been suggested to lead, among others, to severe proliferative defects resulting in aneuploidy and cell death.

Our work demonstrates how MWCNTs display bio-mimetic interactions with microtubules at the molecular level assembling biosynthetic filaments *in vivo*. As a result of the different dynamical behavior of MWCNTs and microtubules, their interaction gives rise to the stabilization of the tubulin polymers. In fact, carbon nanotubes are exceptionally stable, while microtubules are highly dynamic polymers that are continuously undergoing assembly/disassembly cycles *in vivo*. Therefore, these interactions severely perturb microtubule dynamics, disorganizing the centrosomal microtubule pattern in the cell leading to aberrant mitotic spindle assemblies, chromosomal malsegregation between daughter cells, and often to mitotic catastrophe and cell death. Thus, our work gives an consistent explanation aneuploidy and mitotic failure in cells exposed to MWCNTs.

(540b) Differential Analysis Of Single Wall Carbon Nanotubes Cellular Uptake Mechanism

Eldawud, R., West Virginia University
Dong, C., West Virginia University
Sargent, L. M., National Institute for Occupation Safety and Health
Rojanasakul, Y., West Virginia University
Dinu, C. Z., West Virginia University
Originally presented on: 11/6/2013 15:37:00 - 15:59:00

Biomedical applications of Carbon nanotubes (CNTs) are currently being constrained by many uncertainties regarding the cellular mechanisms associated with their uptake and delivery as well as their fate inside the biological systems. In this research, a new approach is used to quantify the cellular mechanisms of a wide library of surface modified CNTs incubated with human lung epithelial cells (BEAS-2B) in real time. This non-invasive technique relies on an electrical cell impedance sensing system (ECIS) used as a proxy to measure morphological changes and cellular interactions upon exposure to different concentrations and functionalities of CNTs. Such measurements are correlated to the cellular behavior under the influence of different parameters like endocytotic inhibitors or electroporation and further complemented by classical microscopic, apoptotic and flow cytometry assays to derive cellular structure-function relationships. Our real time results provide an insight to the underlying cellular mechanisms responsible for the uptake and delivery of CNTs to the biological systems and promises to extend our understanding of such cellular-based interfaces to advance the biomedical applications of CNTs.

(540c) Development of in-Vivo Toxicity Screening Benchmarks for Complex Engineered Nanomaterials

Mahoney, S., Univeristy of Wyoming
Najera, M., Mascaro Center for Sustainable Innovation, University of Pittsburgh
Bai, Q., University of Pittsburgh
Burton, E., University of Pittsburgh
Veser, G., University of Pittsburgh
Originally presented on: 11/6/2013 15:59:00 - 16:21:00

Functional nanomaterials are developing at an ever accelerating pace and are already finding widespread application in consumer products including cosmetics, sporting goods, clothing, and consumer electronics. However, increasing evidence indicates that nanomaterials are showing elevated toxicity, motivating the need to establish appropriate toxicity tests for these emerging technology. Silica nanomaterials have been identified for a wide range of applications including biomedical imaging, drug delivery, catalysis, separations, and sensors, as they constitute a versatile, easily modifiable template that has been established as being non-toxic. However, the rapidly increasing volume of nanomaterials research and production requires development of modern high-throughput *in vivo* nanotoxicity screenings for these nanomaterials. Zebrafish (*Danio rerio*) is rapidly emerging as a well-suited model for such studies due to prolific breeding, embryo tissue transparency, rapid development time, and the relative simplicity of controlled toxin exposure.

Porous silica nano- and microparticles are accepted to be non-toxic, thus leading to the hypothesis that embedding metal nanoparticles in silica could reduce or entirely mitigate nanoparticle toxicity while still providing accessibility of the metal nanoparticle surface and hence maintaining the functionality of the embedded nanoparticle. Three different complex nanostructures are investigated in this work including surface-deposited metal nanoparticles on silica nanoparticle supports, embedded metal nanoparticles in silica and encapsulated metal nanoparticles in silica. In the present work, the toxicity of carefully synthesized Ni-silica nanomaterials with these complex, but well-controlled nanostructures is investigated using zebrafish as the *in vivo* toxicity model.

Comparing the different structures, our results from the *in vivo* zebrafish toxicity study demonstrate that all three silica configurations significantly reduce the toxicity of metal nanoparticles compared to the equivalent dosing of an analogous metal salt. While the engineered nanomaterials showed limited dissolution of metal ions from the structured materials into the media solution, nickel uptake was still detectable at ng levels within the zebrafish larvae tissue. Interestingly, relative to free metal ion concentrations, metal uptake was significantly enhanced for the engineered nanomaterials, suggesting the presence of a "trojan horse" transport mechanism into the fish. All three engineered nanomaterials showed high zebrafish survival rates as compared to the Ni salts, hence necessitating more sensitive means for detection of possible toxic effects. Therefore, the studies were extended to include motility analysis which allowed for a much enhanced sensitivity compared to endpoints of survival and developmental deformations, yielding statistically significant dose-dependent results for the different nanomaterials. Our current work includes extension of these studies to a range of nanoparticle sizes and metals to further investigate the effect of nanostructuring on nanomaterial toxicity.

(540d) NLRP3 Inflammasome Activation Induced By Long Aspect Ratio Engineered Nanomaterials: Role of Oxidative Stress

Sun, B., University of California, Los Angeles
Xia, T., University of California, Los Angeles
Originally presented on: 11/6/2013 16:21:00 - 16:43:00

NLRP3 inflammasome is an intracellular multi-protein complex assembled and activated upon exogenous stimuli including engineering nanomaterials (ENMs). Its activation has been increasingly implicated in nanomaterial-induced toxicity. However, the detailed mechanisms of inflammasome activation induced by ENMs are still not well understood. In this study, we examined the interaction mechanisms of ENMs with NLRP3 inflammasome using a library of long aspect ratio ENMs. We identified that besides the lysosomal damage and cathepsin B release, ROS generated through NADPH oxidase was found to play an important role in long aspect ratio ENM-induced NLRP3 inflammasome activation. The ROS generation could further lead to secondary cytosolic as well as mitochondrial ROS. In the meanwhile, we demonstrated that suppression of oxidative stress could attenuate inflammasome activation *in vitro* and the development of lung fibrosis upon ENM exposure *in vivo*. This study delineates mechanisms of NLRP3 inflammasome activation induced by long aspect ratio ENMs and identifies the role of oxidative stress in ENM-induced chronic lung inflammation and fibrosis.

(540e) Effects of Aerosolized Sub-Micron Particles On The Interfacial Properties of Lung Surfactant Models

Farnoud, A. M., University of Iowa
Fiegel, J., The University of Iowa
Originally presented on: 11/6/2013 16:43:00 - 17:05:00

Pulmonary surfactant plays an important role in lung stability by reducing the surface tension of the alveolar fluid and avoiding alveolar collapse. Changes to the interfacial properties of this surfactant can lead to potential health implications such as increased work of breathing and impaired gas exchange. Inhaled particles that deposit in the alveolar region of the lung come into contact with the pulmonary surfactant and might alter its interfacial properties. A few recent *in vitro* studies have reported surfactant dysfunction following exposure to small particles. However, these studies have been focused on colloidal particle suspensions and little is known about particle-surfactant interactions when dry particles are aerosolized onto surfactant films. The present study seeks to improve the current understanding of the effects of inhaled particles on surfactant function by examining particle-surfactant interactions after exposing surfactant films to aerosolized particles.

Langmuir films of dipalmitoyl phosphatidylcholine (DPPC) and Infasurf (calf lung surfactant extract) were used as lung surfactant models and their interactions with carboxyl-modified polystyrene particles (200 nm) were studied. Surfactant films were spread in a Langmuir trough and their dynamic surface tension was probed using a Wilhelmy plate balance. A Dry Powder Insufflator™ was used to aerosolize dry particles onto surfactant films. Following exposure to particles, surfactant films were subjected to two modes of surface area changes: 1. Surface compression from the surface tension of pure subphase until film collapse and 2. Compression and expansion within the range of physiologically relevant surface tension values to mimic breathing cycles.

Aerosolization of particles on both DPPC and Infasurf films resulted in a shift in the surface area of surfactant collapse and a slight reduction in the minimum surface tension. However, particles induced significant changes (~20 mN/m) in the minimum surface tension of DPPC when aerosolized in the lung relevant surface tension range. In contrast, only minor changes were observed in the maximum surface pressure of Infasurf following aerosolization in the physiologically relevant surface tension range suggesting that Infasurf films are more stable than DPPC in response to particle exposure. Further experiments with different particle-surfactant interaction times showed that particles effects on Infasurf interfacial properties changed as a function of time suggesting that particles affect surfactant function by adsorption of surfactant components. These results provide information on the effects of particles on surfactant function following inhalation, which is of great importance for the fields of pulmonary drug delivery and environmental and occupational health.

(540f) Carbon Nanotubes Induce Invasion of Human Mesothelial Cells Through Matrix Metalloproteinase-2

Lohcharoenkal, W., West Virginia University
Dinu, C. Z., West Virginia University
Stueckle, T., National Institute for Occupational Safety and Health
Wang, L., National Institute for Occupational Safety and Health
Rojanasakul, Y., West Virginia University
Originally presented on: 11/6/2013 17:05:00 - 17:27:00

Carbon nanotubes (CNT) have increasingly been used in a wide variety of applications. However, because of their needle-shape morphology, biopersistence and mode of exposure similar to those of asbestos which is a known carcinogen causing lung mesothelioma, there has been a great concern about their potential carcinogenicity. In this study, we investigated the effect of long-term exposure of CNT on proliferative and invasive properties of human pleural mesothelial cells. We found that sub-chronic exposure of human mesothelial MeT5A cells to low-dose non-cytotoxic concentration of single- and multi-walled CNT (0.02 $\mu g/cm^2$ for 6 months) in culture induced malignant transformation of the cells as indicated by their increased cell invasion and migration. An up-regulation of several key genes known to be important in cell invasion, notably matrix metalloproteinase-2 (MMP-2), was observed in the exposed mesothelial cells as determined by real-time PCR. Western blot analysis of protein expression confirmed the up-regulation of MMP-2 in the exposed cells. Whole genome microarray analysis further indicated the importance of MMP-2 in the invasion gene signaling network of the cells. Knockdown of MMP-2 in the CNT-transformed cells by short-hairpin (sh)RNA stable transfection effectively inhibited MMP-2 activity and expression in these cells as well as their invasive and migratory activities. Preliminary animal studies demonstrated colonization of CNT-transformed cells in the lungs of mice which was absent in mice receiving shMMP-2 knockdown cells. These results indicate MMP-2 as a key regulator of the aggressive phenotype of CNT-transformed cells. This study fortifies the carcinogenic potential of CNT and reveals a novel mechanism that may be important in CNT-induced mesothelioma.

(540g) Neoplastic-Like Transformation Ability of Carbon Nanotubes On Small Airway Epithelial Cells: Linking Toxicogenomic Signatures and Lung Cancer Hallmarks

Stueckle, T., National Institute for Occupational Safety and Health
Mishra, A., National Institute for Occupational Safety and Health
Derk, R., National Institute for Occupational Safety and Health
Meighan, T., National Institute for Occupational Safety and Health
Castranova, V., National Institute for Occupational Safety and Health
Wang, L., National Institute for Occupational Safety and Health
Rojanasakul, Y., West Virginia University
Originally presented on: 11/6/2013 17:27:00 - 17:49:00

Concern for increased risk of CNT-associated lung carcinogenesis has been raised due to asbestos-like high aspect ratio, deep pulmonary deposition, and biopersistance. Studies evaluating long-term human health risks associated with chronic pulmonary CNT exposures compared to other well-characterized particulates (e.g. asbestos) are lacking. To address this knowledge gap, we conducted subchronic *in vitro* exposures of dispersed single walled CNT (D-SWCNT), multi-walled CNT (D-MWCNT) and crocidolite asbestos (ASB) to human small airway epithelial cells (SAEC) to assess their neoplastic transformation potential, an early marker of carcinogenesis. Ultrafine carbon black (D-UFCB) and dispersant-only exposed cells (DISP) served as negative controls. SAEC were exposed for 25 weeks to 0.02 $\mu g/cm^2$ and evaluated for morphological transformation, cancer cell phenotype and subjected to whole genome expression signature profiling. Subchronic *in vitro* exposure to D-SWCNT and D-MWCNT resulted in a more aggressive neoplastic-like transformation phenotype of exposed SAECs compared to asbestos. Furthermore, genome expression signature analysis identified oncogene signaling mechanisms in CNT-exposed SAEC that was substantially different from pro-inflammatory signaling in asbestos-exposed SAEC. Correlation feature selection strategies identified known lung cancer markers and robust subsets of genes that could be used to develop specific gene bio-signatures for CNT-exposed cells. In conclusion, toxicogenomic signature profiling in a CNT neoplastic-like transformed lung cell model identified particle-specific gene markers and known lung cancer markers which can potentially aid in assessing CNT exposure and detection of early disease bio-signatures.

Disclaimer: The findings and conclusions in this abstract are those of the authors and do not necessarily represent the views of the National Institute for Occupational Safety and Health.

(558a) Thermoelectric Properties of Ultra-Long PbSe Hollow Nanofibers

Zhang, M., University of California, Riverside
Park, H., University of California-Riverside
Jung, H., University of California, Riverside
Kim, J.,
Kim, S.,
Lim, J. H., Korea Institute of Materials Science
Choa, Y. H.,
Myung, N., University of California, Riverside

Originally presented on: 11/6/2013 15:15:00 - 15:45:00

Lead selenide (PbSe), a sister material of lead telluride (PbTe), has been less focused on for thermoelectric (TE) applications due to its smaller band gap energy and an expected higher thermal conductivity. Recently, however, an additional notion has been placed to the material, since a TE figure of merit (ZT) up to 1.2 at 850 K was experimentally achieved, approaching the theoretical value of 1.6. In addition, significant progress has been made to enhance the Seebeck coefficient while diminishing the thermal conductivity by using low dimensional PbSe. However, even though a thermal power 3 times greater than that of bulk in PbSe quantum dot superlattices has been reported, only a low power factor was extracted attributing to its high electrical resistivity. One-dimensional (1-D) PbSe, on the other hand, shows a better electrical conductivity with a significant decreased thermal conductivity, but the prevailing difficulty and high cost in a device fabrication hinders its application in a real TE apparatus. Therefore, low dimensional PbSe with superior TE performance is highly demanded, through a cost-effective fabrication process.

While several techniques exist for creating 1-D PbSe nanostructures, electrospinning (ES) has emerged as the most versatile, scalable and cost-effective method to synthesize ultra-long nanofibers (NFs) with controlled diameter and composition. Although various NFs have been synthesized using this technique, limited work has been reported on chalcogenide NFs synthesis due to the incompatibility of the precursors. Therefore, synthesis of PbSe was conducted through ES with the addition of a galvanic displacement reaction (GDR); this is a spontaneous electrochemical process driven by the difference in materials' redox potentials. In this approach, ES is exploited to fabricate ultra-long sacrificial NFs with controlled dimensions, morphology and crystal structures, providing a large materials' data base to tune electrode potentials thereby imparting control over the composition and shape of the nanostructures evolved during GDR. Synthesized samples are designed to be in a bulk scale for easy device fabrication, but reserve the superior properties form nanoscale materials.

In this work, we demonstrated high scalability and cost-effective nanofabrication to synthesize ultra-long hollow PbSe nanofibers by combining ES and a GDR. Control over the diameter, wall-thickness, morphology, composition and crystallinity of the NFs was achieved by tuning the shape and dimension of the sacrificial materials as well as GDR conditions. Temperature dependent electrical resistivity and Seebeck coefficient measurement were performed for the TE properties characterization. Preliminary result showed an excellent Seebeck coefficient of 525 $\mu V/K$ and a power factor of 0.43 $\mu W/K^2m$ at 26 ^oC in the PbSe hollow NF mat.

(558b) Thermal Transport in Nanocrystal Arrays and Self Assembled Monolayers

Ong, W. L., Carnegie Mellon University
Rupich, S., University of Chicago
Majumdar, S., Carnegie Mellon University
Talapin, D. V., University of Chicago
McGaughey, A. J. H., Carnegie Mellon University
Malen, J. A., Carnegie Mellon University
Originally presented on: 11/6/2013 15:45:00 - 16:15:00

The thermal conductivity of nanocrystal arrays (NCAs) is found to be tunable with chemistry and the nanocrystal diameter. Nanocrystals (inorganic crystalline cores 2-10 nm in size encapsulated in organic monolayers) self-assemble from colloidal suspensions into close-packed 3D NCA films. These films exhibit tunable electronic and optical properties that can meet various demands in energy conversion and optoelectronic applications. When coupled with scalable solution-based fabrication processes, these properties make NCAs a cheap and versatile replacement for conventional single crystal semi-conductors and nanostructured materials.

Though electronic transport in NCAs has been studied extensively, little is known, about the nature of thermal transport in these complicated materials. We show that the thermal conductivity of NCAs shows independence from the bulk core thermal conductivity, yet is tunable from 0.1-0.3 W/m-K by increasing the nanocrystal diameter. Effective medium approximations (EMA) modified by incorporating a fintite thermal conductance at the core ligand interface correctly captures the measured diameter-dependent magnitudes and trends. Changing the chemistry of the nanocrystal (i.e., different core and ligand species) enables further control for manipulating the NCA thermal conductivity. To learn more about thermal transport at this interface we have directly studied the thermal conductance of a series of self assembled monolayer (SAM) junctions. By considering various metal-SAM-metal junctions we find that the interface conductance is influenced by both the SAM-metal bond and the overlap of their vibrational states. Molecular dynamics simulations of NCAs and SAMs help us to sort out these effects and how they more generally apply at the organic-inorganic interface.

(558c) Thermoelectric Nanocomposites of Layered Chalogenide Bi2(Te/Se)3 Nanoplateletes and Their Interfacial Effects

Soni, A.,
Xiong, Q.,
Originally presented on: 11/6/2013 16:15:00 - 16:45:00

Thermoelectric nanocomposites of layered chalogenide $Bi_2(Te/Se)_3$ nanoplateletes and their interfacial effects

Ajay Soni[1,2] and Qihua Xiong *[,2,3]

[1] School of Basic Sciences, Indian Institute of Technology Mandi, Mandi, Himachal Pradesh 175001, India

[2] Division of Physics and Applied Physics, School of Physical and Mathematical Sciences, Nanyang Technological University, Singapore 637371, Singapore

[3]Division of Microelectronics, School of Electrical and Electronic Engineering, Nanyang Technological University, Singapore 639798, Singapore

* To whom the correspondence should be addressed. Email address: Qihua@ntu.edu.sg,

Recently the interests in layered chalcogenide materials have grown very rapidly. The compounds of chalcogenides (S, Te and Se) with Bi and Sb are of interests for topological insulators, thermoelectric (TE) applications, while the compounds with transition metals are of interests for optoelectronic applications. The observed fascinating physical properties are arising mainly due to the anisotropic crystal structures, weak van der Waals stacking of layer and quantum confinement effects in atomic thin layers. Historically, chalcogenide materials have been known as good TE materials, especially the family of $(Bi/Sb)_2(Te/Se)_3$ alloys which showed a competitive performances for room temperature TE applications.[1] The efficiency of the TE material is quantified by a dimensionless TE figure of merit, $ZT = S^2s/k$, where S is the Seebeck coefficient, s is the electrical conductivity and k is the thermal conductivity. All three of the physical parameters are strongly dependent on the number of charge carriers (doping) and the arrangements of interfaces/grain boundaries. Thus the interplay of the three physical parameters leads to a very complex situation for engineering of charge and heat carriers in TE materials. For the new class of thermoelectric (TE) nanocomposites,[1-3] while the energy filtering by interfaces and grain boundaries is playing a vital role for enhancing the performances of TE materials, the thermodynamical instability of grain boundaries limits the application range and performance reproducibility.[1] To highlight this theme, we will present our work on the enhanced thermoelectric properties of chemically synthesized 2D n-type $Bi_2(Te/Se)_3$ nanoplatelet composites[2] and the effects of the thermodynamic environment during spark plasma sintering on the TE performance of $Bi_2Te_{2.7}Se_{0.3}$ nanoplatelet composites.[3] Our studies emphasize on a unique coupling charge carriers and phonons and a subtle understanding of the interface crystallization, grain growth and energy filtering from the thermoelectric interfaces.

1. Dresselhaus, M. S., *Adv. Mater.* **2007,** *19*, 1043-1053.

2. Soni, A.; Yanyuan, Z.; Ligen, Y.; Aik, M. K. K.; Dresselhaus, M. S.; Xiong, Q., *Nano Letters* **2012,** *12* (3), 1203-1209.

3. Soni, A.; Yiqiang, S.; Ming, Y.; Yanyuan, Z.; Ligen, Y.; Xiao, H.; Zhili, D.; Aik, M. K. K.; Dresselhaus, M. S.; Xiong, Q., **2012,** 12 (8) 4305.

(558d) Advanced Nanocomposite Fibers for Thermoelectric Energy Harvesting and Motion Sensing

Wu, Y., Purdue University
Finefrock, S., Purdue University
Originally presented on: 11/6/2013 16:45:00 - 17:15:00

In this presentation, we will discuss nanocrystal coated fiber based devices for thermoelectric energy harvesting from waste heat and the conversion of air movement (including wind as well as air flow created by motion). In addition, we will show that these devices are capable to detect an extremely small temeprature fluctuation created by motion, which could find many defense and civilian applications. We will also describe our investigations on the thermal conductivity in the axial direction of a single glass fiber coated with lead telluride (PbTe) nanocrystals using the self-heated 3ω method particularly at low frequency. While prior 3ω measurements have only been demonstrated for high thermal conductivity materials, the present work demonstrates the suitability of the 3ω method for PbTe nanocrystal coated glass fibers where the low thermal conductivity and high aspect ratio result in a significant thermal radiation effect. We simulated the experiment using a finite-difference method that corrects the thermal radiation effect and extract the thermal conductivity of a single glass fiber coated by PbTe nanocrystals to be 0.32-0.46 W/mK in the temperature range 260 K-300 K. This process of including both an experimental approach and numerical modeling could be used to study the intrinsic thermal conductivity of other one-dimensional nanowire or nanotube samples with low thermal conductivity at the single wire/tube level, thus significantly broadening the application of the 3ω measurement technique.

(558e) Development of a Nanocatalytic Microcombustor Power Device

McNally, D.,
Bakrania, S.,

Originally presented on: 11/6/2013 17:15:00 - 17:33:00

With the rising demand for portable power devices, the production of energy through micro-scale combustion of hydrocarbon fuels is becoming increasingly attractive. Microcombustors coupled with thermoelectric devices have the potential to provide a prolonged power source for portable devices as opposed to conventional batteries due to significantly higher energy densities. Recharging then can be easily achieved by simply adding more fuel for such integrated devices, making them desirable for remote operations. Current research is focused on the design of microreactors to sustain combustion and manage temperature gradients within. Previously, our work has successfully demonstrated room temperature ignition and control over catalytic activity of methanol-air mixtures with platinum nanoparticles. For this work, we have developed a microcombustor power generation device to explore the potential application of nanocatalytic combustion. Commercial thermoelectric generators were used to convert the temperature gradient created by combustion directly to electrical power. The overall performance of the device, together with the temperature characterization of the device, will be reported. Key parameters that influence performance and contribute to the modeling efforts were also identified. This work indicates a potential for a highly miniaturized design for such devices. At the same time, our own catalytic combustion research continues to drive the future design of an alternative power source to conventional battery technology.

(558f) Templated Fabrication and Characterization of Thermoelectric Nanowire Arrays - Toward Power-Dense and Efficient Devices

Crosser, L. A., Mainstream Engineering Corporation
Chester, G. E., Mainstream Engineering Corporation
Yelvington, P. E., Mainstream Engineering Corporation
Hill, J. J., Mainstream Engineering Corporation
Originally presented on: 11/6/2013 17:33:00 - 17:51:00

Efficient and power-dense thermoelectric devices would offer significant advantages over other forms of thermal energy conversion and control technologies. A promising route toward enhanced efficiency is through nanostructuring thermoelectric materials, the benefits of which have been predicted and demonstrated by many. However, cost effective assembly that preserves the performance enhancement of the individual nanostructures remains the dominant engineering challenge. Specifically, much attention has been focused on high aspect ratio structures such as nanowires, as they can directionally transport charge carriers and disrupt heat transport without the presence of grain boundaries or excessive interfacial sites. We are developing fabrication and characterization methods for the templated growth and assembly of thermoelectric nanostructures, including advanced low-energy nanofabrication techniques designed to both increase nanostructure anisotropy and enhance aerial power density compared to traditional templated approaches. Empirical and theoretical aspects of the fabrication methods will be highlighted with a focus on maintaining high power density (assembly density) and high efficiency (structure properties & confinement). We will discuss a self-assembly coating procedure which increases radial confinement in nanoporous oxide templates to produce densely arrayed and ultra-low diameter thermoelectric nanowires. Furthermore, we will discuss a method to supplant the loss in aerial power density introduced by the oxide template and its enhancement with respect to overall performance. Thermoelectric property measurement of thin films and anisotropic structures is non-trivial as thermal and electrical contact resistances significantly contribute to the measurement. We will briefly introduce an apparatus that measures thermopower as well as thermal and electrical conductivity designed specifically for thin film or anisotropic structures.

(558g) Discovering Materials With Ultra-Low Work Functions for Thermionics Energy Conversion

Chou, S. H., Stanford University
Voss, J., Stanford University
Vojvodic, A., SLAC National Accelerator Laboratory
Howe, R. T., Stanford University
Abild-Pedersen, F., Technical University of Denmark
Originally presented on: 11/6/2013 17:51:00 - 18:09:00

The work function is one of the most fundamental properties of a metallic surface, important in determining the material's applicability as a contact electrode [1, 2] or electron emitter [3]. Surfaces with ultra-low work functions could therefore improve technologies requiring precise control of contact barriers such as organic and printed electronics [2] or a variety of devices based on electron emission, ranging from fluorescent light bulbs [4, 5] to THz sources [6], thermionic energy converters (TECs) [7], and the photon-enhanced thermionic emission converters (PETEC) [8]. For TEC and PETEC, in particular, discovery of thermally stable materials with work functions of less than 1 eV would allow thermionic conversion of high-temperature (> 500 °C) heat or solar radiation directly to electricity with efficiencies exceeding 50%.

We present a model based on adsorbate atomic orbital interactions which offers new insights into the interfacial phenomena that control the work function [9]. In addition to the model, we will discuss *ab-initio* studies of surfaces with alloyed alkali metal and alkali-earth metal oxides. These adsorbed compound structures can enhance the emissivity of refractory metal substrates commonly used in current electron-emissive devices. By visualizing the electrostatic charge distributions, surface dipole formations, and the corresponding changes in the work function, we can devise design rules for low-work function materials to be incorporated into high-efficiency thermionic energy conversion devices.

[1] Lindell L, Burquel A, Jakobsson F L E, Lemaur V, Berggren M, Lazzaroni R, Cornil J, Salaneck W R and Crispin X 2006 Chem. Mater. **18**, 4246

[2] Zhou Y *et al* 2012 Science **336**, 327

[3] Yamamoto S 2006 Rep. Prog. Phys. **69**, 181–232

[4] Wada M, Gotoh S and Kurumada S 2009 J. Plasma Fusion Res. Series **8**, 1366

[5] Watanabe S, Watanabe T, Ito K, Miyakawa N, Ito S, Hosono H and Mikoshiba S 2011 Sci. Technol. Adv. Mater. **12**, 034410

[6] Snapp J P, Lee J-H, Provine J, Bargatin I, Maboudian R, Lee T H and Howe R T 2012 Sidewall silicon carbide emitters for terahertz vacuum electronics Solid-State Sensor and Actuator Workshop (San Diego, CA, June)

[7] Lee J-H, Bargatin I, Provine J, Clay W, Schwede J, Liu F, Maboudian R, Melosh N, Shen Z-X and Howe R T 2009 Tech. Dig. Power MEMS

[8] Schwede J W *et al* 2010 Nature Mater. **9**, 762

[9] Chou S H, Voss J, Bargatin I, Vojvodic A, Howe R T and Abild-Pedersen F 2012 J. Phys.: Condens. Matter **24**, 445007

(559a) Nanofunctionalized Organic Field-Effect Transistors for Selective, In Situ Biodetection

Hammock, M. L., Stanford University
Knopfmacher, O., Stanford University
Bao, Z., Stanford University

Originally presented on: 11/6/2013 15:15:00 - 15:30:00

Organic field-effect transistors (OFETs) are unique platforms for the detection of chemical and biological species. Such sensors are able to directly transduce an analyte-binding event into an electrical signal, making them highly compatible with sensing platforms requiring a digital readout. In particular, the detection of biologically relevant molecules lends itself to this detection platform because of the inherent charge associated with many biomolecules, which can be detected by the OFET. Additionally, OFETs are more biocompatible than their inorganic counterparts, facilitating their use as biosensors.

OFET sensor applications have historically been limited to the detection of small molecules in the vapor phase, due to the incompatibility of many organic semiconductors with water. Our group first demonstrated the real-time detection of several chemical species in aqueous media using an inherently water-stable organic semiconductor. These sensors were capable of low-voltage operation due to the incorporation of an ultrathin polymer dielectric layer, and were shown to be highly stable operating in aqueous conditions. However, while the limit of detection of these sensors has been demonstrated down to the part-per-billion (ppb) level, the response of the sensors was nonspecific, as the active layer itself displayed a serendipitous sensitivity to a number of small-molecule analytes.

In order to define sensitivity for a particular molecule of interest, a receptor group must be integrated into the device's architecture, preferably by a method that does not damage the OFET itself. We recently developed a novel OFET sensor platform that is capable of stable operation in an aqueous environment while also allowing for the selective detection of a user-defined analyte. Sensitivity for the targeted analyte is engineered by virtue of the ordered array of AuNPs that decorates the OFET's surface. This highly versatile platform is compatible with a large number of available, thiolated receptor groups that can be used to functionalize the AuNPs through the well-known gold-thiol (Au-S) linkage.

We have previously used this platform to demonstrate the highly selective detection of mercury(II) in solution. We have now expanded the sensing capabilities of this platform by demonstrating the selective detection of thrombin to 100 pM. Additionally, using this model system, we systematically investigated the effect of varying the ionic strength and pH of the buffer, as well as the average center-to-center distance between the receptor sites in order to form a more comprehensive picture of biodetection with OFETs. Using these parameters, we have devised a method to tune the dynamic range of our sensors in order to match that necessary for the particular application for which they will be used.

(559b) Peptide Nucleic Acid-Mediated Aggregation of Graphene Oxides and Applications in Detection of DNA Mutations

Kang, T., Yonsei University
Yoon, C., Yonsei University
Jeon, H.,
Kim, B., Hongik University
Yoon, K. A., National Cancer Center
Lee, K., Yonsei University
Originally presented on: 11/6/2013 15:30:00 - 15:45:00

Peptide nucleic acid-mediated aggregation of graphene oxides and applications in detection of DNA mutations

Taegyeong Kang[a], Cheolsang Yoon[a], Hyungjoon Jeon[a], Bumsang Kim[b], Kyong-Ah Yoon[c], and Kangtaek Lee[a,*]

[a] *Department of Chemical and Biomolecular Engineering, Yonsei University, Seoul, Republic of Korea*

[b] *Department of Chemical Engineering, Hongik University, Seoul, Republic of Korea*

[c] *National Cancer Center, Goyang, Gyeonggi, Republic of Korea*

Over the past several years, graphene oxides (GO) has been extensively studied due to its unique properties and potential applications in many fields. Recently, it is found that GO can be folded and aggregated under certain conditions. Herein, we report a simple method to induce aggregation of GOs in aqueous solution using the peptide nucleic acid (PNA). It is well known that single-stranded DNAs can easily adsorb onto the GO surface by non-covalent Â¥g-Â¥g stacking interactions. Therefore, addition of PNA that is a DNA analogue with a neutral polyamide backbone into the GO suspensions is expected to result in adsorption of PNA on the surface of GO, which leads to destabilization and aggregation of GOs. We have investigated aggregation of GOs using optical microscopy, TEM, and quasi-elastic light scattering. Furthermore, PNA-mediated selective aggregation of GO was applied to detect mutations in DNA sequence by using PNA as probe. Using this method, we could successfully detect mutations in DNA samples with high selectivity and sensitivity.

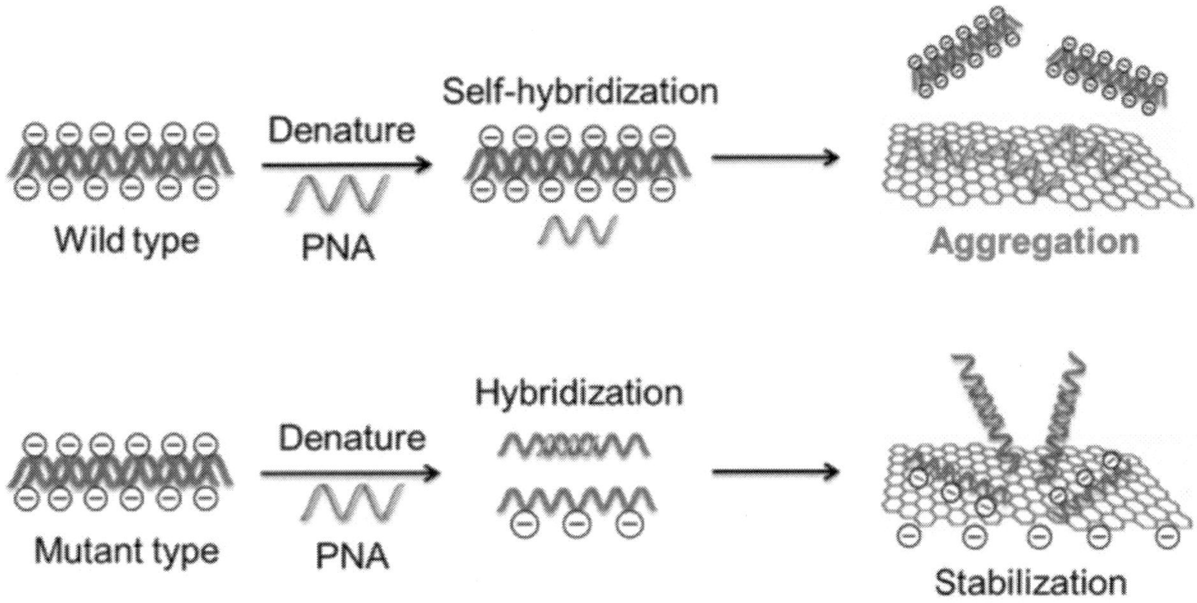

Figure 1. Schematics of PNA-mediated selective aggregation of graphene oxides

(559c) A Graphene-Based Physiometer Array for the Analysis of Single Biological Cells

Paulus, G. L. C., Massachusetts Institute of Technology
Lee, K.,
Reuel, N. F., Massachusetts Institute of Technology
Wang, Q. H., Massachusetts Institute of Technology
Grassbaugh, B.,
Kruss, S., Massachusetts Institute of Technology
Nelson, J.,
Landry, M., Massachusetts Institute of Technology
Zhang, J., Massachusetts Institute of Technology
Mu, B., Georgia Institute of Technology
Kang, J. W.,
Dasari, R.,
Opel, C. F., Massachusetts Institute of Technology
Wittrup, K. D., Massachusetts Institute of Technology
Strano, M. S., Massachusetts Institute of Technology
Originally presented on: 11/6/2013 15:45:00 - 16:00:00

A significant advantage of graphene as a biosensor is its potential to display a continuum of independent and aligned sensors at a specific interface. Here, we demonstrate a nanoscale version of a micro-physiometer – a device that measures cellular metabolic activity from the local acidification rate. Graphene can function as an array of independent pH sensors enabling subcellular and spatiotemporal detection of single biological cells. Raman spectroscopy shows that aqueous protons (H^+) p-dope graphene – in agreement with established doping trajectories, and that graphene displays two distinct pKa values (2.9 and 14.2). The cell physiometer allows micron spatial resolution of cellular acidification, differentiating immunoglobulin (IgG) producing human embryonic kidney (HEK) cells from non-IgG-producing control cells. Graphene also allows single cell counting in a population for mapping phenotypic diversity, variances in metabolic rate, and cellular adhesion. Finally we show the proposed platform can be extended to the detection of other analytes, e.g. dopamine. This work motivates the application of graphene as a unique biosensor array for cellular and sub-cellular interrogation.

(559d) A Gold Nanostructure-Dynamic Light Scattering Tandem for the Detection of Transcription Factors and Micro RNA

Seow, N.,
Yung, L. Y. L., NUS
Tan, Y. N.,
Originally presented on: 11/6/2013 16:00:00 - 16:15:00

Gold nanoparticles (AuNP) functionalized with DNA has shown to be versatile building blocks for the formation of nanostructures, which function as intermediate checkpoints or end-point readouts in diagnostic platforms. The affinity of thiolated DNA for citrate-capped AuNP not just allows their attachment onto AuNPs, but also creates conjugates that have unique physical properties. Through the use of techniques such as electrophoretic mobility gel shift assays, conjugates bearing distinct numbers of DNA can be recovered, and used as probes for further detection.

We have designed conjugates that are complementary to specific miRNA targets, whereupon binding nanostructures are formed. Due to the control of the DNA loading on AuNP, dimers are primarily formed, which is unlike typical aggregation techniques that is without control and lacks a clearly defined structure. This dimer formation process also leads to distinct physical changes such as potential plasmon coupling of the constituent AuNPs, and more directly, an overall increase in the size of the system. To present the successful dimer formation process, we queried the as-formed nanostructures on the dynamic light scattering (DLS) platform. DLS has been well utilized in the characterization of nanoparticles and polymers, and we seek to leverage the DLS readout to show the formation of nanostructures in the presence of the miRNA target. Given the large scattering cross section of AuNPs, and enhanced by the coupled particles in a dimeric arrangement, DLS can distinctly show the increase in the size of the particles in the system, which reflect the binding and detection of the target. The size increase from the nanostructure formation also leads to even more intense signals on the DLS, and this amplification further enhances the sensing capabilities of the technique. With miRNA as the model of study, we present AuNP probes that have successfully detected and distinguished members (let7a, 7f and 7g) of the let7 miRNA family, in a significantly selective and sensitive manner, that is also rapid (DLS readout presented in 5 minutes). Optimization works have also been done to enhance the binding process and stability of the readout. With more works reporting on the role of miRNAs in the pathogenesis of diseases and control of cellular machinery, the detection and study of miRNA has taken on increasing importance, and a simple and direct screening of miRNA in a quantitative manner as shown in our technique presents a viable alternative for miRNA sensing.

In addition, we further developed the nanostructure platform such that dimers are not just endpoints but probes for the detection of transcription factors. With the estrogen receptor (ER) being strongly implicated in breast cancer and treatment considerations, there is strong motivation to study these transcription factors. We had fabricated dimers containing the ER-binding site - ERE, localised within the DNA duplex holding the two AuNPs together. The dimers purified and recovered through gel elecctrophoresis presented a single distinct peak on the DLS. When ER is added, an additional complex peak was observed. And this was observed only for ER on dimers with ERE. When a different DNA sequence was used, or a different protein was queried, little or no complex peak was observed, which validated the selectivity of the technique. A pM detection limit is also an improvement over other AuNP methods.

(559e) Micro-RNA Profiling for Oral Cancer Screening By a Nanocone Optical Fiber Array

Wang, Y., University of Notre Dame
Senapati, S., University of Notre Dame
Stoddart, P., Swinburne University of Technology
Howard, S., University of Notre Dame
Chang, H. C., University of Notre Dame
Originally presented on: 11/6/2013 16:15:00 - 16:30:00

Abnormal experssion of MicroRNAs has been connected to cancer, diabetes, cardiovascular and other chronic diseases. Screening platforms based on miRNA profiling are hence promising early prescreening tests that can significantly improve the survival rate of the more deadly diseases. However, short life time (< 1 hour) and low experssion level of miRNAs pose chanllenges for accurate microRNA profiling, particularly for future transformative home-use personalized medical devices that cannot rely on expensive instrumentation and extensive manual manipulation. Real-time PCR can quantify low number of miRNAs, but can only detect one target at a time, making it impossible to profile a panel of target miRNAs. DNA microarray can detect thousands of targets but only at higher concentrations and takes tens of hours due to diffusion limitation to surface functionalized probes. Both require reverse transciption with expensive instrumentation. Rapid and sensitive multi-target detection and quantification would eliminate reverse-transcription and allow direct miRNA profiling from the sample.

We report a novel conical optical fiber platform, with low-cost LED light sources and miniature cameras in a turn-key instrument, that can profile a large panel of microRNAs down to fentoMolar within 15 minutes, thus allowing direct profiling. The platform relies on unique transport and optical physics at nanoscale geometric singularities. Molecular diffusion to a conic tip with a nanoscale diffusion length and a correspondingly large concentration gradient is much faster than to a planar microarray. Focused optical and plasmonic fields at cone tip also amplifies the fluorescence intensity by 3 orders compared to flat SPR. We recently showed theoretically that plasmonic excitation at a metal cone is broadbanded, encompassing the entire visible spectrum rather than the monochromatic plasmonic frequencey of a gold nanoparticle (Wang, Plouraboue and Chang, Optics Exp, 21, 6609(2013)). A generic multi-spectral platform for SERS, FRET, transmission/absorption, second-harmonic CARS, near-field, interference biosensing can hence be designed based on such unique optical phenomena at geometric singularities. A first prototype with plasmonically enhanced FRET hairpin probes for miRNA profiling is reported here.

Conical array is fabricated by wet etching commercial imaging fiber bundles, with thousands of small fiber cores with optimum spacing for coupling with bulk light. The cone tip has a radius of curvature below 10 nm and can be flattened for near-field applications. A thin gold layer (5 nm) is thermally evaporated onto the array to sustain plasmonics. Hairpin probes with fluorophore tags quenched by the gold film are functionalized to the gold surface. Plasmonically enhanced fluorescent reporting of hybridization occurs when the target molecule breaks the hairpin. Detection of multiple targets in a panel can be achieved by addressing each fiber core optically and using photochemistry to selectively attach different probes. Detailed spectral measurements show broadbanded plasmonic excitation of 3 fluorophores per cone to minimize interference. Intensity tracking in time show a signature constant rapid absorption flux to a conic tip, followed by a slow $t^{-1/2}$ diffusive flux to the substrate. The enhanced tip fluorescence allows detection of the approximately 100 miRNA at the tip in the first region within 15 minutes. Profiling data for miRNA of oral cancer cells from a heterogeneous sample will be shown.

(559f) Towards a Biocompatible Conductive Nanotube Film: An In-Depth Investigation Into Cellular Biocompatibility

Lin, D. W., University of Maryland, Baltimore County
Bao, Z., Stanford University
Originally presented on: 11/6/2013 16:30:00 - 16:45:00

Single-walled carbon nanotubes (SWNTs) have shown promise for use in organic electronic applications including thin film transistors, conducting electrodes, and biosensors. There is a current need to rapidly process SWNTs from solution phase to substrates in order to produce device structures. In terms of SWNT film deposition, previous studies were able to adsorb SWNTs by drop casting, airbrush spray coating, spin coating, vacuum filtration, electrophoretic deposition, and Langmuir-Blodgett deposition. Furthermore, researchers have found that surfaces covalently functionalized with primary amines have been shown to selectively adsorb semiconducting SWNT. However, this and similar techniques are dependent upon environmentally sensitive surface modification techniques. Hence, we explored the potential of substrates modified with physisorbed polymers, poly(L-lysine) (PLL), as a possible alternative methodology. In this work, we detail a number of methods for depositing SWNTs onto various substrate materials using amine-rich PLL and other methods of covalently functionalizing the surface with primary amines. Furthermore, devices were constructed using these methods to observe if cell movement on the surface would elicit changes in the device performance. SWNT adsorption and alignment were characterized by atomic force microscopy (AFM). SWNT surface density was strongly dependent upon the adsorbed concentration of PLL on the surface, spin coating speed, and SWNT solution concentration. Another benefit for using PLL as an adhesion layer was for its biocompatibility with cells. Results from examining mitochondrial hydrogenase activity and Live/Dead fluorescence assay suggest that the PLL SWNTs spin-coat devices exhibited higher biocompatibility with NIH-3T3 fibroblast cells than the drop cast SWNTs devices possibly due to differences of substrate surface roughness. To further elucidate the effect of SWNT roughness on biocompatibility, cell morphology was observed on substrate surfaces of varying SWNT network density using a spray coating method. Additional cells lines with relevant characteristics were also tested for their biocompatibility which included C2C12 myotubes and cardiomyocytes. Furthermore, to observe if cell contractile forces on the device surfaces would elicit a change in electrical performance, 2-terminal resistance measurements were taken at different stages of cell adhesion onto the surface. We envision these conducting biocompatible SWNT networks could potentially be used as biosensors to investigate cell adhesion mechanics or provide a method for cancer diagnostics.

(559g) Prospects for Direct Electron Transfer in Gen-3 Biosensors and Advanced Biofuel Cells With Swnt-Gox Conjugates

Karunwi, O., Center for Bioelectronics, Biosensors and Biochips (C3B)
Guiseppi-Elie, A., Clemson University
Originally presented on: 11/6/2013 16:45:00 - 17:00:00

There is pressing need for the design, development and understanding of abio-bio interfaces that permit direct electron transfer (DET) between oxidoreductase enzymes and metallic, carbonaceous or semiconductor electrodes *[1]*. Generation-3 (Gen-3) biotransduction systems used in biosensors and advanced enzyme biofuel cells may exploit DET enabled by single walled carbon nanotube (SWNT)-enzyme conjugates *[2]*. In this work we explore the use of diazonium salts to covalently immobilize pyrrole monomer onto glassy carbon electrodes (GCE) and the electropolymerization of Py as a method of biofabrication to immobilize SWNT-GOx conjugates onto platinum and glassy carbon electrodes (GCE). Glucose oxidase (GOx) – single walled carbon nanotube supramolecular conjugates (GOx-SWNT) were produced using ultrasonication and ultracentrifugation *[3]* and were shown to suffer very little loss of bioactivity regardless of nanotube fiunctionalization *[4]*. The GOx-SWNT conjugates were immobilized on Φ=100μm Pt microelectrodes (μE) via electropolymerization of pyrrole (Pt|PPy-(GOx-SWNT)) and compared to a typical Gen-1 system wherein GOx was immobilized onto Pt mE via electropolymerization of pyrrole (Pt|PPy-GOx). Electropolymerization times required to achieve 100 mC/cm^2 were 45 min and 26 min for GOx and GOx-SWNT, respectively. The electrodeposited thin films were then overoxidized (OOPPy) (40 cycles, 100 mV/s, -0.2- +1.3 V *vs.* Ag/AgCl, PBS 7.2) to eliminate the role of the conductive PPy layer and allow the effect of the presence of SWNTs to be readily observed. Electrical and electrochemical impedance spectroscopy (EIS) was used to characterize the newly fabricated biotransducers in freshly prepared phosphate buffered saline (1X PBS pH = 7.2) at room temperature in two or three electrode mode, respectively. EIS used non-perturbing 10 mV amplitude over the frequency range 10^{-1} – 10^6 Hz. Multiple scan rate cyclic voltammetry (MSRCV) was performed using 1.0 mM FcCOOH in 1X PBS buffer (pH=7.2). EIS of the Gen-1 (Pt|PPy-GOx) and Gen-3 (PT|PPy-(GOx-SWNT)) biotransducer constructs prepared with increasing electropolymerization charge density caused a dispersion shift towards higher frequencies as well as an increase in the overall impedance magnitude. Both Gen-1 and Gen-3 constructs showed similar behavior for charge densities up to 1 mC/cm^2 while the Gen-3 constructs exhibited a lower impedance magnitude for charge densities > 1 mC/cm^2. MSRCV produced the effective areas of the electrodes from an application of the Randles-Sevcik equation and the literature value of D = 4.51 x 10^{-6} cm^2/s (FcCOOH). There was a ca. 80% reduction of the original electrode area in the case of Pt|OOPPy-GOx compared to a 25% reduction in the case of the Pt|OOPPy-(GOx-SWNT), confirming electrochemical communication between the FcCOOH probe and the underlying Pt electrode. The heterogeneous rate constant k^0 for the electron transfer (ET) reaction for each biotransduction system was obtained from an analysis of the cyclic voltammograms using the methodology described by Nicholson *[5]* and a linearized version of the Nicholson approach developed by Leddy *et al* ***[6]***. The presence of the SWNT within the Pt|OOPPy-(GOx-SWNT) system acts as a conducting medium through the overoxidized ultra-thin film for enhanced electron transfer between the probe molecule in solution and the underlying metallic electrode.

Bioanalytical performance of sensitivity (S), linear dynamic range (LDR) and detection limit [3(SD(blank)/S] as well as apparent enzyme kinetic parameters of $\mathbf{K}_M$, I_{max}, and k_{cat}, were obtained from the dose-response curves for glucose and compared to solution phase behavior.

References:

[1] Karunwi O, Wilson AN, Kotanen C, Guiseppi-Elie A. Engineering the Abio-Bio Interface to Enable More than Moore in Functional Bioelectronics. Journal of The Electrochemical Society. 2013;160:B60-B5.

[2] Guiseppi-Elie A, Lei C, Baughman RH. Direct electron transfer of glucose oxidase on carbon nanotubes. Nanotechnology. 2002;13:559.

[3] Guiseppi-Elie A, Choi S-H, Geckeler K, Sivaraman B, Latour R. Ultrasonic Processing of Single-Walled Carbon Nanotube–Glucose Oxidase Conjugates: Interrelation of Bioactivity and Structure. NanoBiotechnology. 2008;4:9-17.

[4] Karunwi O, Guiseppi-Elie A. Supramolecular glucose oxidase-SWNT conjugates formed by ultrasonication: effect of tube length, functionalization and processing time. Journal of Nanobiotechnology. 2013;11:6.

[5] Nicholson RS. Theory and Application of Cyclic Voltammetry for Measurement of Electrode Reaction Kinetics. Analytical Chemistry. 1965;37:1351-5.

[6] Paul HJ, Leddy J. Direct determination of the transfer coefficient from cyclic voltammetry: isopoints as diagnostics. Analytical Chemistry. 1995;67:1661-8.

(559h) Colorimetric Detection of Ionizing Radiation Using Polypeptide-Templated Gold Nanoparticles

Pushpavanam, K.,
Walker, C. R., Arizona State University
Nair, D. G., Arizona State University
Potta, T., Arizona State University
Sutiyoso, C.,
Sapareto, S.,
Chang, J.,
Rege, K., Arizona State University
Originally presented on: 11/6/2013 17:00:00 - 17:15:00

Ionizing radiation, including gamma rays and X-rays, has numerous applications in nuclear energy, astrophysics, food technology[1], biomedical imaging[2], and cancer therapy[3] However, ionizing radiation causes DNA damage in cells[5] leading to carcinogenesis, and can also cause deleterious effects on the environment[6]. Consequently, the necessity for detecting ionizing radiation has led to development of methods based on fluorescence[7], quantum dots[8] and inorganic films[9]. Most of these detection methods suffer from serious drawbacks of being expensive and hard to fabricate on a large scale. Here, we describe a detection system in which cysteine-containing elastin-like polypeptides (C_nELPs; n=2,12; C=cysteine) were used to reduce colorless metal salts to colored metallic nanoparticles. C_nELPs were used to template gold and silver nanoparticles from metal salts upon exposure to different doses of ionizing radiation. Nanoparticle formation was visualized colorimetrically by the naked eye, and spectroscopically by detecting the plasmonic peak of gold at 520 nm and silver at 420 nm. While gold and silver nanoparticles were formed at doses starting at 175 and 50 Gy, respectively, nanoparticle was not observed when free cysteine's were used, indicating the key role of the polypeptide in this detection system. Transmission Electron Microscopy (TEM) indicated formation of 10-20 nm metallic cores, while dynamic light scattering studies indicated the formation of nanoparticles with hydrodynamic diameters of 80-150nm, following exposure to ionizing radiation. Nanoparticle hydrodynamic diameters reduced with an increase in radiation dose, likely to the loss of integrity of the polypeptide under these conditions. These studies indicate a straightforward colorimetric method for detecting ionizing radiation based on nanoparticle formation from colorless metal salts, and also describes a facile method of nanoparticle formation in which, nanoparticle size can be tailored based on radiation dose and C_nELP type.

REFERENCES

1. Kader, A. A., POTENTIAL APPLICATIONS OF IONIZING-RADIATION IN POSTHARVEST

HANDLING OF FRESH FRUITS AND VEGETABLES. *Food Technology* **1986,** 40, (6), 117-121.

2. Jung, S. Y.; Ahn, S.; Seo, E.; Lee, S. J., Detection of circulating tumor cells via an X-ray imaging technique. *Journal of Synchrotron Radiation* **2013,** 20, 324-331.

3. Chinnaiyan, A. M.; Prasad, U.; Shankar, S.; Hamstra, D. A.; Shanaiah, M.; Chenevert, T. L.; Ross, B. D.; Rehemtulla, A., Combined effect of tumor necrosis factor-related apoptosis-inducing ligand and ionizing radiation in breast cancer therapy. *Proceedings of the National Academy of Sciences of the United States of America* **2000,** 97, (4), 1754-1759.

4. Alrehaily, L. M.; Joseph, J. M.; Musa, A. Y.; Guzonas, D. A.; Wren, J. C., Gamma-radiation induced formation of chromium oxide nanoparticles from dissolved dichromate. *Physical Chemistry Chemical Physics* **2013,** 15, (1), 98-107.

5. Sachs, R. K.; Chen, P. L.; Hahnfeldt, P. J.; Hlatky, L. R., DNA DAMAGE CAUSED BY IONIZING-RADIATION. *Mathematical Biosciences* **1992,** 112, (2), 271-303.

6. Sax, K., THE EFFECT OF IONIZING RADIATION ON PLANT GROWTH. *American Journal of Botany* **1955,** 42, (4), 360-364.

(559i) Chemoresistive Responses of Functionalized SWNTs

Hines, D., The City College of New York
Adebimpe, D.,
Akins, D. L., The City College of New York
Guiseppi-Elie, A., Clemson University
Originally presented on: 11/6/2013 17:15:00 - 17:30:00

There is need for advanced synthetic molecular chemoreceptors in the design and development of all electronic Natural Olfactory Sensor Emulators® (NOSEs)[1]. Carbon nanotubes with their semiconducting properties and relative ease of non-covalent and covalent functionalization are well suited for chemoreceptor use in e-Noses[2]. A series of functionalized SWNTs rationalized on the basis of covalent chemical modification to impart electron-donating or electron-withdrawing functionalities have been synthesized and characterized[3]. These derivatized-nanotube-based materials are designed to serve as chemoreceptors that can facilitate the development of highly selective and sensitive chemical and biological sensor arrays through an "electronic nose" approach which mimics the mammalian olfactory system[4]. Functionalized SWNTs (f-SWNTs) were dispersed in dimethyl formamide (DMF) and cast onto the interdigit space of microlithographically fabricated, pre-cleaned interdigitated microsensor electrodes (IME 1025-M-Pt). Devices were 10 mm line and space with 25 fingers of Pt(100nm)/TiW(20nm) patterned on borosilicate glass. Devices were solvent cleaned and chemically modified with octadecyltrichlorosilane (OTS), cathodically cleaned (20 cycles, 100 mV/s, PBS 7.2), rinsed with isopropyl alcohol and dried in flowing argon. The resulting devices (receptors) were characterized by two–electrode impedance spectroscopy (20 mV p-t-p; 10^{-1} – 10^{6} Hz; RT) and equivalent circuit modeling in various gases/vapors[5]. We compare the EIS profiles when the same chemoresistive f-SWNT layer was cast onto gold vs. platinum devices. We compare the EIS profiles when the same chemoresistive f-SWNT layer was cast onto IME devices that were unmodified or chemically modified with octadecyltrichlorosilane (OTS). The chemo-resistive responses to these same gases/vapors were determined under DC conditions by continuous monitoring following flow injection of gas/vapor.

Keywords: chemoresistors, nanotubes, olfaction, impedance

References:

1. Tchoupo, G.N. and A. Guiseppi-Elie, *On pattern recognition dependency of desorption heat, activation energy, and temperature of polymer-based VOC sensors for the electronic NOSE.* Sensors and Actuators B: Chemical, 2005. **110**(1): p. 81-88.

2. Tasis, D., et al., *Soluble Carbon Nanotubes.* Chemistry – A European Journal, 2003. **9**(17): p.

4000-4008.

3. Yu, C., et al., *Modulating Electronic Transport Properties of Carbon Nanotubes To Improve the Thermoelectric Power Factor via Nanoparticle Decoration.* ACS Nano, 2011. **5**(2): p. 1297-1303.

4. Hirsch, A. and O. Vostrowsky, *Functionalization of Carbon Nanotubes*, in *Functional molecular nanostructures*, A.D. Schlüter, Editor 2005, Springer Berlin Heidelberg. p. 193-237.

5. Yang, L., A. Guiseppi-Wilson, and A. Guiseppi-Elie, *Design considerations in the use of interdigitated microsensor electrode arrays (IMEs) for impedimetric characterization of biomimetic hydrogels.* Biomedical Microdevices, 2011. **13**(2): p. 279-289.

(559j) Hydrophobic Nanofibers With Silver Nanoparticles As a Surface Enhanced Raman Spectroscopy Substrate

Phan, J. C., University of Washington
Tillman, A., University of Washington
Woodrow, K. A., University of Washington
Originally presented on: 11/6/2013 17:30:00 - 17:45:00

Surface enhanced Raman spectroscopy (SERS) has emerged as a platform for ultrasensitive detection of analytes. A key issue in the production of SERS substrates is the production of substrates with high enhancement, reproducibility, and stability. Electrospun polymeric nanofibers incorporating metal nanoparticles have shown potential as a SERS substrate. Existing SERS-active nanofibers have primarily been synthesized with water dissolvable polymers, which are not suitable for detection of biological analytes in aqueous solutions. We show the production of a SERS-active polycaprolactone (PCL) fiber containing silver nanoparticles (AgNP). Electrospinning caused alignment of silver nanoparticles along the axis of the fiber. The PCL-AgNP fiber exhibited at least 10-fold enhancement of the Raman spectra of rhodamine 6G (R6G). The hydrophobic surface of PCL improved signal enhancement of samples dried on the surface, due to the reduced contact area of the droplet and resulting increase in surface concentration. The SERS signal is reproducible across the mat and between electrospinning syntheses. Moreover, we show improved stability of signal enhancement within the fiber compared to silver aggregates. This SERS substrate provides a cost-effective, highly reproducible and stable platform for the sensitive detection of molecules in aqueous solutions.

(559k) Molecule-Terminated, Oxide-Free Silicon Nanowire Field Effect Transistors : Effect of the C-C Bond Nearest to the Surface

Halpern, J. M., Case Western Reserve University
Haick, H., Technion - Israel Institute of Technology
Originally presented on: 11/6/2013 17:45:00 - 18:00:00

We have recently developed silicon nanowire enhanced field effect transistor (FET) devices to improve the sensitivity and diagnostic yield of the current clinical chemiresistors.[1] The silicon nanowire, which has a natural passive oxide layer, can be modified to increase the sensitivity and stability of these devices. Modifications of the device can be made on top of the oxide layer by reacting silanes with oxygen groups;[2] however, we wish to alkylate the nanowire surface without oxygen with the goal of increased sensitivity and selectivity.

We will report on the alkylation of the silicon nanowires (Si NW) on the field effect transistor, as well as the effect of alkylation on the sensitivity and stability of the sensor. The modification is conducted in a three stage reaction mechanism: (1) removal of the oxide layer via HF etch, (2) Si NW Chlorination with a PCl_5 solution, and (3) alkylation with propylmagnesium chloride, 1-propenylmagnesium bromide, and 1-propynylmagnesium bromide. The result is an alkylated monolayer on top of the Si NWs with Si-CH_2-CH_2-CH_3, Si-CH=CH-CH_3, and Si-C≡C-CH_3 terminals respectively. X-ray photon spectroscopy is used to confirm the surface modification, and electrical characterization is used to evaluate the current-voltage profiles; both analyses will verify the success of the surface reaction. Our goal is to achieve complete monolayer coverage with an increase in the device's on/off ratio. We also use standardization gas mixtures (simulated cancer breath samples) to create calibration curves for our sensors and make predictions on the sensitivity and selectivity. We hypothesize that the 1-propenylmagnesium bromide modification (double bond) will be the most stable, while the 1-propynylmagnesium bromide modification (triple bond) will be the most sensitive.

In conclusion, alkylated silicon nanowire field effect transistors could provide a more accurate and sensitive diagnostic breath sensor for evaluating and discriminating various diseases.

References

1. Peng G, Tisch U, Adams O, Hakim M, Shehada N, Broza YY, Billan S, Abdah-Bortnyak R, Kuten A, Haick H. Diagnosing lung cancer in exhaled breath using gold nanoparticles. Nature Nanotechnology 2009;4:669-673.

2. Wang B, Haick H. Effect of functional groups on the sensing properties of silicon nanowires toward volatile compounds. Applied Materials & Interfaces 2013:in press.

(559l) Surface Roughed Pt Nanowires and Their Application As Electrochemical Sensors

Gao, F., University of Massachusetts Lowell
Li, Z.,
Ruan, D.,
Gu, Z., University of Massachusetts Lowell
Originally presented on: 11/6/2013 18:00:00 - 18:15:00

Sensor platforms require significant improvements in sensitivity, specificity and parallelism to meet the future needs in a variety of fields. In this research, platinum nanowires with different surface roughness were fabricated by a controlled galvanostatic electrodeposition method for electrochemical sensors toward hydrogen peroxide detection. The electrochemical behavior has been studied for oxidation of hydrogen peroxide by using cyclic voltammetry and amperometry in phosphate buffer solution. Surface roughness was found to enhance the sensing performance of the Pt nanowire based electrochemical sensor towards hydrogen peroxide, including the higher sensitivity, lower limit of detection and wider linear dynamic range. Furthermore, the effect of the nanowire concentration has been found to play a role in the electrochemical sensitivity. Finally, real sample measurement is studied to quantify the sensing capability based on the surface roughed Pt nanowires.

(630a) Silicon Nanostructures As Efficient Thermoelectric Materials

Yang, P., Unversity of California, Berkeley
Lim, J.,

Originally presented on: 11/7/2013 8:30:00 - 9:00:00

It has been recently reported that thermal conductivity greatly decreases in single crystalline silicon nanostructures without much affecting electrical properties. Although it is qualitatively demonstrated that surface roughness in nanowires and nanohole arrays in silicon membranes (holey silicon) play a significant role in reducing thermal conductivity, the underlying reasons remain unknown and warrant quantitative studies and analysis. In this report, silicon nanowires were controllably roughened and the surface roughness near the length scale of thermal phonon wavelength is shown to interfere with phonon transport below the Casimir limit. Furthermore, a platform to simultaneously measure thermal conductivity, electrical conductivity, and the Seebeck coefficient for holey silicon has been developed to extract ZT directly from a single device. Decoupling of electron and phonon transport associated with their mean free path has been studied at phonon glass electron crystal regime. Lastly, cross-plane thermal conductivity measurement of holey silicon unveiled the multi directional phonon-glass property of holey silicon.

(630b) Toward Physical Models of Thermoelectric Transport Rules At the Organic-Inorganic Interface

Urban, J., Lawrence Berkeley National Laboratories
Originally presented on: 11/7/2013 9:00:00 - 9:30:00

The thermoelectric properties of solution-processed conducting polymer/nanoparticle composite films have been a focus of interest for many groups due to their unique transport mechanisms, enabling new modalities of transport which can surpass common optimization dilemmas in inorganic thermoelectrics. However, many interesting reports have lacked a theoretical framework to guide materials design. Here, I present a few organic/inorganic thermoelectric model systems studied as a function of nanoparticle loading and chemical doping. We use these to develop phenomenological frameworks to guide engineering of this novel material class.

Here I will discuss two different hybrid systems, and the tunability and optimization of their thermoelectric performance: 1.) Poly(3,4 ethylenedioxythiophene):poly(styrenesulfonate) (PEDOT:PSS) and Tellurium nanowire devices, and Polyaniline and nanoparticle-Titania devices. For each of these systems, interesting non-monotonic electrical conductivity is observed at intermediate loadings, suggesting non-effective medium behavior, in contrast to the values of the Seebeck coefficient increase linearly with mass fraction of inorganic. These trends in Seebeck coefficient and electrical conductivity lead to an optimized power factor in the composites that exceed that of the individual components. The surprising electrical conductivity behavior can be explained with a mixing model where carrier transport is primarily through a highly conductive volume of polymer that exists at the nanoparticle-polymer interface. We have developed a general phenomenological model for these systems suggesting that this increase in conductivity results from a structural modification of the polymer, and elucidates a novel route for optimizing transport in nanoscale organic/inorganic composites.

(630c) Thermal and Thermoelectric Transport in Semiconductor Nanowires

Chen, R., University of California, Berkeley

Originally presented on: 11/7/2013 9:30:00 - 10:00:00

Semiconductor nanostructures are ubiquitous in electronic devices. Thermal transport plays an important and sometimes pivotal role in the performance of these devices. For instance, thermal coupling between nano-scale transistors and the surrounding materials would dictate the heat dissipation of the transistors. Thermal conductivity is an important parameter to engineer to enhance energy conversion efficiency of thermoelectric devices. During the past decade, we have witnessed tremendous efforts devoted to the understanding and manipulation of thermal transport properties in semiconductor nanostructures, including thin films, superlattices, nanowires and nano-composites. However, most of these studies have been focused on the classical size regime where the characteristic sizes of nanostructures are greater than dominant phonon wavelength. As the size of nanostructures keeps shrinking and becomes comparable to phonon wavelength, the quantum confinement effect would emerge. Exact picture of phonon transport in this new regime, however, is not fully explored. In this presentation, we use semiconductor nanowires with ultra-small diameters as a model system to explore the quantum confinement effect on thermal and thermoelectric transport. We will present experimental techniques we have developed recently to probe thermal and thermoelectric transport properties in thin nanowires and the new understanding obtained from these experimental studies.

(630d) Nanowire Heterostructure-Based Thermoelectric

Wu, Y., Purdue University
Yang, H., Purdue University
Fang, H.,
Originally presented on: 11/7/2013 10:00:00 - 10:20:00

In this presentation, we will describe our latest progress on the general, reproducible, and scalable synthesis of nanowire heterostructures for potential thermoelectric energy harvesting for low grade heat recovery. Particularly, through a multiple step solution phase process, we can mass produce telluride based thermoelectric nanowire heterostructures with embedded nanoparticles or nanoplates at two ends. In certain cases, significantly improved performance compared to bulk materials has been observed due to the contribution of reduced thermal conductivity through interface phonon scattering and enhanced Seebeck coefficient through energy filtering. The synthesis method could be used to produce nanowires up to kilogram level using our large batch reactor while still maintaining a narrow diameter/length/performance distribution, which could open the door for the real applications of nanostructured thermoelectric materials. These nanowire heterostructures also provide novel platforms in a reproducible way to study new theories in electrical and thermal transport at nanoscale interface inculiding quantum confinement, energy filtering, and phonon scattering.

(630e) Thermoelectric Properties of Bulk Pellets of Unfunctionalized and Functionalized Zn3P2 Nanowires

Brockway, L., Texas A&M University
Vasiraju, V.,
Vaddiraju, S., Texas A&M University
Originally presented on: 11/7/2013 10:20:00 - 10:40:00

Zinc phosphide (Zn_3P_2) nanowire powders, both unfunctionalized and surface-functionalized with organic molecules to passivate the surfaces, were mass produced using reactive vapor transport of phosphorus onto zinc foils in a hot-walled chemical vapor deposition chamber, and hot-pressed into dense pellets for the characterization of their thermoelectric performance. The intent here is to explore the possibility of fabricating thermoelectric modules using inexpensive compound semiconductors. Contrary to expectations, the hot-pressing did not alter the morphology or the dimensions of the nanowires in the pellets, despite the fact that high packing densities (>98%) were achieved in the pellets. It was observed that enhanced flexibility observed in thin nanowires allows for their assembly into highly dense pellets, without any morphological alterations. The elastic nature of the Zn_3P_2 nanowires with diameters less than 50 nm was confirmed using *in-situ* transmission electron microscopy (TEM) studies.

Overall, this experimentation indicated that a 30% reduction in thermal conductivity relative to the bulk is feasible by synthesizing Zn_3P_2 in nanowire format. At the same time, this procedure did not adversely affect the electrical conductivity. In fact, prevention of surface oxidation by functionalizing the nanowires with 1,4-benzenedithiol before their assembly into pellets led to an overall increase in the electrical conductivity of the nanowire pellets. These measurements indicate that it is possible to enhance the thermoelectric performance of materials by synthesizing them in nanowire format. The thermoelectric performance of both unfunctionalized and functionalized Zn_3P_2 nanowire pellets, along with those obtained using by doping them with copper, will be discussed in detail in this presentation.

(630f) Combustion Synthesis of Thermoelectric Oxide Powders

Lin, S., Lamar University
Duruewuru, P., Lamar University
Bonura, J., Lamar University
Originally presented on: 11/7/2013 10:40:00 - 11:00:00

Doped calcium cobaltate powders were prepared by Self-propagating High-temperature Synthesis (SHS). Reactant powders are consolidated to a pellet and ignited by an external heat source. The heat released from the reaction is sufficient to sustain the movement of the reaction front to convert the reactants into the product. TG-DSC analysis was performed to establish the reaction mechanism. Thermocouples and an IR camera were used to measure the temperature history at the pellet centerline and pellet surface during the combustion to determine the velocity of combustion front movement. Products were characterized by XRD for phase purity. Synthesized powders were hot-pressed to form dense pellets and characterized for their thermoelectric properties. Our results show that SHS can economically produce thermoelectric oxides with high ZT values and the thermal conductivity of calcium cobaltate decreases when silver is doped to replace calcium.

(631a) Invited Talk: Controlling Electronic Properties Through Block Copolymer Self-Assembly and Crystallization in Poly(3-alkylthiophenes)

Segalman, R. A., University of California at Berkeley
Originally presented on: 11/7/2013 8:30:00 - 9:00:00

The performance of organic electronic devices depends on the crystalline structure and the morphology at the nanometer length scale. Block copolymers containing a semicrystalline conjugated polymer, such as poly(3-alkylthiophene) (P3AT), can simultaneously address these design considerations by confining crystallization to well-defined nanoscale domains. However, in many conjugated block copolymer systems, crystallinity dominates the thermodynamics of self-assembly such that the melt phase morphology can be significantly distorted. In these cases, periodic domains characteristic of block copolymer self-assembly are not observed, and the morphology at the nanoscale is nearly identical to that of the semicrystalline homopolymer. We show that the rational selection of the alkyl side chain in P3ATs results in control of the thermal transitions over a range of 150C, and that the optoelectronic properties are not significantly altered in the process. Crystallization can be observed in real-time and the resulting UV-vis absorbance and field-effect transistor mobility are correlated to the structural evolution. Interestingly, while the degree of crystallinity increases monotonically over the course of an hour, the hole mobility increases by nearly two orders of magnitude once a threshold value is reached. Additionally, when these poly(3-alkylthiophenes) are incorporated into block copolymers, the decreased melting transition allows for cases in which the strength of segregation is sufficiently high during crystallization to maintain the microphase separated morphology. Phases such as hexagonally-packed cylinders and lamellae are obtained, and it is shown that the crystalline structure of the conjugated polymer is preserved even under confinement. Consequently, the optical bandgap and hole mobility, as determined by time-of-flight measurements, are identical to that of the parent homopolymer.

(631b) A Look Into the Deformation Events in Block Copolymer Modified Epoxies: A SAXS Perspective

Declet-Perez, C., University of Minnesota
Francis, L. F., University of Minnesota
Bates, F. S., University of Minnesota
Originally presented on: 11/7/2013 9:00:00 - 9:20:00

Epoxy resins are an important class of thermosetting polymers. Their high crosslink density provides stiffness, thermal stability and solvent resistance, but also renders these materials very brittle. Addition of small amounts (~5% wt.) of block copolymers (BCPs) significantly increase the facture resistance of epoxy resins. The observed increase in toughness is correlated with the formation of BCP micelles. However, a full description of the mechanisms leading to increased toughness is not available. Here we report the use of simultaneous small angle x-ray scattering (SAXS) and tensile experiments to connect local deformation processes of the BCP micelles to the macroscopic mechanical response of BCP modified epoxies. With this technique we followed the deformation events of rubbery and glassy core BCP micelles. We show that the rubbery core micelles are prone to cavitation and dialation, while the glassy core micelles deforms with constant volume. Finally, we discuss how the onset of cavitation is strongly affected by important characteristics of the epoxy network: its crosslink density and backbone flexibility.

(631c) Ordering of Sphere-Forming Block Polymers

Lee, S., University of Minnesota
Zhang, J., University of Minnesota
Bates, F. S., University of Minnesota
Originally presented on: 11/7/2013 9:20:00 - 9:40:00

Face-centered cubic (FCC) and hexagonally-packed spheres (HEX) are well-established as the most dense arrangements and lowest energy states of order for hard spheres, yet these structures rarely have been observed in block polymers. We recently have identified five different states of ordering in in poly(styrene-*b*-1,4-isoprene-*b*-styrene-*b*-ethylene oxide) (SISO) tetrablock terpolymers that form spherical domains: body-centered cubic (BCC), Frank-Kasper σ-phase, dodecagonal quasicrystals (DQC), Pm-3n (A15), and simple hexagonally packing (sHEX). Some of these sphere-packing symmetries have been documented in dendrimers, another form of soft material that is structurally distinct from linear block polymers, and the σ-phase has been reported in the heavy metal elements uranium and tantalum. Classically, the packing of hard spheres on a crystalline lattice has been modeled using simple pair-potentials such as the Lennard-Jones function. This lecture will consider how surrounding a spherical domain with a corona of polymer blocks modifies interparticle interactions leading to the remarkable compliment of ordered states.

(631d) A Peculiar Layering Structure and Transport Phenomenon of Nano-Blended Perfluoropolyether Thin Films

Chung, P. S., Carnegie Mellon University
Jhon, M. S., Carnegie Mellon University
Originally presented on: 11/7/2013 9:40:00 - 10:00:00

We investigated spreading characteristics of polar/non-polar nano-scale film made of the single component as well as binary blend perfluoropolyether (PFPE) nano films. For single components system, we observed spreading evolves with time in classical diffusive manner for non-polar PFPE Z, while polar PFPE Ztetraol exhibits “layering” structure with characteristic “shoulder” above the spreading edge. Different characterization tool (optical surface analyzer) was used in this research, and our results are consistent with our previous finding. For binary nano blend with Z and Ztetraol, results exhibit ultra-thin layering structure at the spreading edge with rapid proceeding and Ztetraol bottom layer inside the thin film. The qualitative molecular modeling results and description explaining the characteristics of the single component film as well as the peculiar behavior found in the nano blended films will be given.

(631e) Controlling Micro- and Meso- Scaled Pores In Carbon Nanofibers From Immiscible Polymers For Energy Applications

Williams, B., Cornell University
Henderson, K., Cornell University
Yin, J., Cornell University
Joo, Y. L., Cornell University
Originally presented on: 11/7/2013 10:00:00 - 10:20:00

Nanofibers were electrospun from blend solutions of polyacrylonitrile (PAN) and various sacrificial polymers including cellulose diacetate (CDA). The nanofibers were stabilized in air and carbonized in nitrogen to produce micro- and meso- porous nanofibers. The heat treatment converted PAN, a common precursor for carbon fibers, to pure carbon but pyrolyzed the sacrificial polymer to create pores within the nanofiber. Similar to other works, the rapid evaporation of solvent during electrospinning is thought to kinetically trap immiscible polymers in small domains during phase separation caused by evaporating the solvent. In the current study, the electrospinning conditions including temperature, blend ratio, molecular weight, and the presence of a co-solvent were varied to control the pore diameter distribution after carbonization. By varying these conditions, the average pore diameter was varied from 3.4 to 15.6 nm and maintained at least 66 m^2/g of BET surface area. The pore distribution was controlled in a single modal pore distribution as well as in a bimodal distribution. CDA was selectively removed from the uncarbonized nanofibers to further investigate the porous nature of the nanofibers. Nitrogen physisorption and TEM images were used to measure and confirm the pore distributions and surface area. The carbon nanofibers were then tested for their performance as a stand-alone supercapacitor electrode and as a Li-Air cathode which have been reported to have different pore distribution requirements for optimal performance.

(631f) Preparation Of Graphene Based Nanocomposites In Electrospun Polyaniline/Polyethylene Oxide Blends

Moayeri, A., Ecole Polytechnique de Montreal
Ajji, A., National Research Council Canada
Originally presented on: 11/7/2013 10:20:00 - 10:40:00

Electrospinning is a convenient method to produce polymer nanofibers with controlled diameters from tens of nanometers to micrometers. The resulting non-woven fiber mats have high specific surface areas in the range of 1-100 m^2/g. Combining these properties with the high electrical conductivity of intrinsically conductive polymers, conductive electrospun fiber mats are promising for a variety of applications such as electronic, biomedical, sensor and optical fields.

Polyaniline is one of the most studied electrically conductive polymers and it is unique due to its ease of synthesis, environmental stability, and simple doping/dedoping chemistry. It is however relatively difficult to process compared to most other polymers. As is common among intrinsically conductive polymers, it has a fairly rigid backbone due to its high aromaticity. Thus, the elasticity of its solutions is generally insufficient for it to be electrospun directly into fibers. Moreover, polyaniline has poor solubility in common solvents, which further complicates it electrospinnability. To circumvent these problems, it is possible to electrospun polyaniline by blending it with an easily electrospinnable polymer such as polyethylene oxide. However, the presence of an insulator copolymer decreases the fibers conductivity due to dilution of the conducting component. A good strategy for improving the electrical properties of this electrospun fibers blend is to incorporate carbon-based conductive nanofillers into the nanofibers.

Since the discovery of graphene in 2004, it has attracted tremendous research interest. In recent years, graphene nanosheets have been recognized as a promising, cost effective and high quality alternative to carbon nanotubes in composite applications. Graphene is a single-atom-thick, two-dimensional sheet of sp^2-hybrized carbon atoms arranged in a honeycomb crystal structure with exceptionally high strength, surface area, thermal conductivity, and unique electronic conductivity. Therefore, considering the excellent properties of graphene and polyaniline, a highly conductive polyaniline/polyethylene oxide composite with graphene as nanofillers may be obtained.

In this work, we report the preparation of electrospun polyaniline/polyethylene oxide nanofibers fibers filled with non-covalently functionalized graphene sheets. These graphene/polymer nanofibers have the potential to serve in a variety of applications, including conductive wires and sensors.

(631g) Effect of Sulfonation On the Transport Properties of Poly(styrene-isobutylene-styrene) and Poly(styrene-isoprene-styrene) Membranes

Aviles-Barreto, S. L., University of Puerto Rico - Mayaguez Campus
Suleiman, D., University of Puerto Rico, Mayaguez Campus
Originally presented on: 11/7/2013 10:40:00 - 11:00:00

In this study, physical blends of sulfonated and unsulfonated linear poly(styrene-isobutylene-styrene) (SIBS) and radial poly(styrene-isoprene-styrene) (SIS) were casted and analyzed as function of sulfonation level (0-60.8%). Increasing the amount of sulfonic groups improved the ion exchange capacity (IEC) of the membranes, which can be related to their transport properties. To confirm and determine accurate sulfonation levels, SIBS-SIS membranes were characterized using Fourier transform infrared spectroscopy (FTIR) and elemental analysis (EA). Thermogravimetrical analysis (TGA) was performed in order to evaluate the thermal stability, which increases with sulfonation. Absorption limitations and their effect on the membrane's transport properties were investigated through water swelling experiments. Transport properties for direct methanol fuel cell (DMFC) applications (methanol permeability and proton conductivity), were measured and compared to Nafion®. The results suggest that the combination of ionic sulfonic groups with unique morphological arrangements can lead to high selectivities for DMFC applications.

(633a) Design and Development of a Multi-Compartment Human Skin By 3D Printing

Karande, P., Rensselaer Polytechnic Institute
Dai, G., Rensselaer Polytechnic Institute
Lee, V., Rensselaer Polytechnic Institute
Originally presented on: 11/7/2013 8:35:00 - 8:49:00

We demonstrate the potential of 3D bio-printing technology in the generation of a multi-layered human skin tissue model. A modified inkjet printer was used to develop a human skin model in a layer-by-layer approach by printing multiple layers of collagen, fibroblasts and keratinocytes. We were able to print a 13-layered skin structure of thickness 1200μm on both petri-dishes and tissue culture transwell inserts without the presence of a mold or a cast. The skin model was further characterized with H&E and antibody staining, following exposure to air-liquid interface, to initiate the process of self-assembly. Our initial results demonstrate the feasibility of integrating this technology with novel approaches in designing a more physiologically relevant human skin model. This method of generation of skin has applications for modeling in drug discovery, toxicity studies as well as, in designing autologous grafts for burn victims.

(633b) From Carbon Nanomaterials-Based Ultrathin Films to Electronic Skins

Zhang, T., Suzhou Institute of Nanotech and Nanobionics, Chinese Academy of Sciences
Wang, X., Suzhou Institute of Nano-Tech and Nano-Bionics (SINANO), CAS
Xiong, Z.,
Yang, G.,
Cui, Z., Suzhou Institute of Nanotech and Nanobionics, Chinese Academy of Sciences
Originally presented on: 11/7/2013 8:49:00 - 9:03:00

Flexible electronics are an exciting frontier for the next generation devices in both sensing and actuator. Flexible pressure/strain sensors and artificial electronic-skins (E-skins) have attracted increasing massive attention due to their unique advantages in specific potential applications such as sensitive tactile information for future robotic systems minimally invasive surgery and prosthetics. How to improve their flexibility sensitivity transmittance integration and stability is still a key problem for next generation of E-skins.

Free-standing ultrathin films of Single-walled Carbon Nanotubes (SWNTs) and reduced Graphene Oxide (rGO) are promising nanomaterials to meet those needs. Here we present our recent works on the formation of SWNTs and rGO ultrathin films and their potential application in artificial E-skins. By utilizing the high conductivity of SWNTs/rGO network and natural patterned structures on the surface of intertextures of silk, we developed a facile and low-cost method for fabricating patterned PDMS conducting films with uniform microstructured patterns and high transmittance, then constructed the flexible pressure sensor, and demonstrated that it has ultra-sensitivity (detection limit as low as 0.2 Pa) and high stability for detection of super-slight forces. Furthermore we demonstrate that the application of E-skins in monitoring human physiological signals which can broaden their potential applications for diagnosis of diseases and health assessment.

(633c) Highly Efficient Polymer-Based Microrockets and Their Biomedical Applications

Gao, W., University of California, San Diego
Wang, J., University of California, San Diego
Originally presented on: 11/7/2013 9:03:00 - 9:17:00

The motion of synthetic micro/nano objects is of great fundamental and practical interest, and has thus stimulated major research efforts over the past decade in connection to diverse biomedical applications. Here we demonstrated a new polymer-based catalytic tubular microrocket synthesized using a template based electrodeposition method. The effects of different electropolymerized outer layers, including polypyrrole (PPy), poly(3,4-ethylenedioxythiophene) (PEDOT), polyaniline (PANI) and of various inner metal surfaces (Ag, Pt, Au-catalase), upon the movement of such bilayer microtubes are evaluated and compared. These oxygen bubble propelled Pt based microrockets are only 8 μm long, can operate in very low levels of the hydrogen peroxide fuel (down to 0.2%) and achieve a consistently remarkable high speed of 1400 body lengths/s in physiological temperature (the speed record of all artificial nanomotors). These microrockets can also move efficiently in various fuel enhanced biological media such as human serum and can thus serve as an ideal platform for diverse biomedical and environmental applications. For example, lectin modified PANI/Pt microrockets are used for selective bacteria (*E. Coli*) isolation from food, clinical and environmental samples. Poly(3-aminophenylboronic acid) (PAPBA)/Ni/Pt microrockets coupling the selective monosaccharide recognition of the boronic-acid-based outer polymeric layer with the catalytic function of the inner platinum layer. The resulting boronic-acid based microrocket itself provides the 'built in' target recognition capability. 'On-the-fly' capture, transport and release of yeast cells (containing glucose on the cell wall) are illustrated. Molecularly imprinted polymers (MIPs) based microrockets can concentrates the fluorescent-tagged protein target onto the polymer outer layer without the need for additional external functionalization, allowing motion based extraction and isolation of Av-FITC from raw serum and saliva samples along with real-time visualization of the protein loading and transport. Besides the catalytic Pt based microrockets which rely on the hydrogen peroxide fuel, we also demonstrate a new hydrogen bubble propelled micromotor which can be driven by the natural environments (such as human stomach), without any additional chemical fuel. The tubular polyaniline/zinc microrockets display effective autonomous motion in extreme acidic environments (with a speed of over 100 body lengths per second). The ejection of hydrogen bubbles from the exposed zinc inner layer, upon its contact with acid, provides a powerful directional propulsion thrust. The observed speed-pH dependence holds promise for sensitive pH measurements in extreme acidic environments. The attractive guided cargo transport capabilities of these acid driven microrockets hold great promise for practical biomedical applications such as targeted drug delivery.

(633d) The Improved Adhesion Behavior of Fibroblasts On Anodized 316L Stainless Steel

Ni, S., Northeastern Universtiy

Sun, L., Northeastern University

Liu, L., Northeastern University

Webster, T. J., Northeastern University

Originally presented on: 11/7/2013 9:17:00 - 9:31:00

The improved adhesion behavior of fibroblasts on anodized 316L stainless steel

Siyu Ni[a,b], Linlin Sun[b], Luting Liu[b], Thomas J Webster[b]

[a]*College of Chemistry, Chemical Engineering and Biotechnology, Donghua University, Shanghai, China*

[b]*Department of Chemical Engineering, College of Engineering, Northeastern Universtiy, Boston, MA, USA*

E-mail: th.webster@neu.edu, Tel: +1 617 373 8565;

Introduction: It is now recognized that surface nanostructures on biomaterials greatly influence cellular behaviors as well as surface biochemical properties [1, 2]. Nano-featured surfaces might be used as an excellent bioactive interface for implantable materials. Stainless steel is one of the most important alloys for biomedical applications. However, their bioinert characteristic surface is a disadvantage for numerous applications. The aim of this study was to fabricate nano-structured surfaces on 316L stainless steel by an anodization method and examine the effects of these anodized nano-surfaces on the attachment of fibroblasts.

Materials & Methods: 316L stainless steel (2.0 mm °Á 2.0 cm °Á 0.5 mm) (Goodfellow Cambridge Ltd. England) was anodized at 30 V for 4 min in ethylene glycol containing 10 vol. % perchloric acid to create bioactive surfaces. The surface morphology and elemental composition of 316 L after anodization were characterized by scanning electron microscopy (SEM, Hitachi S-4800, Tokyo, Japan) and energy-dispersive X-ray analysis (EDAX). Fibroblasts were seeded on the samples to evaluate initial cellular responses to the nano-structured surfaces. After cell culturing for 4 hours, the attachment of fibroblasts was determined by a methyl thiazolyl tetrazolium (MTT) assay. Smoothly polished 316L was used as control surfaces. Light microscopy and SEM were used to observe the morphology of cells. Experiments were completed in triplicate and repeated at least three times. Data were analyzed statistically using student t-test.

Results and Discussion: A nano-porous structure was successfully fabricated by the anodization procedure. The pore sizes were about 100 nm. After 4 hours of culture, the fibroblasts adhered well on the surfaces of all the stainless steel specimens. SEM images showed that fibroblast exhibited normal morphology and were able to penetrate into pores by long filopodia. MTT tests indicated that the anodization of 316 L stainless steel significantly promoted a higher degree of cell adhesion as compared to the polished 316L stainless steel after 4 hours. ($p<0.05$) (Fig. 1).

Conclusions: The findings in this study showed that anodization significantly improved the initial cell-material interactions on the 316L stainless steel. Although further work is required to determine the long-term effects of the anodized surfaces on the proliferation and differentiation of cells, the results of this study indicated that the anodized surfaces of the 316L stainless steel had better biological activity.

Fig.1 The fibroblast attachment on the anodized surface of 316 L stainless steel using the MTT assay for 4 hours. *$p<0.05$ compared to the smoothly polished 316 L at the same time period

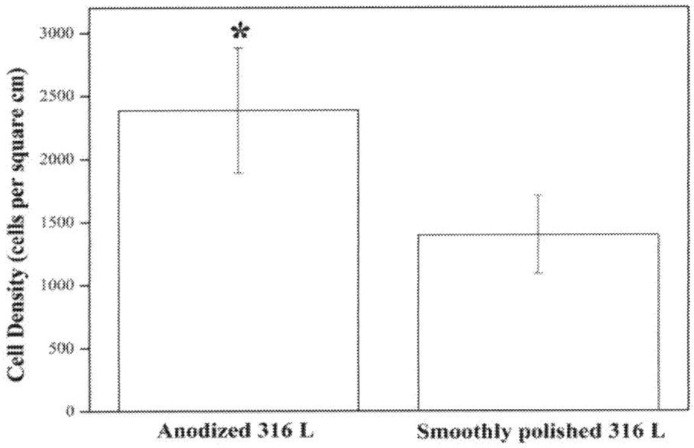

References

1. Mart"ªnez E, Engel E, Planell, JA, Samitier, J. Ann Anat 2009; 191: 126-135.

2. Ercan B, Taylor E, Alpaslan E, Webster TJ. Nanotechnology 2011; 22: 295102

(633e) Flexible Neural Implant From Microfabricated Lbl Nanocomposite

Zhang, H., University of Michigan
Patel, P. R.,
Kotov, N. A., University of Michigan
Originally presented on: 11/7/2013 9:31:00 - 9:45:00

Current neural prosthetic devices induce chronic inflammation due to mismatched mechanical properties with neural tissue and large size of the implants. A flawless integration of tissue/device interface requires a new generation of neural devices fabricated from flexible materials with a combination of record mechanical and electrical performance. In addition to these fundamental challenges, the development of such materials must be combined with their integration with micromanufacturing techniques and innovative methods of implantation. Despite tremendous activity in the area of neural interfaces with carbon nanotubes (CNTs), fully functioning microdevices fabricated from CNTs have not been realized. In this study, we demonstrate the possibility of tissue-integrated devices from a CNT nanocomposite. *In vivo* evaluation demonstrated their functionality by successful registration of brain activity. This study opens the door for different nanocomposites to be used in tissue-mimetic electrodes for long-term brain-machine interface.

(633f) Charge Transfer to Photosystem I Through Hydroxyl-Terminated Alkanethiol SAM Length Modification

Khomami, B., University of Tennessee
Mukherjee, D., Sustainble Energy and Education Research Center (SEERC), The University of Tennessee
Bennett, T.,
Originally presented on: 11/7/2013 9:45:00 - 9:59:00

Photosystem I (PS I) is a supra-molecular photosynthetic protein complex that functions as a biological
photodiode. Upon exposure to light, PS I charge separates between the reaction center (P700 on the lumenal side) and the Fe-S clusters (FA,FX,FB on the stromal side). The highly efficient photo-electrochemical activities of PS I (with 100% quantum efficiency over 54% of the solar spectrum) make it
an ideal candidate for next-generation bio-hybrid electronic devices. Specifically, our interest in using PS
I for future hybrid photovoltaic (PV) device fabrications requires optimal encapsulation of these proteins
onto chemically grafted self-assembled monolayer (SAM) substrates. Our previous results indicate that various experimental parameters alter the surface attachment dynamics of PS I, deposited from colloidal aqueous buffer suspensions onto OH-terminated alkanethiolate/Au SAM substrates, thereby
allowing us to tailor the complex structural arrangements of PS I which affect their electron transfer and
capture pathways. We present surface topographical, combined with electrochemical characterizations
of PS I/Au SAM substrates, to elucidate the controlled monolayer surface attachment-as well as the directional attachment dynamics of PS I to ensure proper photo-activated electronic activities. Specifically, light induced directional electron transfer by surface immobilized PSI is demonstrated via
direct electrochemistry measurements from PS I complexes assembled on varying carbon chain lengths
(C6, C9, C8, C11 alkanethiolate SAM /Au surfaces). The results reveal that the supporting thiol brush
density and chain length, due to their non-optimal packing or large tunneling distance through the SAM,
can act as the critical bottleneck for the electron transfer process from the Au surface to the oxidized
P700+ reaction center of PS I. Similarly, on the stromal side, the electron donation of the reduced

FB- to the surrounding electrolyte solution is enhanced by the presence of a suitable oxidizing agent
or the addition of a soluble electron scavenger. This work provides a detailed and critical in sight into
the optimization of the rate limiting electron transfer processes at the interfaces between the donor surfaces and P700+ as well as the FB- and the electrolyte. In turn , such studies can reveal significant
information regarding the orientation of PS I complexes on various donor substrates.

(633g) Nanoparticle SAMs As Artificial Micro-Environments - Chemistry As Inspired By Biology

Walker, D. A., Northwestern University
Leitsch, E., Northwestern University
Raimondo, C., Northwestern University
Grzybowski, B., Northwestern University
Originally presented on: 11/7/2013 9:59:00 - 10:13:00

One of the canons of organic chemistry has been that a molecule's properties are inherently 'coded' by its chemical structure. This perspective, however, is an over simplification as there are a multitude of examples demonstrating that environmental factors contribute to a molecule's properties equally to, if not more than, the chemical structure. This is best exemplified by living systems – the complexity of human life is created out of just 21 amino acids which are strung together to create proteins and other sophisticated molecular machines. By controlling the environments around this handful of chemical species, and as such controlling their nanoscale interactions, nature has developed an endless list of highly efficient molecular machines.

Nanoparticles provide a useful platform through which we can begin to probe the effects of molecular environments and various types of inter- or intra-molecular interactions. Electrostatic interactions tend to be one of the more versatile interactions at this length-scale as they can be attractive or repulsive, short- or long-ranged, and their magnitudes readily controlled. Here we demonstrate that on-particle self-assembled monolayers can be used to create artificial microenvironments which control the chemical properties of organic molecules. By controlling geometric orientations and electrostatic interactions one can essentially program the pKa of carboxylic acids and redox potentials of organometallics in a manner analogous to proteins and enzymes. These results offer a fundamentally new perspective for synthetic organic chemists and could have substantial implications in the field of synthetic biology. A combination of both experimental and theoretical results will be presented.

(633h) Entrapment of Integral Membrane Proteins in Nanoporous Silica Gels Via Nanolipoprotein Particles

Zeno, W. F., UC Davis
Longo, M. L., University of California, Davis
Risbud, S. H., University of California, Davis
Coleman, M. A.,
Originally presented on: 11/7/2013 10:13:00 - 10:27:00

Immobilization of integral membrane proteins (IMPs) in transparent, nanoporous silica gels has proven to be a challenge, as current and previous techniques utilize liposomes as biological membrane hosts. The instability of liposomes in nanoporous gels is attributed by their size (~150 nm) and altered structure and lipid dynamics upon entrapment within the nanometer scale pores (5-50 nm) of silica gel. This ultimately results in disruption of protein activity. We intend to address these issues by using nanolipoprotein particles (NLPs) as biomembrane hosts. NLPs are discoidal patches of lipid bilayer that are belted by apolipoproteins (~5 nm thickness and 10-15 nm diameter). The IMP-NLP complexes are synthesized in a cell-free environment, which circumvents traditional protein reconstitution in membranes. Bacteriorhodopsin - a robust IMP protein that indicates its proper conformation via distinct purple coloration – will serve as a model IMP for this system. The spectral and physical properties of bacteriorhodopsin-NLPs entrapped within the gel are examined, as well as the phase behavior of the lipids within the NLP, to ensure proper functionality of the system. This bio-inorganic hybrid nanomaterial possesses a variety of viable applications. The success of this work could lead to the development of novel platforms in several areas, including high-throughput drug screening, chromatography, and biosensors.

(633i) Morphological and in Vitro Functional Investigation of Phospholipid-Membrane Associated Protein (PS I)

Niroomand, H., Sustainble Energy and Education Research Center (SEERC), The University of Tennessee
Mukherjee, D., Sustainble Energy and Education Research Center (SEERC), The University of Tennessee
Khomami, B., Sustainble Energy and Education Research Center (SEERC), The University of Tennessee
Sarles, A.,
Venkatesan, G.,
Originally presented on: 11/7/2013 10:27:00 - 10:41:00

Morphological and in vitro functional investigation of phospholipid-membrane associated protein (PS I)

Hanieh Niroomand, Sustainable Energy Education and Research Center (SEERC), Department of Chemical and Biomolecular Engineering, University of Tennessee, Knoxville, TN

Dibyendu Mukherjee, Sustainable Energy Education and Research Center (SEERC), Department of Mechanical, Aerospace and Biomedical Engineering; Department of Chemical and Biomolecular Engineering, University of Tennessee, Knoxville, TN

Bamin Khomami, Sustainable Energy Education and Research Center (SEERC), Department of Chemical and Biomolecular Engineering, University of Tennessee, Knoxville, TN

Membrane protein structural and in-vitro functional studies rely on synthetic membrane mimics (e.g., detergents and liposomes) to stabilize the protein fold and maintain function[1]. Our recent investigations[2] on the cyanobacterial photosynthetic membrane protein complex, Photosystem I (PS I), reveal that the two prototypical non-ionic detergents, n-Dodecyl-β-D-Maltoside (DM) and Triton X-100 (TX-100) can uniquely tailor the solution phase PS I-PS IÂ interactions/aggregations via the Â relative PS I/detergent concentrations[3]. In the present study, we use in-situ dynamic light scattering (DLS) and electron microscopy experiments to investigate the solution-phase aggregation dynamics of PS I complex reconstituted into the phospholipid vesicle diphytanoyl-sn-glycero-3-phosphocholine (DPhPC) for different phospholipid concentrations. Specifically, two well-established transmission electron microscopy (TEM) techniques[4], namely negative staining and direct imaging in cryogenic conditions (cryo-TEM), are utilized. These measurements reveal the complex morphology arising out of the interactions of the phospholipid small unilamellar (SUV) and large unilamellar vesicles (LUV) with the PS I complexes. In this presentation, the shape evolution of PS I with respect to lipid vesicle, and its connection to PS I-PS I as well as PS I-lipid interactions in colloidal solutions are discussed. Furthermore, new approaches for incorporation of the PS I complex into phospholipid liposomes with various head groups and hydrophobic tails are highlighted. Such information is valuable in mimicking the naturally occurring protein-phospholipid interactions in cell membranes. In future, these bio-mimetic systems shall facilitate easy incorporation of these protein-lipid complexes into novel bio-hybrid devices.

1. Dewald A.H., Hodges J.C., Columbus L. Biophys J. 2011; 100(9): 2131?2140.

2. Mukherjee D, May M, Khomami B. J. Colloid Interface Sci. 2011;358(2):477-484.

3. Mukherjee D, May M, Vaughn M, Bruce BD, Khomami B. Langmuir. 2010;26(20):16048-16054.

4. Gennis R.B. *Biomembranes molecular structure and function*. Springer-Verlag, 1989.

(633j) Biomimetic Synthesis of Magnetic Nanocrystals Mediated By Protein Mms6

Liu, X., Iowa State University
Zhang, H., Iowa State University
Parada, G., Iowa State University
Feng, S., Iowa State University
Akinc, M., Iowa State University
Nilsen-Hamilton, M., Iowa State Unversity
Mallapragada, S. K., Iowa State University
Originally presented on: 11/7/2013 10:41:00 - 10:55:00

The synthesis of magnetic nanoparticles with uniform size has attracted a lot of research interest due to their applications in biomedical devices and high-density data storage. Uniform superparamagnetic magnetite nanocrystals, with a size of about 30 nm, were synthesized *in vitro* in the presence of a recombinant Mms6 protein, which is found in the magnetosomes of magnetotactic bacteria. Mms6 was shown to have two phases of iron binding: one high affinity and stoichiometric and the other low affinity, high capacity, and cooperative with respect to iron. Iron can also induce a protein structure change in solution, as shown by small angle X-ray scattering. Nanoparticle fabrication studies showed that the nanoparticles synthesized by the wild type Mms6 differ greatly compared to the mutants, indicating that the number and placement of charged groups in the protein were very critical for its function. We explore the bio-inspired synthesis of magnetite nanoparticles on gold surfaces as well. Self-assembled monolayers on gold were used to study the effect of the surface hydrophobicity on the protein surface aggregates and the resulting magnetite nanoparticle size and morphology. It was shown that the Mms6 formed larger aggregates on hydrophobic surfaces, leading to larger magnetite nanoparticles. These studies are very important in understanding the mechanism by which Mms6 promotes the formation of the uniformly sized nanoparticles under mild conditions, and for using the bio-inspired magnetic materials for different applications.

(633k) Towards Cyberengineering Drug Delivery Vesicles

Stanzione, F., Colorado School of Mines
Hu, M.,
Sum, A. K., Colorado School of Mines
Deserno, M.,
Faller, R., UC Davis
Originally presented on: 11/7/2013 10:55:00 - 11:09:00

We developed a computational framework to understand the assembly conditions and constraints of nanoparticle tethered lipid bilayers which can be used as drug delivery systems. A multi-scale computational approach encompassing atomistic, mesoscale and generic scale molecular dynamics has been used to describe the pertinent interactions on the correct scales. When needed different scale models were mapped onto each other. This approach allowed to determine several principles of interaction as well as design constraints of these systems. The influence of curvature and tethered molecules on the outer lipid membrane was studied in detail.

(635a) Liposomal Co-Encapsulation of 5-Fluorouracil and Doxorubicin for Synergistic Anticancer Activity

Camacho, K. M., Rutgers University
Mitragotri, S., University of California Santa Barbara
Originally presented on: 11/7/2013 8:30:00 - 8:49:00

Combinations of chemotherapy drugs rather than single drugs are clinically used to combat late stage and metastatic cancers, due to their ability to simultaneously attack cancer cells via multiple mechanisms. However, clinical trials show that combination chemotherapy can result in increased side effects and little change in therapeutic efficacy compared to single drug treatments. These reactions can be attributed to both antagonistic interactions (reduction of efficacy upon combination) and separate drug pharmacokinetics, which can lead to drastically different drug concentrations at the tumor site than was originally administered. Here, we investigated drug interactions and developed a dual-drug carrier for controlled and effective combination delivery. Drug pairs were screened for synergistic interactions against the human breast cancer cell line BT474 *in vitro* utilizing the well-established Combination Index method. Antimetabolite 5-fluorouracil and anthracycline doxorubicin were identified as highly synergistic in a specific concentration ratio. It was also found that certain ratios of the same drug pair were antagonistic, demonstrating that poor control of drug ratios can be detrimental to therapeutic efficacy. Multi-drug entrapment in a single vehicle can unify drug pharmacokinetics and ensure that the cancer cells are exposed to the optimal, originally administered synergistic drug ratios. Since both 5-fluorouracil and doxorubicin are water soluble, liposomes were chosen for their encapsulation in the aqueous core. Liposomes which co-encapsulated 5-fluorouracil and doxorubicin in the specific synergistic ratio resulted in enhanced cell death compared to liposomes singly loaded with either drug. Thus, our system for dual-drug delivery provides a method to preserve the potency of the combination throughout blood circulation until reaching the target site, and can ultimately improve the therapeutic efficacy of chemotherapy.

(635b) Robust Nanoscale Liposome Based Coatings On Silk Microspheres Exhibit Exceptionally Low Friction Coefficients for Application in Articular Joints

Zheng, R., Tulane University
Zhan, J., Tulane University
Wang, X., Tufts University
Kaplan, D. L., Tufts University
Pesika, N., Tulane University
John, V. T., Tulane University
Originally presented on: 11/7/2013 8:49:00 - 9:08:00

We present recent results of our work to develop a robust nanoscale liposome based coating on silk microspheres to couple the lubrication properties of highly hydrated phospholipid head groups with the cushioning effects and wear resistance of silk. Such microscale ball bearings with low coefficients of rolling friction have significant potential in injectable systems for joint lubrication and show a high degree of wear resistance even at load levels that are orders of magnitude above those experienced in articular joints. The concept has also been extended to systems of thin silk films covered with a dense packed layer of liposomes which are "captured" and tethered in place using chitosan based biopolymers. Such systems exhibit coefficient of friction values representative of articular joints (0.008 and lower), indicating their potential efficacy as coatings for implant materials. Such synergies between phospholipids to provide the requisite hydration layers for enhanced slippage, and biopolymers to provide the cushioning and wear resistance will be discussed. With current therapeutic options of improving joint comfort being limited, these systems indicate new directions in materials for enhanced biolubrication.

(635c) Thermosensitive Liposomes With Photo-Activated Small Molecule Release

Forbes, N., University of California, Santa Barbara
Zasadzinski, J. A., University of Minnesota
Originally presented on: 11/7/2013 9:08:00 - 9:27:00

A drug delivery system with rapid contents release under continuous-wave near-infrared (NIR) light will enable precise spatial and temporal control of drug release. The drug is encapsulated within a thermosensitive liposome, and rapid drug release is triggered under irradiation by the localized photothermal heating of metallic nanoparticles coupled to the liposome. These nanoparticles heat the liposome membrane through its phase transition at 39-42°C to induce an increase in membrane permeability. A fluorescent dye was encapsulated as a model agent to study dye release kinetics. A small fraction of lysolipid (single-chained phospholipid) enhances the rate of release at the transition. The lysolipid can partition into pre-formed liposomes to enhance membrane permeability, and this partitioning is impacted by bilayer hydrophilicity and phase. Two metallic nanoparticles which absorb NIR light, hollow gold nanoshells and copper sulfide nanoparticles, were compared in terms of their liposomal attachment and heating characteristics. Finally, the anticancer agent doxorubicin was encapsulated within the liposomes and the system was used to deliver doxorubicin to drug-resistant prostate cancer cells (PPC-1) *in vitro*. Near-complete cell killing was observed by the photothermally triggered thermosensitive liposomes at an order of magnitude lower doxorubicin concentration relative to non-irradiated liposomal doxorubicin. The enhanced cell killing at low total concentration is likely due to the rapid release of the doxorubicin, which provides locally higher concentration combined with the thermal damage of the locally increased temperatures. This device provides externally targeted, tumor-specific delivery of small molecule anti-cancer agents from thermosensitive liposomes without requiring bulk tissue hyperthermia, along with expanded control of drug release rate.

(635d) A Microfluidic Platform With Continuous Rapid Cell Trapping and Micro/Nano Electroporation for Single Living Cell Study

Wang, L. J., Nanoscale Science and Engineering Center for Affordable Nanoengineering of Polymeric Biomedical Devices
Li, L., The Ohio State University
Ma, J., University of South Carolina
Wu, Y., The Ohio State University
Gallego-Perez, D., The Ohio State University
Lee, J., Ohio State University
Originally presented on: 11/7/2013 9:27:00 - 9:46:00

A microfluidic platform with continuous rapid cell trapping and micro/nano electroporation for single living cell study

Li-Ju Wang, Lei Li, Junyu Ma, Yun Wu, Daniel Gallego-Perez, James L. Lee

Nanoscale Science and Engineering Center for Affordable Nanoengineering of Polymeric Biomedical Devices, the Ohio State University,

Columbus, Ohio, 43210, United States

Biomedical micro devices have shown incomparable advantages on single living cell study in recent years. Compared to traditional methods, micro chips provide better manipulation of individual cells. Micro devices also show dynamical mimics and precise control of biochemical cellular microenvironment. For single cell study, except for individual cell manipulation and biocompatible micro environment control, efficient and safe delivering of biomolecules is also very important. Single cell micro-electroporation (MEP) has been used to delivery gene/drug into cells. By placing a cell next to a micro aperture or a microchannel, focused electric fields can create temporary pores in cell membranes. Furthermore, nanochannel electroporation (NEP) technology has been developed recently. NEP technology can directly deliver biomolecules into the cell cytosol across nanochannels with highly concentrated electrical fields and much smaller generated pores on cell membranes. Micro/nano electroporation devices has been reported to have superiorities of lower poration voltages, better transfection efficiency, and lower cell mortality comparing to conventional bulk electroporation (BEP) especially for single cell analysis.

MEP/NEP is a highly effective tool to directly delivery gene/drug into single living cells. However, current MEP/NEP devices require complicated operation to place single cells to the electroporation locations, for example, by using optical tweezers or magnetic tweezers. The long-time manipulation increases cell mortality rate. The process is not only time and cost consuming but also has limited throughput at a time. This greatly limits the scale-up and applications of MEP/NEP technologies. Although there are a lot of researches working on microfluidic cell-trapping devices, it is still a challenge to integrate microfluidic cell trapping with micro/nano electroporation in the micro devices.

In this work, an easy-to-use and affordable microfluidic platform combining continuous rapid hydrodynamic cell trapping and micro/nano electroporation gene/drug delivery is presented. The novel device design integrated cell trapping and electroporation channels and was fabricated using soft lithography and was prepared with bio-functionalized surface treatment. A maximum 90% cell trapping rate was achieved by using this device. Each trapped single cell was positioned on the tip of a single microchannel for sequential micro/nano electroporation. After MEP/NEP gene delivery, these cells were able to be incubated inside this biocompatible microenvironment for *in vitro* growth or moved out for further analysis. This newly developed platform was fist tested for human alveolar basal epithelial cells (HBEC) and mouse embryonic fibroblast (MEF) cells by delivering fluorescent-labeled DNA. Image analysis showed high transfection efficiency as well as the higher cell viability which exhibited by calcein AM staining.

(635e) Emergent Properties of Nanosensor Arrays: Applications to Monitoring IgG Affinity Distributions, Weakly-Affined Hypermannosylation, and Colony Selection for Biomanufacturing

Reuel, N. F., Massachusetts Institute of Technology
Strano, M. S., Massachusetts Institute of Technology
Originally presented on: 11/7/2013 9:46:00 - 10:05:00

It is widely recognized that an array of addressable sensors can be multiplexed for the label-free detection of a library of analytes. However, such arrays have useful properties that emerge from the ensemble, even when monofunctionalized. As examples, we show that an array of nanosensors can estimate the mean and variance of the observed dissociation constant (K_D), using three different examples of binding IgG with Protein-A as the recognition site, including polyclonal human IgG (K_D μ = 19 μM, σ^2 = 1000 mM^2), murine IgG (K_D μ = 4.3 nM, σ^2 = 3 μM^2), and human IgG from CHO cells (K_D μ = 2.5 nM, σ^2 = 0.01 μM^2). Second, we show that an array of nanosensors can uniquely monitor weakly-affined analyte interactions *via* the increased number of observed interactions. One application involves monitoring the metabolically-induced hypermannosylation of human IgG from CHO using PSA-lectin conjugated sensor arrays where temporal glycosylation patterns are measured and compared. Lastly, the array of sensors can also spatially map the local production of an analyte from cellular biosynthesis, thus providing a powerful approach to clonal selection. As an example, we rank productivity of IgG-producing HEK colonies cultured directly on the array of nanosensors itself. This study opens new avenues for the use of nanosensor arrays for biomanufacturing applications.

(635f) Rational Design of Polypeptides for the Functionalization of Novel Nanoparticles

VanLang, C. C., Yale University
Chan, W., Stanford University
Lu, Y., Stanford University
Swartz, J. R., Stanford University
Originally presented on: 11/7/2013 10:05:00 - 10:24:00

Bioconjugate chemistry has long been used to combine biomolecules to alter functions and to add properties, especially to larger biomolecular assemblies such as virus-like particles (VLPs). However, there exist several limitations to effectively conjugating two molecules if the available surfaces are sterically hindered or electrostatically repulsive. Furthermore, the reactive moieties must be properly oriented and activated to achieve effective conjugation. Previous strategies to overcome these issues have involved modifying the composition of the surfaces to reduce electrostatic repulsion or adding soluble enhancers to the reaction. However, these strategies often disrupt the structure and functions of the molecule. Alternatively, we can take advantage of the chemical environment by selecting a protein tag that recognizes its target by ion pairing. Using this strategy of "facilitated conjugation", we have designed a series of polypeptides to improve the efficiency of Huisgen Copper (I)-Catalyzed Azide-Alkyne cycloaddition of a variety of ligands including a scFv and a luciferase to the negatively charged surface of the Hepatitis B core VLP. We introduce the concept of facilitated conjugation and evaluate the design rules to further improve conjugation efficiency. We are currently utilizing this strategy to generate functionalized VLPs as vaccines and cancer cell detection.

(635g) Pulmonary Persistence of Intranasally Administered Polyanhydride Nanovaccines

Haughney, S., Iowa State University

Ross, K. A., Iowa State University

Boggiatto, P., Iowa State University

Wannemuehler, M. J., Iowa State University

Narasimhan, B., Iowa State University

Originally presented on: 11/7/2013 10:24:00 - 10:43:00

Polyanhydrides are a class of degradable biomaterials that are suitable for the encapsulation and delivery of subunit vaccines. Nanoparticles comprised of sebacic acid (SA), 1,6-bis-(*p*-carboxyphenoxy)hexane (CPH), 1,8-bis-(*p*-carboxyphenoxy)-3,6-dioxaoctane (CPTEG), and their copolymers have shown the ability to sustain the release and enhanced internalization of fragile protein antigens. Polyanhydride nanovaccines have also demonstrated immunomodulatory capabilities, and more recently, full protection against a lethal bacterial challenge of *Yersinia pestis* was demonstrated 42 weeks post-vaccination using a single intranasal dose.

Intranasal vaccination is an advantageous strategy to induce both systemic and mucosal immunity while avoiding antigen degradation in the first pass metabolism and increasing bioavailability via the large adsorptive surface area. However, it has been shown that soluble antigen alone delivered intranasally does not induce a protective immune response, and therefore, requires the use of adjuvants. Previously, we have demonstrated the favorable deposition, distribution, and persistence of antigen delivered with polyanhydride nanovaccines at early time points after administration, which enhances internalization by antigen presenting cells (APCs) such as macrophages and dendritic cells. While the initial interactions between antigen-loaded nanoparticles and APCs are important in laying a foundation for vaccine efficacy, we hypothesize that the persistence of antigen released from polyanhydride nanoparticles plays a central/key role in the continual recruitment of APCs and the development of high antibody titers with increased avidity. In this work, we followed the persistence of intranasally administered antigen-containing polyanhydride nanoparticles over nine weeks. Utilizing live animal imaging, high throughput microscopy, and flow cytometry, we studied the impact of particle chemistry on the creation of a tunable erosion profile to control antigen release and persistence. Our results demonstrate the impact of persistent antigen encapsulated into polyanhydride nanoparticles on APC interactions in comparison with that of monophosphoryl lipid A (MPLA), and provide new insights into the underlying mechanisms behind the induction of high titer and highly avid antibody upon intranasal administration of nanovaccines.

S.H. and K.A.R. contributed equally to the design and performance of this work.

(635h) Development of Controlled Release Adjuvant Nanoparticle Vaccines for Potent Cell Mediated and Humoral Response

Basto, P., MIT
Alexis, F., Clemson University
Shi, J.,
Moyer, Z.,
Langer, R. S., Massachusetts Institute of Technology
Farokhzad, O. C., Brigham and Women's Hospital - Harvard Medical School
Originally presented on: 11/7/2013 10:43:00 - 11:02:00

In recent years, TLR mechanisms have been elucidated and novel agonists have been developed, yet our generation still has not seen paramount progress in the clinical translation of these agonists due to risks of systemic toxicity and off target effects. We hypothesize that engineering polymeric vaccines to emulate a virus particle with a payload of controlled releasing adjuvant in combination with spatially arranged surface antigen will provide enhanced protection against viral infection and has the potential to serve as an advancement in VLP design. In designing this system, polymeric PLA-PEG nanoparticles are attractive vehicles for vaccine development due to key characteristics of size, biodegradability, biocompatibility, ability of controlled release and stability. We have engineered and synthesized a reproducible nanoparticle delivery system that can efficiently prime B cells and produce potent T cell response through a single packaged dose of R848, a TLR7/8 agonist. The vaccine is able to enhance antigen presentation and co-stimulatory molecules on dendritic cells and subsequently enhanced proliferation of antigen-specific naïve CD4+ and CD8+ cells in vitro. When R848 is conjugated and packaged within a nanoparticle carrier, it was demonstrated that both adjuvant and antigen were able to be localized to antigen presenting cells in vivo as well as a significant reduction in systemic cytokine release. Upon vaccination, our delivery system is able to increase germinal center activity, cell-mediated and humoral response to provide protection in comparison to soluble adjuvant and alum, thereby increasing the potential of bringing novel adjuvants to the clinics.

(690a) Invited Talk: Peptide-Functionalized Nanoparticles for Targeted Delivery of Therapeutics
Kokkoli, E., University of Minnesota
Originally presented on: 11/7/2013 12:30:00 - 13:00:00

The goal of our studies is to develop selective drug delivery systems by incorporating in polymersomes (polymeric vesicles covered with polyethylene oxide) or stealth liposomes (liposomes covered with polyethylene glycol) biomimetic peptide-amphiphiles that recognize and bind, in a specific manner, the targets of choice. Examples will be discussed of different nanoparticle designs that target the alpha(5)beta(1) integrin through the use of a biomimetic peptide, PR_b, that was designed in our group. Our *in vitro* and *in vivo* data will demonstrate that PR_b-functionalized nanovectors can efficiently bind and internalize to cancer cells that over-express the integrin receptor, and release their encapsulated payload (plasmid DNA, siRNA or chemotherapy agents) intracellularly, thus conferring higher levels of cytotoxicity to cancer cells compared to non-functionalized nanoparticles or nanoparticles functionalized with other peptides commonly used in the literature.

(690b) Thermo-Responsive Bottlebrush Polymers

Li, X., Rice University

Pesek, S., Rice University

Hammouda, B.,

Cong, Y., Rice University

Li, Q., Rice University

Verduzco, R., Rice University

Originally presented on: 11/7/2013 13:00:00 - 13:20:00

Bottlebrush polymers are highly branched macromolecules with polymeric side-chains on each repeat unit. The dense grafting of polymeric side-chains results in both backbone and side-chain extension, giving rise to large, highly extended macromolecules with individual molecules exceeding 100 nm in backbone length in some cases. The novel structure of bottlebrush polymers may make them particularly useful in specialty applications, and here we explore the surface and solution properties of model thermoresponsive poly(*N*-isopropylacrilamide) (PNIPAAM) bottlebrush polymers. PNIPAAM macromonomers are prepared through reversible addition-fragmentation chain transfer using a norbornenyl-chain transfer agent, and PNIPAAM bottlebrush polymers are subsequently prepared using a "grafting-through" ring opening metathesis polymerization scheme (Figure 1(a)). This synthetic method gives fully grafted bottlebrush polymers with controlled backbone and side-chain lengths. The resulting PNIPAAM bottlebrush polymers are fully soluble in water after removal of the terminal chain transfer agent, reflecting a disproportionately strong effect of the side-chain end groups on bottlebrush polymer solubility. Small-angle neutron scattering (SANS) measurements of PNIPAAM bottlebrush polymers in D_2O demonstrate that the side-chains collapse as the temperature increases towards the lower critical solution temperature (LCST) (Figure (b)); both the LCST and the collapse of the side-chains is dependent on the side-chain length. Above the LCST, PNIPAAM bottlebrush polymers aggregate, and we find that PNIPAAM bottlebrush polymers with 9 kg/mol side-chains form a lyotropic liquid crystal phase (Figure 1 (e) and (f)), evidenced by low-angle scattering peaks and birefringence under polarized optical microscopy at temperatures above the LCST. Bottlebrush polymers may also be applicable to non-fouling and non-toxic surfaces due to the high density of polymer chains. Bottlebrush thin films with poly(*N*-isopropylacrylamide) (PNIPAM) and 4-acryloylbenzophenone crosslinkable side-chains were prepared by spin casting followed by UV irradiation. Cell attachment studies were conducted for bottlebrush polymers at varying side-chain lengths and solution temperatures.

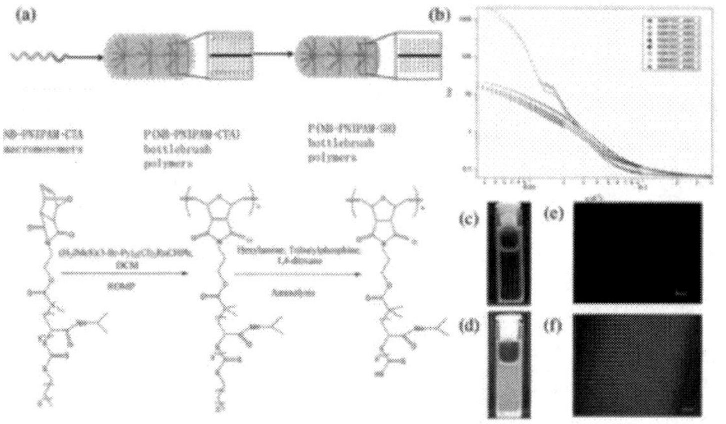

Figure1 Schemetic for the synthesis of P(NB-PNIPAM-SH) (a); SANS data for bottlebrush PNIPAM in D_2O with temperature (b); Photographs of the aqueous solution of P(NB-PNIPAM-SH) taken below LCST (c) and above LCST (d); POM images of P(NB-PNIPAM-SH) aqueous solution taken below LCST (e) and above LCST (f).

(690c) Effect of Molecular Interactions On the Growth, Properties, and Release of Polyelectrolyte Multilayers

Das, B. P., University at Buffalo - The State University of New York (SUNY)
Tsianou, M., University at Buffalo - The State University of New York (SUNY)
Originally presented on: 11/7/2013 13:20:00 - 13:40:00

Multilayered thin films have been the focus of intense research because of their vast array of applications. We investigate the growth and disassembly under different pH conditions of layer-by-layer (LbL) films composed of oppositely charged polyelectrolytes (including poly(diallyldimethylammonium chloride) (PDDA), poly(styrene sulfonate) (PSS), and poly(acrylic acid) (PAA)) and synthetic clays. We monitor the assembly growth by UV-vis spectroscopy, estimate the assembly thickness by scanning electron microscopy (SEM) cross-section images and the surface roughness by atomic force microscopy (AFM). We further examine the charge density of films by Fourier transform infrared (FTIR) spectroscopy, we determine the layer spacing by X-Ray diffraction (XRD), and we evaluate the organic-inorganic content by thermogravimetric analysis (TGA). Our results demonstrate that PDDA-PAA-PDDA-Laponite multilayers display exponential growth aided by internal diffusion of PAA molecules, whereas PDDA-PSS-PDDA-Laponite and PDDALaponite multilayers display linear growth by sequential molecular level deposition. We also illustrate that different assembly regimes are attained by manipulating the deposition pH of PAA solution. The results on the structural characteristics and properties of multilayers formed under different conditions can provide insights into their behavior and sensitivity to external stimuli during various applications.

(690d) Depth-Profiling X-Ray Photoelectron Spectroscopy (XPS) Analysis of Interlayer Diffusion in Nanostructured Polyelectrolyte Multilayers

Gilbert, J. B., Johns Hopkins University
Rubner, M. F., Massachusetts Institute of Technology
Cohen, R. E., Massachusetts Institute of Technology
Originally presented on: 11/7/2013 13:40:00 - 14:00:00

Functional organic thin films often require stratified structures with a distinct purpose for each layer. Applications of stratified films, from the sequential release of therapeutics from a biomedical implant(1) to a polymeric interfacial layer in a dye-sensitized solar cell(2), all demand nanometer level control over the structure. However, due to dynamic diffusion of the species between layers the composition and function of these layers may not be as desired. Analysis of the diffusion of species between layers has proven difficult as other methods require fluorescent labeling of the materials of interest (FRET), require dueterated materials (neutron reflectivity) or have poor z-resolution (confocal microscopy, FTIR). In this talk we will describe high-resolution depth profiling XPS of polymeric nanostructured films using polymer-friendly cluster-ion C_{60} sputtering that was recently published in the *Proceedings of the National Academy of Sciences(3)*. This technique has never before been applied to functional nanostructured films and allows for the direct determination of the composition of polymeric films with ~15nm level resolution.

Applying this technique to a model nanostructured film fabricated using layer-by-layer technology we explore common issues in the field such as the dynamic competition between hydrogen bonding and electrostatic interactions during fabrication, blocking interlayer diffusion, the exchange of film components with a surrounding solution and the kinetics of interlayer diffusion. Using high resolution depth profiling XPS we show that chitosan, a commonly used biopolymer, diffuses readily into a layer comprised of hydrogen bonded polymers with a diffusion coefficient of $1.4*10^{-12}$ cm^2/sec. In addition we demonstrate that the interlayer diffusion of chitosan displaces the hydrogen bonding component poly(ethylene oxide) from the film, thus changing the composition. Finally, we find that a <1nm layer of poly(allylamine hydrochloride) can stop the diffusion of chitosan and thus enables fabrication of highly structured films, each with the desired function. We propose that our results address some common issues of interlayer diffusion and could be applied to multiple fields utilizing organic thin films. Further, the ability to identify and analyze the composition of each layer of a polymeric nanostructured film displays the precision and power of cluster-ion C_{60} depth profiling XPS that could be applied to many other fields.

References:

1. Wood KC, Chuang HF, Batten RD, Lynn DM, Hammond PT (2006) Controlling interlayer diffusion to achieve sustained, multiagent delivery from layer-by-layer thin films. *Proc Natl Acad Sci U S A*

103(27):10207-10212.

2. Kim KH, *et al.* (2012) Interface control with layer-by-layer assembled ionic polymers for efficient low-temperature dye-sensitized solar cells. *Journal of Materials Chemistry* 22(22):11179-11184.

3. Gilbert JB, Rubner MF, Cohen RE (2013) Depth-profiling x-ray photoelectron spectroscopy (xps) analysis of interlayer diffusion in polyelectrolyte multilayers. *Proc Natl Acad Sci U S A* 110(17):6651-6656.

(690e) Directed Self-Assembly of Acrylic Terpolymers With Mixed Block Regimes

Bergman, J. A., Iowa State University
Heinen, J., Iowa State University
Originally presented on: 11/7/2013 14:00:00 - 14:20:00

Obtaining the optimal polymer microstructure with the optimal chemical domains for specific applications is difficult. Because of this, there has been much research into how to manipulate the microstructures of self-assembling polymers such that desired chemical functionalities and microstructures can be simultaneously achieved. The specific microstructure formed in a phase-separated system is determined by minimizing the free energy of the polymer chain conformations and the interfacial energy between the two microdomains. To date, the thermodynamics that define the interfacial curvature, and therefore, the microstructure have largely been tuned by changing polymer composition. Polymers with mixed block regimes offer a relatively new variable for changing the microstructure at a specific composition. For diblock terpolymers, the thermodynamic behavior changes as a third monomer species is copolymerized; differences in self-assembly between the A-b-(B-co-C) architecture, (A-co-C)-b-B architecture, and (A-co-C)-b-(B-co-C) are to be expected. This work explores the differences in these three architectures for an all acrylic system composed of hydroxyethyl acrylate (HEA), methyl acrylate (MA), and octyl acrylate (OA).

Here three diblock architectures, P(HEA)-b-P(OA-co-MA), P(HEA-co-MA)-b-P(OA), and P(HEA-co-MA)-b-P(OA-co-MA) are compared with two triblock architectures, the non-frustrated P(HEA)-b-P(MA)-b-P(OA) and the frustrated P(HEA)-b-P(OA)-b-P(MA). These five architectures were synthesized at four different compositions, where the ratio between the volume fractions HEA and OA remains constant at fHEA/ fOA = 1.74, and the total molecular volume remains constant at approximately 11,600 mL/ mol. The manipulated variable, fMA, was used to explore the effects of altering the enthalpic interaction between polymer blocks. This talk will present results from synchrotron small angle x-ray scattering (SAXS) experiments preformed at Diamond Light Source; the SAXS results characterize the self-assembly of these 20 polymers in the bulk, in water (a selective solvent for HEA), and in THF (a good solvent).

(690f) Mechanical Properties of Poly(styrene-isobutylene-styrene) Membranes As a Function of Sulfonation Level and Counter-Ion Substitution

Padovani, A., University of Puerto Rico
Suleiman, D., University of Puerto Rico, Mayaguez Campus
Negron, A.,
Originally presented on: 11/7/2013 14:20:00 - 14:40:00

Mechanical Properties of Poly(styrene-isobutylene-styrene) Membranes as a Function of Sulfonation Level and Counter-Ion Substitution

Agnes M. Padovani[a,b], David Suleiman[b], and Arnaldo Negrón[b]

Department of Engineering Science and Materials[b], University of Puerto Rico

Mayagüez, Puerto Rico, 00681-9000

Department of Chemical Engineering[b], University of Puerto Rico

Mayagüez, Puerto Rico, 00681-9000

In this study, mechanical properties of sulfonated poly(styrene-isobutylene-styrene) (SIBS) were measured using nanoindentation and dynamic mechanical analysis (DMA) as function of sulfonation level (0-97%) and counter-ion substitution (Mg^{+2}, Ca^{+2}, Ba^{+2}). Nanoindentation results showed that increasing the amount of sulfonic groups did not have a significant effect on the elastic modulus and hardness for the unsubstituted membranes. However, the elastic modulus showed an optimum level with sulfonation around 64% sulfonation. The incorporation of counter-ions improved their elastic modulus and hardness, with ion size playing a significant role on their improvement ($Ba^{+2} > Ca^{+2} > Mg^{+2}$). DMA results showed that sulfonation of the base polymer maintained the glass transition temperature (Tg) of the polyisobutylene (PIB) segment of the membrane at -60°C regardless of sulfonation level. However, the shoulder of the PIB Tg (-30°C), that was perhaps caused by a Rouse type motion, as well as the Tg of PS (105°C) disappeared upon sulfonation. The substitution of inorganic counter-ions maintained the glass transition temperature of the PIB at -60°C regardless of counter-ion exchanged and increased the modulus of the rubbery plateau indicative of a stronger and more thermally stable cross-linked complex formation. Additional unique relaxations were observed for the sulfonated and counter-ion substituted samples, attributed to the stretching/rotation of the S-O bond and the interaction of the cations with the oxygen of the sulfonate group.

(690g) Self-Assembly Based Architecture of 2-D and 3-D Multi-Level Superstructure Nanoarrays

Suresh, V., National University of Singapore
Tan, Y. L., National University of Singapore
Srinivasan, M. P., National University of Singapore
Krishnamoorthy, S., Centre de Recherche Public-Gabriel Lippmann
Originally presented on: 11/7/2013 14:40:00 - 15:00:00

Micro-electro-mechanical systems (MEMS) devices coupled with nanophotonics and plasmonics extend a great deal of promise toward sensing chemical species, metabolites, volatile organic compounds, air pollutants, and noxious gases for application in medicinal equipments, pharmaceuticals and quality control. The plasmonic properties of nanomaterials on surfaces become crucial to sensing and consequently, meticulously-architectured designs are imperative to achieve high levels of sensitivity and selectivity. In an effort to mimic the macroscopic entities such as nature-crafted geometry and patterns on shells, peacock feathers, sun flowers and moth eyes at the nanoscale, we have investigated the design rules that are essential to produce desired architectures that led to the nanofabrication of flower-like patterns in the sub-100 nm regime over macroscopic areas.

We demonstrate here, an entirely self-assembly driven approach to fabricate two-dimensional and three dimensional arrays of binary or ternary flower-like superstructures of various material components. Macroscopic two-dimensional arrays of such superstructures presenting up to three levels of hierarchy are achieved using self-organization of reverse micelles of block copolymer up to the second level and directed electrostatic self-assembly of hard metal nanospheres at the third level. While three-dimensional arrays are fabricated with block copolymer micelles guiding the electrostatic self assembly at the first level of hierarchy which provides the platform for confinement-induced self assembly of block copolymers at the next level. In either case, the hierarchy at every level of the superstructure is meticulously engineered by introducing different material components at different hierarchical levels with ZnO, TiO_2, and Au nanoparticles being prototypes. Such entirely self assembly derived hybrid material architecture that mimic Nature offers rich flexibility in choice of material type, composition and stoichiometries, is unprecedented, and has extensive applications in sensing and plasmonics.

(691a) Enzyme-Based Nanocomposites: Using Nature to Ward Off Emerging Threats

Kane, R. S., Rensselaer Polytechnic Institute
Dordick, J., Rensselaer Polytechnic Institute
Grover, N., RPI
Pangule, R., Rensselaer Polytechnic Institute
Mehta, K., RPI
Mundra, R.,
Paskaleva, E.,
Solanki, K.,
Wu, X.,
Originally presented on: 11/7/2013 12:35:00 - 12:55:00

Nature is unparalleled in its structural and functional diversity. We have taken cues from nature to design nanostructures that are functionalized with biological molecules, complete with tailored selectivities and biocatalytic activities. Herein we highlight our recent efforts to exploit the interface of biology with materials science, with a particular focus on enzyme-nanomaterial composites with a wide range of activities that endow surfaces with self-cleaning properties. In particular, surfaces have been generated with tailored activity against pathogens (e.g., MRSA). Various enzymes have been used, including cell wall lytic enzymes that selectively target pathogens in a manner that is extremely difficult for the pathogen to gain resistance. We have also developed approaches to identify novel cell lytic enzymes. As an example, we will discuss the identification and characterization of a lytic enzyme that is active against bacillus anthracis. Such activity provides a safe and potentially broadly applicable route to eliminating pathogenic microorganisms from common surfaces.

(691b) Enzyme-Based Technologies: Perspectives and Opportunities

Dinu, C. Z., West Virginia University
Campbell, A., West Virginia University
Dong, C., West Virginia University
Wu, N., West Virginia University
Dordick, J. S., Rensselaer Polytechnic Institute
Originally presented on: 11/7/2013 12:55:00 - 13:10:00

Enzyme immobilization is used as a viable alternative to overcome the limitations of enzyme-based applications *in vitro* and to ensure high enzyme activity retention and high operational stability. The choice of immobilization technique is determined by considering both chemical and physical properties of the enzymes and of the support surfaces. As such, immobilization has been achieved by entrapping enzymes into polymer matrices, Langmuir-Blodgett films, solid- or liquid- based membranes, or simply by attachment of enzymes onto solid supports (either by covalent or physical immobilization). This presentation is focused on the current trends in enzyme-based technologies and the research in my group aimed at developing decontamination platforms based on enzymes and capable of neutralizing bacteria, viruses and spores. Various enzyme immobilization strategies are discussed and further insights into the next generation of surface decontamination technologies are provided, outlining the studies that are underway to enable these technologies to be self-sustainable (i.e. operate under ambient conditions without external addition of the enzyme substrate). Further, the interface reactions between biological molecules and nanosupports as well as the methods used for controlling this interface are presented. By "smart design" of the enzyme-nanomaterial-hybrids one could achieve longer shelf life and stability for applications as diverse as biosensors, drug delivery or decontamination platforms.

(691c) Effects of Acid Treatment On Structure, Properties and Biocompatibility of Carbon Nanotubes

Dong, C., West Virginia University
Campbell, A., West Virginia University
Eldawud, R., West Virginia University
Dinu, C. Z., West Virginia University

Originally presented on: 11/7/2013 13:10:00 - 13:20:00

Carbon nanotubes (CNTs) applications in fields as diverse as electronics, optics, nanotechnology, materials science of biomedical industry rely on improved understanding of CNTs physical structures and chemical properties as well as control over their user and environment compatibility. User-controlled strategies that allow changes in physical and chemical properties of CNTs can lead to improved performance of CNTs-based technologies in any of the fields of applications listed above. Herein we carried out a systematic investigation on how acid liquid oxidation treatment changes CNTs physical and chemical properties to lead to nanomaterials with increased biocompatibility. Specifically, by exposing CNTs to strong acid mixtures (sulfuric and nitric acid), we generated user-controlled libraries of CNTs with different characteristics, as recorded and analyzed using Raman energy dispersive x-ray spectroscopy, scanning electron microscopy, energy dispersive spectroscopy, atomic force microscopy, and solubility tests. Systematically characterized CNTs were subsequently tested for their biocompatibility in relation to human epithelial cells (BEAS-2B) or enzymes. Such selected examples are building pertinent relationships between CNTs biocompatibility and their intrinsic properties and suggest that acid oxidation treatment reduces CNTs toxicity. Further, such user-tailored libraries of CNTs with controlled physico-chemical properties can serve as the next generation of nanosupports to provide feasible platforms to be used for biomedical applications or the next generation of biosensors and diagnose tools.

(691d) Enzyme-Nanomaterial Interaction Studies With Application in in Situ Decontaminant Production

Campbell, A. S., West Virginia University
Dong, C., West Virginia University
Meng, F.,
Wu, N., West Virginia University
Dordick, J. S., Rensselaer Polytechnic Institute
Dinu, C. Z., West Virginia University
Originally presented on: 11/7/2013 13:20:00 - 13:30:00

ABSTRACT

Enzyme-Nanomaterial Interaction Studies with Application in *in situ* Decontaminant Production

Alan S. Campbell, Chenbo Dong, Fanke Meng, Nianqiang Wu, Jonathan S. Dordick, Cerasela Zoica Dinu

Enzymes are currently used in industrial catalysis, biological fuel cell production and biosensing. Their high selectivity, specificity and biocompatibility relative to their chemical counterparts recently made enzymes the "green" constituents of coatings with antifouling and decontamination properties. However, before such applications are implemented in large-scale settings, the operational stability and retention of enzymes activities in such coatings needs to be addressed. Herein, an in depth study of the interactions between enzymes and nanomaterials is presented. This systematic study is accomplished using enzymes with varying properties (i.e. size and surface chemistry) and nanosupports with differing characteristics (i.e. surface curvature, charge and aspect-ratio), as well as multiple immobilization techniques (i.e. physical, covalent, or covalent through a linker to bring the enzyme away from the nanosupport). The research will highlight the importance of rational design of the nanomaterial-biological molecule hybrids and the need for control of the bio-nano-interface for achieving most advantageous immobilization conditions that allow high retention of the enzyme structure and functionality. The knowledge gained from this study can further be used to optimize enzyme-nanomaterial interactions in order to maintain finest levels of enzyme activity while enhancing stability and retention upon immobilization. This research can be used for creating the next generation of self-sustainable decontamination coatings and biosensors with improved specificity and efficiency.

(691e) Bioinspired Mechanochemical Nanosystems

Cha, T. G., Purdue University
Pan, J., Purdue University
Robinson, H., Purdue University
Choi, J. H., Purdue University
Originally presented on: 11/7/2013 13:30:00 - 13:45:00

There have been significant efforts in recent years in developing novel nanosystems for bioinspired molecular engineering. We have designed single-wall carbon nanotubes with molecular recognition domain by functionalizing them with nucleic acids. With molecular recognition and self-assembly capabilities of oligonucleotides, we have created nanomechanical systems that convert chemical energy into mechanical motion. We have characterized our nanomechanical systems by using single nanotube fluorescence spectroscopy. Our spectroscopic system is a powerful and unique optical platform that allows one to probe and analyze biomolecular reactions both in ensemble and at single molecule level. The kinetic information of biomolecular reactions was extracted by utilizing the signal transduction from the optical nanomaterials. A simple kinetic model was developed to quantitatively understand the single turnover reaction. The analysis and simulations were performed within the framework of single-molecule kinetics to predict the overall rate of the single turnover reaction. The experimental and theoretical investigation results will be presented.

(691f) Molecular Bioelectronics: Enzyme-Swnt Conjugates for Bioanalytical Biochips

Guiseppi-Elie, A., Clemson University

Originally presented on: 11/7/2013 13:45:00 - 14:00:00

Generation-3 (Gen-3) biosensors that are based on direct electron transfer are reagentless [1] and hold the potential for long term in vivo stability to enable fully implantable continual monitoring of various analytes of a wide variety of physiological conditions and pathologies [2]. Among the novel approaches for establishing a Gen-3 biotransduction mechanism, ultrasonic processing [3] of single walled carbon nanotube-enzyme (***e.g.*** Glucose oxidase, GOx) conjugates [4] was investigated as a function of sonication time (5 and 60 min), SWNT functionalities (-OH and –COOH) and tube length (<1 μm and 5-10 μm) [5]. Using the blue shift ($\Delta\lambda_{max}$) in the absorbance maximum of the FAD-apoenzyme, the change in the sum of α-helix and β-sheet in the circular dichroism spectra, and the change in the k_{cat}/K_M and n of the enzyme conjugate expressed relative to the pristine GOx, were found to show only modest change to the enzyme's structure during supramolecular conjugation. Short SWNTs appeared to support conjugate formation with no egress of FAD for the enzyme; medium SWNT appeared to support conjugate formation moderately while the long and non-functionalized SWNT caused partial denaturation leading to the egress of FAD.

Microdisc electrode arrays (MDEA 5037 Pt) were microlithographically fabricated on borosilicate glass from 10 nm TiW adhesion layer and 100 nm e-gun deposited Pt. The Pt layer was patterned into working, counter and reference electrodes and all electrodes covered by Si_3N_4 which was windowed to reveal a pattern of microdiscs (37 discs, each f=50 μm, A= ???cm^2) over the working electrode and exposed counter and reference electrodes. Biofabrication [6] ***of enzyme amperometric biosensors*** using the process of electropolymerization (Py to PPy at 0.8 V vs. Ag/AgCl) onto the MDEA 5037 in the presence of glucose oxidase (1 mg/ml, DI water, 100 mC/cm^2) resulted in GOx, GOx-ferrocene monocarboxylic acid and GOx-SWNT biotransducers that were responsive to glucose and that allowed parallel evaluation of the relative efficacy of Gen-1, Gen-2 and Gen-3 mechanisms of transduction. The bioanalytical performance of sensitivity (S), linear dynamic range (LDR) and detection limit [3(SD(blank)/S] as well as apparent enzyme kinetic parameters of $\mathbf{K}_M$, I_{max}, and k_{cat}, were obtained from the dose-response curves for mutarotated glucose solutions (0.0 mmol/L to 13.7 mmol/L in PBS 7.2) at RT and compared to solution phase behavior. There was a good linear relationship (R^2=0.99) between glucose concentration and response from 2.0 mM – 9.0 mM. The sensitivity was 10^{-3} mA/mM and the detection limit was 0.002 mA/mM.

Keywords: Biosensors, Bioimmobilization, Nanotubes, Direct Electron Transfer, Glucose-SWNT

References:

[1] Guiseppi-Elie A, Lei C, Baughman RH. Direct electron transfer of glucose oxidase on carbon nanotubes. Nanotechnology. 2002;13:559.

[2] Kotanen CN, Moussy FG, Carrara S, Guiseppi-Elie A. Implantable Enzyme Amperometric Biosensors. Biosensors and Bioelectronics. 2012:(in press).

[3] Guiseppi-Elie A, Choi S-H, Geckeler KE. Ultrasonic processing of enzymes: Effect on enzymatic activity of glucose oxidase. Journal of Molecular Catalysis B: Enzymatic. 2009;58:118-23.

[4] Guiseppi-Elie A, Choi S-H, Geckeler K, Sivaraman B, Latour R. Ultrasonic Processing of Single-Walled Carbon Nanotube–Glucose Oxidase Conjugates: Interrelation of Bioactivity and Structure. NanoBiotechnology. 2008;4:9-17.

[5] Karunwi O, Guiseppi-Elie A. Supramolecular glucose oxidase-SWNT conjugates formed by ultrasonication: effect of tube length, functionalization and processing time. Journal of Nanobiotechnology. 2013;11:6.

[6] Karunwi O, Wilson AN, Kotanen C, Guiseppi-Elie A. Engineering the Abio-Bio Interface to Enable More than Moore in Functional Bioelectronics. Journal of The Electrochemical Society. 2013;160:B60-B5.

(691g) Protein-Based Bioelectronic Devices

Lederman, D., West Virginia University
Jett, J.,
Bostick, C.,
Li, D.,
Gannett, P.,
Tracy, T., University of Kentucky
Originally presented on: 11/7/2013 14:00:00 - 14:15:00

After billions of years of evolution, many proteins have developed the ability to transfer charge in a controlled and stable fashion from the environment to the protein's active site. Prime examples are heme-containg proteins. Exploiting this property in multifunctional (e.g., capable of sensing and computation) electronic circuitry could result in devices that are biocompatible and have a wide range of applications. In this talk I will describe recent efforts to make electronic devices out of proteins, and in the process, learn about the fundamental mechanism responsible for electron transfer in biological systems. One example is an attempt to fabricate single electron transistors (SETs) using myoglobin, a heme protein found in muscle fibers. SETs were fabricated by depositing the protein on nanometer scale metal junctions which were broken via electromigration on n-Si/SiO2 substrates used as back gates. By performing control experiments on blank junctions and apo-myoglobin (myoglobin without heme group), signals were observed that indicate single electron conductivity in a few devices. In another example, cytochrome P-450 proteins were immobilized on gold pillars less than 30 nm in diamter, and their electrical conductance was measured using conducting probe atomic force microscopy (CP-AFM). Systematic changes in the conductance of these proteins (in dry conditions) when bound to different small molecules were found to mirror the effects of these molecules on the known metabolic activity of the protein. These experiments show that it may be feasible to make advanced biocompatible electronic devices in the near future.

This work was supported by the National Science Foundation (grant EPS-1003907) and the National Institutes of Health (GM-086891).

(691h) Fundamentals and Applications of Nanomaterial-Biomolecule Interactions

Asuri, P., Santa Clara University

Originally presented on: 11/7/2013 14:15:00 - 14:30:00

Recent research has illustrated the significant influence of nanomaterials on the structure and function of a variety of biomolecules and biological species. These interactions impact applications ranging from biotransformations and diagnostics to drug delivery, disease therapy and development of novel biomaterial nanocomposites. However, many challenges remain before the potential of nanomaterials can be fully realized, including understanding and controlling the nanomaterial-biomolecule interface. This talk will focus on our work developing a fundamental understanding on how nanomaterials affect the fate and function of proteins, biopolymers, and cells. Topics will include enzyme stability on carbon nanotubes, development of efficient viral nanoparticles that infect human stem cells, effect of silica nanoparticles on biomaterial stiffness, and role of particle size on cellular nanotoxicity.

(691i) Biomimetic Two-Dimensional Crystals of Photosynthetic Proteins for Membrane Based Energy Production

Saboe, P. O., Pennsylvania State Univerity
Lubner, C. E., Pennsylvania State Univerity
McCool, N. S., Pennsylvania State Univerity
Vargas-Barbosa, N. M., Pennsylvania State Univerity
Golbeck, J. H., Pennsylvania State University
Kumar, M., The Pennsylvania State University
Originally presented on: 11/7/2013 14:30:00 - 14:45:00

It has recently been shown that a molecular wire can successfully tether a hydrogenase enzyme (H_2ase) and a photosynthetic protein complex, Photosystem I (PSI) to integrate hydrogen generation with effective solar energy conversion. The light harvesting component, PSI is a robust membrane-bound protein complex with the ability to provide the chemical potential energy needed to reduce protons. The H_2ase-PSI construct can generate significant amounts of hydrogen when in solution with an electron donor. Studies to increase hydrogen production rates by the attachment of the H_2ase-PSI construct to an electrode are currently focusing on single particle approaches. In order to achieve maximized electron transfer to immobilized PSI, high protein density and correct orientation on the surface are required. Incorporating PSI in a membrane environment mimics its native state and allows for control over the protein orientation, providing favorable conditions for electron transfer. End group modification of lipids is an immobilization technique of bilayers to minimizing the distance between the membrane and surface. PSI will be incorporated with thiolated end group modified lipids and attached to a gold electrode for cathodic photocurrent measurements. Protein incorporation into lipid membranes has been determined by TEM. Results currently show mosaic crystallization of the PSI trimers in unmodified end group lipids. Increasing the stability, orientation and efficiency of PSI based membranes can be achieved by using tailored amphiphilic block copolymers (BCPs) as a substitute for modified lipids. We have created lipid and polymer membranes with a high density of PSI molecules and have characterized their activity using time-resolved optical spectroscopy and photocurrent measurements.

(691j) A Plant Nanobionic Approach to Enhance Solar Energy Conversion of Extracted Chloroplasts Using Spontaneously Assembled Nanoparticles

Giraldo, J. P., Massachusetts Institute of Technology
Faltermeier, S., Georgia Institute of Technology
McNicholas, T. P., Massachusetts Institute of Technology
Boghossian, A. A., Massachusetts Institute of Technology
Reuel, N. F., Massachusetts Institute of Technology
Hilmer, A. J., Massachusetts Institute of Technology
Sen, F., Massachusetts Institute of Technology
Brew, J., Massachusetts Institute of Technology
Landry, M., Massachusetts Institute of Technology
Strano, M. S., Massachusetts Institute of Technology
Originally presented on: 11/7/2013 14:45:00 - 15:00:00

The interface between plant organelles and non-biological nanostructures has the potential to impart the former with new and enhanced functions, and unlike photosynthetic proteins, has been relatively unexplored for energy applications. This nanobionic approach can yield isolated chloroplasts – plant organelles responsible for converting CO_2 and solar energy to glucose – that are more stable to reactive oxygen species *ex vivo*, possess enhanced solar conversion efficiencies, are able to produce glucose, and allow real time information exchange via embedded nanosensors for free radicals. We show that cationic or anionic single walled carbon nanotubes (SWNT) have the ability to passively transport and irreversibly localize within the lipid envelope of extracted plant chloroplasts from *Spinacia oleracea*. This mechanism extends to poly (acrylic acid) nanoceria (PAA-NC) and their corresponding nanotube complexes (SWNT-NC). Concentration of superoxide can be suppressed 59% by assembling PAA-NC inside chloroplasts. SWNT-NC promote photoactivity 3.1 times above controls while PAA-NC extend glucose production from 0.5 to 88 hours. SWNT chloroplast complexes also allow fluorescent reporting of nitric oxide generation. Nanobionic engineering of extracted plant organelles may contribute to developing biomimetic light-harvesting materials with regenerative properties, enhanced efficiency and prolonged lifetimes.

(692a) Controlled Plasmonic Coupling of Multifunctional Gold Nanoparticles for Photothermal Cancer Treatment

Sotiriou, G. A., Harvard University
Starsich, F.,
Pratsinis, S. E., ETH Zurich
Originally presented on: 11/7/2013 12:30:00 - 12:50:00

Plasmonic nanoparticles show great potential to the in-vivo photothermal ablation of tumors because of their efficient near-IR light-to-heat conversion. Here, we show that by finely tuning the plasmonic Au nanoparticle coupling of those nanoparticles the optical absorption can be also finely controlled, broadening it to the near-IR spectral region. Hybrid plasmonic-superparamagnetic nanoparticles are encapsulated in-situ during their gas-phase synthesis by a nanothin, smooth silica coating with a precise control over the coating thickness [1,2,3]. This fine control of the interparticle distance within each agglomerate enables such hybrid bionanoprobes to efficiently convert light-into-heat upon their near-IR irradiation and facilitates their employment as efficient therapeutic agent in photothermal heat tumor ablation. The feasibility of such multifunctional [4] gold nanoparticles as efficient plasmonic photothermal agents is demonstrated by destroying breast cancer cells in vitro upon near-infrared laser irradiation.

References

[1] Sotiriou G. A., Sannomiya T., Teleki A., Krumeich F., Vörös J. and Pratsinis S. E. Non-toxic dry-coated nanosilver for plasmonic biosensors. *Adv. Funct. Mater.* **20**, 4250-4257 (2010).

[2] Sotiriou G. A., Hirt A. M., Lozach P. Y., Teleki A., Krumeich F. and Pratsinis S. E. Hybrid, silica-coated, Janus-like plasmonic-magnetic nanoparticles. *Chem. Mater.* **23**, 1985-1992 (2011).

[3] Sotiriou G. A., Franco D., Poulikakos D. and Ferrari,A. Optically Stable Biocompatible Flame-Made SiO_2-Coated Y_2O_3:Tb^{3+} Nanophosphors for Cell Imaging. *ACS Nano* **6**, 3888-3897 (2012).

[4] Sotiriou G. A., Biomedical Applications of Multifunctional Plasmonic Nanoparticles. *WIREs Nanomed. Nanobiotechnol.* **5**, 19-30 (2013).

(692b) Synthesis, Characterization, Cellular Internalization and Mitochondrial Targeting of Triphenylphosphonium-Conjugated PAMAM Dendrimer Nanocarriers

Bielski, E., Wayne State University
Brown, M.,
Della Porta, K.,
da Rocha, S. R. P., Wayne State University
Originally presented on: 11/7/2013 12:50:00 - 13:10:00

Lung cancer is the leading cause of cancer related death worldwide with over 1.3 million deaths annually. Less than 15% of those diagnosed with lung cancer have an overall prognosis greater than 5 years. Therefore, new strategies to fight lung cancer have tremendous potential in improving the health and survival rate of a large number of patients. Mitochondria play vital roles in energy production within cells, are key to eukaryotic cell survival, in the generation of reactive oxygen species, in the regulation of calcium homeostasis, and importantly, in the regulation of programmed cell death. Mitochondrial dysfunction is related to a range of diseases including cancer. A change in the cellular bioenergetics is a crucial marker of cancer, as well as resistance to apoptosis, invasiveness, malignancy, and decreased autophagy. Many of these factors have been linked to mitochondrial dysfunction, making mitochondria vital intracellular targets for anticancer therapies.

Delocalized lipophilic cations (DLCs), such as triphenylphosphonim cations (TPP), are the most widely used mitochondrial targeting agents due to their delocalized positive charge, which allows them to pass through plasma and mitochondrial membranes in a potential driven process, showing 100-1000 fold selective accumulation into the cell cytosol and mitochondrial compartments. DLCs also accumulate more readily into tumor cells as they possess more negative plasma and mitochondrial membrane potentials. Therefore, the use of DLCs such as TPP may offer unique strategies to deliver numerous anticancer agents preferentially to cancerous cells, and can thus be very effective in maximizing drug effects while reducing side effects. The combination of DLCs with drug nanocarriers, such as poly(amido amine) (PAMAM) dendrimers, may afford further opportunities for the cellular and intracellular targeting and controlled release of therapeutics to treat lung cancer.

In this work, we report i) the synthesis and characterization of triphenylphosphonim cations (TPP) to PAMAM dendrimers, which are potential nanocarriers of anticancer agents, ii) the cellular internalization and colocalization of such nanocarriers in A549 alveolar epithelial adenocarcinomal cell line using flow cytometry and confocal microscopy, respectively, and iii) the effect of PEG-TPP conjugation to PAMAM dendrimers on cellular internalization and colocalization in A549 cells.

This study is also relevant in the delivery of other agents to the mitochondria for the treatment of mitochondrial disorders related to other diseases such as diabetes, neurodegenerative, neuromuscular disorders, and inherited mitochondrial diseases not necessarily associated with cancer.

(692c) Localized Sustained Delivery of siRNA for the Treatment of Diabetic Ulcers

Castleberry, S., Oklahoma State University
Hammond, P. T., Massachusetts Institute of Technology
Originally presented on: 11/7/2013 13:10:00 - 13:30:00

Wound healing is a complex and highly synchronized process of multiple biological pathways, consisting of an assortment of signaling molecules including many growth factors, which evolves over time with the progression of the wound. Development of drug delivery system that can locally modify cell behavior on the basic level of gene transcription provides a powerful tool to alter the pathology of wound healing. RNA interference is a highly specific gene silencing process mediated through the use of short interfering RNAs (siRNA) that has the potential to be an extremely powerful therapeutic tool in medicine.

Impaired wound healing is a major problem in diabetes, leading to the development of more than 750,000 diabetic foot ulcers and 70,000 lower extremity amputations per year in the USA alone. Recent work in understanding the molecular mechanisms underlying poor wound resolution in diabetic patients have shown increased expression of metalloproteinase 9 (MMP-9) correlates with poor outcomes. Small interfering RNA (siRNA) has emerged as a powerful tool in post-transcriptional gene silencing. In this work we describe the development and implementation of a polymer-based thin film that successfully delivered siRNA for the knockdown of MMP-9 expression within the wounds of genetically diabetic mice. This reduction resulted in significantly accelerated wound healing and increased collagen deposition within the wound bed.

Developing the capability to effectively deliver siRNA locally and in a way that addresses the complications common in diabetic wound healing would significantly advance the current state of diabetic ulcer treatment. This work presents an opportunity to substantially improve the treatment of patients and outcomes.

(692d) Design of a Microinjection Device for Injection of in Situ Gelling Hydrogels for Ophthalmic Drug Delivery

Campbell, S. B., McMaster University

Yang, J.,

Wu, W. I., McMaster University

Selvaganapathy, P. R., McMaster University

Hoare, T. R., McMaster University

Originally presented on: 11/7/2013 13:30:00 - 13:50:00

Design of a microinjection device for injection of *in situ* gelling hydrogels for ophthalmic drug delivery

Scott Campbell*, Jun Yang**, Wen-I Wu**, P. Ravi Selvaganapathy**, and Todd Hoare*

**Department of Chemical Engineering, McMaster University, Hamilton, ON, Canada*

****Department of Mechanical Engineering, McMaster University, Hamilton, ON, Canada*

E-mail: campbesb@mcmaster.ca | hoaretr@mcmaster.ca

INTRODUCTION: The primary causes of vision loss in developed nations are diseases associated with the posterior eye, such as age-related macular degeneration, diabetic retinopathy, posterior uveitis, and retinitis due to glaucoma.[1–3] Thus, the development of materials that can effectively treat such ailments would be highly beneficial. However the posterior region of the eye is a particularly difficult target tissue due to anatomic and physiologic limitations.[2] Treatments involving injections to the back of the eye have proven quite successful, but the required frequent injections greatly increases the risk of complications over time.[2] Hydrogels present a potential solution to this issue, as they can facilitate prolonged delivery of a drug to the eye and thus limit the number of required injections for effective treatment.[4] Furthermore, unlike most drug release vehicles, hydrogels can be designed to effectively match the refractive index of the vitreous humour and be injectable *in situ*, exploiting the rapid chemical reaction of complementary groups grafted to two polymers (such as aldehyde-mediated crosslinking of alcohol-, amine-, or hydrazide-functionalized polymers or thiol-mediated crosslinking with vinylsulfone or acryloyl groups) to form the crosslink structure.[5,6] Crosslinked networks may also find applications as vitreous replacement materials for patients suffering from opacification of the vitreous, commonly observed upon aging.

We have developed several *in situ*-gelling injectable hydrogels using aldehyde-hydrazide chemistry to produce hydrolytically degradable hydrazone crosslinks.[7,8] When polymers with these two functional groups come into contact, they rapidly form a hydrogel, entrapping the desired drug to be released in the process. These hydrogel materials have promising tunable drug delivery characteristics for a variety of drugs, can have refractive indices similar to that of the vitreous, and slowly degrade in physiological conditions.[7] However, to assess the *in vivo* capabilities of these hydrogels in mouse, rabbit, and ultimately human eyes, very small amounts of each reactive material (1 – 10 μL) must be injected. While this is routinely done with single component systems, the additional requirement of mixing the two reactive polymers upon injection (to prevent premature gelation) poses a significant additional challenge with administering these injectable hydrogels to the point that no suitable injection system currently exists. In response, this work describes the fabrication of a novel microinjector that: (1) effectively mixes the two precursor polymers from separate microchannels upon injection; (2) controllably and precisely injects volumes in the 1 - 10 μL range through a narrow gauge needle suitable for ophthalmic applications; and (3) rapidly injects these materials to prevent gelation, and blockage within the needle. The developed dual-barrel microinjection system is characterized in terms of its optimal channel length, the precision of the injected volumes, the properties of the ejected polymer solutions/hydrogels after injection, and its use in initial *in vivo* studies.

EXPERIMENTAL: The *in situ*-injectable hydrogels are typically prepared by the mixing a hydrazide-functionalized polymer (A-Polymer) and an aldehyde-functionalized polymer (B-Polymer) (both have a MW <32 kDa) upon injection using a double barrel syringe (Figure 1). The double-barrelsyringe has two compartments: one with hydrazide functionalized NIPAM and the other with aldehyde functionalized dextran in PBS solutions. Upon injection, these two solutions interact in the mixing channel and needle, forming hydrazone bonds that crosslink the polymer to form a hydrogel.

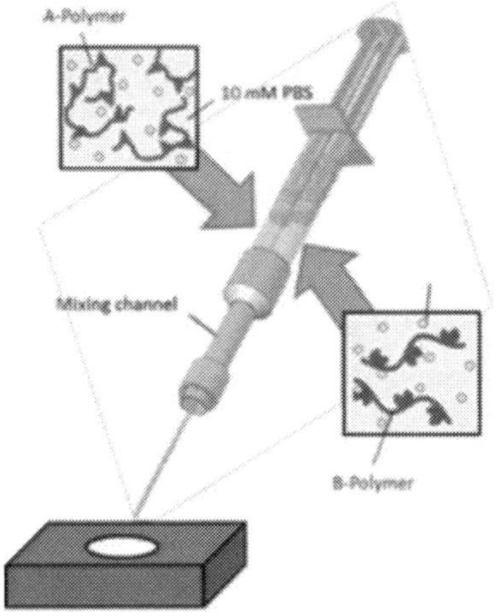

Figure 1: Typical hydrogel fabrication process using a double barrel syringe.

The microfluidic injector was designed to operate analogous to this macroscale system. The A- and B-polymer solutions are contained within separate reservoirs that are connected to the same pneumatic pressure source. Upon the application of a given pressure pulse, the polymers will travel through a microfluidic channel of a given length. To effectively mix the two polymer components, a staggered herringbone mixer originally described by Stroock et al. (2002) will be utilized which generates transverse flows to induce microvortexes and promotes mixing.[9] After navigating through the mixing channel, the mixed polymer solutions will be ejected from the microfluidic injector via a 30 ½ (or higher) gauge needle, which are typically used for *in vivo* ocular injections. A simplified schematic of the microinjection device is shown in Figure 2.

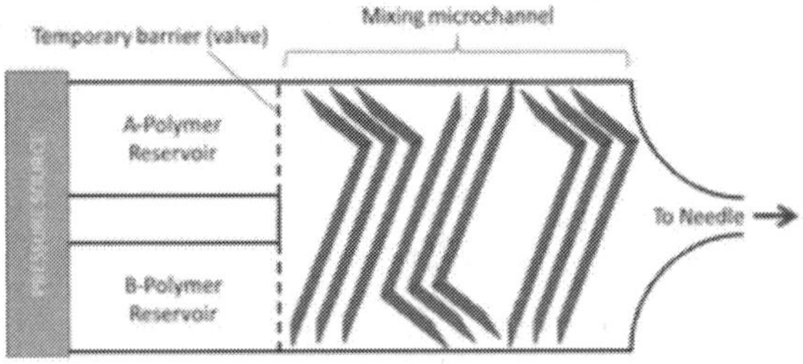

Figure 1: Schematic of the microinjector design. A pressure source propels the polymer solutions from their reservoirs into a herringbone mixing microfluidic channel before their emission from a needle.

RESULTS: The initial prototype is shown in Figure 3, which depicts the initial device design and the staggered herringbone structure. The channels have a width of 200 μm and a height of 100 μm with the herringbone grooves having a height of 23 μm and spacing of 50 μm.

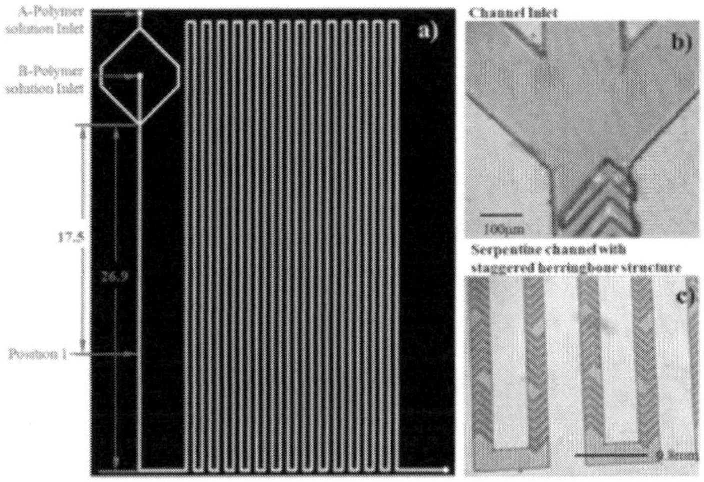

Figure 3: (a) A schematic of the device design (all units in mm) with the polymer solution inlets labelled; (b) a microscope image of the channel inlet; (c) a microscope image showing the serpentine structure with staggered herringbone grooves in the channels.

A syringe pump was set to generate an equal flow rate at both the inlets. DI water mixed with methylene blue dye was flowed in the central inlet while DI water was flown in the other two inlets. A comparison of mixing in microchannels with and without the herringbone design was made, as shown in Figure 4. It shows that complete mixing occurs at 17.5 mm in the channel with the herringbone design. On the other hand in the case of a simple microchannel the two streams are still delineated from each other and the mixing is only due to diffusional effects. The degree of mixing was shown to increase with channel length as well.

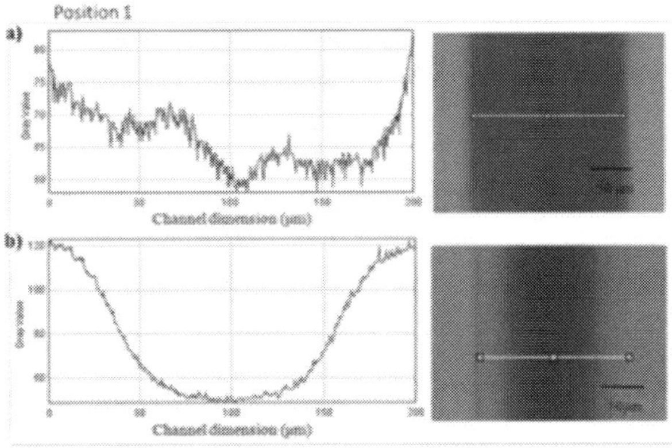

Figure 4: Mixing distribution over channel width measured via grayscale value for channels (a) with and (b) without staggered herringbone structures.

Next, the mixing channel design was used with polymer A and B solution to determine whether effective mixing occurs. The interaction of two polymer solutions in these channels can be observed as the two polymer solutions are transparent and become opaque upon gel formation. As shown in Figure 5, the herringbone design promotes faster mixing and gel formation in these microchannels.

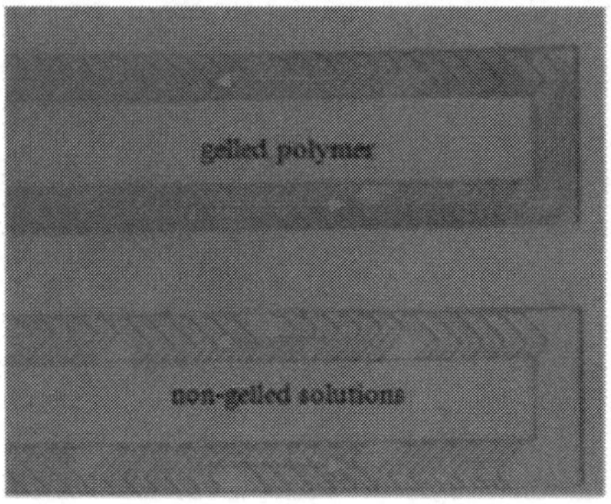

Figure 5: The determination of when the material has formed a hydrogel within the channel as a function of length (the direction of flow is in indicated by the red arrows).

The point of gelation can be used to determine the placement of the outlet. The optimal length will depend on the polymer solution viscosities, the polymer concentration in the initial solutions, the applied flow rate, and the gelation rate of the polymers (which depends on the degree of functionalization and polymer properties). Tests on the final material after it is ejected from the needle at varying degrees of mixing will be discussed, aiming to determine the optimal channel length for a given flow rate. For a chosen solution, the structure of the channels will be altered to lower the channel lengths and allow for the insertion of neddles to the channel outlet. The level of control over the injected volume will be reported, along with the preliminary *in vivo* testing of the device and their comparison with conventional injections.

CONCLUSIONS: The development of a device that could inject precise small amounts of in-situ gelling hydrogel precursors allows for the *in vivo* assessment of injectable hydrogels as ocular drug delivery materials, enabling potentially significant improvements over current therapies in terms of treating diseases of posterior eye.

REFERENCES: (1) Myles, M. E.; Neumann, D. M.; Hill, J. M. *Advanced Drug Delivery Reviews* 2005, *57*, 2063–79. (2) Sheardown, H. *Future Medicinal Chemistry* 2012, *4*, 2123–2125. (3) Fitzpatrick, S. D.; Jafar Mazumder, M.; Lasowski, F.; Fitzpatrick, L. E.; Sheardown, H. *Biomacromolecules* 2010, *11*, 2261–2267. (4) Patenaude, M.; Hoare, T. *Biomacromolecules* 2012, *13*, 369–378. (5) Van Tomme, S. R.; Storm, G.; Hennink, W. E. *International Journal of Pharmaceutics* 2008, *355*, 1–18. (6) Overstreet, D. J.; Dutta, D.; Stabenfeldt, S. E.; Vernon, B. L. *Journal of Polymer Science Part B: Polymer Physics* 2012, *50*, 881–903. (7) Patenaude, M.; Hoare, T. *ACS Macro Letters* 2012, *1*, 409–413. (8) Campbell, S. B.; Patenaude, M.; Hoare, T. *Biomacromolecules* 2013, *14*, 644–653. (9) Stroock, A. D.; Dertinger, S. K. W.; Ajdari, A.; Mezic, I.; Stone, H. a; Whitesides, G. M. *Science* 2002, *295*, 647–651.

ACKNOWLEDGEMENTS: This research is funded by the Natural Sciences and Engineering Research Council of Canada (NSERC), the Ontario Graduate Fellowship, and the J.P. Bickell Foundation (Medical Research Grant).

(692e) Aortic Smooth Muscle Cell Benchmark Force Measurement

Koons, B., Virginia Tech

Phillippi, J. A.,

Gleason, T. G.,

Nain, A. S., Virginia Tech

Originally presented on: 11/7/2013 13:50:00 - 14:10:00

Aortic Smooth Muscle Cell Benchmark Force Measurement

Brian Koons[1]

Julie A. Phillippi[2]

Thomas G. Gleason[2]

Amrinder S. Nain[1,3]

[1]Department of Mechanical Engineering, Virginia Tech, Blacksburg, VA, USA

[2] Department of Cardiothoracic Surgery, University of Pittsburgh, Pittsburgh, PA, USA

[3]School of Biomedical Engineering and Sciences, Virginia Tech, Blacksburg, VA, USA

Introduction

Rupture of aortic aneurysm is the thirteenth leading cause of death in the United States and is a clinical emergency. The main reason for the high death rate is that the condition is commonly asymptomatic, until the aorta ruptures, leading rapidly to death. An aortic aneurysm is characterized by an enlargement of the aorta's diameter. Although vessel diameter varies among individuals, it can be related to body surface area, and aortic diameter larger than 30 mm is concerning and potentially life threatening [1]. Aortic aneurysm can be either fusiform, affecting the entire circumference of the vessel, or saccular, where only a portion of the circumference is enlarged. The dilatation is due to mechanical weakening of the blood vessel wall, initiated by biological pathways that are distinct for aneurysms of the thoracic and abdominal aorta. Current expert clinical opinion dictates surgical intervention to replace the aneurysmal aorta with a synthetic graft when the maximal orthogonal aortic diameter reaches 50-55 mm. Unfortunately, the aneurysmal aorta has been known to dissect or rupture at lower aortic diameters and accurate risk assessments for patients is challenging and lacking. Hence, there is a great clinical need for enabling technologies to improve rupture risk profiles, increased knowledge of the biological mechanisms mediating the pathophysiology and new medical therapies for patients affected by aneurysm. We have engineered a model system used to measure forces exerted by migratory primary human aortic smooth muscle cells (HASMC). This technology could be used as a benchmark to determine putative differences in cell forces exerted by diseased HASMCs. Using our previously reported STEP (Spinneret based Tunable Engineered Parameters) non-electrospinning fiber fabrication platform [2], aligned polymeric fibers were deposited in multiple perpendicular layers and forces exerted by migratory cells were calculated through the deflection of fibers in the elastic limit using Euler beam mechanics. This platform will provide fresh insights in single-cell force generation in normal vs.

diseased cells and can potentially be used to determine efficacy of novel therapeutics to treat patients at risk for aortic catastrophe.

Materials and Methods

STEP manufactured nanofibers were suspended in a highly aligned and oriented fashion on polystyrene frames. Various concentrations of polystyrene dissolved in xylene was used to fabricate fibers of 250 nm and 2 μm, respectively. The two fiber types were oriented perpendicular to one another and fused at intersections to engineer a nanofiber mesh where both fiber diameters coexist within the same 2-D geometric plane. The fibers were coated with the extracellular matrix protein fibronectin (2 μg/ml) to increase cell attachment (seeding density 20,000 cells/ml). HASMCs were isolated from non-aneurysmal patients with approval of the IRB and with informed patient consent. Time-lapse phase contrast microscopy and forces measurements were conducted using Carl Ziess microscope and Axiovision software. As cells migrated on the fibers, fiber deflections were observed and measured to calculate the cell forces using Euler beam theory.

Results

Cellular force generation was found to be dependent upon fiber structural stiffness (N/m). The cells were unable to deflect the stiffer 2 μm diameter fibers. However, large deflections, ranging from 2 to 12 μm, occurred on the less stiff 250 nm fibers. Therefore, the less stiff fibers can be modeled as deformable beams where the stiff fibers at their edge create a simply supported boundary condition. Preliminary results with HASMC exhibited a force generation of 0.01 nN at the least stiff regions along the fiber and up to 0.8 nN near the stiffest regions of the fiber, as shown in Figure 1. These measurements provide the benchmark values for non-diseased HASMC lines and can be used to compare force generation by cells from diseased patients.

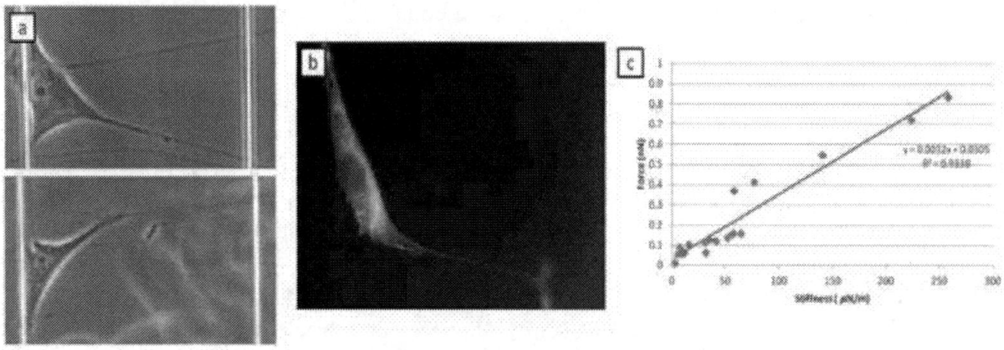

Figure 1. (a) Healthy HASMC deflecting 250 nm fiber segments between stiff boundaries, (b) immunocytochemistry for the cytoskeletal proteins F-actin (red) and zyxin (green) while a healthy HASMC

is deflecting a fiber, and (c) healthy HASMC force generation on fibers of various stiffness.

Discussion

The STEP fiber fabrication technique provides a novel biophysical platform for determining the force generation of single migratory HASMCs. The method may be used to establish a new force-based hypothesis for aneurysm development, progression and risk for rupture and will help test the efficacy of novel therapeutics. It is expected that future studies on diseased cells will show a decrease in cell force generation for HASMC isolated from aneurysmal patients when compared with healthy HASMC. This enabling technology can later be applied for the study of several other pathologies.

References

[1] Sakalihasan, N Limet, R Defawe, O D. Abdominal aortic aneurysm. Lancet. 2005

[2] Nain AS, Sitti M, Jacobson A, Kowalewski T, Amon C. Dry Spinning Based Spinneret Based Tunable Engineered Parameters (STEP) Technique for Controlled and Aligned Deposition of Polymeric Nanofibers. Macromolecular Rapid Communications. 2009;30:1406-12.

(692g) Synthesis and Characterization of Drug-Polysaccharide Nanoparticles for Oral Drug Delivery

Davis, R. M., Virginia Tech
Mazumder, S., Virginia Tech
Pereira, J., Virginia Polytechnic Institute and State University
Edgar, K. J., Virginia Tech
Originally presented on: 11/7/2013 14:30:00 - 14:50:00

For the treatment of many diseases such as tuberculosis (TB) and Human Immunodeficiency Virus (HIV), it is important to formulate drugs so that they are released in a controlled and sustained manner. Polymer-encapsulated drug nanoparticles are particularly interesting due to the enhanced properties that result from interactions between the polymer carrier and the drug and from a reduction in particle dimensions. Oral administration of therapeutic agents represents by far the easiest and most convenient route of drug delivery but it is challenging to design delivery systems optimizing the chemical and morphological stability of a drug in the gastrointestinal (GI) tract so that a desirable pharmacokinetic profile may be attained. Dissolution of drugs is quite often the rate-limiting step which controls the bioavailability of the drug. Recent work has demonstrated the great potential for using polysaccharides to form complexes with drugs that can result in enhanced oral drug solubility and, thus, enhanced bioavailability due to the disruption of drug crystallinity. This work concerns the synthesis and characterization of nanoparticles comprised of drugs and polysaccharides selected for their potential use in oral drug delivery.

Nanoparticles comprised of drugs encapsulated in carboxymethyl cellulose acetate butyrate (CMCAB) were produced by rapid precipitation and the formulation process and properties of the nanoparticles were investigated. An antibacterial drug – rifampicin used to treat TB - and less soluble anti-viral drugs – ritonavir and efavirenz which are used to treat HIV - were selected as model drugs for this study. The purposes of this research were two-fold. First, the methodology for producing drug-polymer nanoparticles with well-defined particle size distributions was developed. Second, the factors affecting drug loading and release properties were investigated. The nanoparticles were formed under turbulent conditions in a multi-inlet vortex mixer. Particle diameters ranged between 150-400 nm with narrow polydispersity indices as measured by dynamic light scattering. The rifampicin drug loading efficiencies ranged from 14-40% whereas, for the less soluble antiviral drugs, drug loading efficiencies were as high as 85%. X-ray diffraction and differential scanning calorimetry experiments established that the drugs in the drug-polymer nanoparticles were mostly amorphous which is important for increasing the effective solubility and hence bioavailability of the drugs. Dissolution studies in a pH 6.8 buffer at 37°C were consistent with this, showing an increase in the solubility of the amorphous drug nanoparticles compared to crystalline forms of the drugs.

(706a) Photo-Chemical Tuning of the Dirac Point in CVD-Grown Graphene

Baltazar, J., Georgia Tech
Sojoudi, H., Georgia Tech
Paniagua, S., Georgia Tech
Lawson, R. A., Georgia Institute of Technology
Graham, S.,
Tolbert, L. M., Georgia Institute of Technology
Henderson, C. L., Georgia Institute of Technology
Originally presented on: 11/7/2013 12:30:00 - 12:48:00

Graphene, a two-dimensional sp^2 hybridized carbon lattice, has attracted significant interest due to its distinctive electrical and mechanical properties including its nearly linear energy dispersion relation that results in electric field induced generation of electrons and holes which travel as massless Dirac fermions with high velocities giving rise to its high charge carrier mobility.

These electrical properties have spurred research directed at modifying graphene for use in a variety of electronic, optoelectronic, and sensor technologies. In order to fully utilize its potentials, it is critical to manipulate its electronic properties while maintaining its pristine state. Intentional doping of graphene is such an effective method to modify the electrical properties of graphene, while preserving its electronic properties.

Here we demonstrated the use of TPS-F and NBC as photo-chemical dopants respectively. The neutrality point of graphene field effect transistors (GFET) was tuned significantly by controlling the light exposure time. Furthermore, the charge transfer between both molecules and graphene was analyzed via Raman Spectroscopy, X-Ray Photoelectron Spectroscopy (XPS) and Ultraviolet Electron Spectroscopy (UPS).

(706b) Synthesis of Graphene and Novel Carbon Nanostructures By Induction Heating of Nanocrystalline 3C-SiC Particles At Atmospheric Pressure

Miettinen, M., Fine Particle and Aerosol Technology Laboratory, Dept. of Environmental Science, University of Eastern Finland

Hokkinen, J.,

Torvela, T., University of Eastern Finland

Karhunen, T., University of Eastern Finland

Pfüller, C.,

Ramsteiner, M.,

Tapper, U.,

Jokiniemi, J., VTT Fine Particles

Lähde, A., University of Eastern Finland

Originally presented on: 11/7/2013 12:48:00 - 13:06:00

Cubic silicon carbide (3C-SiC) has been shown to be a suitable substrate for the growth of graphene (Suemitsu et al. 2009). For advanced applications, the uniform graphene film on an insulating substrate is required. Thermal decomposition of SiC at atmospheric pressure has been proposed to improve the homogeneity of the formed graphene layers compared to vacuum conditions (Emtsev et al. 2009). A considerable advantage of using SiC-based methods to grow graphene is that it is not required to transfer the as-grown graphene onto another insulating substrate. We have studied the post-processing of previously manufactured nanocrystalline 3C-SiC particles by induction heating at atmospheric pressure (Miettinen et al. 2011, Miettinen et al. 2013). Three precursor powders with increasing crystallinity were synthesised at 1000 °C, 1200 °C and 1400 °C by atmospheric pressure chemical vapor synthesis and induction temperatures from 1900 °C to 2600 °C were used.

Growth of carbon sheets and spherical carbon particles was observed (Fig. 1). The spherical carbon particles turned out to be new carbon structures, i.e. carbon nanoflowers (CNFs). Curved graphite layers grew from the Si-C core, which decreased in size with increasing induction temperature and eventually disappeared above induction temperature of 2200 °C. To our knowledge the formation of this kind of structures has not been reported previously. The growth of CNFs may proceed through the cap formation mechanism proposed by Wang et al. (2007). The existence of C-C σ-bonds perpendicular to the SiC surface and the persistence of C-C π-bonds results in a repulsion in the cap center between two adjacent dangling bond sites which is responsible for the formation of a stable curved structure. The growth mechanism was probably induced by simple thermal decomposition of the SiC surface followed by the evaporation of Si. The number of the carbon layers in the nanoflower structure decreased with increasing temperature from 11 at induction temperatures below 2200 °C to 7 at the induction temperature of 2600 °C with the precursor powder synthesized at 1000 °C. The number of layers decreased also with increasing crystallinity of the precursor material and 5 layers were detected with the most crystalline precursor at the induction temperature of 2050 °C. The carbon sheets were more abundant when the precursor particles of higher degree of crystallinity (SiC-1400) were used. The SAED analysis indicated crystallinity of the carbon sheets and the X-ray diffraction analysis verified the presence of crystalline carbon. Raman spectrum displayed sharp G and 2D bands with intensity ratio 2D/G over four which is typical for single-layer, undoped graphene (Ferrari 2007). A very low intensity D band in this spectrum indicates a highly ordered structure.

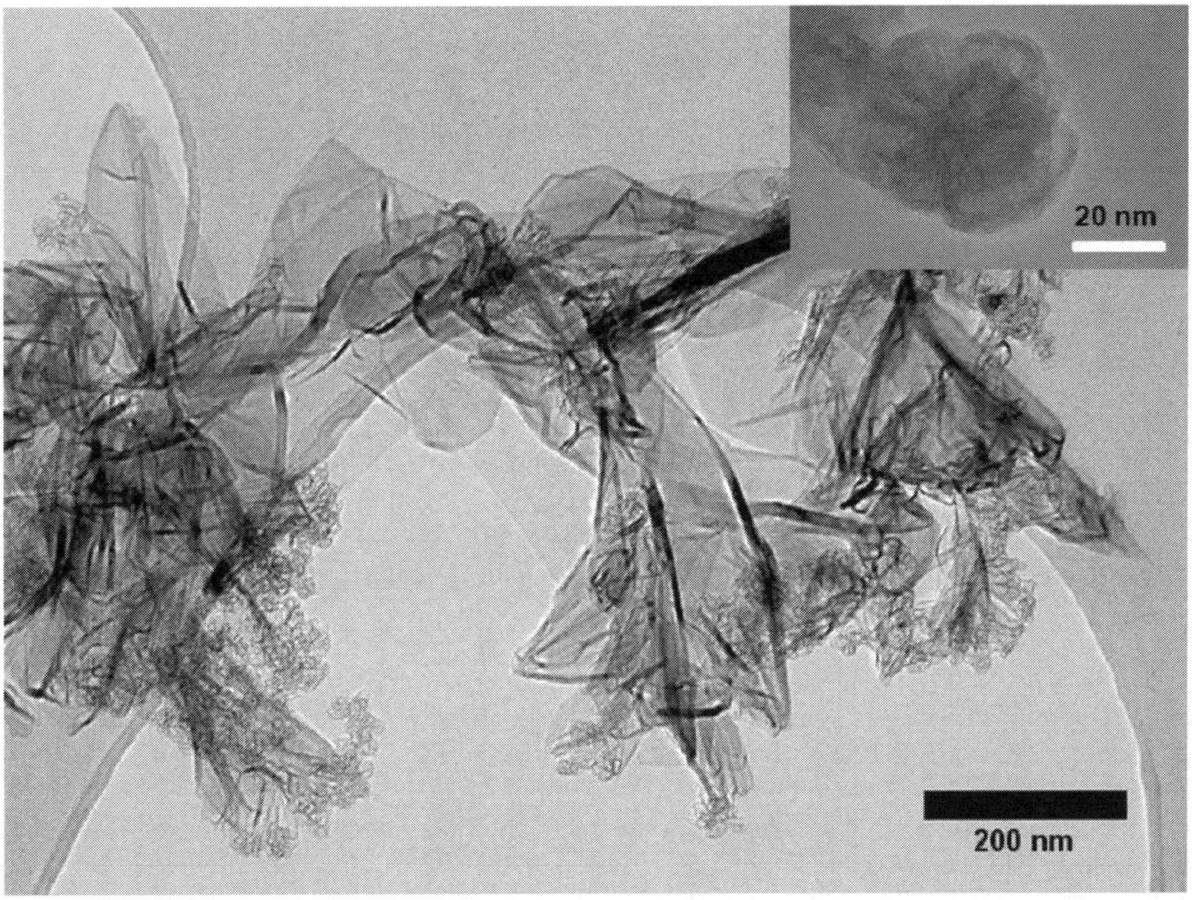

Fig. 1 Transmission electron microscopy images of the produced carbon sheets and nanoflowers (inserted is a larger magnification of the nanoflower)

The method presented is easily scalable and opens up the possibility to produce the advanced carbon structures that can be exploited as precursors for numerous applications, e.g. deposition of graphene films or in biomedical applications.

This work was supported by the strategic funding of the University of Eastern Finland under the NAMBER spearhead project.

Emtsev KV, Bostwick A, Horn K, Jobst J, Kellogg GL, Ley L, McChesney JL, Ohta T, Reshanov SA, Röhrl J, Rotenberg E, Schmid AK, Waldmann D, Weber HB, Seyller T (2009) Nat Mater 8:203-207

Ferrari AC (2007) Solid State Communications 143:47-57

Miettinen M, Johansson M, Suvanto S, Riikonen J, Tapper U, Pakkanen TT, Lehto VP, Jokiniemi J, Lähde A (2011) J Nanopart Res 13:4631-4645

Miettinen M, Hokkinen J, Karhunen T, Torvela T, Pfüller C, Ramsteiner M, Tapper U, Jokiniemi J, Lähde A (2013) J Nanopart Res under review

Suemitsu M, Miyamoto Y, Handa H, Konno A (2009) e-J Surf Sci Nanotech 7:311-313

Wang Z, Irle S, Zheng G, Kusunoki M, Morokuma K (2007) J Phys Chem C 111:12960-12972

(706c) Micro-Nanoscale Tuneable Reduction of Graphene/Graphite Oxide Via Focused Ion Beam Irradiation

Lobo, D. E., Monash University
Fu, J., Monash University
Gengenbach, T., CSIRO
Majumder, M., Rice University
Originally presented on: 11/7/2013 13:06:00 - 13:24:00

Graphene has risen to prominence as a versatile and scientifically intriguing material with unparalleled potential for application in electronics. Consequently, control of its properties and its fabrication are key research focuses toward realisation of this potential. An intriguing and innovative route to property-controlled synthesis of graphene is spatially patterned reduction of graphene oxide (GO) which allows fabrication of conductive reduced-graphene oxide (rGO) pathways into an otherwise electrically insulating (GO) material with a focussed ion beam (FIB). Dual FIB-SEM systems are a common feature of all modern microscopy centres therefore the importance and relevance of using FIB irradiation to achieve this goal is rising in prominence. FIB irradiation also has a number of particular advantages including direct imprinting of complex micron-scale shapes and sub-20nm lines of rGO in insulating films/flakes of GO. FIB irradiation can also be controlled such that degrees of reduction can be obtained, including amorphisation of the sample. This tuneable reduction method can be of significance in the quest towards wholly carbon based electronics and allows us to control not only the chemical makeup of the substrate but also its porosity, roughness and most importantly its electrical conductivity.

FIB technology gives us the ability to control each of these parameters across a wide range of length scales and to miniaturise the entire process which is crucial to fabricate nanoscale devices, parameters which are also important in the creation of future all-carbon super-capacitors. Thus, the ability to control patterned features varying in size from ~10's nm to ~100's of μm, in a mask-less and efficient manner, can be established as a crucial step towards graphene electronic devices.

We present proof-of-concept experimental results of the controlled reduction[1]and amorphisation process including a detailed study of the regimes of reduction and amorphisation of GO flakes with respect to the incident ion dosage.

1 Lobo, D. E., Fu, J., Gengenbach, T. & Majumder, M. Localized Deoxygenation and Direct Patterning of Graphene Oxide Films by Focused Ion Beams. *Langmuir* **28**, 14815-14821, doi:10.1021/la303369m (2012).

(706d) Atmospheric-Pressure Controllable Synthesis of Heteroatom-Doped Carbon Nanotubes

Chiang, W. H., National Taiwan University of Science and Technology
Li, T. J.,
Ho, K., National Taiwan University
Originally presented on: 11/7/2013 13:24:00 - 13:42:00

Recent theoretical [1] and experimental [2] studies have suggested that heteroatom-doped carbon nanomaterials such as carbon nanotube (CNTs) and graphene as novel materials with exceptional properties for applications including nanoelectronics, energy storage, fuel cells, and electrochemical sensing. However, current synthesis methods of heteroatom-doped carbon nanomaterials usually involve complicated vacuum systems, making it difficult to enable industrial-scale production [1]. Consequently, the development of atmospheric-pressure and controllable synthesis of heteroatom-doped carbon nanomaterials will lead to important advances on both scientific studies and innovation applications.

Here we demonstrate an atmospheric-pressure carbonthermic method.to produce boron-doped CNTs (BCNTs) with tunable boron atomic concentration ranging from 0.4 to 21.1 atomic percentage (at%). Systematic X-ray diffraction and X-ray photoelectron spectroscopy characterizations indicated that boron atoms were successfully doped into the sp^2graphene lattice of CNTs. Detailed high-resolution transmission electron microscopy electron energy loss spectroscopy (HRTEM-EELS) revealed that boron substitution doping mainly occurred on the CNT surface. Electrochemical characterization suggested that the as-produced BCNTs exhibit superior electrocatalytic activity for dopamine (DA) sensing than the pristine CNTs and conventional glassy carbon electrodes. Extensive electrochemical analysis confirmed a significant enhancement of 25% in the amperometric response for BCNTs with 2.1 at% boron atomic concentration than pristine CNTs, suggesting its potential as efficient catalysts for electrochemical detection of DA. It is also noteworthy from a practical point of view that the developed atmospheric-pressure synthesis method is amenable to industrial-scale production since it avoids the need for a vacuum system.

Reference

1. P. Ayala et al., Rev. Mod. Phys., **82**, 1843 (2010).
2. L. Yang et al., Angew. Chem. Int. Ed., **50**, 7132 (2011).

(706e) CVD Growth of Continuous Carbon Nanotube Arrays in Segregated Reactant Flow

Chester, G. E., Mainstream Engineering Corporation
Nicol, K. T., Mainstream Engineering Corporation
Hill, J. J., Mainstream Engineering Corporation
Originally presented on: 11/7/2013 13:42:00 - 14:00:00

Scalable and low-cost production of ultra-long carbon nanotubes (CNTs) will enable rapid insertion of these materials in electronic and structural technologies. In chemical vapor deposition (CVD) systems, precise control over reactor flow and thermal characteristics as well as uniformity and composition of the catalyst is needed. Furthermore, premature CNT growth termination often occurs due to amorphous carbon deposition or structural aggregation. Mainstream Engineering is developing a method for continuous production of ultra-long CNTs via a structure-directed downward growth mechanism. A novel CNT growth reactor has been designed which is designed to segregate CVD gas flow and mitigate some causes of growth termination. Another focus of this work is to grow long, straight and continuous CNT arrays/fibers at low temperature. Low-temperature growth presents challenges associated with catalyst treatment, composition and growth conditions. The system is also designed to structure-direct catalyst sites of high curvature without regard to substrate/catalyst surface energy, enabling the seeded process. We will discuss the current development as well as the technological roadmap to low-cost and large-scale manufacturing of continuous CNTs.

(706f) Sulfur Doped Co/SiO2 Catalysts for Chirally Selective Synthesis of Single Walled Carbon Nanotubes

Wang, H., Nanyang Technological University
Wei, L., Nanyang Technological University
Chen, Y., Nanyang Technological University
Originally presented on: 11/7/2013 14:00:00 - 14:18:00

Electronic and optical properties of single-walled carbon nanotubes (SWCNTs) are correlated with their geometric structures, i.e. chirality index (n,m). Thus, many potential applications require the use of monochiral SWCNTs. Chirally selective synthesis of nanotubes is the ultimate goal in SWCNT production, which
remains a great challenge.In this work, we demonstrated a method to convert three types of Co/SiO_2 catalysts, which are either inactive for the SWCNT growth or only selective to small-diameter nanotubes, into chirally selective catalysts to grow SWCNTs enriched with large-diameter (9,8) tubes (up to 40.5%) by doping catalysts with S. Catalysts were characterized by temperature programmed reduction. Based on experimetnal results, we also proposed in the mechanism that S atoms near Co atoms assist the formation of Co nanoparticles which are selective to (9,8) tubes. Moreover, H^+ ions may react with CoO_x to form well dispersed Co hydrosilicate and surface Co silicate on SiO2, which increases the selectivity to SWCNTs.

Reference: Chem. Commun., 2013, 49, 2031--2033

(706g) Amplification of SWNT Retaining n, m Distribution

Ren, F., Yale University
Wang, H., Nanyang Technological University
Wei, L., Nanyang Technological University
Haller, G. L., Yale University
Chen, Y., Nanyang Technological University
Pfefferle, L. D., Yale University
Originally presented on: 11/7/2013 14:18:00 - 14:36:00

Single-walled carbon nanotubes (SWNTs) have enormous potential for application in electronic applications such as field effect transistors, but the mixture of metallic and semiconducting nanotubes in bulk limits the reproducible production of these devices. Herein we present a method of amplifying a particular (n, m) SWNT sample in a manner scalable for industrial regrowth. The starting materials were two batches of SWNTs with different n, m distribution. The SWNTs were shortened using two different methods, liquid-phase oxidative/mechanical cutting and solid-state reaction cutting. The shortened SWNTs were then impregnated with cobalt catalysts and uniformly deposited onto fumed silica to form regrowth seeds. The regrowth seeds were pretreated by hydrogen followed by ethanol pyrolysis at different temperature initiating regrowth. After regrowth, the highest amplification factor achieved is 4X, as determined by thermogravimetric analysis. Photoluminescence excitation spectroscopy indicated that the SWNT-templated regrowth retained the (n, m) distribution from the original nanotube types. X-ray absorption spectroscopy was used to probe the state of the cobalt catalysts. This method requires little post-synthesis cleaning and produces high quality SWNTs.

(706h) Diffusion Mechanisms of Carbon During Single-Walled Carbon Nanotube Growth On Supported Metal Nanoparticles: The Effect of Temperature

Gomez-Ballesteros, J. L., TAMU
Gomez-Gualdron, D. A., Texas A&M University
Balbuena, P. B., Texas A&M University
Originally presented on: 11/7/2013 14:36:00 - 14:54:00

A highly controlled synthesis of single-walled carbon nanotubes requires a clear understanding of the mechanisms involved. However, understanding the role and mechanism of carbon diffusion and dissolution during chemical vapor deposition synthesis still remains a challenge. Our contribution addresses this issue by examining the individual trajectories of carbon atoms in reactive molecular dynamics simulated growth of single-walled carbon nanotubes on supported Ni nanoparticles. We identify carbon atoms participating in surface diffusion, bulk diffusion, and potential carbide formation by observing the dynamic evolution of the system at fixed nanoparticle size and adhesion energy of the nanoparticle to the substrate, at different temperatures, 750, 900 and 1000 K. Nanoparticle saturation with carbon and changes in the catalytic regime determine the transitions between induction, nucleation, and growth stages. From our results at 1000K, we found that due to the low-energy barrier for surface diffusion, nucleation may occur simultaneously with dissolution before saturation is reached. Induction and nucleation are governed preeminently by bulk diffusion, while growth is by surface diffusion. Additionally, Ni/C ratios were found to be compatible with stoichiometries of known Ni carbides. Here we examine the effect of lowering the synthesis temperature that may have a significant impact in the nanoparticle morphology and consequently on the carbon solubility inside the nanoparticle.

(720a) Unified Pressure and Electroosmotic Flow Enhancement Mechanism in Carbon Nanotube Membranes

Hinds, B. J. III, University of Kentucky

Originally presented on: 11/7/2013 15:15:00 - 15:40:00

Carbon nanotubes have three key attributes that make them of great interest for novel membrane applications 1) atomically flat graphite surface allows for ideal fluid slip boundary conditions and extremely fast flow rates 2) the cutting process to open CNTs inherently places functional chemistry at CNT core entrance for chemical selectivity and 3) CNT are electrically conductive allowing for electrochemical reactions and application of electric fields gradients at CNT tips. In general, the transport mechanisms through CNT membrane are a) ionic diffusion is near bulk expectation with no enhancement from CNT b) gas flow is enhanced by ~1-2 order of magnitude due to specular reflection off of flat graphitic surface c) and pressure driven flux of a variety of solvents (H2O, hexane, decane ethanol, methanol) are 4-5 orders of magnitude higher than conventional Newtonian flow [*1*] due to atomically flat graphite planes inducing nearly ideal slip conditions. Nearly all applications require chemical selectivity in what is allowed to pass across the membrane. However the act of placing selective functional chemistry at pore entrance or along the core of CNTs, dramatically/completely eliminates the enhanced flow effects by eliminating the near perfect slip boundary condition[*2*]. Needed is a mechanism to pump chemicals through the pore where selective chemistry is. This is routinely achieved in protein channels where permeates are accelerated through regions of precise functionality. The CNT membrane, with tips functionalized with charged molecules, is a nearly ideal platform to induce electro-osmotic flow with high charge density at pore entrance and a nearly frictionless surface for the propagation of plug flow. Through diazonium electrochemical modification we have successfully bound anionic surface charge to CNT tips and along CNT cores. High electro-osmotic flows of 0.16 cm/s-V at are seen by the pumping of neutral caffeine molecules. Improvements in electroosmotic power efficiency of 25-112 fold are seen in CNTs compared to conventional nanoporous materials with atomically rough interfaces [*3*]. Use of the electro-osmotic phenomenon for responsive/programmed transdermal drug delivery devices is discussed with the voltage gated delivery of clonidine and nicotine across CNT membrane at therapeutically useful fluxes [*4*]. In small diameter SWCNTs ion mobilities are seen to be ~6 fold enhanced due to induced electroosmotic flow. Electroosmotic flow enhancements of 10,000 fold are seen [*5*] and are consistent with pressure driven flow enhancements.

1 "Nanoscale hydrodynamics: Enhanced flow in carbon nanotubes" Majumder, M.; Chopra, N.; Andrews, R; Hinds, B.J *Nature* **2005**, 438, 44.

2 'Mass Transport through Carbon Nanotube Membranes in three different regimes: ionic diffusion, gas, and liquid flow' Mainak Majumder, Nitin Chopra, B.J. Hinds *ACS Nano* **2011** 5(5) 3867-3877

3 'Highly Efficient Electro-osmotic Flow through Functionalized Carbon Nanotubes Membrane' Ji Wu, Karen Gerstandt, Mainak Majunder, B.J. Hinds, *RCS Nanoscale* **2011** 3(8) 3321-28

4 " Programmable transdermal drug delivery of nicotine using carbon nanotube membranes" J. Wu, K.S. Paudel, C.L. Strasinger, D. Hamell, Audra L. Stinchcomb, B. J. Hinds *Proc. Nat. Acad. Sci.* **2010** 107(26) 11698-11702.

5 "Electrophoretically Induced Aqueous Flow through sub-Nanometer Single Walled Carbon Nanotube Membranes" Ji Wu, Karen Gerstandt, Hongbo Zhang, Jie Liu, and Bruce. J. Hinds *Nature Nano* **2012** DOI: 10.1038/NNANO.2011.240

(720b) Improving the Separation Performance of Nanocomposite Membrane By Aligning Carbon Nanotubes Into Polymeric Matrix

Zhao, H.,
Zhou, Z., Zhejiang University
Zhang, L., Zhejiang University
Chen, H. L., Zhejiang University
Originally presented on: 11/7/2013 15:40:00 - 16:00:00

The nanocomposite membranes containing carbon nanotubes (CNTs) have shown better separation performance than the bare membranes. For example, several studies have demonstrated that a higher permeation flux can be obtained by incorporating functionalized CNTs in the polymeric membrane. However, the improvement in separation performance is limited due to the disordered state of CNTs in the nanocomposite membranes.

In this work, a facile method, hot-press combined with peel-off (HPPO), is established, through which CNTs can be successfully vertically aligned on the polymeric membrane substrate. Shear force and mechanical stretch are proposed to be the main forces to align the tubes perpendicular to the substrate surface during the peel-off process. The TEM images show that the alignment of CNTs keeps their orientation in a thin hybrid membrane by coating polymer casting solution. The stable alignment of CNTs by HPPO greatly contributes to the improved separation performance of the nanocomposite membranes. The CNTs facilitate the permeation of components due to their nature molecular channel and the microvoids between the polymer matrix and tubs.

(720c) Translating Carbon Molecular Sieves (CMS) Dense Film Gas Transport Performance to Asymmetric Hollow Fiber Membranes

Bhuwania, N., Georgia Institute of Technology
Koros, W. J., Georgia Institute of Technology
Williams, P. J., Shell International Exploration and Production (US) Inc
Originally presented on: 11/7/2013 16:00:00 - 16:20:00

Fundamental studies on Carbon Molecular Sieves (CMS) dense film membranes have shown enhanced gas separation ability based on its bimodal pore distribution. The simplified structure of CMS membrane is generally assumed to compose a 'slit-like' dual-pore morphology consisting of micropore (~6-20Å) and ultra-micropore (<6Å). The nano-size pore distribution is obtained by the thermal degradation of polymer precursors under controlled atmospheres. For industrial application, it is important to translate this high separation performance in hollow fiber membrane configuration. In our talk, we will discuss an approach to overcome the challenge in maintaining the asymmetric morphology of CMS hollow fiber membranes. To achieve the superior gas separation performance and asymmetric morphology in final CMS hollow fibers, we use the sol-gel reaction between organic-alkoxy silane and moisture (V-Treatment). CMS V-Treated hollow fibers membranes are evaluated for both pure and mixed gas feeds.

(720d) Slip and Spin-Coupling: Foundations of the Flow Enhancement of Fluids in Graphene Slit Channels and Carbon Nanotubes With Important Applications in Nanoscale Fluid Pumping

Todd, B. D., Swinburne University of Technology
Kannam, S.,
De Luca, S.,
Hansen, J.,
Daivis, P. J., RMIT University
Originally presented on: 11/7/2013 16:20:00 - 16:45:00

We present both theoretical and simulation studies that clearly demonstrate the importance of several non-classical phenomena that are fundamental features of fluid flow under extreme confinement. These are: (1) the prevalence of slip and (2) the strong coupling of molecular spin to linear translational momentum for molecularly structured fluids. In the first of these, we utilize a newly developed equilibrium based model[1] to accurately predict the slip velocity and slip lengths of systems such as water or methane flowing in graphene nanochannels and carbon nanotubes[2-4]. We demonstrate that traditional nonequilibrium molecular dynamics simulations of such systems are far less efficient and accurate than the easily implemented model we propose. Our results confirm large flow enhancements that are accurately quantified and compared to existing experimental, theoretical and simulation data in the literature. Next, we show that ignoring the coupling of spin angular momentum to linear translational motion of a highly confined fluid can lead to significant over-estimation of the predicted flow rates using conventional Navier-Stokes treatments. By including spin-coupling into the extended Navier-Stokes equations, hydrodynamic prediction is seen to be very accurate down to length scales of a few atomic diameters[5]. We also demonstrate how this knowledge, coupled with our knowledge of slip, can be used to pump molecular fluids such as water via non-intrusive application of a rotating electric field[6-8].

[1]J.S. Hansen, B.D. Todd and P.J. Daivis. Phys. Rev. E **84**, 016313 (2011).

[2]S. K. Kannam, B.D. Todd, J.S. Hansen and P.J. Daivis. J. Chem. Phys. **135**, 144701 (2011).

[3]S. K. Kannam, B.D. Todd, J.S. Hansen and P.J. Daivis. J. Chem. Phys. **136**, 024705 (2012).

[4]S. K. Kannam, B.D. Todd, J.S. Hansen and P.J. Daivis. J. Chem. Phys. **138**, 094701 (2013).

[5]J.S. Hansen, J.C. Dyre, P.J. Daivis, B.D. Todd and H. Bruus. Phys. Rev. E **84**, 036311 (2011).

[6]J. D. Bonthuis, D. Horinek, L. Bocquet, and R. R. Netz, Phys. Rev. Lett. **103**, 144503

(2009).

[7]J.S. Hansen, H. Bruus, B.D. Todd, and P.J. Daivis. J. Chem. Phys. **133**, 144906 (2010).

[8]S. De Luca, B.D. Todd, J.S. Hansen and P.J. Daivis, J. Chem. Phys. **138**, 154712 (2013)

(720e) Coupling Electrodialysis With a Novel Carbon Nanotube Topology Structure for Desalination: An Idea Demonstration

Chen, Q., Tsinghua University
Kong, X., Tsinghua University
Lu, D., Tsinghua University
Liu, Z., Tsinghua University
Originally presented on: 11/7/2013 16:45:00 - 17:05:00

Carbon nanotube (CNT) has been one of the research spotlights in recent years. By virtue of the several orders of magnitude faster water flow than macroscopic hydrodynamics and the impressive anti-fouling properties, aligned CNT membrane has become a super-star candidate for the next-generation reverse-osmosis (RO) desalination membrane materials. However, the shadow casting over the promising application of CNTs lies in its lack of intrinsic water/ion selectivity. Despite many significant research attempts by using steric hindrance or electrostatic interactions, researchers failed to achieve high-flux and high-selectivity at the same time, either because of the difficulty to synthesize small-diameter CNTs or the reduction of water flux once entrance of CNTs is functionalized. To obtain both high selectivity and flux, we design a novel membrane, which is filled with honeycomb-shaped CNT, and demonstrate the feasibility of desalination with high flux and selectivity by using molecular dynamics simulation. In operation, an external electric field vertical to the flow direction is applied to serve as the driving force for the spatial segregation of ions and water. In our honeycomb-CNT model, the diameters of all the CNTs are identical, ranging from 1.3 nm to 2.0 nm in different simulations. Each separation stage contains two Y-junctions, with a unit length as long as 5 nm. There are four CNT entrances and outlets on the graphene layer which functions as the membrane surface, resulting in the pore density around 6.7×10^{12} pores/cm^2. Concentrated streams, consisting of Na^+ and Cl^-, flow through the two outlets near anode and cathode, respectively, while the dilute streams flow through the two outlet streams in the middle. The ion concentration of the incoming stream is set to be 1-2 mol/L and the hydrostatic pressure at different side of the membrane is 400 MPa to enable more permeation events observed in 60 ns simulation. Under the electric field of 0.2-0.6 V/nm, desalination is successfully realized: the ion permeation ratio, defined as the ratio between ion concentrations of the dilute and the incoming streams, is close to 0%, indicating pure water is obtained in the middle outlet. In addition, the selectivity, defined as the ratio between ion concentrations of the concentrate and the dilute streams, can be as high as 5.5 ± 1.1. For better understanding of separation mechanism, further analysis was carried out. It is demonstrated that, first of all, low ion flux in the dilute streams is attributed to both steric hindrance and electric-field driven migration, which is consistent with our initial guess. Secondly, free energy barrier for ions to move upwards or downwards at the Y-junction is approximately 5 kcal/mol, which is the major driving force for the spatial distribution of ions. Thirdly, the importance of vertical ion transport is further demonstrated in the aligned CNT model, in which the lack of vertical connectivity leads to poor ion rejection performance even if the vertical electric field is applied. Finally, faster water transport in the concentration streams is subjected to examination, which probably resulted from the electro-osmosis effect. In conclusion, this hybrid approach can solve problems caused by both traditional electrodialysis (ED) and CNT membrane. On the one hand, CNT membrane is equipped with additional driving force for high selectivity. On the other hand, the lower ion concentration inside CNT membrane will significantly reduce the energy consumption of ED desalination. Notably, our model is not far from experimental practices. For example, techniques to synthesize sub-2-nm or Y-shaped CNTs are currently available, and applying electric field is much easier than charge modification. Therefore, our simulation provides an insight into the design of a more complex CNT membrane to achieve desalination.

(720f) Simulations and Experiments of Carbon Nanotube-Polyamide Nanocomposite Membranes for Water Desalination

Chen, H. Y., University of Pittsburgh
Chan, W. F., Virginia Polytechnic Institute and State University
Marand, E., Virginia Tech
Johnson, J. K., University of Pittsburgh
Originally presented on: 11/7/2013 17:05:00 - 17:30:00

We present a combined simulation and experimental study of zwitterion functionalized carbon nanotube (CNT) membranes for water desalination. We constructed model CNT membranes for the simulation studies by embedding four CNTs between graphene sheets to prevent fluid flow around the outside of the CNTs. Each end of each CNT was functionalized with one or more zwitterion groups. The model membrane was placed a periodically repeated saltwater box in order to study the flux of water and salt ions through the membrane. A pressure drop of about 200 MPa was applied across the membrane in order to generate a high water flux over the course of the simulation. The ion rejection ratio was found to be 100% when two zwitterions were affixed to the ends of (20,0) nanotubes, which are 1.56 nm in diameter. We found that five zwitterions per tube end was required to completely block ion transport in a 2.03 nm diameter (26,0) CNT. In contrast, the rejection ratio for non-functionalized CNTs and functionalized CNTs with five COOH groups at each tube end was found to be 0% and 20%, respectively. We have synthesized asymmetric carbon nanotube/polyamide membranes with varying loadings of CNTs and assessed these membranes for water desalination in experiments. The flux of water was found to be increased from 6.8 to 28.7 gallons per square foot per day as the weight fraction of zwitterion functionalized CNTs in the polyamide membrane was increased from 0 to 20 wt%. Importantly, the ion rejection ratio also increased from 97.6% to 98.6% with increasing weight fraction of functionalized CNTs. Additional simulation tests revealed the ion blocking mechanism is almost exclusively due to a steric hindrance, with a minor contribution due to electrostatic repulsion. Theoretical predictions indicate that an ideal CNT/polymer membrane having a loading of 20 wt % CNTs has a maximum flux of about 20000 gallons per square foot per day at the conditions of our experiments.

*** Formatting Issue - Best Available Paper/Graphic ***

(720g) Explaining Ultra-High Water Flow Rates Observed in Carbon Nanotube Membranes
Mattia, D.,
Calabrò, F.,
Originally presented on: 11/7/2013 17:30:00 - 17:50:00

Explaining ultra-high water flow rates observed in carbon nanotube membranes

Davide Mattia,1* Francesco Calabr?2

1Department of Chemical Engineering, University of Bath, UK

2 DIEI, Universit? di Cassino, Italy

d.mattia@bath.ac.uk, www.bath.ac.uk/nanotech

Liquid flow through nanoporous media has been studied in disciplines as diverse as membrane science, soil permeability and cell physiology. In all these fields, though, the emphasis has been on the macroscopic outcome, often neglecting the effects of intermolecular forces and liquid-pore wall physicochemical interactions on liquid behaviour. The availability of carbon nanotubes (CNTs) as a platform for mass transport studies at the nanoscale has generated a large number of experimental and molecular dynamics studies which have shown unexpected fluid behaviours compared to the macroscale. The most striking one has been the observation of several orders of magnitude higher liquid flow velocities compared to the no-slip Haagen-Poiseuille equation. ADDIN EN.CITE ADDIN EN.CITE.DATA
3C456E644E6F74653E3C436974653E3C417574686F723E4D61747469613C2F417574686F723E3C596561
723E323030383C2F596561723E3C5265634E756D3E3932353C2F5265634E756D3E3C446973706C617954
6578743E3C7374796C652066616365D22737570657273637269707422E312C20323C2F7374796C653E
3C2F446973706C6179546578743E3C7265636F72643E3C7265632D6E756D6265723E3932353C2F726563
2D6E756D6265723E3C666F726569676E2D6B6579733E3C6B657920617070D22454E222064622D69643D
227A737461323272736E64357A706665737A326F70646477777474663273786673667830392 23E393235
3C2F6B65793E3C2F666F726569676E2D6B6579733E3C7265662D74797065206E616D653D224A6F75726E
616C2041727469636C65223E31373C2F7265662D747970653E3C636F6E7472696275746F72733E3C6175
74686F72733E3C617574686F723E4D61747469612C20446176696465 3C2F617574686F723E3C61757468
6F723E476F676F7473692C20597572793C2F617574686F723E3C2F617574686F72733E3C2F636F6E7472
696275746F72733E3C7469746C65733E3C7469746C653E5265766965773A207374617469632061 6E6420
64796E616D6963206265686176696F72206F66206C697175696473 20696E736964652063617262 6F6E20
6E616E6F74756265733C2F7469746C653E3C7365636F6E646172792D7469746C653E4D6963726F666C75
6964696373 20616E64204E616E6F666C7569646963733C2F7365636F6E646172792D7469746C653E3C2F
7469746C65733E3C706572696F646963616C3E3C66756C6C2D7469746C653E4D6963726F666C75696469
637320616E64204E616E6F666C7569646963733C2F66756C6C2D7469746C653E3C616262722D313E4D69
63726F666C7569642E204E616E6F666C7569642E3C2F616262722D313E3C2F706572696F646963616C3E
3C70616765733E3238392D3330353C2F70616765733E3C766F6C756D653E353C2F766F6C756D653E3C6E
756D6265723E333C2F6E756D6265723E3C65646974696F6E3E31372F30342F323030383C2F6564697469
6F6E3E3C6B6579776F7264733E3C6B6579776F72643E6E616E6F666C7569646963733C2F6B6579776F72
643E3C6B6579776F72643E636172626F6E206E616E6F74756265733C2F6B6579776F72643E3C6B657977
6F72643E66696C6C696E673C2F6B6579776F72643E3C2F6B6579776F7264733E3C64617465733E3C7965
61723E323030383C2F796561723E3C2F64617465733E3C63616C6C2D6E756D3E303932353C2F63616C6C
2D6E756D3E3C75726C733E3C72656C617465642D75726C733E3C75726C3E687474703A2F2F64782E646F
692E6F72672F31302E313030372F7331303430342D3030382D303239332D353C2F75726C3E3C2F72656C
617465642D75726C733E3C2F75726C733E3C656C656374726F6E69632D7265736F757263652D6E756D3E
31302E313030372F7331303430342D3030382D303239332D353C2F656C656374726F6E69632D7265736F
757263652D6E756D3E3C2F7265636F72643E3C2F436974653E3C436974653E3C417574686F723E576869
7462793C2F417574686F723E3C596561723E323030373C2F596561723E3C5265634E756D3E3737333C2F
5265634E756D3E3C7265636F72643E3C7265632D6E756D6265723E3737333C2F7265632D6E756D626572
3E3C666F726569676E2D6B6579733E3C6B657920617070D22454E222064622D69643D227A7374613232
72736E64357A706665737A326F706464777777474663273786673667830392 23E3737333C2F6B65793E3C
2F666F726569676E2D6B6579733E3C7265662D74797065206E616D653D224A6F75726E616C2041727469
636C65223E31373C2F7265662D747970653E3C636F6E7472696275746F72733E3C617574686F72733E3C
617574686F723E576869747462792C204D2E3C2F617574686F723E3C617574686F723E517569726B652C20
4E2E3C2F617574686F723E3C2F617574686F72733E3C2F636F6E7472696275746F72733E3C7469746C65
733E3C7469746C653E466C75696420466C6F7720696E20436172626F6E204E616E6F747562657320616E
64204E616E6F70697065733C2F7469746C653E3C7365636F6E646172792D7469746C653E4E6174757265

*** Formatting Issue - Best Available Paper/Graphic ***

204E616E6F746563686E6F6C6F67793C2F7365636F6E646172792D7469746C653E3C2F7469746C65733E
3C706572696F646963616C3E3C66756C6C2D7469746C653E4E6174757265204E616E6F746563686E6F6C
6F67793C2F66756C6C2D7469746C653E3C616262722D313E4E61742E204E616E6F2E3C2F616262722D31
3E3C2F706572696F646963616C3E3C70616765733E38372D39343C2F70616765733E3C766F6C756D653E
323C2F766F6C756D653E3C6B6579776F7264733E3C6B6579776F72643E4E616E6F666C75696469633C2F
6B6579776F72643E3C6B6579776F72643E6C697175696420666C6F773C2F6B6579776F72643E3C6B6579
776F72643E666C6F773C2F6B6579776F72643E3C6B6579776F72643E6E616E6F747562653C2F6B657977
6F72643E3C6B6579776F72643E4E616E6F706970653C2F6B6579776F72643E3C6B6579776F72643E7761
73686275726E2065717561746696F6E3C2F6B6579776F72643E3C6B6579776F72643E426F73616E717565
742065717561746696F6E3C2F6B6579776F72643E3C2F6B6579776F7264733E3C64617465733E3C796561
723E323030373C2F796561723E3C2F64617465733E3C63616C6C2D6E756D3E303737333C2F63616C6C2D
6E756D3E3C776F726B2D747970653E7265766965773C2F776F726B2D747970653E3C75726C733E3C2F75
726C733E3C2F7265636F72643E3C2F436974653E3C2F456E644E6F74653EFF00 1, 2
3C456E644E6F74653E3C436974653E3C417574686F723E4D61747469613C2F417574686F723E3C596561
723E323030383C2F596561723E3C5265634E756D3E3932353C2F5265634E756D3E3C446973706C617954
6578743E3C7374796C6520666163653D22737570657273637269707422E312C20323C2F7374796C653E
3C2F446973706C6179546578743E3C7265636F72643E3C7265632D6E756D6265723E3932353C2F726563
2D6E756D6265723E3C666F726569676E2D6B6579733E3C6B657920617070D22454E222064622D69643D
227A73746132327736E64357A706665737A326F706464777774746632737866736678303922 3E393235
3C2F6B65793E3C2F666F726569676E2D6B6579733E3C7265662D74797065206E616D653D224A6F75726E
616C2041727469636C65223E31373C2F7265662D747970653E3C636F6E7472696275746F72733E3C6175
74686F72733E3C617574686F723E4D61747469612C20446176696465C2F617574686F723E3C61757468
6F723E476F676F7473692C20597572793C2F617574686F723E3C2F617574686F72733E3C2F636F6E7472
696275746F72733E3C7469746C65733E3C7469746C653E5265766965773A207374617469632616E6420
64796E616D69632062656861766696F72206F66206C6971756964732096E73696465206361726626F6E20
6E616E6F7475626573C2F7469746C653E3C7365636F6E646172792D7469746C653E4D6963726F666C75
696469637320616E64204E616E6F666C7569646963733C2F7365636F6E646172792D7469746C653E3C2F
7469746C65733E3C706572696F646963616C3E3C66756C6C2D7469746C653E4D6963726F666C75696469
637320616E64204E616E6F666C7569646963733C2F66756C6C2D7469746C653E3C616262722D313E4D69
63726F666C7569642E204E616E6F666C7569642E3C2F616262722D313E3C2F706572696F646963616C3E
3C70616765733E3238392D3330353C2F70616765733E3C766F6C756D653E353C2F766F6C756D653E3C6E
756D6265723E333C2F6E756D6265723E3C65646974696F6E3E31372F30342F323030383C2F6564697469
6F6E3E3C6B6579776F7264733E3C6B6579776F72643E6E616E6F666C7569646963733C2F6B6579776F72
643E3C6B6579776F72643E636172626F6E206E616E6F7475626573C2F6B6579776F72643E3C6B657977
6F72643E66696C6C696E673C2F6B6579776F72643E3C2F6B6579776F7264733E3C64617465733E3C7965
61723E323030383C2F796561723E3C2F64617465733E3C63616C6C2D6E756D3E303932353C2F63616C6C
2D6E756D3E3C75726C733E3C72656C617465642D75726C733E3C75726C3E687474703A2F2F64782E646F
692E6F72672F31302E313030372F7331303430342D3030382D303239332D353C2F75726C3E3C2F72656C
617465642D75726C733E3C2F75726C733E3C656C656374726F6E69632D7265736F757263652D6E756D3E
31302E313030372F7331303430342D3030382D303239332D353C2F656C656374726F6E69632D7265736F
757263652D6E756D3E3C2F7265636F72643E3C2F436974653E3C436974653E3C417574686F723E576869
7462793C2F417574686F723E3C596561723E323030373C2F596561723E3C5265634E756D3E3737333C2F
5265634E756D3E3C7265636F72643E3C7265632D6E756D6265723E3737333C2F7265632D6E756D626572
3E3C666F726569676E2D6B6579733E3C6B657920617070D22454E222064622D69643D227A7374613232
72736E64357A706665737A326F706464777774746632737866736678303922 3E3737333C2F6B65793E3C
2F666F726569676E2D6B6579733E3C7265662D74797065206E616D653D224A6F75726E616C2041727469
636C65223E31373C2F7265662D747970653E3C636F6E7472696275746F72733E3C617574686F72733E3C
617574686F723E5768697462792C204D2E3C2F617574686F723E3C617574686F723E517569726B652C20
4E2E3C2F617574686F723E3C2F617574686F72733E3C2F636F6E7472696275746F72733E3C7469746C65
733E3C7469746C653E466C75696420666C6F7720696E20636172626F6E204E616E6F747562657320616E
64204E616E6F70697065733C2F7469746C653E3C7365636F6E646172792D7469746C653E4E6174757265
204E616E6F746563686E6F6C6F67793C2F7365636F6E646172792D7469746C653E3C2F7469746C65733E
3C706572696F646963616C3E3C66756C6C2D7469746C653E4E6174757265204E616E6F746563686E6F6C
6F67793C2F66756C6C2D7469746C653E3C616262722D313E4E61742E204E616E6F2E3C2F616262722D31
3E3C2F706572696F646963616C3E3C70616765733E38372D39343C2F70616765733E3C766F6C756D653E
323C2F766F6C756D653E3C6B6579776F7264733E3C6B6579776F72643E4E616E6F666C75696469633C2F
6B6579776F72643E3C6B6579776F72643E6C697175696420666C6F773C2F6B6579776F72643E3C6B6579
776F72643E666C6F773C2F6B6579776F72643E3C6B6579776F72643E6E616E6F747562653C2F6B657977
6F72643E3C6B6579776F72643E4E616E6F706970653C2F6B6579776F72643E3C6B6579776F72643E7761
73686275726E2065717561746696F6E3C2F6B6579776F72643E3C6B6579776F72643E426F73616E717565
742065717561746696F6E3C2F6B6579776F72643E3C2F6B6579776F7264733E3C64617465733E3C796561
723E323030373C2F796561723E3C2F64617465733E3C63616C6C2D6E756D3E303737333C2F63616C6C2D
6E756D3E3C776F726B2D747970653E7265766965773C2F776F726B2D747970653E3C75726C733E3C2F75

*** Formatting Issue - Best Available Paper/Graphic ***

726C733E3C2F7265636F72643E3C2F436974653E3C2F456E644E6F74653E0000

These experimental results have generated an intense debate about the underlying physical phenomena and whether such flow can be explained within the continuum model on which traditional fluid mechanics is based or whether there is a size threshold below which it no longer applies (Figure 1). These results also hold great promise for filtration applications in general and water desalination in particular. ADDIN EN.CITE
<EndNote><Cite><Author>Lee</Author><Year>2011</Year><RecNum>1294</RecNum><DisplayTex
t><style Article">17</ref-type><contributors><authors><author>Lee, Kah
Peng</author><author>Arnot, Tom C.</author><author>Mattia,
D.</author></authors></contributors><titles><title>A review of reverse osmosis
membrane materials for desalination--Development to date and future
potential</title><secondary-title>Journal of Membrane
Science</secondary-title></titles><periodical><full-title>Journal of Membrane
Science</full-title><abbr-1>J. Membr. Sci.</abbr-1><abbr-2>J Membr
Sci</abbr-2></periodical><pages>1-22</pages><volume>370</volume><number>1-2</number>
<keywords><keyword>Desalination</keyword><keyword>Reverse osmosis
(RO)</keyword><keyword>Membrane material</keyword><keyword>Membrane
performance</keyword><keyword>Nano-materials</keyword></keywords><dates><year>2011</
year></dates><isbn>0376-7388</isbn><call-num>1294</call-num><urls><related-urls><url
>http://www.sciencedirect.com/science/article/B6TGK-51SPS08-2/2/6f8473a0951f30842ad2
aace7de13864</url></related-urls></urls><electronic-resource-num>DOI:
10.1016/j.memsci.2010.12.036</electronic-resource-num></record></Cite></EndNote>3

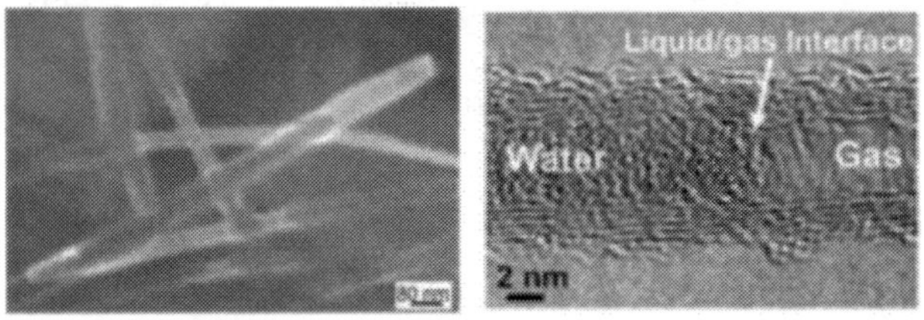

Figure 1. Water plugs in 40-60 nm and 2-5 nm diameter carbon nanotubes, respectively, showing a transition from a continuum to a sub-continuum behaviour.
ADDIN EN.CITE ADDIN EN.CITE.DATA
3C456E644E6F74653E3C436974653E3C417574686F723E4D61747469613C2F417574686F723E3C596561
723E323030363C2F596561723E3C5265634E756D3E3634393C2F5265634E756D3E3C446973706C617954
6578743E3C7374796C6520666163653D22737570657273637269707422E342C20353C2F7374796C653E
3C2F446973706C6179546578743E3C7265636F72643E3C7265632D6E756D6265723E3634393C2F726563
2D6E756D6265723E3C666F726569676E2D6B6579733E3C6B657920617070D22454E222064622D69643D
227A737461323272736E64357A706665737A326F706464777774746632737866736678303922E363439
3C2F6B65793E3C2F666F726569676E2D6B6579733E3C7265662D74797065206E616D653D224A6F75726E
616C2041727469636C65223E31373C2F7265662D747970653E3C636F6E747269627574746F72733E3C6175
74686F72733E3C617574686F723E4D61747469612C20442E3C2F617574686F723E3C617574686F723E52
6F7373692C204D2E20502E3C2F617574686F723E3C617574686F723E4B696D2C20422E4D2E3C2F617574
686F723E3C617574686F723E4B6F726E6576612C20472E3C2F617574686F723E3C617574686F723E4261
752C20482E482E3C2F617574686F723E3C617574686F723E476F676F7473692C20592E3C2F617574686F
723E3C2F617574686F72733E3C2F636F6E747269627574746F72733E3C7469746C65733E3C7469746C653E
456666656374206F6620477261706869746973A6174696F6E206F6E20746865205765747474162696C6974
7920616E6420456C656374726963616C20436F6E647563746976697479206F66204356442043617262626F
6E204E616E6F747562657320616E642046696C6D733C2F7469746C653E3C7365636F6E646172792D7469
746C653E4A6F75726E616C206F662050687973696963616C204368656D697374727920423C2F7365636F6E
646172792D7469746C653E3C2F7469746C65733E3C706572696F646963616C3E3C66756C6C2D7469746C
653E4A6F75726E616C206F662050687973696963616C204368656D697374727920423C2F66756C6C2D7469
746C653E3C616262722D313E4A2E20506879732E204368656D2E20423C2F616262722D313E3C2F706572
696F646963616C3E3C70616765733E39383530202D393835353C2F70616765733E3C766F6C756D653E31
31303C2F766F6C756D653E3C6E756D6265723E32303C2F6E756D6265723E3C6B6579776F7264733E3C6B
6579776F72643E616E6E65616C696E673C2F6B6579776F72643E3C6B6579776F72643E7765747474696E67
3C2F6B6579776F72643E3C6B6579776F72643E4356443C2F6B6579776F72643E3C6B6579776F72643E6E
616E6F747562653C2F6B6579776F72643E3C6B6579776F72643E636F6E647563746976697479793C2F6B65
79776F72643E3C6B6579776F72643E77617465723C2F6B6579776F72643E3C6B6579776F72643E677261
70686974697A6174696F6E3C2F6B6579776F72643E3C2F6B6579776F7264733E3C64617465733E3C7965
61723E323030363C2F796561723E3C2F64617465733E3C63616C6C2D6E756D3E303634393C2F63616C6C
2D6E756D3E3C75726C733E3C2F75726C733E3C2F7265636F72643E3C2F436974653E3C436974653E3C41
7574686F723E4D61747469613C2F417574686F723E3C596561723E323030373C2F596561723E3C526563
4E756D3E3836393C2F5265634E756D3E3C7265636F72643E3C7265632D6E756D6265723E3836393C2F72
65632D6E756D6265723E3C666F726569676E2D6B6579733E3C6B657920617070D22454E222064622D69
643D227A737461323272736E64357A706665737A326F706464777774746632737866736678303922E38

*** Formatting Issue - Best Available Paper/Graphic ***

36393C2F6B65793E3C2F666F726569676E2D6B6579733E3C7265662D74797065206E616D653D22436F6E
666572656E63652050726F63656564696E6773223E31303C2F7265662D747970653E3C636F6E74726962
75746F72733E3C617574686F72733E3C617574686F723E4D61747469612C20442E3C2F617574686F723E
3C617574686F723E526F7373692C204D2E20502E3C2F617574686F723E3C617574686F723E59652C2048
2E3C2F617574686F723E3C617574686F723E476F676F7473692C20592E3C2F617574686F723E3C2F6175
74686F72733E3C2F636F6E7472696275746F72733E3C7469746C65733E3C7469746C653E3C7374796C65
20666163653D226974616C69632220666F6E743D2264656661756C7422207369A653D223130302523E
496E207369747520466C7569643C2F7374796C653E3C7374796C6520666163653D226E6F726D616C2220
666F6E743D2264656661756C7422207369A653D22313030252223E2053747564696573206F6E20436172
626F6E204E616E6F747562657320776974682044696D657465727320526167696E672066726F6D20
3120746F20353030206E6D3C2F7374796C653E3C2F7469746C653E3C7365636F6E646172792D7469746C
653E357468204941534D452F575345415320496E7465726E6174696F6E616C20436F6E666572656E6365
206F6E20466C756964204D656368616E69637320616E64204165726F64796E616D6963733C2F7365636F
6E646172792D7469746C653E3C7465727469617279D7469746C653E4D617468656D617469637320616E
6420436F6D7075746572732696E20536369656E636520616E6420456E67696E656572696E673C2F7465
7274696172792D7469746C653E3C2F7469746C65733E3C70616765733E3239372D3239393C2F70616765
733E3C6B6579776F7264733E3C6B6579776F72643E436172626F6E206E616E6F747562653C2F6B657977
6F72643E3C6B6579776F72643E6E616E6F666C75696469963733C2F6B6579776F72643E3C6B6579776F72
643E666C7569642074726616E73706F72743C2F6B6579776F72643E3C6B6579776F72643E77657474696E
673C2F6B6579776F72643E3C6B6579776F72643E636F6E746163742616E676C653C2F6B6579776F7264
3E3C6B6579776F72643E73757266616365206368656D69737472793C2F6B6579776F72643E3C2F6B6579
776F7264733E3C64617465733E3C796561723E323030373C2F796561723E3C2F64617465733E3C707562
2D6C6F636174696F6E3E417468656E732C20477265656563653C2F7075622D6C6F636174696F6E3E3C6973
626E3E3937382D3936302D383435372D39392D37203C2F6973626E3E3C63616C6C2D6E756D3E30383639
3C2F63616C6C2D6E756D3E3C75726C733E3C2F75726C733E3C2F7265636F72643E3C2F436974653E3C2F
456E644E6F74653E0000 4, 5
3C456E644E6F74653E3C436974653E3C417574686F723E4D61747469613C2F417574686F723E3C596561
723E323030363C2F596561723E3C5265634E756D3E3634393C2F5265634E756D3E3C446973706C617954
6578743E3C7374796C6520666163653D22737570657273637269707422E342C20353C2F7374796C653E
3C2F446973706C6179546578743E3C7265636F72643E3C7265632D6E756D6265723E3634393C2F726563
2D6E756D6265723E3C666F726569676E2D6B6579733E3C6B657920617070D22454E222064622D69643D
227A737461323272736E64357A706665737A326F7064647777747466327378667366783039223E363439
3C2F6B65793E3C2F666F726569676E2D6B6579733E3C7265662D74797065206E616D653D224A6F75726E
616C2041727469636C65223E31373C2F7265662D747970653E3C636F6E7472696275746F72733E3C6175
74686F72733E3C617574686F723E4D61747469612C20442E3C2F617574686F723E3C617574686F723E52
6F7373692C204D2E20502E3C2F617574686F723E3C617574686F723E4B696D2C20422E4D2E3C2F617574
686F723E3C617574686F723E4B6F726E6576612C20472E3C2F617574686F723E3C617574686F723E4261
752C20482E482E3C2F617574686F723E3C617574686F723E476F676F7473692C20592E3C2F617574686F
723E3C2F617574686F72733E3C2F636F6E7472696275746F72733E3C7469746C65733E3C7469746C653E
456666656374206F662047726170686974697A6174696F6E206F6E207468652057657474616269C6974
7920616E6420456C656374726963616C20436F6E647563746976697479206F66204356442043617262F
6E204E616E6F747562657320616E642046696C6D733C2F7469746C653E3C7365636F6E646172792D7469
746C653E4A6F75726E616C206F662050687973696361C204368656D69737472792042C2F7365636F6E
646172792D7469746C653E3C2F7469746C65733E3C706572696F646963616C3E3C66756C6C2D7469746C
653E4A6F75726E616C206F662050687973696361C204368656D69737472792042C2F66756C6C2D7469
746C653E3C616262722D313E4A2E20506879732E204368656D2E20423C2F616262722D313E3C2F706572
696F646963616C3E3C70616765733E39383530202D393835353C2F70616765733E3C766F6C756D653E31
31303C2F766F6C756D653E3C6E756D6265723E32303C2F6E756D6265723E3C6B6579776F7264733E3C6B
6579776F72643E616E6E65616C696E673C2F6B6579776F72643E3C6B6579776F72643E7765747474696E67
3C2F6B6579776F72643E3C6B6579776F72643E4356443C2F6B6579776F72643E3C6B6579776F72643E6E
616E6F747562653C2F6B6579776F72643E3C6B6579776F72643E636F6E647563746976697479C2F6B65
79776F72643E3C6B6579776F72643E77617465723C2F6B6579776F72643E3C6B6579776F72643E677261
70686974697A6174696F6E3C2F6B6579776F72643E3C2F6B6579776F7264733E3C64617465733E3C7965
61723E323030363C2F796561723E3C2F64617465733E3C63616C6C2D6E756D3E303634393C2F63616C6C
2D6E756D3E3C75726C733E3C2F75726C733E3C2F7265636F72643E3C2F436974653E3C436974653E3C41
7574686F723E4D61747469613C2F417574686F723E3C596561723E323030373C2F596561723E3C526563
4E756D3E3836393C2F5265634E756D3E3C7265636F72643E3C7265632D6E756D6265723E3836393C2F72
65632D6E756D6265723E3C666F726569676E2D6B6579733E3C6B657920617070D22454E222064622D69
643D227A737461323272736E64357A706665737A326F7064647777747466327378667366783039223E38
36393C2F6B65793E3C2F666F726569676E2D6B6579733E3C7265662D74797065206E616D653D22436F6E
666572656E63652050726F63656564696E6773223E31303C2F7265662D747970653E3C636F6E74726962
75746F72733E3C617574686F72733E3C617574686F723E4D61747469612C20442E3C2F617574686F723E
3C617574686F723E526F7373692C204D2E20502E3C2F617574686F723E3C617574686F723E59652C2048

***** Formatting Issue - Best Available Paper/Graphic *****

2E3C2F617574686F723E3C617574686F723E476F676F7473692C20592E3C2F617574686F723E3C2F6175
74686F72733E3C2F636F6E7472696275746F72733E3C7469746C65733E3C7469746C653E3C7374796C65
20666163653D226974616C69632220666F6E743D2264656661756C74222073697A653D2231303025223E
496E2073697475520466C7569643C2F7374796C653E3C7374796C6520666163653D226E6F726D616C2220
666F6E743D2264656661756C74222073697A653D2231303025223E2053747564696573320696E20436172
626F6E204E616E6F747562657320776974682044696D65746572732052616E67696E672066726F6D20
3120746F20353030206E6D3C2F7374796C653E3C2F7469746C653E3C7365636F6E646172792D7469746C
653E357468204941534D452F575345415320496E7465726E6174696F6E616C20436F6E666572656E6365
206F6E20466C756964204D656368616E69637320616E64204165726F64796E616D6963733C2F7365636F
6E646172792D7469746C653E3C74657274696172792D7469746C653E4D617468656D617469637320616E
6420436F6D707574657273320696E20536369656E636520616E6420456E67696E656572696E673C2F7465
7274696172792D7469746C653E3C2F7469746C65733E3C70616765733E3239372D3239393C2F70616765
733E3C6B6579776F7264733E3C6B6579776F72643E436172626F6E206E616E6F747562653C2F6B657977
6F72643E3C6B6579776F72643E6E616E6F666C7569646963733C2F6B6579776F72643E3C6B6579776F72
643E666C75696420747261616E73706F72743C2F6B6579776F72643E3C6B6579776F72643E77657474696E
673C2F6B6579776F72643E3C6B6579776F72643E636F6E74616374420616E676C653C2F6B6579776F7264
3E3C6B6579776F72643E737572666163652063686656D697374727793C2F6B6579776F72643E3C2F6B6579
776F7264733E3C64617465733E3C796561723E323030373C2F796561723E3C2F64617465733E3C707562
2D6C6F636174696F6E3E417468656E732C204772656563653C2F7075622D6C6F636174696F6E3E3C6973
626E3E3937382D3936302D383435372D39392D37203C2F6973626E3E3C63616C6C2D6E756D3E30383639
3C2F63616C6C2D6E756D3E3C75726C733E3C2F75726C733E3C2F7265636F72643E3C2F436974653E3C2F
456E644E6F74653E0000 The micrograph on the left has been obtained using an environmental SEM by preferential condensation of water inside a CNT. The micrograph on the right was obtained using high resolution TEM and shows the coexistence of water liquid and vapor in a capped CNT.

The attempts made so far to explain these phenomena have focused on the slip length at the wall and its possible dependencies on a wide range of parameters, from contact angle to roughness, viscosity and shear rate. All these solutions are somewhat unsatisfying when slip values in the order of tens of micrometers for membranes with CNTs with diameters of only few nanometers are obtained. Moreover, no model has yet been capable of explicitly relating the flow enhancement with solid-liquid molecular interactions occurring at the channel walls.

[illegible]

Here a model capable of explaining the observed dependence of the flow enhancement on the membrane channels' characteristic dimensions and surface wettability is proposed. The model can accurately predict the effect of each component for membrane channels and nanotubes of different sizes, wall surface chemistry and structure. Comparison with the experimental and modelling data available shows good agreement (Figure 2). ADDIN EN.CITE <EndNote><Cite><Author>Mattia</Author><Year>2012</Year><RecNum>1472</RecNum><DisplayText><style
db-id="zsta22rsnd5zpfesz2opddwwttf2sxfsfx09">1472</key></foreign-keys><ref-type
name="Journal Article">17</ref-type><contributors><authors><author>Mattia,
D.</author><author>Calabr?,
F.</author></authors></contributors><titles><title>Explaining high flow rate of
water in carbon nanotubes via solid-liquid molecular
interactions</title><secondary-title>Microfluidics and
Nanofluidics</secondary-title></titles><periodical><full-title>Microfluidics and
Nanofluidics</full-title><abbr-1>Microfluid.
Nanofluid.</abbr-1></periodical><pages>1-6</pages><keywords><keyword>work of
adhesion</keyword><keyword>surface
diffusion</keyword><keyword>slip</keyword><keyword>solid-liquid
interaction</keyword></keywords><dates><year>2012</year></dates><publisher>Springer
Berlin /
Heidelberg</publisher><isbn>1613-4982</isbn><call-num>1472</call-num><urls><related-urls><url>http://dx.doi.org/10.1007/s10404-012-0949-z</url></related-urls></urls><electronic-resource-num>10.1007/s10404-012-0949-z</electronic-resource-num></record></Cite></EndNote>6 The proposed model is based on a two-phase model for water flow in cylindrical channels with reduced viscosity at the wall, , and a non-zero velocity at the wall which is solely a function of solid-liquid interactions (ex. graphitic carbon and water) and the applied pressure. By solving the Haagen-Poiseuille equation with this velocity at the wall term, the following expression for the fluid

*** Formatting Issue - Best Available Paper/Graphic ***

flow enhancement, , at the wall is obtained:

where and are the tube/channel/pore length and radius, respectively; is the thickness of the annular region and is the liquid bulk density; is the surface diffusion term arising from the chemical potential generated by the pressure gradient, ADDIN EN.CITE <EndNote><Cite><Author>Ruckenstein</Author><Year>1983</Year><RecNum>1314</RecNum><DisplayText><style no-slip boundary condition of hydrodynamics</title><secondary-title>Journal of Colloid and Interface Science</secondary-title></titles><periodical><full-title>Journal of Colloid and Interface Science</full-title><abbr-1>J. Colloid Interface Sci.</abbr-1><abbr-2>J Colloid Interface Sci</abbr-2></periodical><pages>488-491</pages><volume>96</volume><number>2</number><keywords><keyword>slip</keyword><keyword>surface diffusion</keyword></keywords><dates><year>1983</year></dates><isbn>0021-9797</isbn><call-num>1314</call-num><urls><related-urls><url>http://www.sciencedirect.com/science/article/B6WHR-4CT7DM2-NX/2/a2dbcea1f67474c1626147d2f0f2fb1c</url></related-urls></urls><electronic-resource-num>10.1016/0021-9797(83)90050-4</electronic-resource-num></record></Cite></EndNote>7 and is the work of adhesion which represents the strength of the solid-liquid interaction at the tube/channel/pore wall. The two latter terms can be measured independently and at the macroscale for any solid-liquid couple.

Two recent publications have offered further validation of the proposed model: First, a molecular dynamics paper of water flow in CNTs of different lengths has found a linear dependence of the flow enhancement with the former. ADDIN EN.CITE <EndNote><Cite><Author>Nicholls</Author><Year>2011</Year><RecNum>1358</RecNum><DisplayText><style Jason</author></authors></contributors><titles><title>Water transport through (7,7) carbon nanotubes of different lengths using molecular dynamics</title><secondary-title>Microfluidics and Nanofluidics</secondary-title></titles><periodical><full-title>Microfluidics and Nanofluidics</full-title><abbr-1>Microfluid. Nanofluid.</abbr-1></periodical><pages>257-264</pages><volume>12</volume><number>1-4</number><keywords><keyword>Physics and Astronomy</keyword></keywords><dates><year>2011</year></dates><publisher>Springer Berlin / Heidelberg</publisher><isbn>1613-4982</isbn><call-num>1358</call-num><urls><related-urls><url>http://dx.doi.org/10.1007/s10404-011-0869-3</url></related-urls></urls><electronic-resource-num>10.1007/s10404-011-0869-3</electronic-resource-num></record></Cite></EndNote>8

Second, water flow measurements in hydrophilic alumina nanochannels with diameters ranging from 20 nm to 100 nm have shown to be well represented by the above equation. For the smallest channels tested, the water permeability is more than double than the theoretical prediction using the Haagen-Poiseuille equation. Though such an enhancement is significantly smaller than what has been measured in hydrophobic carbon nanotubes, it clearly shows that flow enhancement and water slippage occur on hydrophilic surfaces as well, contrary to existing theoretical models (Figure 2, red empty circles ○). ADDIN EN.CITE <EndNote><Cite><Author>Lee</Author><Year>2012</Year><RecNum>1486</RecNum><DisplayText><style Peng</author><author>Leese, Hannah</author><author>Mattia, Davide</author></authors></contributors><titles><title>Water flow enhancement in hydrophilic nanochannels</title><secondary-title>Nanoscale</secondary-title></titles><periodical><full-title>Nanoscale</full-title></periodical><keywords><keyword>slip</keyword><keyword>flow enhancement</keyword><keyword>Hagen-poisuille</keyword><keyword>AAO</keyword></keywords><dates><year>2012</year></dates><publisher>The Royal Society of Chemistry</publisher><isbn>2040-3364</isbn><call-num>1486</call-num><urls><related-urls><url>http://dx.doi.org/10.1039/C2NR30098B</url></related-urls></urls></record></Cite></EndNote>9

***** Formatting Issue - Best Available Paper/Graphic *****

Finally, the above equation has been successfully used to model molecular dynamics simulations of water flow in 2nm diameter silicon carbide (SiCNTs, yellow diamonds ♦) and boron nitride (BNNTs, blue squares ■) nanotubes.10

Figure 2. Normalized flow enhancement as a function of nanotube radius (dashed curve). Literature: ○ - (Thomas and McGaughey, Nano Lett, 2009); □ - (Holt et al., Science, 2006); × - (Thomas et al., Int J Therm Sci, 2010); ▲ - (Majumder et al., Nature, 2005); ♦ - (Du et al., Langmuir, 2011); ■ - (Whitby and Quirke, Nano Lett, 2007). Literature values (open symbols for MD and full symbols for experimental) are reported as a range between minimum and maximum possible values. ○ Enhancement values for water flow in alumina nanopores; ♦ SiCNTs; ■ BNNTs; Adapted from ref.s ADDIN EN.CITE ADDIN EN.CITE.DATA

3C456E644E6F74653E3C436974653E3C417574686F723E4D61747469613C2F417574686F723E3C596561
723E323031323C2F596561723E3C5265634E756D3E313437323C2F5265634E756D3E3C446973706C6179
546578743E3C7374796C6520666163653D227375706572736372697074223E362C20393C2F7374796C65
3E3C2F446973706C6179546578743E3C7265636F72643E3C7265632D6E756D6265723E313437323C2F72
65632D6E756D6265723E3C666F726569676E2D6B6579733E3C6B6579206170703D22454E222064622D69
643D227A737461323272736E64357A706665737A326F7064647777747466327378667366783039223E31
3437323C2F6B65793E3C2F666F726569676E2D6B6579733E3C7265662D74797065206E616D653D224A6F
75726E616C2041727469636C65223E31373C2F7265662D747970653E3C636F6E7472696275746F72733E
3C617574686F72733E3C617574686F723E4D61747469612C20442E3C2F617574686F723E3C617574686F
723E43616C616272C3B22C20462E3C2F617574686F723E3C2F617574686F72733E3C2F636F6E74726962
75746F72733E3C7469746C65733E3C7469746C653E4578706C61696E696E67206869676820666C6F7720
72617465206F6620776174657220696E20636172626F6E206E616E6F74756265732076696120736F6C69
64E280936C6971756964206D6F6C6563756C617220696E746572616374696F6E733C2F7469746C653E3C
7365636F6E646172792D7469746C653E4D6963726F666C75696469637320616E64204E616E6F666C7569
646963733C2F7365636F6E646172792D7469746C653E3C2F7469746C65733E3C706572696F646963616C
3E3C66756C6C2D7469746C653E4D6963726F666C75696469637320616E64204E616E6F666C7569646963
733C2F66756C6C2D7469746C653E3C616262722D313E4D6963726F666C7569642E204E616E6F666C7569
642E3C2F616262722D313E3C2F706572696F646963616C3E3C70616765733E312D363C2F70616765733E
3C6B6579776F7264733E3C6B6579776F72643E776F726B206F66206164686573696F6E3C2F6B6579776F
72643E3C6B6579776F72643E7375726661636520646966667573696F6E3C2F6B6579776F72643E3C6B65
79776F72643E736C69703C2F6B6579776F72643E3C6B6579776F72643E736F6C69642D6C697175696420
696E746572616374696F6E3C2F6B6579776F72643E3C2F6B6579776F7264733E3C64617465733E3C7965
61723E323031323C2F796561723E3C2F64617465733E3C7075626C69736865723E537072696E67657220
4265726C696E202F2048656964656C626572673C2F7075626C69736865723E3C6973626E3E313631332D
343938323C2F6973626E3E3C63616C6C2D6E756D3E313437323C2F63616C6C2D6E756D3E3C75726C733E
3C72656C617465642D75726C733E3C75726C3E687474703A2F2F64782E646F692E6F72672F31302E3130
30372F7331303430342D3031322D303934392D7A3C2F75726C3E3C2F72656C617465642D75726C733E3C
2F75726C733E3C656C656374726F6E69632D7265736F757263652D6E756D3E31302E313030372F733130
3430342D3031322D303934392D7A3C2F656C656374726F6E69632D7265736F757263652D6E756D3E3C2F
7265636F72643E3C2F436974653E3C436974653E3C417574686F723E4C65653C2F417574686F723E3C59
6561723E323031323C2F596561723E3C5265634E756D3E313438363C2F5265634E756D3E3C7265636F72
643E3C7265632D6E756D6265723E313438363C2F7265632D6E756D6265723E3C666F726569676E2D6B65
79733E3C6B6579206170703D22454E222064622D69643D227A737461323272736E64357A706665737A32
6F7064647777747466327378667366783039223E313438363C2F6B65793E3C2F666F726569676E2D6B65
79733E3C7265662D74797065206E616D653D224A6F75726E616C2041727469636C65223E31373C2F7265
662D747970653E3C636F6E7472696275746F72733E3C617574686F72733E3C617574686F723E4C65652C
204B61682050656E673C2F617574686F723E3C617574686F723E4C656573652C2048616E6E61683C2F61
7574686F723E3C617574686F723E4D61747469612C2044617669643C2F617574686F723E3C2F617574
686F72733E3C2F636F6E7472696275746F72733E3C7469746C65733E3C7469746C653E57617465722066
6C6F7720656E68616E63656D656E7420696E20687964726F70686F626963206E616E6F6368616E6E656C
733C2F7469746C653E3C7365636F6E646172792D7469746C653E4E616E6F7363616C653C2F7365636F6E
646172792D7469746C653E3C2F7469746C65733E3C706572696F646963616C3E3C66756C6C2D7469746C
653E4E616E6F7363616C653C2F66756C6C2D7469746C653E3C2F706572696F646963616C3E3C6B657977
6F7264733E3C6B6579776F72643E736C69703C2F6B6579776F72643E3C6B6579776F72643E666C6F7720
656E68616E63656D656E743C2F6B6579776F72643E3C6B6579776F72643E486167656E2D706F69737569
6C6C653C2F6B6579776F72643E3C6B6579776F72643E41414F3C2F6B6579776F72643E3C2F6B6579776F
7264733E3C64617465733E3C796561723E323031323C2F796561723E3C2F64617465733E3C7075626C69
736865723E54686520526F79616C20536F6369657479206F66204368656D69737472793C2F7075626C69
736865723E3C6973626E3E323034302D333336343C2F6973626E3E3C63616C6C2D6E756D3E313438363C
2F63616C6C2D6E756D3E3C75726C733E3C72656C617465642D75726C733E3C75726C3E687474703A2F2F

*** Formatting Issue - Best Available Paper/Graphic ***

64782E646F692E6F72672F31302E313033392F43324E523330303938423C2F75726C3E3C2F72656C6174
65642D75726C733E3C2F75726C733E3C2F7265636F72643E3C2F436974653E3C2F456E644E6F74653E00
5200 6, 9
3C456E644E6F74653E3C436974653E3C417574686F723E4D61747469613C2F417574686F723E3C596561
723E323031323C2F596561723E3C5265634E756D3E313437323C2F5265634E756D3E3C446973706C6179
546578743E3C7374796C6520666163653D22737570657273637269707422E362C20393C2F7374796C65
3E3C2F446973706C6179546578743E3C7265636F72643E3C7265632D6E756D6265723E313437323C2F72
65632D6E756D6265723E3C666F726569676E2D6B6579733E3C6B657920617070D22454E222064622D69
643D227A737461323272736E64357A706665737A326F706464777777474663273786673667830392 23E31
3437323C2F6B65793E3C2F666F726569676E2D6B6579733E3C7265662D74797065206E616D653D224A6F
75726E616C2041727469636C65223E31373C2F7265662D747970653E3C636F6E7472696275746F72733E
3C617574686F72733E3C617574686F723E4D61747469612C20442E3C2F617574686F723E3C617574686F
723E43616C616272C3B22C20462E3C2F617574686F723E3C2F617574686F72733E3C2F636F6E74726962
75746F72733E3C7469746C65733E3C7469746C653E4578706C61696E696E67206869676820666C6F7720
72617465206F662077617465722069E20636172626F6E206E616E6F7475626573207669612073686F6C69
64E280936C697175696420696D6F6C6563756C617220696E7465726163746696F6E733C2F7469746C653E3C
7365636F6E646172792D7469746C653E4D6963726F666C7569646963320616E64204E616E6F666C7569
646963733C2F7365636F6E646172792D7469746C653E3C2F7469746C65733E3C706572696F646963616C
3E3C66756C6C2D7469746C653E4D6963726F666C75696469637320616E64204E616E6F666C7569646963
733C2F66756C6C2D7469746C653E3C616262722D313E4D6963726F666C7569642E204E616E6F666C7569
642E3C2F616262722D313E3C2F706572696F646963616C3E3C70616765733E312D363C2F70616765733E
3C6B6579776F7264733E3C6B6579776F72643E776F726B206F66206164686573696F6E3C2F6B6579776F
72643E3C6B6579776F72643E737572666163652064696666757369 6F6E3C2F6B6579776F72643E3C6B65
79776F72643E736C69703C2F6B6579776F72643E3C6B6579776F72643E736F6C69642D6C697175696420
696E746572616374696F6E3C2F6B6579776F72643E3C2F6B6579776F7264733E3C646174657333E3C7965
61723E323031323C2F796561723E3C2F64617465733E3C7075626C69736865723E537072696E67657220
4265726C696E202F2048656964656C626572673C2F7075626C69736865723E3C6973626E3E313631332D
343938323C2F6973626E3E3C63616C6C2D6E756D3E313437323C2F63616C6C2D6E756D3E3C75726C733E
3C72656C617465642D75726C733E3C75726C3E687474703A2F2F64782E646F692E6F72672F31302E3130
30372F7331303430342D3031322D303934392D7A3C2F75726C3E3C2F72656C617465642D75726C733E3C
2F75726C733E3C656C656374726F6E69632D7265736F757263652D6E756D3E31302E313030372F733130
3430342D3031322D303934392D7A3C2F656C656374726F6E69632D7265736F757263652D6E756D3E3C2F
7265636F72643E3C2F436974653E3C436974653E3C417574686F723E4C65653C2F417574686F723E3C59
6561723E323031323C2F596561723E3C5265634E756D3E313438363C2F5265634E756D3E3C7265636F72
643E3C7265632D6E756D6265723E313438363C2F7265632D6E756D6265723E3C666F726569676E2D6B65
79733E3C6B657920617070D22454E222064622D69643D227A737461323272736E64357A706665737A32
6F706464777777474663273786673667830392 23E313438363C2F6B65793E3C2F666F726569676E2D6B65
79733E3C7265662D74797065206E616D653D224A6F75726E616C2041727469636C65223E31373C2F7265
662D747970653E3C636F6E7472696275746F72733E3C617574686F72733E3C617574686F723E4C65652C
204B61682050656E673C2F617574686F723E3C617574686F723E4C656573652C2048616E6E61683C2F61
7574686F723E3C617574686F723E4D61747469612C20446176696465653C2F617574686F723E3C2F617574
686F72733E3C2F636F6E7472696275746F72733E3C7469746C65733E3C7469746C653E57617465722066
6C6F7720656E68616E63656D656E7420696E20687964726F7068696C6963206E616E6F368616E6E656C
733C2F7469746C653E3C7365636F6E646172792D7469746C653E4E616E6F7363616C653C2F7365636F6E
646172792D7469746C653E3C2F7469746C65733E3C706572696F646963616C3E3C66756C6C2D7469746C
653E4E616E6F7363616C653C2F66756C6C2D7469746C653E3C2F706572696F646963616C3E3C6B657977
6F7264733E3C6B6579776F72643E736C69703C2F6B6579776F72643E3C6B6579776F72643E666C6F7720
656E68616E63656D656E743C2F6B6579776F72643E3C6B6579776F72643E486167656E2D706F69737569
6C6C653C2F6B6579776F72643E3C6B6579776F72643E41414F3C2F6B6579776F72643E3C2F6B6579776F
7264733E3C64617465733E3C796561723E323031323C2F796561723E3C2F64617465733E3C7075626C69
736865723E54686520526F79616C20536F6369657479206F66204368656D69737472793C2F7075626C69
736865723E3C6973626E3E323034302D333336343C2F6973626E3E3C63616C6C2D6E756D3E313438363C
2F63616C6C2D6E756D3E3C75726C733E3C72656C617465642D75726C733E3C75726C3E687474703A2F2F
64782E646F692E6F72672F31302E313033392F43324E523330303938423C2F75726C3E3C2F72656C6174
65642D75726C733E3C2F75726C733E3C2F7265636F72643E3C2F436974653E3C2F456E644E6F74653E00
0000

References:

ADDIN EN.REFLIST 1. D. Mattia and Y. Gogotsi, Microfluid. Nanofluid., 2008, 5, 289-305.

2. M. Whitby and N. Quirke, Nat. Nano., 2007, 2, 87-94.

*** Formatting Issue - Best Available Paper/Graphic ***

3. K. P. Lee, T. C. Arnot and D. Mattia, J. Membr. Sci., 2011, 370, 1-22.

4. D. Mattia, M. P. Rossi, B. M. Kim, G. Korneva, H. H. Bau and Y. Gogotsi, J. Phys. Chem. B, 2006, 110, 9850 -9855.

5. D. Mattia, M. P. Rossi, H. Ye and Y. Gogotsi, in 5th IASME/WSEAS International Conference on Fluid Mechanics and Aerodynamics, Athens, Greece, 2007, pp. 297-299.

6. D. Mattia and F. Calabr?, Microfluid. Nanofluid., 2012, 1-6.

7. E. Ruckenstein and P. Rajora, J. Colloid Interface Sci., 1983, 96, 488-491.

8. W. Nicholls, M. Borg, D. Lockerby and J. Reese, Microfluid. Nanofluid., 2011, 12, 257-264.

9. K. P. Lee, H. Leese and D. Mattia, Adv Math Lett, 2012.

10. K. Ritos, D. Mattia, F. Calabr? and J. Reese, 2013.

(734a) Ultrasmall Metal Nanoparticles With Molecular-Like Properties: Synthesis and Applications

Xie, J., National University of Singapore

Originally presented on: 11/7/2013 15:15:00 - 15:50:00

In very recent years, ultrasmall metal nanoclusters with core sizes smaller than 2 nm have emerged as a new direction in nanoparticle research due to their discrete and size dependent electronic structures and molecular-like properties, such as HOMO-LUMO transitions in optical absorptions, quantized charging, and strong luminescence. Synthesis of high quality metal nanoclusters in sufficiently large quantities is necessary for establishing reliable size-property relationships and exploring potential applications. This presentation will summarize our recent progress in the development of novel synthetic strategies for the production of monodisperse metal nanoclusters. The preparation of monodisperse metal nanoclusters is viewed as engineerable process where both the precursors (input) and their conversion chemistry (processing) may be rationally designed to achieve the desired outcome – monodisperse metal nanoclusters (output). We will demonstrate several efficient strategies for tailoring the precursor and the conversion process, and present our understanding of the processes involved. We will also highlight some application developments of metal nanoclusters in important sustainability topics including sensors for environment and human health, and biomedical applications.

(734b) Stability of Phosphonate-Functionalized Iron Nanoparticles

Greenlee, L. F., National Institute of Standards & Technology
Rentz, N. S., National Institute of Standards & Technology
Wilson, S. J., National Institute of Standards & Technology
Originally presented on: 11/7/2013 15:50:00 - 16:10:00

Iron nanoparticles are of interest in fields such as water treatment and alternative energy, due to their reactive properties and low cost. When combined with other metals, iron-metal nanoparticles can act as catalysts for a diverse set of reactions. In solution-based nanoparticle synthesis, organic stabilizers play a critical role in the properties of the nanoparticles, including size, morphology, composition, and colloidal stability. While polymers are often used in nanoparticle synthesis, chelator-type molecules can also be used as nanoparticle stabilizers. In this study, we characterize iron nanoparticles synthesized in water in the presence of a phosphonate chelator, amino tris(methylene phosphonic acid) (ATMP) for a range of molar ratios of ATMP to iron. An increase in the molar ratio from 0.05 to 0.8 decreases nanoparticle size from approximately 150 nm to less than 10 nm. Colloidal ATMP-stabilized iron nanoparticles are stable over a broad range of pH values, and zeta potential measurements were used to evaluate colloidal stability. Diffraction analysis indicates that ATMP-stabilized iron nanoparticles may have a nano-crystallite structure, with potential regions of amorphous iron. Characterization results of ATMP-stabilized iron nanoparticles are compared to results obtained previously for carboxymethyl cellulose (CMC)-stabilized iron nanoparticles. Quartz crystal microbalance measurements are used to evaluate the effect of surface functionalization on oxidation rate of the iron nanoparticles, and nanoparticle stability during storage is evaluated.

(734c) Magadiite Silylated With Sulfur-Functional Organosilanes: Investigation of Structure and Interlayer Accessibility

Mao, Y., University of South Carolina
Li, S., University of South Carolina
zur Loye, H. C., University of South Carolina
Ploehn, H. J., University of South Carolina
Originally presented on: 11/7/2013 16:10:00 - 16:30:00

Sodium magadiite (Na-MGD) is a prospective active filler for elastomer applications. In this work, Na-MGD was functionalized with sulfur-containing silane coupling agents, either bifunctional (bis (triethoxysilylpropyl) polysulfide, Si69) or monofunctional (3-mercaptopropyltriethoxysilane, MPTES) using cetyltrimethylammonium bromide modified MGD (CTA-MGD) as the intermediate. During grafting, the organosilanes displace CTA^{+} cations from CTA-MGD surface, leading to a decrease in the interlayer spacing as measured by XRD; however, some CTA^{+} cations remain in the MGD interlayer. The amounts of organosilane molecules grafted on MGD surface indicate that MPTES leads to a higher grafting density than Si69. Having functionalized the MGD interlayer surfaces, the accessibility of the sulfur-functional interlayer was evaluated using squalene, a model compound for natural rubber. Various sulfur-functional MGD materials (CTA-MGD, Si69-MGD, and MPTES-MGD) were reacted with pure liquid squalene. FTIR results show that new Si-O-C bonds are formed after reaction of squalene with sulfur-functional MGDs, but not with CTA-MGD. TGA and elemental analysis were used to quantify the amount CTA+, silane, and squalene in the products. XRD results demonstrate that squalene can access the interlayer sulfur only when the silane graft density is relatively low. This work shows the value of using squalene as a model compound for characterizing the structure and properties of sulfur-functional MGD prior to preparation of MGD-based elastomer composites.

(734d) Polymer-Gold Composite Particles Via Pickering Emulsion Polymerization: Synthesis, Characterization and Application

Zhang, M.,
Dai, L., Arizona State University
Phelan, P.,
Originally presented on: 11/7/2013 16:30:00 - 16:50:00

A one-step Pickering emulsion polymerization route was successfully applied to synthesize core-shell polystyrene-gold composite particles. Interestingly, in addition to core-shell structured polystyrene-gold composite particles, we found asymmetric polystyrene-gold nanocomposite particles are also produced, where a single gold nanoparticle is attaching on each polystyrene substrate. It is found these asymmetric particles are formed via "seeded-growth" mechanism. The core-shell and asymmetric polystyrene-gold composite particles proves to be efficient catalysts as they successfully catalyzed Rhodamine B reduction reaction. Further, PNIPAAM were successfully incorporated to produce "smart" polystyrene-PNIPAAM-gold composite particles. Without the aid of cross linker, the synthesized particles have up to 72% size decrease percentage. The polystyrene-PNIPAAM-gold asymmetric particles are found to have tunable optical properties due to thermo-responsiveness of PNIPAAM. It is found that absorption/scattering intensity could be adjusted significantly and the refractive index of substrates of the asymmetric particles influence their SPR properties. These tunable optical properties show great potential for biomedical and optical applications.

(734e) Synthesis and Characterization of Tin Selenide Nanocrystals Using Air-Stable Precursors

Qi, Y., University of Massachusetts
Mountziaris, T. J., University of Massachusetts
Originally presented on: 11/7/2013 16:50:00 - 17:10:00

Near-infrared (NIR) absorbing semiconductor nanocrystals, also known as quantum dots (QDs), have unique properties, such as size tunable bandgap, and potential applications in photothermal therapy, photovoltaics, and infrared absorbing glass coatings or interlayers. SnSe QDs are attractive as NIR materials, because they do not contain a heavy metal, such as Cd, Hg or Pb. The synthesis of SnSe colloidal QDs from organometallic precursors that are highly toxic and pyrophoric has been reported in the literature (e.g., Franzman et al., *J. Am. Chem. Soc.*, 2010, 132 (12), pp 4060–4061; Baumgardner et al., *J. Am. Chem. Soc.*, 2010, 132 (28), pp 9519–9521). Our study focuses on the development of a new synthesis method for SnSe QDs that employs an air-stable tin(II) chloride-oleylamine (OLA) complex and selenium powder dissolved in trioctylphosphine (TOP) as precursors. SnSe QDs with average diameter between 1.7 and 2.5 nm were synthesized by injecting the precursors in a hot coordinating solvent consisting of OLA and TOP. The effects of reaction conditions on the growth kinetics and final morphology of the QDs were investigated. A high OLA concentration in the growth solvent was found to be necessary for activating the reaction forming SnSe nuclei. On the other hand, TOP was found to hinder SnSe nucleation and, at increased concentrations, to yield larger particles. The relative amounts of precursors, OLA and TOP in the growth mixture control the growth kinetics and the aggregation of the QDs. The evolution of QD size and morphology as function of reaction time for different relative amounts of precursors, OLA and TOP has been studied using TEM. The formation of aggregates has been studied as function of reaction conditions and methods of controlling their size and shape have been identified. Growth conditions that provide precise control over the final particle size and prevent particle aggregation have been identified. The ultimate objective of this study is the development of a practical method for manufacturing SnSe QDs using readily available, air-stable precursors.

(734f) Asymmetric Functionalization of Shape-Anisotropic Polymer Nanoparticles

Guignard, F., University of Fribourg
Lattuada, M., ETH Zurich
Originally presented on: 11/7/2013 17:10:00 - 17:30:00

In the research in the field of nanoparticle in the last years, much attention has been devoted to the synthesis of anisotropic nanoparticles, also called Janus Nanoparticles. Anisotropicity may be provided in different ways, usually either in terms of non-homogeneous surface functionalization, or as compartmentalized internal structure. These special features render the nanoparticles much more interesting than their isotropic counterparts, especially regarding their self-assembling properties.

In this work, a two-step emulsion based polymerization has been applied to synthesize shape-anisotropic dumbbell nanoparticles, which size and morphology can be tuned by changing the process parameters. Narrowly distributed surfactant-free polystyrene nanoparticles are used as seeds. They are coated with a random copolymer of styrene and 3-trimethyloxyxilyl propyl acrylate (MPS), bringing silane groups on the surface of the seeds. The particles are then swollen with a monomer solution before a second polymerization. Due to the hydrophilic shell on the surface of the nanoparticles, the newly formed polymer is bulging out of the seeds, giving birth to dumbbell-like nanoparticles. The resulting nanoparticles are not only anisotropic in shape, but also in term of surface chemistry. The first bulb is bearing silane groups coming from the MPS, while the second is only made of polystyrene. This specific feature can be used for further asymmetric functionalize the dumbbells. Silane chemistry can be applied, which will result in the functionalization of the first hemisphere only. Superparamagnetic iron oxide nanoparticles (SPIONS) can be coated by surfactant having silane group, offering the possibility to attach SPIONS only to one bulge. The same approach has been used for platinum nanocrystals, leading to particles that can be used as Janus nanomotors in H2O2 solution.

(738a) Disorder Imposed Limits of Mono- and Bilayer Graphene Electronic Modification Using Covalent Chemistry

Shih, C. J., Massachusetts Institute of Technology
Wang, Q. H., Massachusetts Institute of Technology
Jin, Z., Massachusetts Institute of Technology
Paulus, G. L. C., Massachusetts Institute of Technology
Blankschtein, D., Massachusetts Institute of Technology
Strano, M. S., Massachusetts Institute of Technology
Originally presented on: 11/7/2013 15:15:00 - 15:31:00

A central question in graphene chemistry is to what extent chemical modification can control an electronically accessible band gap in monolayer and bilayer graphene (MLG and BLG). Density functional theory predicts gaps in covalently functionalized graphene as high as 2 eV, while this approach neglects the fact that lattice symmetry breaking occurs over only a prescribed radius of nanometer dimension, which we label the S-region. Therefore, high chemical conversion is central to observing this band gap in transport. We use an electrochemical approach involving phenyl-diazonium salts to systematically probe electronic modification in MLG and BLG with increasing functionalization for the first time, obtaining the highest conversion values to date. We find that both MLG and BLG retain their relatively high conductivity after functionalization even at high conversion, as mobility losses are offset by increases in carrier concentration. For MLG, we find that band gap opening as measured during transport is linearly increased with respect to the ID/IG ratio but remains below 0.1 meV in magnitude for SiO2 supported graphene. The largest transport band gap obtained in a suspended, highly functionalized (ID/IG = 4.5) graphene is about 1 meV, lower than our theoretical predictions considering the quantum interference effect between two neighboring Sregions and attributed to its population with midgap states. On the other hand, heavily functionalized BLG (ID/IG = 1.8) still retains its signature dual-gated band gap opening due to electric-field symmetry breaking. We find a notable asymmetric deflection of the charge neutrality point (CNP) under positive bias which increases the apparent on/off current ratio by 50%, suggesting that synergy between symmetry breaking, disorder, and quantum interference may allow the observation of new transistor phenomena. These important observations set definitive limits on the extent to which chemical modification can control graphene electronically.

(738b) Toughness, Self-Healing, and Network Formation in Pristine Graphene/Polyacrylamide Gels

Das, S., Texas Tech University
Irin, F., Texas Tech University
Ma, L.,
Hedden, R., Texas Tech University
Green, M., Texas Tech University
Originally presented on: 11/7/2013 15:31:00 - 15:47:00

Enhancement of toughness in nanomaterial-based hydrogels is a critical metric for the various engineering applications of these gels. Here pristine graphene/polyacrylamide (PAM) hydrogels are synthesized via in-situ polymerization of acrylamide monomer in PAM-stabilized graphene dispersions. In-situ polymerization leads to the uniform dispersion of the graphene sheets in the hydrogel. The pristine graphene sheets interact with the elastic chains of the hydrogel through physisorption and permit gelation in the absence of any chemical cross-linker. This study represents the first report of pristine graphene as a physical cross-linker in a hydrogel. The properties of the graphene-polymer hydrogel are characterized by viscoelastic measurements and compressive tests, revealing an increase in the storage modulus and toughness of the hydrogels compared to the chemically cross-linked PAM analogs. The physically cross-linked graphene hydrogels also exhibit excellent self healing properties. These graphene/PAM hydrogels serve as effective precursors for aerogels which become electrically conductive when graphene sheets form an interconnected network inside the network of the aerogel skeleton. Monomer infusion into the aerogel scaffold leads to the formation of ultralow percolation threshold graphene composites. The backfilling of the additives creates a conductive nanocomposite with a very low filler loading, without any network disruption.

(738c) Segregation of Single-Walled Carbon Nanotubes to Liquid-Liquid Interfaces

Hoagland, D. A., University of Massachusetts
Feng, T., University of Massachusetts Amherst
Russell, T. P., University of Massachusetts Amherst
Originally presented on: 11/7/2013 15:47:00 - 16:03:00

The interfacial assembly of carbon nanotubes (CNTs) is needed for many of their applications, and we have developed a new process for driving this assembly at water-organic liquid-liquid interfaces. Our interest in such interfaces was motivated by a desire to find a practical, continuous process for the preparation of CNT membranes, one that would exploit interfacial polymerization to prepare membrane films containing the interface-segregated CNTs. Acid treatment of single-walled CNTs (SWCNTs), which can reduce their lengths to hundreds of nanometers, decorates the tubes with carboxyl functionalities that make them water-dispersible. When an aqueous SWCNT suspension is placed in contact with an organic liquid containing hydrophobic solutes functionalized with tertiary amine(s), carboxyl-amine bonding drives the CNTs to the liquid-liquid interface, as demonstrated by a dramatic lowering of interfacial tension. With amine-terminated polystyrene (PS-NH2) in toluene, for example, the interfacial segregation of SWCNTs can drop the interfacial tension of the water-toluene interface from ~35 mN/m to ~10 mN/m, a change sufficient for the preparation of stable emulsions; the segregation to the interface was further verified by fluorescence microscopy of labeled SWCNTs. To understand the assembly better, interfacial tension for these systems was examined as a function of SWCNT and PS-NH2 concentration as well as PS-NH2 molecular weight and SWCNT degree of carboxylation. Unexpectedly, the interfacial tension fell sharply over an extremely narrow window of SWCNT concentration. Further study showed the same behavior when the SWCNT suspension was replaced by a poly(acrylic acid) solution. These trends suggest that pH of the aqueous phase is a controlling variable, the SWCNT segregation sensitive to the degree of protonation of the carboxyl groups. The sharp SWCNT concentration dependence is thereby a consequence of SWCNT impact on pH. Unfortunately, microscopic observation of dried emulsion droplets indicated that the segregated SWCNTs, while trapped in a film, do not have a preferential orientation relative to the interface.

(738d) Materials for Exploring New Electronic and Optical Behaviors in SWCNT Systems

Duque, J. G., Los Alamos National Lab
Hamilton, C., Los Alamos National Lab
Gupta, G., Los Alamos
Doorn, S., Los Alamos National Laboratory

Originally presented on: 11/7/2013 16:03:00 - 16:19:00

Single-walled carbon nanotubes (SWCNTs) have unique mechanical, electronic, and optical properties which make them an ideal material for optoelectronics, nanophotonics, and sensing applications. However, development of SWCNT mesomaterials for optoelectronics and nanophotonics has been especially challenging due to the high sensitivity of the optical and electronic properties to environmental interactions. I'll present an innovative approach developed at LANL by our team to obtain highly fluorescent SWCNT-mesocomposites. Here we report a general route to prepare silica gels, films, and monolithic nanocomposites doped with individual surfactant stabilized SWNTs with optimum photoluminescence (PL) properties. The approach provides a rationale for tailoring SWCNT photophysical response via interfacial chemistry. I will also show some results on sensing and energy transfer processes obtained using our mesomaterials.

These unique solid platforms of fluorescent SWCNTs will enable ultrasensitive optical sensors and optoelectronics devices as well as basic research in areas like low temperature exciton dynamics, exciton-phonon coupling, and energy transfer for photovoltaic applications.

(738e) Optimization of a Benign Supercritical Carbon Dioxide Treatment for Carbon Nanotube Nanocomposites

Quigley, J., Virginia Tech
Herrington, K., Virginia Tech
Qian, C., Virginia Tech
Gaddam, P., Virginia Tech
Baird, D., Virginia Tech
Originally presented on: 11/7/2013 16:19:00 - 16:35:00

In this work, a benign method of deagglomerating commercially available carbon nanotubes (CNT) using supercritical carbon dioxide ($scCO_2$) was optimized. Twin screw extrusion is the current method employed to disperse and reduce these agglomerates in polymer matricies, but results in significant destruction of the CNT in applications where aspect ratio retention is desired (up to 70% decrease in length). Thermodynamic principles were employed to discover that the optimal soak temperatures and pressures corresponded to a $scCO_2$ fluid density of ~750 kg/m^3 for the current apparatus. Soak time and CNT/$scCO_2$ ratio were also studied to acquire a more comprehensive understanding of the system. Transmission electron microscopy was utilized to determine the nanotube lengths both before and after $scCO_2$ treatment of Baytube C150 P and Nanocyl NC-7000. The best case expansions resulted in a decrease in CNT bulk density from 150 kg/m^3 (Baytube) and 50 kg/m^3 Nanocyl) to 5 kg/m^3 and 1 kg/m^3 respectively. It was discovered that the $scCO_2$treatment did not damage the nanotubes. Inclusion of the CNT into polymer may result in improved conductivity, mechanical properties, and reduced percolation threshold.

Acknowledgments: The authors would like to thank Bayer MaterialScience for funding this research. Also thanks to Dr. Erdogan Kiran for thoughtful advice in designing the experiments for pressure and temperature optimization.

(738f) Influence of the Surfactant On the Supramolecular Structure of the Single Walled Carbon Nanotube-Polyvinyl Pyrrolidone System: Consequences for Swnt Aqueous Dispersion

Yu, T.,
Davis, T.,
Herrera, J., The University of Western Ontario
Originally presented on: 11/7/2013 16:35:00 - 16:51:00

Influence of the Surfactant on the Supramolecular Structure of the Single Walled Carbon Nanotube-Polyvinyl pyrrolidone System: Consequences for SWNT Aqueous Dispersion.

Tennison Yu, Taylor J. Davis, and Jose E. Herrera

Department of Chemical and Biochemical Engineering, University of Western Ontario

London, Ontario, Canada, N6A 5B9

Obtaining stable aqueous dispersions is one of the main challenges hindering a widespread and effective use of single-walled carbon nanotubes (SWNT) in many applications. Although it has been recognized that their versatility makes them an extremely attractive material, the unique molecular structure that gives SWNTs their unmatched mechanical and electronic properties is also responsible for strong attractive forces between the nanotubes themselves. These are the result of van der Waals interactions, which are an inherent consequence of their carbon sp^2 hybridization network. This, combined with extremely high aspect ratios and flexibility, causes SWNTs to adhere strongly into tightly bundled ropes. In these bundles, SWNTs do not possess to nearly the same usefulness of their linearized unbundled equivalents. Thus, in order to take advantage of their mechanical strength effectively, SWNTs must be debundled and linearized[1]. Although several strategies have been suggested as a means to overcome these challenges, non-covalently coat nanotubes using amphiphilic molecules such as surfactants have gained significant attention in the recent years.[2]

In this contribution we will present our results exploring the use of an aqueous dispersant system containing cetyl trimethylammonium bromide (CTAB) and polyvinyl pyrrolidone (PVP). Although individually these molecules are poor dispersers of SWNTs, they show a synergic effect when combined. We have probed the reasons for this observation using a battery of spectroscopic and microscopic characterization techniques. Our data suggests that the observed synergistic effect is linked to the formation of stable supramolecular structures: "a 2-dimensional dispersion" as opposed to "one dimensional" dispersion systems, based on a single surfactant, traditionally used. We propose our approach as a novel alternative for nanotube aqueous dispersion, potentially allowing for an unlocking of SWNT's full capabilities.

[1] Biswas C, Kim KK, Geng HZ, Park HK, Lim SC, Chae SJ, et al. Strategy for high concentration nanodispersion of single-walled carbon nanotubes with diameter selectivity. The Journal of Physical Chemistry C. 2009;113(23):10044-51.

[2] Shin JY, Premkumar T, Geckeler KE. Dispersion of Single-Walled Carbon Nanotubes by Using Surfactants: Are the Type and Concentration Important? Chemistry-A European Journal. 2008;14(20):6044-8.

(738g) Controlling the Chemistry of Graphene

Hernandez, S., Naval Research Laboratory
Lock, E., Naval Reserach Laboratory
Bezares, F.,
Tsoi, S.,
Robinson, J.,
Caldwell, J.,
Stine, R.,
Tamanaha, C.,
Reinecke, T.,
Sheehan, P.,
Walton, S.,

Originally presented on: 11/7/2013 16:51:00 - 17:07:00

Graphene has attracted a wealth of technological interest due to its unique structural and electronic properties. Given its pure two dimensional nature, adsorbates have a strong impact on these properties and so chemical modification provides the opportunity to control the characteristics of graphene. Control over the spatial distribution of chemical moieties provides an even greater functionality in that the properties can be manipulated locally, offering a range of opportunities for subsequent surface procedures focused on band gap engineering, device fabrication, and sensor applications.

Global and spatially-controlled chemical functionalization of graphene using electron beam generated plasmas to produce well-controlled patterned chemistries and chemical gradients on the surface of graphene will be demonstrated. The resulting chemical, structural, and electrical properties of the functionalized graphene as they originate from oxygen, fluorine, and nitrogen functionalities will be shown. Finally, the utility of these surfaces in device and sensing applications will be discussed.

This work is supported by the Naval Research Laboratory Base Program.

(738h) Guided Assembly of Graphene Oxide in Aqueous Solution

Godfrin, M., Brown University
Guo, F., Brown University
Chakraborty, I., University of Rhode Island
Heeder, N., University of Rhode Island
Shukla, A., University of Rhode Island
Bose, A., University of Rhode Island
Hurt, R., Brown University
Tripathi, A., Brown University
Originally presented on: 11/7/2013 17:07:00 - 17:23:00

Graphene oxide is a chemically modified, hydrophilic form of graphene that can be processed in water to yield new carbon-based films, fibers, and folded particles. The atomic thinness of GO along with its micron-scale lateral dimensions make it a flexible sheet-like colloid with an extreme aspect ratio. Graphene oxide solutions have been observed to phase separate into ordered regions under shear when salt is introduced into the solution. We interpret this assembly process as the formation of vorticity bands based on a log-rolling mechanism. Here, we investigate multiple processing parameters that have an effect on the formation of the ordered regions. The characteristics of these regions have been found to be a function of GO concentration, salt valency and concentration and gap confinement; the control of these parameters allows for the manipulation of the characteristics of the regions of separated GO phase, a desirable ability in nanomanufacturing.

(738i) Surfactant Removal and Purification of Single-Wall Carbon Nanotubes

Soule, K., Rochester Institute of Technology
Rossi, J. E., Rochester Institute of Technology
Cox, N., Rochester Institute of Technology
Merrill, A., Rochester Institute of Technology
Mastrangelo, T. L., Rochester Institute of Technology
Cress, C., Naval Research Center
Landi, B. J., Rochester Institute of Technology
Originally presented on: 11/7/2013 17:23:00 - 17:39:00

Single-wall carbon nanotubes (SWCNTs) are attractive for a variety of applications including novel nanoelectronic devices, interconnects, advanced wires and cables, and solar cells. Until recently, the use of SWCNTs in such applications has been limited due to the presence of mixed chiralities, electronic-types, and diameters in as-produced materials, however, advances in post-synthesis processing has allowed for successful SWCNT separation. Utilizing processes such as column chromatography and density gradient ultracentrifugation (DGU), as-produced SWCNTs have been separated into their respective metallic and semiconducting species. These techniques rely on the use of surfactants, commonly sodium dodecyl sulfate (SDS) and sodium deoxycholate (DOC), to exploit differences in the surface chemistry or buoyant density of the SWCNTs, respectively. In an effort to fully probe the intrinsic physical properties of electronic-type-separated SWCNTs, complete surfactant removal must be accomplished.

It had previously been reported that filtration followed by copious water washing was sufficient to remove bulk surfactant from SWCNT materials, however, this did not address the residual surfactant adsorbed to the surface of the SWCNTs. More recently, acidification and thermal oxidation processes have been used to aid in further surfactant removal, however, a universal method for surfactant removal and determination of SWCNT purity is required. In the current study, surfactant removal was systematically examined by employing a variety of acids, bases, and organic solvents to 10 μg/mL SWCNT dispersions in either a 2 wt% solution of SDS or DOC. Upon mixing of the desired reagent, the SWCNT dispersions were bath sonicated, immersed in an ice bath, and observed over the course of 48 hours. The reagents that interrupted the SWCNT/surfactant interaction were selected to allow the SWCNTs to be easily filtered from the bulk solution. Based on a qualitative analysis of approximately 20 reagents, five reagents yielded optimal results, including acetone, acetonitrile, N,N-dimethylacetamide, cyclohexylpyrrolidine, and ethanol. The ratio of reagent:SWCNT dispersion was investigated to minimize the amount of solvent needed to precipitate the SWCNTs. Subsequently, the SWCNTs were filtered into bulk papers containing 2 mg of SWCNTs with an areal density of approximately 1 mg/cm^2. The SWCNT papers were rinsed with the appropriate solvent, thermally oxidized, and were characterized via thermogravimetric analysis (TGA), Raman spectroscopy, optical absorbance spectroscopy, and scanning electron microscopy (SEM). Additionally, electrical characterization was conducted using 4-point probe measurements and the van der Pauw method. The purity of the treated SWCNT samples were quantified in relation to control samples that were rinsed from the surfactant only with water, and to traditionally purified SWCNT papers, which had not been exposed to surfactants. Results demonstrating the successful removal of SDS and DOC from SWCNT dispersions will be discussed.

(739a) Separation of Small Diameter Single-Wall Carbon Nanotube Species Via Aqueous Two-Phase Extraction

Fagan, J. A., National Institute of Standards and Technology
Khripin, C., National Institute of Standards and Technology
Silvera Batista, C., National Institute of Standards and Technology
Hight Walker, A.,
Zheng, M., National Institute of Standards and Technology
Originally presented on: 11/7/2013 15:15:00 - 15:31:00

Aqueous two-phase extraction (ATPE) was recently demonstrated [1] to enable the separation of semiconducting from metallic species of single-wall carbon nanotubes (SWCNTs) as well as partitioning of several small diameter nanotubes species across the separated phases. In this presentation we will describe the use of a dextran – polyethylene glycol (PEG) aqueous two-phase system along with a separation scheme of varying surfactant concentrations to enable isolation of high purity specific small diameter SWCNT species. Separation by ATPE is rapid and robust, with a remarkable tunability that allows isolation of many single nanotube chiralities at high purity. The rapidity and robust open opportunity for scale-up and cost reductions beyond competitor technologies. Choice of surfactant(s), temperature, polymer concentrations, and the addition of small molecule salts can all be used to tune the exact partitioning of single SWCNT species between the two phases. Details of the separation, and properties of the separated fractions will be presented

[1] Khripin, C.Y.; Fagan, J.A.; Zheng, M. *J. Am. Chem. Soc.* **2013,** http://dx.doi.org/10.1021/ja402762e.

(739b) High-Fidelity Single Column Separation of Single Walled Carbon Nanotubes Using Agarose Gel

Clar, J. G., University of Florida
Bonzongo, J. C. J., University of Florida
Ziegler, K. J., University of Florida
Originally presented on: 11/7/2013 15:31:00 - 15:47:00

For their inclusion in novel electronic devices, the post synthesis separation of single walled carbon nanotubes (SWCNTs) is of great interest. While several methods have been utilized to achieve these separations with various levels in purity, the most promising technique to produce large-scale separations is the selective adsorption of SWCNTs onto an agarose gel stationary phase. Recently, several researchers have reported successful modification to the process in order to achieve further separation of the semiconducting fractions by (n,m) type. While these changes are a significant step in SWCNT processing, they are difficult for development on an industrial scale due to the use of harsh chemical treatments (pH adjustments) and temperature gradients that may be difficult to control. In addition, many of these approaches require multiple columns in series, causing substantial dilution of the collected fractions, which increases the amount of surfactant required and may require the fractions to be concentrated before use. Here we report a method for both type and chirality separations of sodium dodecyl sulfate (SDS) suspended SWCNTS at a single temperature. Using our understanding of the mechanism behind selective adsorption of SWCNTs onto agarose, we have developed high-fidelity separations based on a single elution profile in a single column. The results confirm that the surfactant orientation on the SWCNT surface is an important parameter that affects selective adsorption during separation. The design of a simple, single-column technique to separate SWCNTs by (n,m) type is a major step in realizing their full potential.

(739c) Point of Zero Charge of Single-Walled Carbon Nanotubes

Kanaan, S., Yale University
Ren, F., Yale University
Pfefferle, L., Yale University
Originally presented on: 11/7/2013 15:47:00 - 16:03:00

It has been shown through theoretical models that the point of zero charge (PZC) of single- walled carbon nanotubes (SWNT) changes with the size of the tubes. By modifying post-growth treatments, we show that different chiralities/diameters have different PZC and could allow us to separate bulk SWNT samples. In our lab, the SWNT are synthesized with a Co catalyst incorporated in an MCM41 matrix [1]. In the earlier work, highly dispersed size controllable Co clusters were produced by reduction treatment of Co-MCM41. Using the Co-MCM41 catalysts and performing the reaction at the different temperatures allows tuning over a range of different diameter/chiral tubes with small diameter distributions. The product SWNT has to be treated with harsh acids, a traditional method of functionalizing the surface of the SWNT for potential applications. The problem of this method is that the small diameter (<1nm) SWNTs are easy to be consumed in such a harsh conditions, therefore, this kind of post-treatment will not be favorable because it changes the distribution of the tubes we are trying to separate. We refine the method to functionalize the SWNT by using a more mild treatment that utilizes ozone concentrated water, according to fluorescence measurement, the small tubes survives and the distribution remains unchanged.

We have shown that the functionalization of SWNT using ozone gives comparable results as with nitric acid treatment. By modifying the functionalization treatment method, we can still maintain the same surface charge characteristics but maintain PZC characteristics. As shown in the figure below, the PZC of ozone treated tubes is actually lower than that of nitric acid treated and could be modified by annealing. This information allows for a less hazardous treatment and can be used in functionalizing SWNT surfaces.

[1] Loebick, C. Podila, R., Reppert, J., Chudow, J. Ren, F., Haller, G., Rao, A., Pfefferle, L. *J. Am. Chem. Soc.*, 7, 11127 (2010).

(739d) A Quantitative Theory of Adsorptive Separation for the Electronic Sorting of Single-Walled Carbon Nanotubes

Jain, R., Massachusetts Institute of Technology
Tvrdy, K., Massachusetts Institute of Technology
Han, R.,
Ulissi, Z., Massachusetts Institute of Technology
Strano, M. S., Massachusetts Institute of Technology
Originally presented on: 11/7/2013 16:03:00 - 16:19:00

High surface area adsorbents, including amide baring surfaces, allow for separation of single walled carbon nanotubes (SWNT) through selective adsorption and desorption of different *(n,m)* chiral indices. The mechanistic dependence on surfactant concentration and a full quantitative model of the elution order have not yet been developed. We develop a quantitative theory that describes the separation order and find kinetic rate constants of each SWNT chirality *(n,m)* based on the surfactant-induced, areal charge density, which ranges from 0.41 e^-/nm for *(7,3)* in 17mM sodium dodecyl sulfate (SDS) to 3.32 e^-/nm for *(6,5)* in 105mM SDS. Adsorption onto the amide baring support is balanced by short distance hard-surface and long distance electrostatic repulsive SWNT/substrate forces, the latter of which we postulate is strongly dependent on surfactant concentration and ultimately leads to gel-based single-chirality semiconducting SWNT separation. These molecular-scale properties are utilized to derive bulk-phase, forward adsorption rate constants for each SWNT chirality. The resulting theory quantitatively describes the experimental elution profiles of 15 unique SWNT chiralities as a function of anionic surfactant concentration between 17 and 105mM, as well as phenomenological observations of the impact of varying preparatory conditions such as extent of ultrasonication and ultracentrifugation. SWNT elution order and separation efficiency are primarily driven by the morphological change of SDS surfactant wrapping on the surface of the nanotube, mediated by SWNT chirality and the ionic strength of the surrounding medium. This work provides a foundational understanding for high purity, preparative scale separation of as-produced SWNT mixtures into isolated, single-chirality fractions.

(739e) Low-Temperature Transfer-Free Synthesis of Large-Area Unfolding Graphene Films By Electrospray On Super-Hydrophilic Glass

Wang, L. J., Nanoscale Science and Engineering Center for Affordable Nanoengineering of Polymeric Biomedical Devices
Li, L., The Ohio State University
Ouyang, X.,
He, H., The Ohio State University
Wu, Y., The Ohio State University
Yu, J.,
Lee, J., Ohio State University
Originally presented on: 11/7/2013 16:19:00 - 16:35:00

Low-temperature transfer-free synthesis of large-area unfolding graphene films by electrospray on super-hydrophilic glass

Li-Ju Wang, Lei Li, Xilian Ouyang, Hongyan He, Yun Wu, Jianfeng Yu, L. James L Lee*

Nanoscale Science and Engineering Center for Affordable Nanoengineering of Polymeric Biomedical Devices, the Ohio State University,

Columbus, Ohio 43210, United States

Abstract

Graphene as a new material is widely studied in optoelectronics, energy storage, composites and biomedical applications because graphene possesses a unique combination of extraordinary electrical, optical, thermal and mechanical properties and high specific surface area. However the synthesis of large area graphene film is a challenge work and greatly limiting the application of graphene. The most commonly used synthesis method of graphene films is via chemical vapor deposition (CVD) at elevated temperature (~1,000 °C) on polycrystalline metal surfaces and later transferred from metal support to desirable substrates. The high growth temperature imposes limitation on the choice of substrates and results in the sequential transfer process. During the transfer process, wrinkled or rippled graphene frequently forms cracks or tears.

Solution-based synthesis of graphene film, such as filtration method, is an alternative way used to overcome the high-temperature limitation and to avoid the transfer process. However, the soft sheet-like graphene tend to compress and aggregate together because of strong van der waals attraction. Once the graphene dispersions are dried, the sheet-like graphene not only aggregate to form coffee-ring but also forming three-dimensional (3D) crumpled ball-like structure. Therefore, fabrication of large-area thin graphene films is still a big challenge due to preventing aggregation (coffee ring) and making uniform unfolding films in a scalable and inexpensive method.

In this study, we present an effective and scalable transfer-free synthesis of large-area unfolding graphene thin films by electrospraying graphene sheets directly on the super-hydrophilic heated glass at relatively lower temperature (100-150°C). When the graphene nanosheets were directly electrospray bombarded onto the super-hydrophilic and preheated glass, the strong shear force unfolded the crumpled graphene and the fast water evaporating simultaneously applies capillary forces to spread and pin graphene sheets on glass surface. This fabrication method was able to avoid aggregation (coffee-ring) and to make the uniformly graphene film. Efficiently, for the large-area (4.4cm×4.4cm) graphene film deposition, only 1ml graphene solution (~0.4 mg/ml) is required to fabricate transparent glass with 80% transparency and ~2kΩ/sq resistance. This low-temperature transfer-free approach enables us to fabricate large-area unfolding graphene coatings with few amount of solution and this method has great potential for large scale and low cost applications.

(739f) Strain-Sensing Application of Natural Rubber Composites Reinforced With Different Carbon Nano-Fillers

He, Q., Lamar University
Wei, S., Lamar University
Guo, Z., Lamar University
Originally presented on: 11/7/2013 16:35:00 - 16:51:00

Carbon nano-fillers including carbon nanotubes and carbon nanofibers reinforced natural rubber/carbon black composites were fabricated through a dissolution-mixing approach. The as-prepared natural rubber composites were utilized for strain-sensing application through monitoring the dynamic relationship between the electrical conductivity and the room temperature mechanical properties such as elongation / compression stress. Specifically, a real time correlation has been demonstrated through simultaneously recording the changes of electrical conductivity and stress upon repeated elongation / compression cycles. The morphology of different carbon nanofillers in the natural rubber matrix was studied by scanning electron microscopy, and thermal properties were investigated through thermal gravimetric analysis and differential scanning calorimetry. Both two fillers were observed to increase the electrical conductivity, storage and loss modulus of their reinforced natural rubber/carbon black composites. Significant difference in the electrical conductivity - strain sensing performance through using different carbon nanofillers was observed and the mechanism was also discussed.

(739g) Thermodynamic Theory for Performance of Graphene Oxide Based Pickering Emulsions

Creighton, M. A., Brown University
Ohata, Y.,
Chakraborty, I., University of Rhode Island
Miyawaki, J.,
Bose, A., University of Rhode Island
Hurt, R. H., Brown University
Originally presented on: 11/7/2013 16:51:00 - 17:07:00

Fine particles can assemble at aqueous-organic interfaces and stabilize droplets to form what is known as a Pickering emulsion. Graphene-based nanomaterials are of potential interest for this application because a large fraction of their atoms reside at the interface. This allows the stabilization of emulsions at very low mass doses relative to conventional isometric particles. For atomically thin sheets such as monolayer graphene oxide, every atom lies on the interface and interacts with both liquid phases, and stabilization energies can be orders of magnitude higher than spherical nanoparticles at equal mass loading. Experiments show that graphene oxide is an effective stabilizer for a range of organic-in-water emulsions, but few-layer graphene with pristine (hydrophobic) surfaces is much less effective. This trend is consistent with the thermodynamic theory, which provides a basis for optimizing the surface chemistry of a graphene-based material to achieve the highest free energy of stabilization. Other potential advantages of graphene-based emulsion stabilizers include tunable surface chemistry, elastic conformal coverage of curved surfaces, and unfavorable detachment from the interface.

(739h) Properties of Bismaleimides Composites With Hyperbranched Polysilane Grafted Graphene

Yan, H., School of Science, Northwestern Polytechnical University, 127# Youyi West, Shaanxi, Xi'an, 710072, China
Feng, Y., School of Science, Northwestern Polytechnical University
Zhang, J., School of Science, Northwestern Polytechnical University
Li, T., School of Science, Northwestern Polytechnical University
Originally presented on: 11/7/2013 17:07:00 - 17:23:00

Properties of Bismaleimides Composites with Hyperbranched Polysilane Grafted Graphene

Hongxia Yan[1],Yichen Feng, Junping Zhang, Tingting Li

Department of Applied Chemistry, School of Science, Northwestern Polytechnical University, Xi'An, 710072, China

Abstract: Graphene is a kind of novel two-dimensional nanocomposite. the special single atomic layer structure gives it distinctive physical properties, such as excellent mechanical properties, good electrical and thermal conductivity properties, outstanding tribological property and so on. BMI resin is a high-performance material, which has been widely used in aerospace and electronic industries due to its outstanding thermal, mechanical, electrical, and tribological properties. However, pure BMI resin is brittle and has poor processing characteristic. Therefore, in this paper, graphene were used as filler to modify bismaleimide (BMI) which pre-polymer containing allyl phenyl ether and a low viscosity. To solve the problem of graphene's dispersion in BMI, graphene oxide are grafted by hyperbranched polysilane (HBPSi-GO), and The changes in GO surface morphology and chemistry conditions are characterized by Transmission electron microscopy (TEM) and X-ray photoelectron spectroscopy (XPS). The HBPSi-G/BMI composites are prepared by module casting method. The effect of content of HBPSi-GO on the mechanical properties and tribological properties of the composites are investigated, while the wearing mechanisms of the materials are also researched by SEM. The results show that, the suitable addition of HBPSi-GO can largely enhance both the mechanical properties and tribological properties of BMI resin. The severe wear of BMI resin is also converted from fatigued wear to adhesive wear with the addition of HBPSi-GO.

Key words: Graphene; Bismaleimide; Hyperbranched; polysilane; tribology; wear

---

Corresponding author: Tel.: +86 2988431657; fax: +86 29 88431655.

E-mail: hongxiayan@nwpu.edu.cn

This work was financially supported by the Aerospace Science Foundation of China (2011ZF53064)

(739i) Functionalized Carbon Nanotube Nanocomposite Membranes for Water Desalination: Experimental Study

Chan, W. F., Virginia Polytechnic Institute and State University
Chen, H. Y., University of Pittsburgh
Marand, E., Virginia Tech
Johnson, J. K., University of Pittsburgh
Originally presented on: 11/7/2013 17:23:00 - 17:39:00

We have fabricated a novel nanocomposite membrane for water desalination application by combining functionalized single-walled carbon nanotubes (SWNTs) with traditional thin film polyamide (PA). We then characterized this type of membrane using reverse osmosis technique with different salt/de-ionized water mixtures as the feed solutions in order to investigate the water permeation flux and salt rejection rate. The SWNTs were functionalized with zwitterionic functional groups (Z-SWNTs), which act as molecular gatekeepers at the entrance of SWNTs to enhance the blockage for salt ions. The nanocomposite membrane was synthesized with 26wt% of Z-SWNTs embedded in PA skin layer introduced by interfacial polymerization. Vacuum filtration was applied before the polymerization to align the SWNTs by flow-induced orientation. Cross-sectional image of the nanocomposite membrane taken by scanning electron microscopy (SEM) showed semi-aligned Z-SWNTs on top of a porous support covered by a thin PA film with an overall thickness of approximately 250nm. The nanocomposite membranes had significant improvement in permeation flux when the embedded Z-SWNTs content increased from 0 to 26wt%. There was almost three-fold increase in water flux and a slight increase in salt rejection for sodium ion. Experimental study also showed that Z-SWNTs/PA membrane had higher salt rejection rate when the cation concentration in feed solution increased from 50 to 2000ppm. This result demonstrated that separation mechanism is dominated by size exclusion over Donnan charge exclusion due to the absence of electrostatic screening effect from increasing solution ionic strength. Z-SWNTs/PA membrane also experienced higher salt rejection rate towards cations with larger outer hydration shell. For example, the rejection for potassium cation (K^+), whose hydrated radius is 3.31 Å, was approximately 5% lower than the rejection rates for sodium, Na^+, whose hydrated radius is 3.58 Å and magnesium, Mg^{2+},with a hydration radius is 4.38 Å . These results confirm that the separation mechanism of the functionalized carbon nanotubes is dominated by size exclusion rather than by charge exclusion. These results furthermore agree with our simulation study in which zwitterions offer steric hindrance at the tip of SWNTs to separate salt ions, rather than electrostatic repulsion.

(750a) Hydrogen Production From Methanol and Formic Acid Reactions Over Atomically-Dispersed Precious Metal (Au and Pt) Catalysts

Yi, N., Tufts University
Saltsburg, H., Tufts University
Flytzani-Stephanopoulos, M., Tufts University
Originally presented on: 11/7/2013 15:15:00 - 15:35:00

Steam reforming of methanol (SRM), which produces the highest amount of hydrogen among all hydrogen-producing reactions using methanol as reactant, is one of the significant catalytic processes to produce hydrogen in different scales [1]. But methanol is only one option among possible liquid fuel carriers for a hydrogen-based fuel cell. As is the case with introducing any novel industrial application in competitive markets, other possible alternatives, such as formic acid, must be fully considered [2]. From the fundamental viewpoint, it is interesting to compare these two liquids in terms of the pathways involved and the selectivity of different catalysts to hydrogen. Moreover, if precious metals are to be used as catalysts, it is highly desirable to explore efficient approaches that decrease the amount of precious metals used while maintaining high activity at close-to- ambient temperatures for electrocatalytic applications. Here we investigated the potential of two candidate catalysts: dispersed sub-nm gold on ceria and sodium-promoted platinum on silica for catalytic hydrogen production through steam reforming of methanol and decomposition of formic acid reactions[3,4,5].

1. Palo, R. M., Dagle, R. A., Holladay J. D. *Chem. Rev.*107, 3992 (2007).

2. Yu, X., Pickup, P. G. *J. Power Sources*182, 124 (2008).

3. Yi, N., Si, R., Saltsburg, H., Flytzani-Stephanopoulos, M. *Appl.Catal. B* 109, 87 (2010).

4. Yi, N., Si, R., Saltsburg, H., Flytzani-Stephanopoulos, M. *Energy Environ. Sci.* 3, 831 (2010).

5. Yi, N., Saltsburg, H., Flytzani-Stephanopoulos, M. *ChemSusChem* 6, 816 (2013).

(750b) Structure Sensitivity in Pt Nanoparticle Catalysts for Hydrogenation of 1,3-Butadiene: In Situ Study of Reaction Intermediates Using SFG Vibrational Spectroscopy

Michalak, W. D., Lawrence Berkeley National Laboratory
Krier, J., University of California
Komvopoulos, K., University of California
Somorjai, G. A., University of California, Berkeley
Originally presented on: 11/7/2013 15:35:00 - 15:55:00

The product selectivity during 1,3-butadiene hydrogenation on monodisperse, colloidally synthesized, Pt nanoparticles was studied under reaction conditions with kinetic measurements and *in situ* sum frequency generation (SFG) vibrational spectroscopy. Four products are formed at 75 °C: 1-butene, cis-2-butene, trans-2-butene, and n-butane. Ensembles of Pt nanoparticles with average diameters of 0.9 and 1.8 nm exhibit a ~30% and ~20% increase in the full hydrogenation products, respectively, as compared to Pt nanoparticles with average diameters of 4.6 and 6.7 nm. Methyl and methylene vibrational stretches of reaction intermediates observed under working conditions using SFG were used to correlate the stable reaction intermediates with the product distribution. Kinetic and SFG results correlate with previous DFT predictions for two parallel reaction pathways of 1,3-butadiene hydrogenation. Hydrogenation of 1,3-butadiene can initiate with H-addition at internal or terminal carbons leading to the formation of 1-buten-4-yl radical (metallocycle) and 2-buten-1-yl radical intermediates, respectively. Small (0.9 and 1.8 nm) nanoparticles exhibited vibrational resonances originating from both intermediates, while the large (4.6 and 6.7 nm) particles exhibited vibrational resonances originating predominately from the 2-buten-1-yl radical. This suggests each reaction pathway competes for partial and full hydrogenation and the nanoparticle size affects the kinetic preference for the two pathways. The reaction pathway through the metallocycle intermediate on the small nanoparticles is likely due to the presence of low-coordinated sites.

(750c) Design of Catalysts At the Subnanometer to Nanometer Scale: Tuning Performance Via Size, Composition, Doping, Support and Assembly

Vajda, S., Argonne National Laboratory

Originally presented on: 11/7/2013 15:55:00 - 16:15:00

The elucidation of the size/composition/shape/structure and function relationship, the effect of support along with the determination and control of the nature of the catalytic particles under reaction conditions are instrumental for addressing fundamental aspects of catalysis on the way to the design of new classes of catalytic materials. Highly uniform particles on technologically relevant supports are prerequisites for such studies.

Our experimental studies are based on 1) fabrication of technologically relevant supports, 2) physical size- and composition selected cluster deposition with atomic precision control, 3) *ex situ* and *in situ* microscopies and 4) *in situ* synchrotron X-ray characterization of cluster size/shape and oxidation state under realistic working conditions, combined with mass spectroscopy analysis of the reaction products. The experimental studies are complemented with DFT calculations. Using model size-selected clusters, this contribution will outline the applicability of the outlined approach on example of selective C-H bond activation in the oxidative and non-oxidative dehydrogenation of cyclohexane and cyclohexene. Select examples will be presented on coupling studies of model and practical catalysts.

Oxidative and non-oxidative dehydrogenation of cyclohexane on subnanometer Co_4O_x and $Co_{27}O_x$ clusters and nanometer size Co_3O_4 and $(Au)Co_3O_4$particles. In the absence of oxygen, the clusters became active around 300 °C, with the larger clusters being about 2x as active as the smaller ones. The clusters were active already around 100 °C – with somewhat different selectivities for the two cluster sizes. The performance of the subnanometer Co_nO_x clusters will be compared with that of various size nanometer Co_3O_4 particles prepared by wet-chemical methods - bare as well as Au-doped. Notably, the activity of cobalt clusters under oxidative conditions makes them as potential candidates to replace Pt-based dehydrogenation catalysts, while the spectrum of investigated cobalt oxide clusters is also offering a tunable selectivity and activity with particle size, doping and oxidation state.

Dehydrogenation of cyclohexene on subnanometer $Co_{27}O_x$clusters and their nanoscale assemblies: Increased activity through assembly under reaction conditions and the effect of the support and reaction conditions on catalyst performance. 27-atom cobalt clusters were deposited on TiO_2, ZnO, Al_2O_3, MgO and ultrananocrystalline diamond (UNCD). While the activity and selectivity of TiO_2, ZnO and Al_2O_3 supported clusters were comparable, MgO-supported clusters were 2 to 3 times more active under oxygen rich conditions. *In situ* data reveal that a formation of a fluxional ~2-3 nm structure of a $Co_xMg_yO_z$ composition is responsible for the dramatic increase in activity. Under oxygen lean conditions, UNCD-supported clusters possessed the highest activity as well as highest toward selectivity towards dehydrogetanted C6 products. Along with the dehydrogenation of cyclohexane, these results indicate extraordinary propensities of sub-nm Co_nO_x clusters in low-temperature activation of O_2.

References

1. "Oxidative Dehydrogenation of Cyclohexane on Cobalt Oxide (Co_3O_4) NanoparticlesThe Effect of Particle Size on Activity and Selectivity", by E. C. Tyo, C. Yin, M. Di Vece, Q. Qian, S. Lee, B. Lee, S. Seifert, R. E. Winans, R. Si, B. Ricks, S. Goergen, M. Rutter, B. Zugic, M. Flytzani-Stephanopoulos, Z. Wang, R. E. Palmer, M. Neurock, and S. Vajda, *ACS Catal.* 2, p. 2409–2423 (2012)

2. "Support-Dependent Performance of Size-Selected Subnanometer Cobalt Cluster-Based Catalysts in the Dehydrogenation of Cyclohexene", S. Lee, M. Di Vece, B. Lee, S. Seifert, R. E. Winans and S. Vajda, *Chem. Cat. Chem.* 4, p. 1632-1637 (2012)

3. "Oxidative Dehydrogenation of Cyclohexene on Size Selected Subnanometer Cobalt Clusters: Improved Catalytic Performance via Evolution of Cluster-Assembled Nanostructures", S. Lee, M. Di Vece, B. Lee, S. Seifert, R. E. Winans and S. Vajda, *Phys. Chem. Chem. Phys.*, **14**, p. 9336 - 9342 (2012)

4. L. Curtiss, J. Greeley, M. Flytzani-Stephanopoulos, G. Haller, S. Lee, M. Neurock, R. Palmer, L. Pfefferle, Q. Qian, R. Schlögl, D. Teschner, E. Tyo, S. Vajda, R. Winans et al, papers in preparation

5. "Subnanometre Platinum Clusters as Highly Active and Selective Catalysts for the Oxidative Dehydrogenation of Propane", S. Vajda, M. J. Pellin, J. P. Greeley, C. L. Marshall, L. A. Curtiss, G. A. Ballentine, J. W. Elam, S. Catillon-Mucherie, P. C. Redfern, F. Mehmood and P. Zapol, *Nat. Mater.* 8, 213-216 (2009)

(750d) Morphology Dynamics of Precious Metal Catalysts for Use in Steam Reformation of Oxygenated Fuels

Crowley, S., The City College of New York
Castaldi, M. J., City College of New York
Originally presented on: 11/7/2013 16:15:00 - 16:35:00

The advancement in liquid fuel production from biological sources will result in oxygenated fuel compositions. While oxygen is beneficial for combustion purposes, its presence within the chemical makeup of the fuel is detrimental to the fuel's energy density. In understanding how oxygenated fuel may be reformed to remove the oxygen functional groups, we have been exploring ethanol reforming as a test case. Recently, initial results show for the first time that the oxidation state of catalytic Rh nanoparticles changes dynamically under reaction conditions with ethanol and that the extent of Rh oxidation appears to control catalyst activity and the rate of deactivation. Catalyst deactivation is one of the major challenges faced in fuel reforming and there exists a need not only to determine the primary causes, but model them in order to take preventative measures. Through the use of common analytical techniques such as CO chemisorption and X-ray Diffraction (XRD), the surface morphology of the catalysts has been examined. Each of three formulations was calcined at temperatures ranging from 550°C to 950°C. In addition, X-ray Diffraction (XRD) has been performed in order to detect any conformational changes in the crystal structure of the catalysts as a function of calcination temperature. The extent of sintering of the precious metal catalysts as well as any changes in oxidation state, catalyst poisons adsorbed to the catalyst, and burrowing of the precious metals into the support are suspected to result in lower catalytic activity. This presentation will correlate and explain the active metal distribution and catalytic activity as well as the time dependence of the distribution during the reaction. In this way, we can create a complete profile of the molecular composition of the catalyst surface, which could lead to better overall performance. It is anticipated that these results will provide an important basis for DFT and mechanistic modeling to further elucidate the underpinnings of the observed performance.

(750e) Morphological, Electronic, and Catalytic Properties of Graphene-Supported Pt Nanoclusters

Fampiou, I., University of Massachusetts Amherst
Ramasubramaniam, A., University of Massachusetts Amherst
Originally presented on: 11/7/2013 16:35:00 - 16:55:00

Graphene has received considerable recent attention as a support material for fuel-cell electrodes due to its high surface area and excellent electrical conductivity. Experiments indicate that graphene-supported Pt nanoparticles exhibit greater stability and better electrocatalytic activity as compared to their counterparts on traditional carbon supports, or even bulk Pt electrodes.

In this study we use a combination of empirical-potential molecular-dynamics calculations and first-principles density functional theory calculations to investigate the structural, electronic, and catalytic properties of Pt nanoclusters on graphene supports. We present detailed thermodynamic and electronic structure analyses of Pt clusters on both defective and defect-free graphene substrates and elucidate the role of support point defects in altering the electronic structure of the supported clusters. Defects in graphene supports are found to enhance the metal-support interaction thereby increasing stability towards catalyst sintering. The strong Pt-C interactions are also accompanied by significant charge redistribution between the cluster and the substrate, which alters the catalytic properties of the cluster.

As a model reaction, we investigate CO adsorption and oxidation on graphene-supported Pt clusters. We show that clusters bound to defects in graphene could potentially exhibit increased tolerance toward CO poisoning. We examine the activity of the graphene-supported Pt clusters for the CO oxidation reaction and obtain estimates for CO-oxidation kinetics. Reaction kinetics on supported Pt nanoclusters are comparable with bulk Pt and, on average, faster than on unsupported clusters.

Overall, our results suggest promising avenues for controlling the dispersion and catalytic activity of Pt nanoclusters on graphene supports via defect engineering.

(750f) Nanoparticle SAMs As Molecular Rulers: Two Cu(I) Catalytic Centers Are Required for 'Click' Catalysis

Walker, D. A., Northwestern University
Browne, K. P., Northwestern University
Grzybowski, B., Northwestern University
Originally presented on: 11/7/2013 16:55:00 - 17:15:00

The ‘click’ reaction has become an exceedingly popular coupling reaction in organic chemistry, but the mechanism by which the reaction operates remains unknown. It has been suggested that two Cu(I) catalytic centers are required for the mechanism, but to date there is no experimental evidence for this assumption. Here we employ on particle mixed self-assembled monolayers (mSAMs) to control the relative spacing between catalytic centers. This technique allows us to (i) experimentally prove that two catalytic centers are necessary for the ‘click’ reaction to occur and (ii) measure the critical separation at which a catalytically active pair of Cu(I) ions become inactive. This technique, beyond providing useful insights into the mechanism of the ‘click’ reaction, develops a platform for which other catalytic diads and triads can be investigated. A mixture of experimental and theoretical results will be resented.

(750g) Ni-Based Catalysts Synthesized By Atomic Layer Deposition for Dry Reforming of Methane

Gould, T. D., University of Colorado at Boulder
Lubers, A. M., University of Colorado at Boulder
Medlin, J. W., University of Colorado Boulder
Weimer, A. W., University of Colorado at Boulder
Falconer, J. L., University of Colorado
Originally presented on: 11/7/2013 17:15:00 - 17:35:00

Supported Ni catalysts with particle diameters of approximately 2.5 nm were synthesized by atomic layer deposition (ALD). These catalysts were evaluated for dry reforming of methane activity and stability. The Ni ALD catalysts exhibited a higher activity per gram of metal than catalysts prepared by incipient wetness, and performed slightly better than Ni catalysts in the literature that were promoted with Mg or Pt. Additionally, the Ni ALD catalysts exhibited higher long-term activity than other catalysts for up to 72 h at 600°C. When the Ni ALD catalysts were modified with Pt ALD, their activity increased, sintering decreased, and resistance to coking was enhanced. Modifying the incipient wetness Ni catalysts with 1 cycle of Pt ALD increased their activity per gram of metal by more than a factor of 2. Adding 1 cycle of Pt ALD to the Ni ALD catalysts doubled the activity per gram of metal of the Ni ALD catalyst, resulting in the highest activity per gram of metal of all the samples investigated. Additionally, modifying the Ni ALD catalyst with Pt ALD decreased the loss of activity due to sintering. The Ni ALD catalysts deactivated by ~50% of their initial activity due to sintering within the first 6 h, but the Pt ALD-modified catalysts only deactivated by 2% over 6 h at 600°C. The nature of bimetallic interaction was probed via CO-TPD and temperature-programmed reduction. Additionally, the structure of the as-prepared samples and heat treated samples were analyzed by HRTEM.

(751a) Molecular Mobility in Ultrathin Polymer Films

Priestley, R., Princeton University

Originally presented on: 11/7/2013 15:15:00 - 15:45:00

Central to the advancement of many technologies is the miniaturization of functional devices to the nanometer length scale. As polymers continue to play a prominent role in material solutions in meeting the challenges of reducing size, they are increasingly being utilized at length scales that are approaching the dimensions of the unperturbed macromolecule. Furthermore, with decreasing the confining dimension to the nanoscale an increasingly larger fraction of molecules are in direct contact with solid surfaces. As a consequence, in ultrathin polymer films, molecular mobility can differ significantly from the bulk. Here we present a non-contact method to measure the viscoelastic properties of supported thin polymer films. The approach is based on the blow-off method originally developed to measure viscosity in monolayer films of molecular liquids. The method entails flowing gas under laminar flow conditions through a narrow channel in order to generate a well-define shear stress on a properly placed sample and subsequently measuring the deformation. Our design is modernized in numerous ways, including, the ability to simultaneously control the gas and substrate temperature. In combination with hole- and edge- dewetting experiments, the new blow-off features will allow for the measurement of time- and temperature-dependent viscoelastic properties of thin polymer films.

(751b) Nanoconfined Polymerization Reaction Kinetics and Thermodynamics

Zhao, H., Texas Tech
Begum, F., Texas Tech University
Gao, S., Texas Tech Univeristy
Simon, S. L., Texas Tech University
Originally presented on: 11/7/2013 15:45:00 - 16:05:00

The behavior of materials confined at the nanoscale has been of considerable interest over the past several decades. Recent work in our laboratory has focused on the influence of nanoconfinement on polymerization kinetics, as well as the effects of synthesis under nanoconfinement on polymer properties. In the case of step-growth polymerization of cyanate esters and free radical polymerization of methyl methacrylate, we find enhanced reactivity with decreasing nanopore size. On the other hand, in the ring-opening polymerization of dicyclopentadiene, we find reduced reactivity and the presence of a side reaction not present in the bulk system. In sulfur polymerization, the ring/chain equilibrium has also been found by other researchers to shift towards less polymer in confinement. All of the results can generally be explained by competition between changes in local packing and diffusivity under confinement coupled with entropic and surface effects. In this presentation, we detail the ability to describe nanoconfined reaction kientics using models that incorporate changes in diffusivity and entropy upon nanoconfinement using scaling from the literature.

(751c) Glassy Behavior of Polymers Confined in Unique Geometries

Zhang, C., Princeton University
Priestley, R., Princeton University
Originally presented on: 11/7/2013 16:05:00 - 16:25:00

Polymers confined to the nanometer length-scale are important in applications ranging from active layers in photovoltaic cells to membranes in separation technologies to vehicles in drug delivery. As shown extensively in the past, the glass transition temperature (T_g) (and its associated dynamics) of confined polymers can deviate substantially from the bulk. Due to ease of processing, most of these studies have focused on the 1-dimensional thin film geometry, while few have extended investigations to unique geometries and to higher dimensionalities. Here, we utilize a 3-dimensional nanoparticle geometry to explore confinement-induced effects on the T_g and the dynamic fragility of polymers. In the case of bare polystyrene (PS) and poly(4-methylstyrene) (P4MS) nanoparticles, the T_g is observed to decrease systematically with decreasing nanoparticle diameter. On the contrary, by capping the nanoparticles with a rigid silica shell, *i.e.*, a polymer/silica core-shell geometry, the T_g remains invariant as a function of diameter. Furthermore, using variable cooling rate differential scanning calorimetry, we measure the isochoric fragility (m_v) of silica-capped PS and P4MS nanoparticles, which decreases significantly with confinement. In the case of P4MS, we show that the ratio of the isobaric fragility to the isochoric fragility, *i.e.*, m_p/m_v, remains constant with diameter, indicating that the relative effects of thermal activation and volume contribution on the dynamics near the glass transition remain unchanged under confinement. Lastly, we utilize a novel confining geometry, *i.e.*, thin polymer films coated onto silica nanoparticles, to extend our understanding of confinement effects on the T_g and the dynamic fragility. We explain the observed T_g-confinement and fragility-confinement results in terms of interfacial effects perturbing glass transition dynamics.

(751d) Self-Assembled Protein Structures Are Altered By Underlying Fluctuations

Cordella, N., Stanford University
Lampo, T., Stanford University
Heilshorn, S. C., Stanford University
Spakowitz, A. J., Stanford University
Originally presented on: 11/7/2013 16:25:00 - 16:45:00

Clathrin proteins stabilize the budding and curvature of cell membranes that are characteristic of endocytosis, an essential process by which cells internalize nutrients and other cargo. Clathrin's robust ability to self-assemble *in vitro* into a variety of structures depending on experimental conditions also makes it a promising tool for templating inorganic nanostructures. Observations of clathrin aggregation on membranes *in vivo* show the formation of two distinct structures, dependent on membrane conditions [Saffarian *et al.*, *PLoS Biology*, 2009, **7**, e1000191]. Specifically, large, flat "plaques" are observed predominantly where the underlying membrane is adhered to a rigid substrate, while short-lived, curved "pits" are found on all membrane surfaces. These findings indicate that cellular transport may be modulated by subtle physical changes in the mechanical properties of cell membranes, and have important implications for systems self-assembling on flexible substrates in general. We develop a simplified physical model of clathrin proteins coupled to a membrane. This model is used in Monte Carlo simulations and analytical theory to show the effect of membrane fluctuations on the phase behavior of clathrin lattices. Crystalline, plaque-like lattices are stabilized when membrane tension is much higher than its physiological value, thereby suppressing thermal membrane fluctuations. Inversely, large membrane fluctuations give rise to disordered, fluid clathrin phases. Our observations are quantified through correlation functions of lattice order. We also predict the response of clathrin assemblies to indentations of the membrane similar to those felt during endocytosis. Fluid lattices, brought about by membrane fluctuations at low tension, are shown to be more capable of coating these indentations than crystalline lattices, which typically result in tearing of the lattice and generation of void spaces localized to the indentation. These findings illustrate how the mechanical properties of membranes dictate the nature of their associated protein aggregates, thereby explaining experimental observations regarding disparate clathrin phases on different cell surfaces. This useful insight on the role that physical properties of the surrounding environment have in self-assembly on surfaces can result in better understanding of biological processes and guidance of bio-templated nanotechnology.

(751e) Tuning the Domain Size of Block Copolymers for Directed Self-Assembly Using Polymer Blending

Lawson, R. A., Georgia Institute of Technology
Peters, A., Georgia Institute of Technology
Ludovice, P. J., Georgia Institute of Technology
Henderson, C. L., Georgia Institute of Technology
Originally presented on: 11/7/2013 16:45:00 - 17:05:00

Implementation of block copolymers (BCPs) in a directed self-assembly (DSA) scheme for industrial lithography introduces a series of scientific and engineering challenges that have not been significantly addressed in previous block copolymer phase separation studies. One fundamental engineering challenge that underlies the entire DSA field is the development of methods for producing large scale quantities of BCPs with low, controlled polydispersity and well controlled total and individual block molecular weights. Since the domain size D of a BCP is proportional to the degree of polymerization N, the pattern size and pitch of DSA patterns are strongly dependent on BCP molecular weight. In current optical photoresists, molecular weight variation in polymer batches mainly affects photospeed and is typically overcome using blending of polymer resins from different batches. Generally, this polymer blending in optical resists has relatively little effect on other resist performance characteristics such as CD control or line edge roughness (LER). In contrast, current DSA efforts are driving for the synthesis and use of extremely low polydispersity (i.e. PDI<1.1) polymers. This perceived need and demand for low PDI polymers for DSA processes is in direct conflict with such polymer blending strategies that could potentially be used mitigate polymer batch-to-batch variation and achieve tight matching of CD and pitch targets for DSA materials. Synthetic control of polymer molecular weight is still very challenging, and molecular weight variations of up to 10% are not uncommon in current DSA polymer synthesis. For reference, a 10% variation in molecular weight for a low PDI block copolymer used for DSA lithography would result in more than a 6% change in domain size. Therefore, there is a strong motivation and need to develop and understand methods for fine-tuning the domain size of DSA BCPs to meet the CD and pitch specifications that will be required for practical implementation of DSA processes. This study investigates two methods of fine-tuning the domain size of a specific batch of BCP through blending of the BCP with another polymer, either 1. a set of similar homopolymers (HPs) or 2. a similar BCP of different molecular weight. Each method was investigated and compared using a coarse grained molecular dynamics simulation. For BCP-HP blends, the domain size increases as the amount of HP increases because the HPs tend to slightly swell the BCP domains. For blends of different molecular weight BCPs, two different scaling regimes were identified; one regime is majority large chains and the other regime is majority small chains. Using the simulation results, design heuristics were developed for both methods of blending which should guide experimental attempts to fine tune domain size.

(751f) Annealing Polymer Nanocomposite Fibers and Films Via Photothermal Heating: Effects On Overall Crystallinity and Spherulite Density

Viswanath, V.,
Maity, S.,
Bochinski, J., North Carolina State University
Clarke, L. I., North Carolina State University
Gorga, R. E., North Carolina State University
Originally presented on: 11/7/2013 17:05:00 - 17:25:00

Metal nanoparticles embedded within polymeric systems can be made to act as localized heat sources thereby aiding in-situ polymer processing. This is made possible by the surface plasmon resonance (SPR) mediated photothermal effect of gold nanoparticles, wherein incident light absorbed by the nanoparticle generates a non-equilibrium electron distribution which subsequently transfers this energy into the surrounding medium, resulting in a temperature increase in the immediate region around the particle. The current research demonstrates this effect in polymer nanocomposite systems, electrospun nanofiber mats and thin films, which have been annealed at temperatures above the glass transition and below melting. A non-contact temperature measurement technique utilizing embedded fluorophores has been used to monitor the average temperature within samples. The effect of annealing methods (conventional and plasmonic), annealing conditions (temperature and duration) and cooling mechanisms on the morphology, crystallinity, and mechanical properties of polymeric nanocomposite systems will be discussed. The specificity of plasmonic heating coupled with the inside-outside approach of annealing presents a unique tool to thermally process polymers.

(751g) Molecular Simulation Studies of Polydispersity Effects On The Morphology of Polymer Nanocomposites

Martin, T. B., Clarkson University
Jayaraman, A., University of Colorado at Boulder
Originally presented on: 11/7/2013 17:25:00 - 17:45:00

In this talk we will present molecular simulation studies elucidating polydispersity effects in polymer grafted nanoparticle based polymer nanocomposites. Past literature has shown that, in polymer nanocomposites containing monodisperse polymer (brush) grafted particles in a monodisperse polymer matrix, when the graft and matrix polymers are chemically similar, the grafted particles are dispersed (aggregated) in the polymer matrix when the grafted polymer chains are much longer (shorter) than the matrix polymer chains. We have shown in recent work [1,2] that polydisperse grafted polymers can stabilize dispersions of polymer grafted nanoparticles in a polymer matrix even if the average grafted polymer chain length is less than the matrix chain length. This is because the polydispersity in grafted polymers relieves monomer crowding in the grafted polymer layer on the particle, and improves wetting of the grafted layer by the matrix chains. In this talk, we will present our recent work extending the understanding of polydispersity effects in nanocomposite by focusing on polydispersity in matrix chains. In the case of a bidisperse matrix, consisting of short and long chains, with monodisperse polymer grafted particles, we find that the grafted polymer layer is preferentially wet by the short matrix chains, even when the short matrix chains are longer or higher in molecular weight than the graft polymer chains. These results suggest that polydispersity can be a useful design parameter to improve filler dispersion in polymer nanocomposites.

[1] Martin, T. B.; Dodd, P. M.; Jayaraman, A.; Polydispersity for Tuning the Potential of Mean Force between Polymer Grafted Nanoparticles in a Polymer Matrix. Phys. Rev. Lett. 2013, 110 (1), 018301.

[2] Martin, T. B.; Jayaraman, A.; Polydisperse homopolymer grafts stabilize dispersions of nanoparticles in a chemically identical homopolymer matrix: an integrated theory and simulation study. Soft Matter 2013. DOI: 10.1039/C3SM00144J

(779a) Characterization of Functional Carbon Nanotubes By Titration

Zhang, Z., Yale University
Haller, G. L., Yale University
Originally presented on: 11/8/2013 8:30:00 - 8:45:00

As we all known, since its discovery, CNT has been an attractive supporting material of catalyst. Before it can be used for catalyst supports, we must functionalized it first. However the functional groups on the carbon nanotubes can be very complicated, including hydroxyl groups, carbonyl groups, carboxyl groups and so on. To characterize the functional groups can be difficult. All of the traditional methods have their problems. Some are not capable for quantitive measurement like RI, TGA, TPD, Raman, TEM. Some need a lot of sample and time consuming like Chemical Derivatization, Bohem titration. Or some is very subjective due to analysis process like XPS and XAFS. Here we introduce using the simple acid-base titration to characterize the functionalized carbon nanotubes. By using principle component analysis of the titration data, we can calculate the density of certain functional groups. The result is compared with XANES and XPS for both C and O edges.

(779b) Cooperative Catalysis By Acid-Base Bifunctional Graphene

Zhang, F., Tianjin University
Zhang, G.,
Fan, X., Tianjin University
Li, Y., Tianjin University
Originally presented on: 11/8/2013 8:45:00 - 9:00:00

Cooperative catalysis by Acid-Base Bifunctional Graphene

Graphene is a promising candidate to disperse or immobilize catalytically active species for heterogeneous catalysis. Despite various graphene-based nanocatalysts have been reported, the applications of graphene supported nanocatalysts limit to individual reaction. In another aspect, Acid-base bifunctional catalyst is of particularly interest, but also has great challenges in synthesis and applications. In this study, the acid-base bifunctional graphene was prepared by modifications of graphene oxide (GO) with organic amine and sulfonic acid groups through consequent silylanization and free-radical addition, and catalytic organic amine and sulfonic acid groups could be homogeneously introduced to the surfaces of graphene sheets with high densities.. The successful bifunctionalization of GO was confirmed through a series of Characterizations, including fourier transform infrared spectroscopy (FTIR), the X-ray photoelectron spectroscopy (XPS), Scanning electron microscopy (SEM) and Raman spectroscopy. It is found that the obtained acid-base bifunctional graphene show extraordinary performances in a typical one-pot acid-base reaction sequence called Deacetalization--Nitroaldol reaction sequence, which are probably due to the high densities of both acidic and basic functional groups, as well as the unique two-dimensional structure of the graphene supports. Such kind of graphene-based bifunctional catalysts used in acid-base one-pot reaction sequences can reduce cost and waste water by saving separation/purification steps and solvents/reagents, as well as increase the yield by avoiding the separation of intermediate products.

(779c) MnO2-Functionalized Graphene Nanosheets Supported Pt Nanoparticles With Excellent Performance for Electrooxidation of Methanol

Huang, H.,
Wang, X., Nanjing University of Science and Technology
Originally presented on: 11/8/2013 9:00:00 - 9:15:00

1. Introduction

Direct methanol fuel cells (DMFCs) as promising alternative power sources have attracted considerable attention due to their high theoretical energy conversion efficiency, low pollutant emission, and convenient storage and transport of liquid fuel. Platinum is the best-known precious metal catalyst that exhibits significant electrocatalytic activity for methanol oxidation. However, given the limited resources and high cost of platinum, it is very necessary to improve the utilization efficiency of noble metals for the practicality and commercialization of DMFCs. In this regard, hybrid electrode materials consisting of metal Pt incorporated into other elements have been explored. Among them, MnO_2 is of particular interest due to their excellent proton conductivity, low density, and good permeation. On the other hand, graphene, a single layer of graphite, has been a hot topic in this area because of its unique physicochemical properties, such as extremely high specific surface area and superior electronic conductivity. Therefore, the interaction of Pt nanoparticles (NPs), MnO_2 and graphene sheets (GS) is promising to simultaneously possess respective merits of each component, which is highly desirable and technologically important for improving the catalytic performance of Pt catalysts. Herein, we report a simple and cost-effective strategy for the synthesis of a ternary Pt/MnO_2/GS composite as an advanced electrocatalyst for high-performance DMFCs.

2. Methods

As shown in Figure 1, graphene oxide (GO) was prepared from powdered flake graphite by a modified Hummer's method. With the introduction of $KMnO_4$ aqueous solution into the GO dispersion system, the carbon atoms of graphene framework would reduce MnO_4^- *in situ* to form MnO_2 in a water-containing system at room temperature. Thereafter, $Pt(NO_3)_2$ was added to the MnO_2/GO nanosheets colloid suspension under stirring. During the solvothermal reaction, both of the reduction of graphene oxide and Pt NPs loading can be achieved, leading to the formation of Pt/MnO_2/GS composite. Pt/GS, Pt/XC-72 (commercial carbon black, Vulcan XC-72) and Pt/MnO_2/XC-72 were also synthesized for comparison.

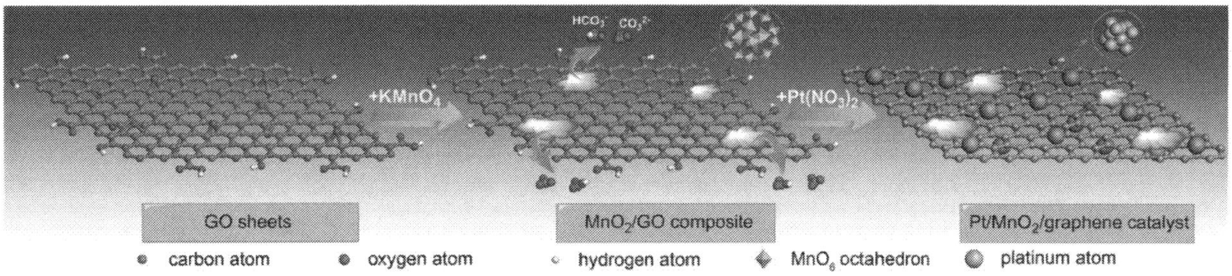

Figure 1 Illustrations of the synthesis of Pt/MnO_2/graphene catalyst.

3. Results and discussion

X-ray photoelectron spectra (XPS) was used to determine the composition of the as-prepared composites. As shown in Figure 2a-b, the intensity of oxygen-containing groups such as C-OH (286.1 eV), C-O-C (287.6 eV), H-O-C=O (288.9 eV) in Pt/MnO_2/GS was obviously reduced during the solvothermal reaction, while the peak at 284.5 eV, C-C bond, became predominant. The significantly lower O/C ratio in the composite can be linked to the much faster rate of charge transfer, which is favorable for the oxidation of methanol. XPS analysis also shows that there are different Pt oxidation states in the catalyst. It can be seen from Figure 2c that the Pt 4f signal of Pt/MnO_2/GS represents the sum of two pairs of doublets: the most intense doublet (71.4 and 74.7 eV) is due to metallic Pt and the second doublet (72.0 and 76.0 eV) can be ascribed to the +2 oxidation state of Pt. In addition, the Mn 2p XPS spectrum of the Pt/MnO_2/GS composite exhibits two characteristic peaks at 642.2 eV and 654.0 eV, corresponding to the Mn $2p_{3/2}$ and Mn $2p_{1/2}$ spin-orbit peaks of MnO_2, respectively (Figure 2d). Thus it can be concluded that MnO_2, graphene and Pt coexist in the prepared nanohybrid.

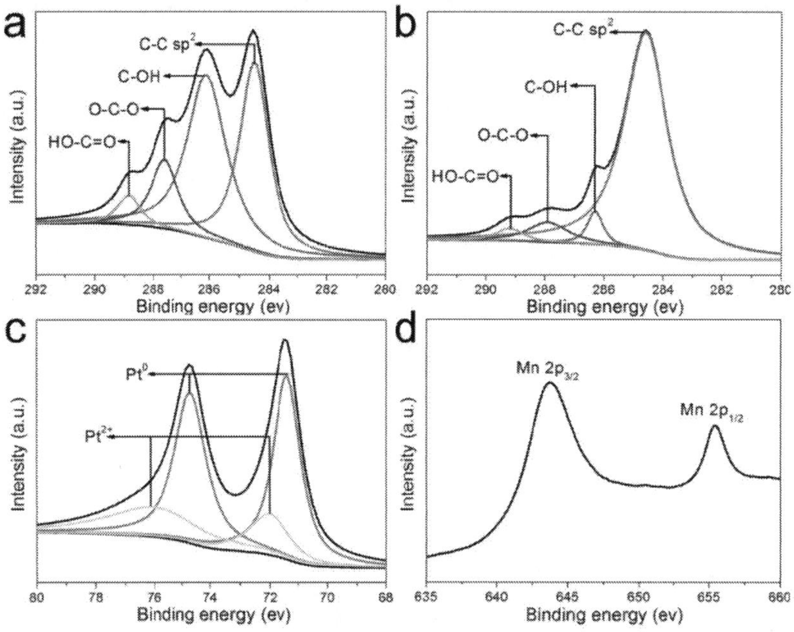

Figure 2 C 1s core-level XPS spectra of (a) graphite oxide and (b) Pt/MnO_2/GS; (c) Pt 4f and (d) Mn 2p core-level XPS spectra of Pt/MnO_2/GS.

Figure 3 depicts transmission electron microscopy (TEM) images of the MnO_2/GO, Pt/MnO_2/GS, Pt/GS and Pt/XC-72 nanohybrids. As shown in Figure 3a, it was found that the almost transparent GO sheets were thinly covered with well-dispersed layered MnO_2. No obvious stacking of these MnO_2 lamellas was observed, mainly due to the fact that the formed MnO_2 coating could function as a barrier preventing further access of the MnO_4^- ions to the functionalized carbon surface. Afterwards, MnO_2-functionalized GO sheets were used to load Pt NPs for producing ternary Pt/MnO_2/GS composite, as displayed in Figure 3b. The most striking feature is that the ultrafine Pt metal NPs with diameters less than 3 nm were uniformly distributed on the surface of MnO_2/GS sheets. Lattice fringes of both the (111) plane in face-centered cubic Pt (0.22 nm) and the (400) plane in MnO_2 (0.24 nm) are shown in the high-resolution TEM (HRTEM) image (the inset of Figure 3b), which correlate well with the known data. In Figure 3c-d, it is clearly seen that Pt NPs were also deposited on GS and XC-72 supports, however, their distributions became much broader than those on MnO_2/GS surfaces and the aggregation of some Pt NPs even took place. The statistical size analysis of Pt NPs indicates that the average particle sizes are about 1.7 nm, 3.6 nm and 4.7 nm for Pt/MnO_2/GS, Pt/GS and Pt/XC-72, respectively (Figure 3e). The findings can be explained in terms of the reasons that the presence of hydrous MnO_2 can create large hydrophilic regions on the surface of GO, which can facilitate the diffusion of Pt^{2+} ions and effectively prevent the metal NPs from agglomeration.

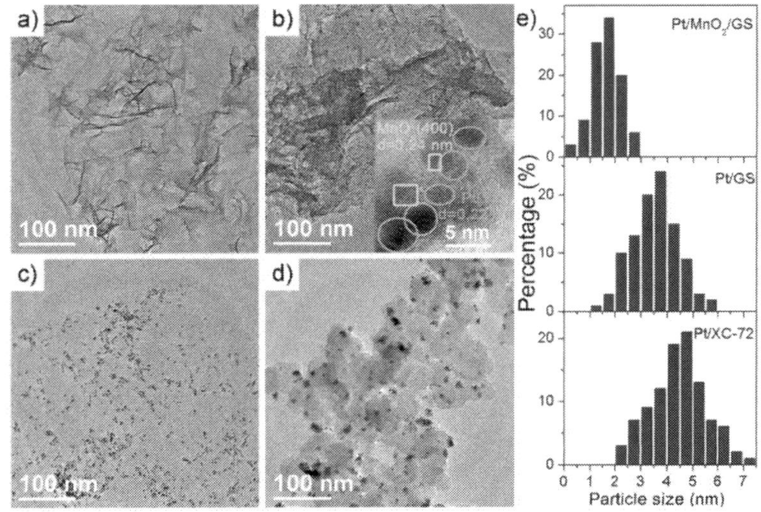

Figure 3 TEM images of (a) MnO_2/GO, (b) Pt/MnO_2/GS, (c) Pt/GS and (d) Pt/XC-72. The inset of (b) is an HRTEM image of Pt/MnO_2/GS. (e) The histograms of Pt particle size distribution of different carbon-based samples.

To explore the potential applications, the as-obtained Pt/MnO_2/GS hybrid was tested as a nanoelectrocatalyst for oxidation of methanol to evaluate its catalytic activity. By measuring the charge collected in the hydrogen adsorption/desorption region after double-layer correction, the electrochemically active surface area (ECSA) of Pt/MnO_2/GS was calculated to be as high as 103.2 $m^2\ g^{-1}$, more than 2 times as large as that for Pt/GS (49.1 $m^2\ g^{-1}$), nearly 3 times the Pt/XC-72 (38.5 $m^2\ g^{-1}$) and 2 times the Pt/MnO_2/XC-72 (54.4 $m^2\ g^{-1}$) catalyst (Figure 4a), demonstrating that the Pt/MnO_2/GS electrocatalyst not only possesses a larger number of catalytically active sites available, but also is electrochemically more accessible, which are essential for the electrocatalytic reactions. This remarkable improvement can be primarily attributed to the concerted effect of MnO_2 nanolamellas and graphene sheets and the high dispersion of small-sized Pt NPs on modified carbon sheets. Figure 4b presents the cyclic voltammograms (CVs) of the electrodes coated with the Pt/MnO_2/GS, Pt/GS, Pt/XC-72 and Pt/MnO_2/XC-72 catalysts in a 1 M H_2SO_4 solution containing 2 M CH_3OH. It can be seen that the Pt/MnO_2/GS catalyst exhibits an ultrahigh forward peak current density of 1224 mA mg^{-1}, which is significantly higher than the one obtained from Pt/GS (357 mA mg^{-1}), Pt/XC-72 (225 mA mg^{-1}), Pt/MnO_2/XC-72 (435 mA mg^{-1}).

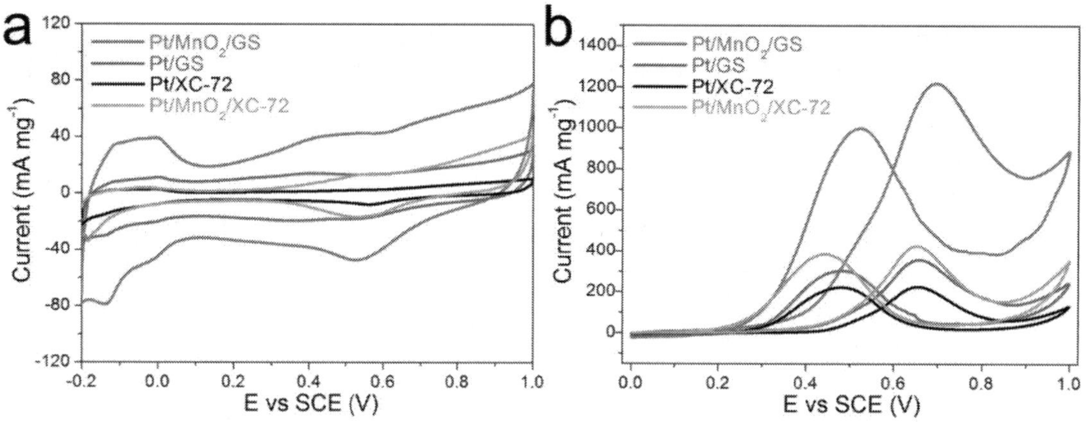

Figure 4 (a) CVs of Pt/MnO_2/GS, Pt/GS, Pt/XC-72 and Pt/MnO_2/XC-72 in 1 M H_2SO_4 at 20 mV s^{-1}. (b) CVs of Pt/MnO_2/GS, Pt/GS, Pt/XC-72 and Pt/MnO_2/XC-72 in 1 M H_2SO_4 and 2 M methanol at 20 mV s^{-1}.

Based on the results above, it can be deduced that the rational combination of Pt NPs, MnO_2 and graphene into a sophisticated 2D architecture is conducive to the development of high Pt utilization catalysts. First, the small size and good dispersion of Pt NPs on the modified carbon sheets can offer higher surface-areas that host more accessible active sites, thus achieving an extremely high electrocatalytic activity; Second, hydrous MnO_2 coating on graphene surface can facilitate the diffusion of electrolyte and also allow the transportation of the poisoning species and protons (H^+) from the Pt sites during the reaction; Third, graphene sheets in the ternary composite act as not only the support for the deposition of Pt NPs but also the electronic conductive channels, leading to a significant decrease in charge-transfer resistance and effectively improving the electrochemical kinetics.

4. Conclusions

In summary, a facile and cost-effective approach has been developed to disperse Pt NPs on MnO_2-functionalized graphene sheets. Graphene could be amazingly used as both a green reductant in synthesis of MnO_2 and an ideal substrate for growing and anchoring Pt NPs with a very narrow size distribution. Most importantly, the rationally designed Pt/MnO_2/GS ternary electrocatalyst exhibits ultrahigh electrocatalytic activity, significantly outperforming the Pt/GS, Pt/XC-72 and Pt/MnO_2/XC-72 catalysts. We believe this design concept can be extended to the fabrication of other novel nanocomposites containing noble metals, such as Pd, Au and Ag, opening up a brand new avenue for a large spectrum of device applications.

(779d) Surface Energy of Graphene Determined By Intrinsic Friction Microscopy

Krajina, B. A.,
Kocherlakota, L. S., University of Washington
Overney, R. M., University of Washington
Originally presented on: 11/8/2013 9:15:00 - 9:30:00

Surface Energy of Graphene Determined by Intrinsic Friction Microscopy

Brad Krajina, Lakshmi S. Kocherlakota , RenÃ© M. Overney

Due to its exceptional properties, research in graphene is actively and widely pursued for applications in electronics, electro-optics, photovoltaics, and sensing. A critical and surprisingly still outstanding property is graphene's surface energy and its transition towards bulk graphite. The reason for this shortcoming is the challenge to obtain surface energies on materials that possess surface inhomogneities, such as step edges, or thickness variations, an aspect that is of particular importance for graphene. While the graphene sheet homogeneity and size could be improved significantly by chemical routes over the initial mechanical cleaving process, oxidation effects and follow up treatments add additional complications for reaching consensus on the surface energy.

In this light, we explored the surface energy of "virgin" graphene as obtained through mechanical cleaving of highly ordered pyrolytic graphite (HOPG) on the sub-inhomogeneity scale. We utilized a scanning scattering methodology that extracts an energetic signature from the coupling events of an atomic force microscopy (AFM) tip with the thermally active modes within the interface between AFM tip and sample. This method dubbed "intrinsic friction microscopy" (IFA) has been used in the past on organic systems to analyze the energetics of mobile groups.[1,2] It is here for the first time employed on an inorganic material, resulting in an energy value for HOPG that is in correspondence with the Hamaker constant of 1.49×10^{-19} J known for HOPG and SiO_2 (AFM tip material). Subsequent analysis of single layer graphene reveal a decrease in the Hamaker constant by a factor of two, which leads via the Liftshift Theory of Van der Waals interaction applied to metal-insulator interfaces to an equivalent drop in the surface free energy. This paper discusses also the dependence on the surface energy on the number of graphene layers and identifies a critical graphene/HOPG thickness of around 5 nm, beyond which the HOPG bulk value for the surface energy is restored.

(1) Sills, S.E.; Gray, T.; Overney, R.M., J. Chem. Phys. (2005), **123**, 134902.

(2) Knorr, D.B.; Gray, T.O.; Overney, R.M., J. Chem. Phys. (2008), **129**, 074504.

(779e) Water-Induced Interactions Between Boron-Doped Carbon Nanotubes: A Multiscale Simulation Study

Zhang, Z., University of Alabama
Turner, C. H., University of Alabama

Originally presented on: 11/8/2013 9:30:00 - 9:45:00

We have deployed molecular dynamics (MD) to investigate the hydration, the water-induced interactions, and the debundling behavior of boron-doped single-walled carbon nanotubes (B-CNTs) within aqueous solutions. Partial charges on the atom sites are fit to electrostatic potentials (derived from first-principles calculations), which are used to capture the polarization effects within the B-CNTs, and this model is then used in the MD simulations. These charges are fit with the DDEC method, which is proven to reproduce correct electrostatic potential of periodic nanotubes.

To clarify the water-induced interactions of B-CNTs, the potential of mean force (PMF) of two and three solvated B-CNTs are evaluated by MD over a range of different temperatures. The water density profiles, orientations, and hydrogen-bonds are also studied to shed light on the hydration behavior of the B-CNTs. The same properties of pristine CNTs are also investigated as a benchmark to project the effects of B-doping on the water-CNT interactions. Enhanced B-CNT-water interactions and water-induced interactions between B-CNTs are observed at more prolonged distance, as compared to pristine CNTs. These simulations provide important insight into the effects of substitutional doping of CNTs on their dispersion behavior, which is a critical issue for engineering composite materials and other CNT/graphene-based system.

(779f) Structural, Electronic, and Mechanical Properties of Superlattices of Interlayer-Bonded Domains in Twisted Bilayer Graphene

Muniz, A. R., Federal University of Rio Grande do Sul
Machado, A., Federal University of Rio Grande do Sul
Maroudas, D., University of Massachusetts Amherst
Originally presented on: 11/8/2013 9:45:00 - 10:00:00

We present a comprehensive computational analysis of the atomic and electronic structure and properties of a novel class of carbon nanostructures, formed due to interlayer covalent sp^3 C-C bonding in twisted bilayer graphene as a result of controlled chemical functionalization (hydrogenation or fluorination). Depending on the twist angle and local stacking of layers, these nanostructures are superlattices of diamond-like nanocrystals or caged fullerene-like configurations embedded within the bilayer. According to first-principles density functional theory calculations, the electronic behavior of these sp^2/sp^3 hybrid configurations ranges from semi-metallic, characterized by linear dispersion around the K point in the first Brillouin zone (Dirac cones), to semi-conducting with electronic band gaps from a few meV to ~1.2 eV, to insulating with band gaps up to ~4 eV; the band gap of the semiconducting configurations can be tuned by controlling the density of interlayer C-C bonds and the chemical functionalization pattern. The electronic character of the superstructures depends on the symmetry and periodicity of the superlattices and on the type of chemisorbed species (H or F atoms).

We have also conducted a systematic study of the mechanical response of these superlattices based on molecular-dynamics simulations of tensile deformation and shear loading tests according to a reliable interatomic bond-order potential. We have found that the mechanical properties of these superstructures can be precisely tuned by controlling the fraction of sp^3-hybridized C-C bonds in the material through the extent of chemical functionalization. Their Young modulus and tensile strength decrease moderately compared to those of pristine bilayer graphene, but remain superior to those of most conventional engineering materials. However, their interlayer shear modulus increases strongly compared to that of pristine bilayer graphene and monotonically with the fraction of sp^3-hybridized C-C bonds.

(779g) Structural, Morphological, Electronic, and Mechanical Properties of Irradiated Single-Layer Graphene

Carpenter, C., University of Massachusetts Amherst
Ramasubramaniam, A., University of Massachusetts Amherst
Maroudas, D., University of Massachusetts Amherst
Originally presented on: 11/8/2013 10:00:00 - 10:15:00

Defect engineering of graphene has been studied extensively, both experimentally and computationally, aimed at approaches for tuning graphene's electronic, optical, and mechanical properties. Ion-beam and electron-beam irradiation have been employed to introduce in the honeycomb lattice of graphene defects, such as vacancies, reconstructions with pentagon and heptagon rings, as well as larger-scale defects such as two-dimensional (2D) holes or voids. Recent experimental studies also have reported defect-induced amorphization of graphene after high electron-beam exposure. High-resolution imaging of the resulting amorphized graphene demonstrated that such irradiation-induced amorphized graphene remains a coherent single-layer 2D structure that consists of a random patch of polygons and also contains 2D holes.

In this presentation, we address some fundamental questions on the structure and properties of irradiated graphene, including how the defect-induced structure and morphology depend on the inserted defect concentration, the determination of the onset of defect-induced amorphization, and the effects of the defect density and amorphization transition on the graphene's electronic and mechanical properties. Our study is based on isothermal-isobaric molecular-dynamics (MD) simulations according to a reliable bond-order potential for the description of interatomic interactions in graphene and using large supercells that contain thousands of C atoms. Random distributions of single vacancies were introduced into the graphene honeycomb lattice at a given concentration and the response of the irradiated graphene sheets was explored systematically over a broad range of defect concentration based on a variant of simulated annealing. The analysis of the simulation results for the relaxation of the irradiated graphene samples included detailed structural and morphological characterization of the relaxed configurations based on calculation of the radial distribution function, structural order parameters, and RMS surface roughness; the relaxed configurations constitute computer models of irradiated graphene. Finally, we used density-functional tight-binding calculations to compute the electronic density of states of the structurally relaxed configurations.

We predict that the onset of the amorphization transition occurs at an inserted vacancy concentration between 5 and 10%, with the transition becoming less abrupt with vacancy concentration as temperature increases, and determine the surface roughness as a function of defect concentration. We find that the computed amorphized configurations are in good agreement with those observed in recent experiments of high-dose exposure of graphene to 100-keV electron irradiation. We also find that the electronic density of states of vacancy-amorphized graphene is characterized by introduction of localized states near the Fermi level of perfect single-layer graphene, consistent with those arising from random vacancy insertion in graphene at low concentrations. Our findings set the stage for exploratory work on the functionalization of defect-amorphized graphene toward useful band-gap engineering and tuning of electronic transport and optical properties.

We have also determined the mechanical properties of irradiated single-layer graphene sheets as a function of inserted vacancy concentration, x. Specifically, based on MD simulations of uniaxial deformation tests at constant strain rate and temperature, we have computed as a function of x the tensile strength, toughness, and failure strain of the irradiated graphene sheets. We find that the vacancy-induced crystalline-to-amorphous transition is accompanied by a brittle-to-ductile transition in the failure response of irradiated graphene sheets for x values over the range 8-12%. While point defects and larger voids appreciably degrade the strength of pristine graphene, we find that even heavily-damaged samples (~20% vacancies) exhibit tensile strengths of ~30 GPa, in significant excess of those typical of engineering materials. Our results suggest that defect engineering of graphene is feasible without losing its desirable mechanical properties.

(779h) Modeling Surfactant-Assisted Stabilization of Carbon Nanotubes and Graphene in Ionic Surfactant Aqueous Solutions: Coarse-Grained Molecular Dynamics Simulations and Modified DLVO Theory

Shih, C. J., Massachusetts Institute of Technology
Lin, S., Massachusetts Institute of Technology
Strano, M. S., Massachusetts Institute of Technology
Blankschtein, D., Massachusetts Institute of Technology
Originally presented on: 11/8/2013 10:15:00 - 10:30:00

Dispersion of single-walled carbon nanotubes (SWCNTs) and graphene in ionic surfactant aqueous solutions is a very important route to further purify, self-assemble, and functionalize these nanomaterials for electronic, biological, and sensor applications. In order to optimize the yield and colloidal stability, it is essential to understand the mechanisms responsible for the concentration-dependent adsorption of surfactant molecules on carbon nanomaterials, as well as the effects of surfactant micellization and surface geometry of the carbon nanomaterials. Molecular dynamics (MD) simulations are a powerful tool to investigate the interactions between surfactants and carbon nanomaterials at the molecular level. However, the length and time scales associated with the systems considered here are far beyond current computational capabilities using conventional all-atomistic MD simulations. With this in mind, we have used "coarse-grained" MD simulations to successfully simulate micellization and adsorption on SWCNTs or graphene in one simulation box. Specifically, the adsorption density of surfactant molecules was quantified as a function of the bulk surfactant concentration. Based on the simulated results, we developed a model that describes the adsorption density as a function of surfactant concentration and the geometry of the SWCNTs and graphene. In combination with a modified DLVO theory of colloid stability, for the first time, we successfully modeled the surface electric potential, the intertube (or intersheet) potential energy profile, and the energy barrier height. These results allow us to predict the optimized surfactant concentration needed to disperse different SWCNTs and graphene in aqueous media. We believe that the development of advanced methods to separate SWCNTs and graphene in aqueous media will be greatly facilitated by the molecular-level insight and theoretical analyses presented here.

(779i) The Evolution of the Physical and Electronic Structure of Bilayer Graphene Upon Chemical Functionalization

Wang, Q. H., Massachusetts Institute of Technology
Shih, C. J., Massachusetts Institute of Technology
Paulus, G. L. C., Massachusetts Institute of Technology
Strano, M. S., Massachusetts Institute of Technology
Originally presented on: 11/8/2013 10:30:00 - 10:45:00

Bilayer graphene is increasingly attractive for electronic applications because of its electric field tunable band structure and band gap. However, while the chemical functionalization of monolayer graphene has been widely explored to date, there is much less known for bilayer graphene. Of particular interest in bilayer graphene is the use of chemical functionalization to generate an external electric field to open a bandgap. In this paper, we show the covalent and non-covalent functionalization of bilayer graphene using an electrochemical process with aryl diazonium salts, and we use Raman spectroscopic mapping and conductive atomic force microscopy (cAFM) to study the resulting changes in the physical and electronic structures. We report that under strong covalent functionalization, the G peak of bilayer graphene splits into two components that broaden in width and shift in position with increasing functionalization, the D peak forms with four components, and the 2D peak decreases in intensity in the $2D_{12}$ and $2D_{11}$ components but broadens in width in the $2D_{22}$ and $2D_{21}$ components. These changes suggest a complex interplay between the physical and electronic structures of bilayer graphene. We report for the first time these distinct changes observed in the Raman spectra as a function of degree of covalent and non-covalent functionalization, which can provide a guide for further studies of the chemistry of bilayer graphene.

(779j) Using Reactive Force Fields to Model Adatom Domains in Fluorinated Graphenes

Junkermeier, C., Penn State University
Srinivasan, S., Penn State University
van Duin, A. C. T., Pennsylvania State University
Originally presented on: 11/8/2013 10:45:00 - 11:00:00

In the last decade graphene has been studied intensively due to its high electrical conductivity, strength, and elasticity. In application graphene suffers as it is a zero band gap semiconductor and is weak chemically interactions. Surface chemical functionalization is one method to modify the properties of graphene such that it could be used in transformative technologies. Fluorine has been a common choice, for examining the effects of surface functionalization on mechanical and electrical properties of graphene. Experimental observation has shown that graphene can have domains with what appears to be highly ordered fluorine adsorption. Though the mechanisms by which ordered domain formation happens are not fully understood.

To date, (ReaxFF based) fluorocarbon reactive forcefields have been produced using sets of small fluorocarbon molecules. While this technique been used with some success in describing fluorinated graphitic materials, it has failed to reproduce some of the mechanical and dynamic properties seen in higher levels of theory. Fluorinated graphene structures produced using density functional theory (DFT) and reported previously in references [1-3] were used to refine the force field parameters and to show that they better reproduces the mechanical and dynamical properties of fluorinated graphitic materials.

The new forcefield parameters are applied to a combination of climbing-image nudged elastic band (CI-NEB), molecular dynamics, and parallel replica dynamics (PRD) calculations to study the how fluorine domains form and what occurs at domain boundaries. We will discuss our latest results and what they may mean for producing stable, high quality fluorographenes for use in technological applications.

1. Jeremy T. Robinson, James S. Burgess, Chad E. Junkermeier, Stefan C. Badescu, Thomas L. Reinecke, F. Keith Perkins, Maxim K. Zalalutdniov, Jeffrey W. Baldwin, James C. Culbertson, Paul E. Sheehan and Eric S. Snow. "Properties of Fluorinated Graphene Films" Nano Lett., 10:3001-3005 (2010)
2. Maxim Zalalutdinov, Jeremy T. Robinson, Chad E. Junkermeier, Jim Culbertson, Thomas L. Reinecke, Rory Stine, Paul E. Sheehan, Brian H. Houston, and Eric S. Snow. "Engineering Graphene Mechanical Systems" Nano Lett. 12:4212-4218 (2012)
3. Chad E. Junkermeier, Stefan C. Badescue, and Thomas L. Reinecke. "Highly Fluorinated Graphene" (arXiv preprint arXiv:1302.6878

(783a) Proteotoxic Stress Induced By Magnetic Fluid Hyperthermia (MFH) Enhances Bortezomib Cytotoxicity

Alvarez, M., University of Puerto Rico
Castillo, A., University of Puerto Rico
Rinaldi, C., University of Puerto Rico
Torres-Lugo, M., University of Puerto Rico
Originally presented on: 11/8/2013 8:30:00 - 8:55:00

Clinical studies have demonstrated the effectiveness of combined hyperthermia and anticancer drug treatment. However, challenges related to effective heat transfer have limited its clinical application. The use of magnetic fluids to induce hyperthermia (MFH) has been an attractive treatment to overcome these challenges; however, the mechanism by which MFH enhances anticancer drugs cytotoxicity has not been explored in depth. In this work, studies of mechanisms involved in the synergistic potentiation of bortezomib in combination with magnetic fluid hyperthermia (MFH) were performed. Our hypothesis was that MFH can produce significant protein damage inducing cell death. To test the hypothesis, studies between hyperthermia and the proteasome inhibitor bortezomib for various cell lines (MDA-MB 468, Caco-2, and A2780) were conducted and results were compared with those of hot water hyperthermia. Aggresome formation and microtubule disruption as a function of recovery time in MCF-7 cells were done. Results show that there is a synergistic potentiation between hyperthermic treatment and bortezomib in all tested cell lines, but, MFH induced cell cytotoxicity in sensitive and resistant cell lines significantly higher than HWH. Studies of aggresome formation and microtubule disruption indicated that MFH induces higher proteotoxic stress that hot water hyperthermia. This result suggests that this is one of the mechanisms by which MFH enhaces bortezomib cytotoxicity.

(783b) Synthesis and Characterization of Peptide-Conjugated Iron Oxide Nanoparticles for Hyperthermia Applications

Kruse, A. M., University of Kentucky
Meenach, S. A., University of Rhode Island
Hilt, J. Z., University of Kentucky
Anderson, K. W., University of Kentucky
Originally presented on: 11/8/2013 8:55:00 - 9:20:00

One of the current challenges in the delivery of systemic nanoparticle systems in cancer therapy applications is the lack of effective tumor penetration. In this research, a nanoparticle system composed of an iron oxide core with a crosslinked dextran coating functionalized with the tumor homing peptide, CREKA, has been developed to overcome this limitation by homing to and penetrating into tumor tissue. The iron oxide core allows for particle heating upon exposure to an alternating magnetic field (AMF) while the dextran coating stabilizes the particles in suspension and decreases the cytotoxicity. The overall goal of this study was to develop and optimize these systems for effective hyperthermia treatment applications. Hyperthermia, the heating of tissue between 42 and 45°C, has been shown to enhance the effects of radiation and chemotherapeutics, but current methods of hyperthermia often result in severe side effects due to lack of localization and overheating of tissue. Magnetically mediated hyperthermia provides the opportunity for localized heating, however, this method is currently limited by the lack of particle penetration into tumor tissue. It is hypothesized that this specific particle system can enhance particle accumulation at the tumor site, providing a high enough concentration to induce localized hyperthermia conditions. The particles were characterized for size, stability, biocompatibility, and heating capabilities. The particles were stable in PBS and media over at least twelve hours, had a hydrated diameter of 50 nm, and generated enough heat to raise solution temperatures well into the hyperthermia range when exposed to an AMF. The biocompatibility of the particles was analyzed through cytotoxicity studies on A549 lung cancer cells and NIH 3T3 fibroblasts. These studies were completed for both high particle concentrations with low exposure time and low particle concentrations with high exposure time, and the results determined that the particles have low cytotoxicity over both time frames and concentrations. Fibrinogen clots were used to determine the binding affinity of CREKA. The equilibrium dissociation constant was determined for CREKA alone and for CREKA-conjugated particles. The binding affinity of CREKA conjugated iron oxide nanoparticles to fibrinogen gels is significantly greater than the binding affinity of dextran coated iron oxide nanoparticles. The binding of CREKA conjugated iron oxide nanoparticles to multicellular spheroids was also evaluated. Results demonstrated that this novel system has the potential to be used for effective localized hyperthermia treatment of tumors.

(783c) Embolic Polylactide-Co-Glycolide Microspheres for Image-Guided,Transcathter Delivery of Sorafenib to Hepatocellular Carcinoma

Chen, J., Northwestern University

Shea, L. D., Northwestern University

Larson, A.,

Originally presented on: 11/8/2013 9:20:00 - 9:45:00

Introduction: Hepatocellular carcinoma (HCC) is the most common form of primary liver cancer and most patients are not eligible for potentially curative treatments (resection, transplant) as the cancer is often diagnosed during later stages. Therefore, these patients undergo palliative treatments such as transcatheter therapies. These therapies involve having doctors place a catheter in the femoral artery and using image guidance to guide it to tumor feeding branches of the hepatic artery. Once the catheter is placed, chemotherapy or radiation therapy is given along with embolic agents to help contain the therapy locally to the tumor as well as to prevent blood flow to the tumor to help starve it. The purpose of our studies is to develop a drug delivery system to allow for sorafenib to be delivered as part of a transcatheter therapy via an embolic polymer microsphere bead. Additionally, we have added an iron oxide component to serve as a contrast agent to allow for magnetic resonance imaging (MRI) of the delivery.

A more recent therapy for HCC is sorafenib (brand name Nexavar), which is a multikinase inhibitor that was approved by the FDA in 2007. It is currently formulated as an oral tablet, and therefore, the systemic administration is associated with severe side effects such as diarrhea, hand foot syndrome (severe rashes of the hands and feet), and hypertension. Therefore, patients do not tolerate sorafenib well. Re-formulating sorafenib for delivery using transcatheter techniques would improve patient tolerance. Furthermore, it would also allow for increased does at the tumor and an added embolic effect to improve therapy.

Methods: We have chosen polylactide-co-glycolide (PLG) as our platform and we employed a single emulsion/solvent evaporation method to fabricate our microspheres to co-encapsulate the sorafenib and a hydrophobic iron oxide. We then characterized the size and morphology using microscopy and Image J software analysis. Sorafenib and iron oxide loading were characterized with high performance liquid chromatography and inductively coupled plasma mass spectroscopy, respectively. Sorafenib and iron oxide release were also monitored using similar methods. The contrast effect these microspheres have under MRI was monitored *in vitro* using a 7 Tesla MRI scanner. Lastly, the microspheres were injected to tumor feeding arteries transcather in a rabbit VX2 animal model to mimic clinical delivery. The rabbit was then imaged in the same 7 Tesla MRI scanner to visualize the distribution of the microspheres. After imaging, the rabbit was sacrificed and sections of the liver, tumors, stomach and lungs were harvested and submitted for histology with Prussian blue staining for iron to confirm the microsphere biodistribution.

Results: The resulting microspheres were of an average diameter of 22.7 microns. Iron oxide content and sorafenib content were 0.209% w/w) and 14.0% (w/w) respectively. The release study indicated slow burst release whereas 2.27% of the loaded sorafenib released within 3 days. *In vitro* MRI imaging showed that increasing concentrations of the microspheres from 0-20 mg/mL caused a reduction in T2* from 111 to 8.13 ms. Lastly, after infusion of the microspheres into the rabbit, we were able to depict the microspheres under MRI as areas of darkening near the tumors. Prussian blue histology to confirm microsphere biodistribution indicated that the tumors showed high staining at the tumors and minimal staining in the normal liver, stomach and lungs (stomach and lungs are of concern as microsphere distribution to those tissues can cause problems).

Conclusion: These rigorous characterization studies show that embolic PLG microspheres can co-encapsulate sorafenib and iron oxide. These microspheres can also be delivered locally to liver tumors through a catheter to ultimately produce a therapeutic response.

(783d) Stimuli Responsive Poly (ethylene glycol) (PEG) Hydrogel Coated Magnetic Iron Oxide Nanoparticles for Targeted Drug Delivery

Nazli, C., KOC UNIVERSITY
Demirer, G. S., Koc University
Yagci Acar, F. H., KOC UNIVERSITY
Kizilel, S., Koc University
Originally presented on: 11/8/2013 9:45:00 - 10:10:00

One of the major limitations of nanoparticle-based cancer therapeutics is non-specific distribution of drugs in the body. Undesirable side effects and resistance of cancer cells to drugs are some of the additional limitations of a successful cancer therapy. Magnetic iron oxide nanoparticles (MIONPs) has demonstrated great potential for simultaneous imaging and targeted delivery of therapeutic agents into tumor site. The aim of this study was synthesis and characterization of enzymatically responsive poly (ethylene glycol) (PEG) hydrogel coated MIONPs for cancer cell targeted drug delivery. RGDS short peptide as a targeting ligand and GGGPQGIWGQGK peptide as matrix metalloprotease (MMP) responsive degradable peptide sequences were used for the design of functionalized PEG hydrogel coating. MIONPs could be internalized into cancer cells due to RGDS functionalized PEG hydrogel coating, and degradable sequence within PEG hydrogel structure allows for the accelerated release of therapeutic drugs into cytoplasm. Intracellular uptake studies via Prussian blue staining and inductively coupled plasma optical emission spectrometry with HeLa cells demonstrated that the protease responsive and RGDS-functionalized PEG hydrogel-coated MIONPs were superior compared to bare or only PEG hydrogel-coated MIONPs. In addition, *in vitro* release of doxorubicin from protease responsive PEG hydrogel coated nanoparticles was characterized, and viability of cancer cells exposed to this nanoparticle system has been studied. We find that both RGDS and protease responsive coating play a role in enhancing nanoparticle internalization into HeLa cells and lead to reduced cancer cell viability. This approach may be beneficial for enhanced cancer drug efficiency and for reduced the systemic side effects of loaded drugs.

(783e) Characterizing Spions Permeability By Using An in Vitro Blood-Brain Barrier Model

Shi, D.,
Sun, L., Northeastern University
Bhattacharya, S.,
Nayar, S.,
Webster, T. J., Northeastern University
Originally presented on: 11/8/2013 10:10:00 - 10:35:00

Abstract: In the present study, using murine brain endothelioma cells, an in vitro blood-brain barrier model was set up and confirmed. Confirmation of the blood-brain barrier model was completed by examining the permeability of FITC-Dextran at increasing exposure times in serum-free medium and comparing such values with values in the literature. After such confirmation, the permeability of eleven different magnetic nanoparticle samples (SPIONs) was determined by this blood-brain barrier model. Through these experiments, SPIONs that would not pass through the blood brain barrier were identified for MRI usage and SPIONs which would pass through the blood brain model were identified for drug delivery usage.

Keywords: nanoparticles, blood-brain barrier, permeability

Introduction

The blood-brain barrier is known as a "wall" that separates somatic circulating blood from the cerebrospinal fluid in central nervous system (CNS). Consisting of tight junctions that lie between endothelial cells of capillaries in CNS, it keeps large or hydrophilic molecules and microorganisms from passing into the cerebrospinal fluid and therefore maintains the microenvironment. [1]

However, previous research shows that small lipid-soluble-molecule that has a molecular weight less than 600 Da or, 200 nm in diameter, can be transported through blood-brain barrier. Therefore it requires that nanoparticles, as effective pharmaceuticals, are able to undergo successfully transportation through the blood-brain barrier. [2]

Among several nanoparticles, superparamagnetic iron oxide nanoparticles (SPIONs) have been discovered as a contrast agent which can potentially increase image contrast and improve MRI sensitivity and specificity. Moreover, SPIONs can also be used to track brain activity and detect cellular receptors that presented in brain diseases. On the other hand, if the SPIONs inside blood-brain barrier have not been efficiently cleared, the accumulation of those nanoparticles could be toxic and harmful to normal brain function in a long term. [3]

Since there are concerns about both the application of delivering drugs across the blood-brain barrier and the accumulation of SPIONs in the body, this study set up an *in vitro* blood-brain barrier model to test new SPIONs' ability to pass through blood-brain barrier.

Materials And Methods [4]

<>Material Characterization

According to a patented process entitled "A biomimetic process for the synthesis of aqueous ferrofluids for biomedical applications" [5], the synthesis of ferrofluids involved incubation of a ferrous/ferric salt solution in phosphate-buffered saline supplemented with the additives of interest using ammonium hydroxide under highly alkaline conditions. After synthesis, ferrofluids were centrifuged for a stability test. Dynamic light scattering was used to characterize hydrodynamic diameter and zeta potential was used to characterize the charge of each sample. <>Confirmation of the blood-brain barrier model

Characterization was carried out through fluorescent sugar transport to confirm the blood-brain barrier model. For the blood-brain barrier model, murine brain endothelioma cells (b.End3) were used, and cultured in Dulbecco's modified Eagle medium (DMEM) with 10% fetal bovine serum (FBS) and 1% penicillin-streptomycin (P/S). In the next step, 6.5 mm Transwell®-COL collagen-coated 0.4 µm pore polytetrafluoroethylene membrane inserts were combined with 24-well plates for the model. A hemocytometer and a light microscope were used to determine the concentrations of cells. After diluted down to 10^5 cells per mL in DMEM + 10% FBS + 1% penicillin-streptomycin, cells in the previous cell culture section were pipetted onto the inserts. Then the well plates with inserts were incubated with daily changed medium in the outer well until confluency was reached. After reaching the confluency, medium should then be replaced by 1:1 DMEM/Ham's F12 with 1% penicillin-streptomycin for the next 24, 48, 72 and 96 hours respectively.

To confirm the blood-brain barrier model, the permeability of the inserts would be tested in triplicate. After each time point was reached, inserts should be transferred to wells that contain 600 µL of Hank's Balanced Salt Solution (HBSS) before adding fluorescein isothiocyanate-labeled dextran (FITC-Dextran). Immediately after the transfer, the permeability of the inserts was tested in triplicate using 10 µg/mL FITC-Dextran. Positive controls were set up by adding 100 µL of FITC-dextran solution with 600 µL of HBSS and negative controls were set up by adding 700 µL of HBSS only.

After 2 hours incubation, 100 µL were taken from each well and transferred to a black 96-well plate with a clear bottom. A fluorescent plate reader was excited at 490/20 nm and measured at 528/20 nm. Data were converted into relative fluorescence Unit (RFU) from the reader and relative permeability was then determined: <>Experimental samples in the blood-brain barrier model

The previously described *in vitro* blood-brain barrier model was used to test the permeability of the various nanoparticles, among which three of them were solid powder (IONP) containing bovine serum albumin (BSA), polyvinyl alcohol (PVA), collagen, and graphene and the remaining eight samples were ferrofluids (FF) containing BSA, collagen, graphene, glutamic acid, and four concentrations (0.5%, 1.5%, 2.0%, and 2.5%, w/w) of PVA.

The inserts were transferred into a 24-well plate that containing 600 µL of HBSS in each well. Nanoparticles were diluted 1:19 with HBSS to lower the concentration. Next, followed the permeability experiment talked above, the inserts were exposed to the nanoparticles for 2 hours. After 2 hours, 500 µL would be pipetted from each well into a 20 mL scintillation vial and the inserts were removed.

To determine their iron concentration, aqua regia was used to convert IONPs into ionic iron. Then inductively-coupled plasma atomic emission spectroscopy (ICP-AES) was used to determine the iron concentration.

Results and Discussion

<> Material Characterization

The DLS results showed that hydrodynamic diameters of nanoparticles range from 20 nm to 200 nm. The results of zeta potential showed the lowest absolute zeta values for IONP with collagen is -5.36 mV, for ferrofluids with PVA and glutamic acid is -3.38 mV, and for ferrofluids with collagen is -5.39 mV. The highest absolute zeta values were from IONP with graphene and ferrofluids with graphene.

<>Confirmation of the blood-brain barrier model

The value of the experimental blood-brain barrier model was in accordance with values in the previous articles, indicating that the model was successfully established. First, the decline, which was about 30%, in permeability was similar with the decline described by Brown et al. [6] Furthermore, the decline of permeability with increasing exposure to serum-free medium also indicated the success of a tightening of the barrier model. <>Experimental samples using blood-brain barrier model

Permeability of each sample has been tested and further experiments were carried out to test whether one would be able to set up parameters to moderate permeability of nanoparticles by using zeta potential, size, or surface modification. Results showed that the highest permeability was obtained from ferrofluids (FF) with collagen (1.5% collagen-FF) and the lowest permeability was obtained from IONP with collagen (1.5% collagen-IONP). These results suggest that for nanoparticles that need to be delivered through blood-brain barrier, FF should be coated with collagen and, on the other hand, IONP should be coated with collagen to keep nanoparticles from penetrating the blood-brain barrier.

<>Conclusion

An *in vitro* model of blood-brain barrier was established using b.End3 cells. As the permeability decreased with increasing exposure to serum-free medium, the model was confirmed by comparing the permeability trend of FITC-dextran in serum-free medium with previous research. With the successfully established model, the permeability of eleven nanoparticle samples was then carefully examined. Results showed that in order to be delivered through blood-brain barrier, FF should be coated with collagen and IONP should be coated with collagen to avoid penetration.

Acknowledgment

The authors would like to thank the Northeastern University for funding this research.

<>References

[1] Hamilton RD, Foss AJ, Leach L (2007). "Establishment of a human *in vitro* model of the outer blood–retinal barrier". Journal of Anatomy 211 (6): 707-16.

[2] Pardridge, W. M. (1995). "Transport of small molecules through the blood-brain barrier: biology and methodology." Advanced Drug Delivery Reviews 15(1-3): 5-36.

[3] Berry CC, Curtis ASG. Functionalisation of magnetic nanoparticles for application in biomedicine. J

Phys D Appl Phys. 2003; 36:198-206.

[4] Dan Hoff, Lubna Sheikh, Thomas J Webster. Comparison of ferrofluid and powder iron oxide nanoparticle permeability across the blood-brain barrier.International Journal of Nanomedicine. 2012; 2012:7-1.

[5] Nayar S, Sinha A, Pramanick AK, inventors. A biomimetic process for the synthesis of aqueous ferrofluids for biomedical applications. Application number 0672DEL2010. March 22, 2010.

[6] Brown RC, Morris AP, O'Neil RG. Tight junction protein expression and barrier properties of immortalized mouse brain microvessel endothelial cells. Brain Res. 2007; 1130:17–30.

(783f) In Vivo Assessment of Temporal and Spatial Behavior of Carboxyl Methyl Dextran Iron Oxide Nanoparticles in Breast Cancer Model

Court, K., University of Puerto Rico Mayaguez Campus
Santiago, L., University of Puerto Rico Mayaguez Campus
Latorre, M., University of Puerto Rico Mayaguez Campus
BohÃ³rquez, A., University of Florida
Mora, E., (3) University of Puerto Rico Comprehensive Cancer Center, San Juan, Puerto Rico
Juan, E., University of Puerto Rico Mayaguez Campus
Rinaldi, C., University of Puerto Rico
Torres, M., University of Puerto Rico Mayaguez Campu
Originally presented on: 11/8/2013 10:35:00 - 11:00:00

Hyperthermia consists of heating an organ or tissue to temperatures between 41 to 46°C, which induces damage to cancer cells and triggers cell death and/or cell protective mechanisms. Magnetic fluid hyperthermia is being proposed to deliver heat at the desired area using magnetic nanoparticles. The main challenge in this area is to reach the necessary nanoparticle concentration in the tumor in order to be able to reach the necessary temperatures when administered through tail vein injection, to achieve this goal the temporal and spatial properties of the particles need to be assessed. For this purpose, iron oxide nanoparticles functionalized with carboxyl methyl dextran with various degrees of negative charge have been developed. Particles were designed to have different amounts of carboxyl groups per chain (30, 23 and 5 COOH). The objective of this work is to assess the in vivo temporal and spatial behavior of the particles described herein to assess the effect of particle charge on tumor accumulation given that previous work from in vitro uptake experiments with the MCF-7 cell line showed that higher uptake was obtained when higher degrees of negative charge were applied. For this purpose, in vivo pharmacokinetics and biodistribution experiments were conducted. Pharmacokinetic experiments were performed by injecting nanoparticles to cannulated Wistar rats. Each animal was administered a dose of 58mg Fe/kg. Iron concentration in plasma was measured throughout a 24 hour period at various time intervals. Results demonstrated that nanoparticle half life was significantly affected and was inversely proportional to nanoparticle charge (higher charge, lower half life). Values ranged from 16 min for those nanoparticles with higher amounts of carboxyl groups to 4 hours for the 5COOH nanoparticles. Biodistribution experiments were conducted in Nu/Nu mice inoculated subcutaneously in the torso with MCF-7 cells and after 6 weeks experiments were performed. Particles were injected through tail vein injection at 29 mg Fe/kg of nanoparticles per mice. Mice were dissected after 6 hours and iron concentration was determined in liver, spleen, kidney and tumor using a spectrophotometric technique. Higher particle uptake in the tumor was observed for those particles with the lowest charge. These results indicate that in vivo temporal and spatial behavior can be significantly affected by surface properties, therefore nanoparticles must be optimized for these purposes.

(786a) Exceptional Activity for Methane Oxidation With Catalysts Prepared By Modular Assembly of Subunits

Cargnello, M., University of Pennsylvania
Originally presented on: 11/8/2013 8:30:00 - 8:50:00

Exceptional activity for methane oxidation with catalysts prepared by modular assembly of subunits

M. Cargnello,[1,2] J. J. Delgado JaÃ©n,[3] J. C. HernÃ¡ndez Garrido,[3] K. Bakhmutsky,[4] T. Montini,[1]
J. J. Calvino Gamez,[3] R. J. Gorte,[4] P. Fornasiero[1]

[1] *Department of Chemical and Pharmaceutical Sciences, ICCOM-CNR, Consortium INSTM, University of Trieste, via L. Giorgieri 1, 34127 Trieste, Italy*

Â [2] Department of Chemistry, University of Pennsylvania, 231 S. 34th Street, Philadelphia, PA 19104, USA

[3] *Departamento de Ciencia de los Materiales e IngenierÃa MetalÃºrgica y QuÃmica InorgÃ¡nica, Facultad de Ciencias, Universidad de CÃ¡diz, Campus RÃo San Pedro, 11510 Puerto Real, CÃ¡diz (Spain)*

[3] *Department of Chemical and Biomolecular Engineering, University of Pennsylvania, 311A Towne Building, 220 S. 33rdÂ Street, Philadelphia, PA 19104, USA*

There is a critical need for improved methane-oxidation catalysts to both reduce emissions of methane, a green-house gas that will soon be regulated, and significantly enhance the efficiency of power generation in gas turbines. Presently available, emissions-control catalysts are notoriously ineffective at reducing CH_4 in exhaust streams. High-temperature combustion also results in the emission of toxic nitrogen oxides (NO_x) and CO. Combustion of CH_4 promoted by heterogeneous catalysts could utilize the available energy of methane at lower temperatures, increasing efficiency and limiting emissions by drastically reducing the required temperatures. Materials for this application must also be catalytically and mechanically stable at high reaction and flame temperatures. PdO_x is recognized as one of the best catalysts for catalytic CH_4 oxidation. Unfortunately, Pd-based catalysts tend to deactivate, both due to the loss of active surface by sintering and due to the transformation into the less active but thermodynamically favored metallic Pd phase at temperatures above 600 Â°C. Although alumina is a commonly employed support, both experimental and theoretical studies reveal that ceria (CeO_2) can improve the catalytic activity of supported Pd due to the stabilization of PdO_x species. Materials that could simultaneously enhance the performance of Pd-based catalysts at low temperatures and limit deactivation at elevated temperatures would greatly improve the viability of catalytic CH_4 combustion processes.

In this contribution, we report on a novel, hierarchical design of core-shell type catalysts that is inspired by the concepts of supramolecular chemistry. Two active building blocks, Pd and CeO_2, are prepared separately, then self-assemble and organize in solution to form supramolecular core-shell structures held together by metal ion-ligand coordination chemistry. We exploit the pre-organization of the functionalized Pd@CeO_2 core-shell structures to disperse *single units* onto a modified, catalytically inert alumina carrier. Transmission Electron Microscopy (TEM) investigations demonstrate that it is indeed possible to deposit single structures where the metal-promoter interaction is maintained even after severe thermal treatments at temperatures up to 850 Â°C. The special configuration of the hierarchical catalyst gives rise to exceptionally high and stable performance for the catalytic combustion of methane with reduced amounts of Pd and ceria. The particular geometry appears to stabilize the active phase of the catalyst, not only preventing agglomeration of palladium oxide particles during the catalytic reaction but also preventing the PdO_x from being transformed to Pd at its usual transition temperature.

(786b) Shape Dependence of Oxygen Reduction Activity On Ag Nanoparticles

Van Cleve, T., University of Michigan
Linic, S., University of Michigan
Originally presented on: 11/8/2013 8:50:00 - 9:10:00

To mitigate negative economic and environmental consequences arising from our dependence on petroleum, there is a definite need to develop alternative energy systems such as low temperature fuel cells. Low temperature fuel cells generate electrical energy and water by reacting hydrogen and oxygen gases. It has been well established that the majority of the energy losses originates from the oxygen reduction reaction at the cathode. Currently, fuel cells are limited by large kinetic overpotential, limited stability with cycling, and high material cost of platinum electrodes. With recent advances in anion exchange membranes, cheaper materials have become available as potential cathodes in low temperature alkaline fuel cells.

In particular, we are investigating Ag nanoparticles as alternative electrocatalysts for oxygen reduction. Different particle geometries have different distributions of surface facets, which have distinct chemical and catalytic properties. Specifically, we are interested in examining differences in activity between nanoparticles terminated with differing proportions of 100 and 111 facets. In this work, silver nanoparticles of several geometries are synthesized using several colloidal routes prior to being deposited onto a Vulcan XC72R support. Kinetic current and electrochemical surface area measurements are obtained using linear sweep and cyclic voltammetry with a rotating disc electrode in a three electrode electrochemical cell. Silver nanospheres are shown to have slightly higher mass and specific kinetic currents compared to nanocubes. Our results are consistent with previous work on silver single crystal electrodes, which show the Ag 111 surface is more active than Ag 100.[1]

REFERENCES

1. B.B. Bliznanc, P.N. Ross, and N.M. Markovic, J Phys Chem B 110 (2006) 4735.

(786c) Photo-Enhanced Reactive Desulfurization Via Gold-ZnO Nanoparticles

Behl, M., University of Illinois at Urbana-Champaign
Jain, P. K., University of Illinois at Urbana-Champaign

Originally presented on: 11/8/2013 9:10:00 - 9:30:00

Increasingly stringent regulations on the sulfur-content of fuels require continuous development of improved sulfur-removal technology. Technologies involving high temperature adsorptive removal of organo-sulfur impurities from raw fuel sources are widely practiced. Adsorption-based technologies that rely on metal oxides often require higher temperature to accelerate the chemical conversion, which often results in sorbent volatilization, chemical reduction, loss of sorbent surface area and physical attrition. It would be highly desirable to devise a low-temperature *catalytic adsorbent*, which can provide accelerated reaction together with higher capacity of metal oxides.

Here, we show that it is possible to use visible light to achieve room temperature reactive desulfurization by using gold-zinc oxide nano-heterostructures as adsorbents. In preliminary studies, we found that, when exposed to green light Au-loaded ZnO nanoparticles achieve room temperature catalytic decomposition of model organo-sulfur compounds (thioacetamide), which is accompanied by reactive transformation of zinc oxide to zinc sulfide. Without light, the system had to be heated to 373 K to cause the desired conversion. Photo-excited hot-electrons from gold are likely responsible for this photocatalytic process. In order to shed light on this potential mechanism, we are investigating the performance of the heterostructures with different excitation wavelengths, electron/hole scavengers, and other metals. Analysis is being carried out via X-ray diffraction, mass-spectrometery, and analytical electron microscopy. We are particularly interested in probing the possible plasmonic origin of the photocatalytic effect and synergy between the metal and oxide.

(786d) Synthesis and Characterization of Metal Carbide and Nitride Nanoparticle Catalysts for Bio-Oil Deoxygenation

Ruddy, D. A., National Renewable Energy Laboratory
Wang, J., National Renewable Energy Laboratory
Schaidle, J., The University of Michigan
Blackburn, J., NREL
Hensley, J., National Renewable Energy Laboratory
Originally presented on: 11/8/2013 9:30:00 - 9:50:00

Pyrolysis bio-oil is a promising intermediate for biomass-to-fuels processes because this liquid oil holds the potential to supply diesel and jet fuels, in addition to many valuable chemical intermediates. Elemental analysis of bio-oil reveals that despite its dark brown viscous appearance, it is compositionally different from petroleum. Bio-oils contain relatively little S and N, but have a high O content, typically near 40 wt% and as high as 50 wt%. This oxygen is present in the 400 compounds across a variety of functionalities, including acids, aldehydes, esters, alcohols, ethers, ketones, phenolics, sugars, and furans. Due to these stark chemical differences from petroleum, pyrolysis oil cannot be directly used as a transportation fuel nor can it be "dropped-in" to existing petroleum refinery processes. The oil must first be upgraded, however, efficient deoxygenation has proven to be especially challenging. Therefore, deoxygenation remains a critical obstacle to enabling transportation fuel production from pyrolysis oil at the industrial scale, and there is a continuing need for new catalytic processes and chemistries. This presentation will focus on our recent advancements in the syntheses of transition metal carbides and nitrides for use as deoxygenation catalysts. Both high-temperature tube-furnace synthetic methods (>600 °C) and lower temperature solution-phase methods (<400 °C) have been employed to prepare nanoparticle carbide and nitride materials with diameters ranging from 5 – 100 nm. Notably, size control was achieved in the high-temperature synthesis of Mo_2N by varying the gas flow rate, and average particle sizes of 5 or 50 nm can be selected. The structural characterization of Mo, W, and V (and mixed-metal) carbide/nitride will be presented, including in-situ experiments to observe compositional changes under flowing reactive gases (e.g., nitride-to-carbide). The carbide/nitride active phases were dispersed on silica for characterization relevant to deoxygenation conditions. The acidity of the catalysts was probed using FTIR spectroscopy with pyridine adsorption to differentiate and quantify Lewis acid sites and Bronsted acid sites on the catalysts. Finally, hydrogen activation is an essential parameter for deoxygenation catalysis, and the hydrogen activation chemistry of these catalysts will be presented and compared.

(786e) Cu3(BTC)2 Metal-Organic-Framework As a Solid Acid Catalyst for the Production of Biodiesel

Nambo, A., University of Louisville
Carreon, M. A., University of Louisville
Originally presented on: 11/8/2013 9:50:00 - 10:10:00

Biodiesel is a fuel derived from the transesterification of fats and oils that has similar properties to that of diesel produced from crude oil and can be used directly to run existing diesel engines or as a mixture with crude oil diesel. The main advantages of using biodiesel is that it is biodegradable, can be used without modifying existing engines, and produces less harmful gas emissions such as sulfur oxide. [1]. The transesterification reaction can be catalyzed by acid or base solids, the advantages of using solid acid catalyst are: (1) they are insensitive to free fatty acids (FFA) content, (2) esterification and transesterification occurs simultaneously, (3) eliminate the washing step of biodiesel, (4) easy separation of the catalyst from the reaction medium, resulting in lower product contamination level, (5) easy regeneration and recycling of catalyst and (6) reduce corrosion problem, even with the presence of acid species. Currently, biodiesel research is focused on exploring new and sustainable solid acid catalysts for transesterification reaction.[2]

In recent years, metal-organic frameworks (MOFs) have received considerable attention as potentially valuable gas storage, sensing, separation and catalytic materials. They are composed of an organic linker, for example a dicarboxylic acid and a transition metal complex or cluster, both units being building blocks that are designed to assemble a three-dimensional open framework. In the present work a highly active and selective Lewis acid catalyst [4], copper-benzentricarboxi metal-organic-framework Cu-MOF was prepared via hydrothermal treatment and used as catalyst for the transesterification of tryglicerides to produce biodiesel.

(786f) Hydrolysis of Glucans Grafted On Inorganic Oxides: Role of OH-Defect Sites

Katz, A., University of California, Berkeley
Gazit, O., University of California, Berkeley
Originally presented on: 11/8/2013 10:10:00 - 10:30:00

We describe the first example of OH-defect-site catalysis in water, with the hydrolysis of grafted glucans on silica and alumina. Catalytic comparisons between these two systems demonstrate that the key feature that controls hydrolysis activity is OH-defect site density on the surface, both during glucan anchoring as well as during hydrolysis catalysis. The latter enforces a high local density of defect sites, which act as an array to activate glycosidic oxygens that are constrained to be near the surface for hydrolysis catalysis. This result has significant repercussions on understanding other weak-acid site catalysts for glucan hydrolysis such as unfunctionalized carbon.

(786g) A Highly Selective Sn-Doped PMO Catalyst for the Isomerization of Glucose to Fructose

Crisci, A., University of Wisconsin-Madison

Alamillo, R., University of Wisconsin-Madison

Coller, D.,

Gallo, A., University of California

Scott, S. L., University of California, Santa Barbara

Dumesic, J. A., University of Wisconsin-Madison

Originally presented on: 11/8/2013 10:30:00 - 10:50:00

Title: A Highly Selective Sn-Doped PMO Catalyst for the Isomerization of Glucose to Fructose

Anthony J. Crisci*, Ricardo Alamillo, Daniel H. Coller, Alessandro Gallo, Susannah L. Scott, and James A. Dumesic

The potential platform chemical, 5-hydroxymethylfurfural (HMF), can be produced selectively from fructose by acid-catalyzed dehydration. Fructose itself is produced by isomerization of glucose. Some efforts to convert glucose directly to HMF using a combination of a high-boiling solvent (e.g., DMSO, ionic liquids) and a metal chloride catalyst (e.g., $CrCl_2$, $PrCl_3$, $NdCl_3$, $DyCl_3$, $YbCl_3$) have resulted in high HMF yields, but product isolation remains a challenge, adding to potential production costs. To produce HMF efficiently, the catalyst(s) must isomerize glucose to fructose, then dehydrate the fructose to HMF, all in a reaction medium from which the products and unconverted reactants are easily separated.

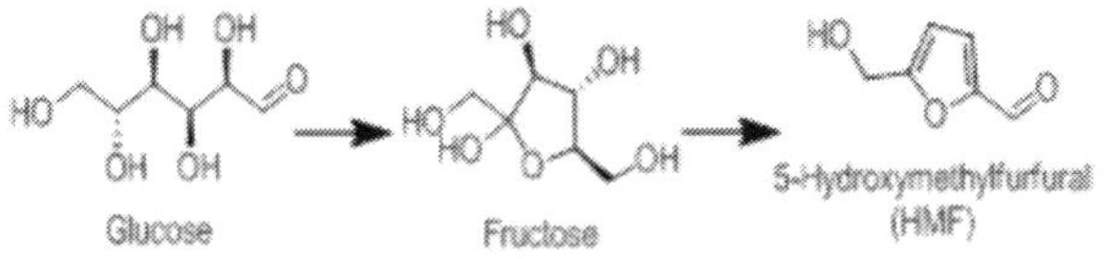

Figure 1. Reaction pathway for the direct production of HMF from glucose by first, glucose isomerization to fructose, followed by fructose dehydration to HMF.

Heterogeneous Lewis acid catalysts may provide an efficient pathway to HMF. These catalysts can be easily coupled with additional catalysts for use in a single pot reaction or in a packed bed reactor for the continuous production of HMF from glucose (to ease catalyst separation). A small number of solid Lewis acid catalysts have shown activity for glucose isomerization in an aqueous solution. Recently, Sn-β zeolite was reported to isomerize glucose to similar fructose yields as the enzyme used industrially. Sn-β zeolite has also been applied in tandem with HCl (dehydration catalyst) for the direct conversion of glucose to HMF. However, the process required a biphasic solvent system to achieve a moderate selectivity (6% single phase and 72% bi-phasic).

In addition to being difficult and time consuming to synthesize, post-synthesis functionalization of crystalline Sn-β zeolite with organosilanes (e.g., MPTMS, APTMS) is limited to silanols generated by surface defects. Similar to Sn-β, Lewis acidity can be incorporated into ordered mesoporous silicas through the addition of metal chlorides during the sol-gel synthesis. These Sn-modified (PMOs) are quick to synthesize (2 days), easily modified with additional functionalities, and potentially scalable. A series of phenylene-bridged periodic mesoporous organosilicas (PMOs) were synthesized and doped with various amounts of Sn (from ca. 0.5 to 6 wt%). A Sn-PMO (0.5 wt%) containing alkylsulfonic acid groups was also functionalized. The materials were extensively characterized (elemental analysis. XRD, TEM/STEM, N_2 physisorption, solid-state NMR, IR, EXAFS and TGA) and glucose isomerization reaction studies were conducted. Reaction studies were also performed using Sn-β zeolite, and a Sn-SBA-15 for comparison.

Each reaction was ran for 2 hours in a 2 wt% glucose solution in a mono-phasic low boiling 4:1 THF:water (w:w) solvent at 413 K. Both the Sn:glucose molar ratio (1:100) and the total catalyst mass (diluted with Sn-free PMO) were held constant. PMO with isolated Sn sites (0.4 wt%) converted 28 % of the glucose to fructose with a selectivity of 47 %. However, as the Sn-loading increased, both glucose conversion and selectivity to fructose decreased, compared to the 0.4 wt% Sn-PMO. Sn-β zeolite under these conditions fully converted glucose, but did not yield any fructose and possessed a low selectivity to HMF (7%). The reaction time was reduced to compare results at the same conversion levels to that of the Sn-PMOs. Between 80 and 90% glucose conversion, the selectivity to fructose is 30-50%. Once glucose is completely consumed, the desired products (fructose and HMF) react and the corresponding selectivity drops significantly and rapidly. In contrast, the 0.4 % Sn-PMO maintains its selectivity to desired products around 40% even at higher conversions (> 85%). Sn-PMOs are capable of selectively isomerizing glucose to fructose in THF/Water without rapidly consuming glucose. The bifunctional PMO (with Lewis and Brønsted acid sites) at 30 % conversion had a selectivity to fructose and HMF of 27 and 14 %, respectively. With double the reaction time, the HMF selectivity improved to 37 % while fructose selectivity decreased to 4 % at 52 % conversion. HMF selectivity is expected to improve with the optimization of the Lewis: Brønsted acid ratio and the reaction conditions. Sn-PMOs are capable of selectively isomerizing glucose to fructose in THF/Water without rapidly consuming glucose. Through the incorporation of alkylsulfonic acid sites, reasonable selectivity to HMF from glucose is achieved without the use of costly and/or separation unfriendly solvents.

(794a) Formation of Nanotubes By Catanionic Mixtures of Drug-Amphiphiles

Lin, Y. A., Johns Hopkins University
Cheetham, A. G., Johns Hopkins University
Ou, Y. C., Johns Hopkins University
Cui, H., Johns Hopkins University
Originally presented on: 11/8/2013 8:30:00 - 8:45:00

Self-assembly of therapeutic agents into well-defined architectures not only creates a new means for the development of self-delivering anticancer drugs or molecular probes, but the self-assembly process itself also provides new insight for the rational design of micro- or nanostructures with increased complexity. Tubular structures, in particular, have drawn much research interest in various biomedical and electronical areas, owing to new properties associated with their hollow nature. There is a great deal of literature that reports the co-assembly of oppositely charged amphiphilic molecules into bilayer-based structures, such as vesicles, icosahedra, nanodiscs, nanobelts, and nanotubes. It is generally believed that the formation of ion pairs by oppositely charge surfactants alters the packing geometry as a result of charge screening. In this context, we report our use of catanionic mixtures (CAMs) of drug amphiphiles containing a 36% fixed drug loading to construct tubular structures. We attempted to elucidate the potential factors that contribute to the construction of such interesting tubular morphologies.

(794b) Investigation of Self-Assembled Peptide Amphiphile Micelles for Targeting Early Stage Atherosclerotic Plaques

Drews, L. B., University of California, Berkeley
Chung, E. J., University of Chicago
Wonder, E., Institute for Molecular Engineering
Tirrell, M., University of Chicago
Originally presented on: 11/8/2013 8:45:00 - 9:00:00

Atherosclerosis, the leading cause of death in the United States, is characterized by the formation of plaques in the arterial wall, which progress through multiple stages. If left untreated, these plaques can become unstable and rupture, leading to potentially fatal consequences. Delivery of therapeutic drugs to these plaques using vehicles targeted directly to markers located on the plaque could allow for development of new therapies for atherosclerosis. Our group has shown the ability to target fibrous cap formation, characteristic of late stage plaques, using self-assembled peptide amphiphile (PA) micelles [1]. Detection of plaques at an earlier stage, however, could lead to the potential for treatment prior to these plaques becoming unstable. Here, we discuss micelles that are designed to bind to vascular cell adhesion molecule-1 (VCAM-1), an inflammatory marker of early stage atherosclerotic plaques. PAs are formed by conjugating a hydrophobic, lipid tail to a hydrophilic peptide headgroup. PAs under investigation in this study are composed of a peptide headgroup that recognizes and binds to VCAM-1, a 2,000 molecular weight polyethylene glycol (PEG) spacer, and a diC_{18} hydrophobic tail to form a PA and resulting micelle termed DSPE-PEG-VCAM.

Dynamic light scattering (DLS) of DSPE-PEG-VCAM micelles shows the hydrodynamic diameter to be approximately 20 nm. The large PEG spacer helps to confer a large headgroup and sphere formation is both predicted and observed using negative stain TEM. The critical micelle concentration (CMC), or the concentration at which micelles begin to form, was found to be approximately 60 μM as determined by the DPH method [2]. Finally, circular dichroism shows that the peptide in solution as well as DSPE-PEG-VCAM, retain similar secondary structures, shown by fitting the resulting spectra (peptide: 33±1% beta sheet, 67±1% random coil; PA: 28±1% beta sheet, 72±1% random coil).

Human umbilical vein endothelial cells (HUVEC) and aortic endothelial cells (AEC) were shown to remain viable in the presence of DSPE-PEG-VCAM micelles. Furthermore, *in vitro* results have shown localization of DSPE-PEG-VCAM micelles to aortic endothelial cells using confocal microscopy with micelles fluorescently labeled with both fluorescein and Cy7. This may indicate binding between the micelle and the VCAM-1 marker and these results will be further discussed by examining the binding specificity of DSPE-PEG-VCAM micelles for AECs versus control micelles. Specifically, the effects of micelle composition and PA formulation with the ability of the micelle to associate with AECs will be discussed. Future work will determine where DSPE-PEG-VCAM PAs localize within AECs as well as assessing the ability to deliver a hydrophobic payload to both cells in an *in vitro* setting and atherosclerotic plaques *in vivo*.

[1] Peters, D, et al. Targeting Atherosclerosis by Using Modular, Multifunctional Micelles. *PNAS*. 2009; **106**: 9815-9819.

[2] Chattopadhyay, A and London, E. Fluorimetric Determination of Critical Micelle Concentration Avoiding Interference from Detergent Charge. *Analytical Biochemistry*. 1984; **139**: 408-412.

(794c) Engineering Large Self-Assembled Amyloid Fibers

Ridgley, D., Virginia Tech
Barone, J. R., Virginia Tech
Originally presented on: 11/8/2013 9:00:00 - 9:15:00

Amyloid fibers are robust biomaterials capable of self-assembling from a variety of proteins. It is generally accepted that any protein can assemble into an amyloid structure. However, few systems have been able to assemble at physiological conditions or to very large length scales. Here we introduce a system that utilizes peptide mixtures to form large micrometer sized amyloid fibers at near physiological conditions.[1] A short, hydrophobic template peptide interacts with an α-helical adder protein to cooperatively self-assemble into an amyloid fiber. By altering the *in vitro* solution conditions and peptides in the mixture it is possible to engineer a fiber with a tailored morphology (rectangular or cylindrical cross-section) and modulus (0.1-2.5 GPa).[2] Atomic force microscopy (AFM) shows fiber self-assembly occurs in four stages: 1) peptides self-assemble into elementary protofibrils (width=15-30 nm), 2) protofibrils laterally and vertically stack into fibrils (width=60-120 nm) that 3) aggregate into large fibrils (width=150-480 nm), and upon drying 4) large fibrils assemble into amyloid fibers (width=10-20 μm).[3] The type of peptides in the mixture can be varied to initiate morphological and property differentiation after stage 2. Thus, the hierarchical self-assembly, morphology, and rigidity of the fiber can be designed and predicted based on overall peptide properties in the mixture. In recent work, we have demonstrated that the fiber morphology and properties can be encoded at the genetic level by re-programming cells to extracellularly express template and adder peptides.

References

1 Ridgley, D. M., Ebanks, K. C. & Barone, J. R. Peptide Mixtures Can Self-Assemble into Large Amyloid Fibers of Varying Size and Morphology. *Biomacromolecules* **12**, 3770-3779, doi:10.1021/bm201005k (2011).

2 Ridgley, D. M., Claunch, E. C. & Barone, J. R. The effect of processing on large, self-assembled amyloid fibers. *Soft Matter* **8**, 10298-10306, doi:10.1039/c2sm26496j (2012).

3 Ridgley, D. M. & Barone, J. R. Evolution of the Amyloid Fiber over Multiple Length Scales. *ACS nano* **7**, 1006-1015, doi:10.1021/nn303489a (2012).

(794d) The Role of Spacers On the Self-Assembly of ssDNA Aptamer-Amphiphiles Into Micelles and Nanotapes

Pearce, T. R., University of Minnesota
Waybrant, B., University of Minnesota
Kokkoli, E., University of Minnesota
Originally presented on: 11/8/2013 9:15:00 - 9:30:00

Biopolymers formed from nucleic and amino acid building blocks have been used to create functional supramolecular and multi-dimensional structures that rely on extensive intermolecular interactions to produce biocompatible materials with structural order on the nanometer scale for applications in electronics, biosensing, and tissue engineering. Single-stranded DNA (ssDNA) is often used to create these materials as the DNA bases provide an intrinsic molecular code that can be exploited to allow for programmed assembly of structures based upon Watson-Crick base-pairing. However, engineering complex materials and devices from biopolymers alone requires careful design to ensure that the intrinsic forces responsible for organizing them are capable of creating the desired products. Additional control over supramolecular assembly can be achieved by chemically modifying the biopolymers with hydrophobic moieties to create amphiphilic molecules, which adds the hydrophobic effect to the list of contributing forces that help dictate the self-assembly process. This strategy has been widely utilized to form peptide-amphiphiles consisting, typically, of hydrocarbon chains conjugated to hydrophilic peptide headgroups. But despite widespread research utilizing ssDNA for bionanotechnology applications, ssDNA-amphiphiles have not been studied as thoroughly as peptide-amphiphiles. In this work we created a set of ssDNA aptamer-amphiphiles composed of hydrophobic tails, hydrophilic ssDNA aptamer headgroups and different spacer molecules linking these groups together. Through the use of cryo-transmission electron microscopy (cryo-TEM), small angle x-ray scattering (SAXS), and circular dichroism (CD), we found that the aptamer-amphiphiles can assemble into a variety of structures depending on the spacer used. For the first time we demonstrate the presence of self-assembled aptamer-amphiphile nanotapes, and show that the choice of the spacer used in the design of aptamer-amphiphiles influences the self-assembly and secondary structure of the aptamer-amphiphiles.

(794e) An Implicit Solvent Coarse-Grained Model for Design and Characterization of Bio-Nanostructured Soft Materials

Aydin, F., Rutgers University
Dutt, M., Rutgers University
Originally presented on: 11/8/2013 9:30:00 - 9:45:00

Via the use of coarse-grained Molecular Dynamics simulation technique, we design and characterize multi-component bio-nanostructured soft materials. Our objective is to use the self-assembly of amphiphilic lipid molecules to generate a stable vesicle through the use of suitable implicit solvent coarse-grained interaction potentials. The amphiphilic lipid molecules are composed of a hydrophilic head group and two hydrophobic tails. We investigate the effect of temperature on the stability of the morphology and the mechanical properties of these lipid vesicles. We find the lipid bilayer vesicles to maintain their morphology with changes in the shape and mechanical properties with temperature. We have extended the model to study the self-assembly of two species of phospholipids to form stable two-component lipid vesicles. We find that we can tune the degree of phase segregation between the two lipid species through their distinct effective chemical specificity and molecular architecture. The phase segregation process is characterized by computing the pure species domain populations, and is shown to influence the structure, and mechanical properties of the lipid vesicle. We further demonstrate the combined effect of temperature and phase segregation on the morphology and mechanical properties of the lipid vesicle. These results can be used to design new soft materials for applications in medicine, sensing and energy.

(794f) Self-Assembly of Complex Nucleic Acid Nanostructures From Single-Stranded RNA Tiles

Green, A. A., Harvard University
Silver, P., Harvard Medical School
Collins, J. J., Howard Hughes Medical Institute, Boston University
Yin, P., Harvard Medical School
Originally presented on: 11/8/2013 9:45:00 - 10:00:00

Programmed self-assembly using nucleic acids is a powerful method for producing nanostructures with precisely defined shapes and sizes. Recently, a new paradigm has emerged for constructing such structures using short DNA oligonucleotides or tiles. This single-stranded tile (SST) approach utilizes multiple nucleic acid strands each containing four binding sites that are complementary to the four nearest neighbors of the tile in the specified structure. This simple yet versatile construction motif has been successfully applied to fabricating complex one-, two-, and three-dimensional DNA architectures. SST nanostructures however have so far been limited to DNA, leaving RNA with its distinct structural, chemical, and biological properties completely unexplored as an SST building material. Here, we describe how the SST paradigm can be successfully applied to the assembly of complex RNA-based molecular architectures. These RNA SST structures are constructed from enzymatically synthesized RNA tiles that are subsequently pooled and assembled through thermal annealing in a one-pot process. We have used this approach to construct one-dimensional RNA nanoribbons and nanotubes with prescribed widths and diameters, respectively. Furthermore, we have assembled complex two-dimensional lattices with well-defined dimensions composed of up to 95 unique RNA tiles. These RNA nanostructures offer distinct biochemical properties compared to their DNA counterparts. Notably, we have found that RNA SST assemblies can be successfully processed by recombinant human dicer enzyme into siRNAs and offer the potential to encode dozens of siRNAs within a single nanostructured delivery vehicle.

(794g) DNA Origami Based Fabrication of Reconfigurable Nanostructures

Chen, H., Purdue University
Choi, J. H., Purdue University
Originally presented on: 11/8/2013 10:00:00 - 10:15:00

DNA origami is a relatively new technique that uses hundreds of short "staple" DNA oligonucleotides to direct the folding of a single stranded DNA scaffold into a predefined structure. Such folded DNA structures possess nanometer precision and can be used as high precision templates for nanofabrication. Here, we show that the chirality of the nano-helices can be readily controlled by rationally design the complementary DNA strands that link adjacent DNA origami monomers together. The strategy we present here offers a facile way to create nano helices and tune their unique properties of these chiral structures. We present the use of DNA origami as a template for arranging nano objects into helical configuration. Unlike previous works that uses single DNA origami as template, we assemble multiple DNA origami monomers into larger nano-helices whose dimensions approach the wavelength of visible light. Such larger structures are capable to template more nano objects with larger dimensions and can lead to stronger chiral properties. Most importantly, we present a novel strategy which can be used to reversibly reconfigure the geometry of hierarchically assembled DNA structures.

(794h) Synthesis of Highly Stable Self-Assembled Virus-Like Particles

Lu, Y., Stanford University
Swartz, J. R., Stanford University
Originally presented on: 11/8/2013 10:15:00 - 10:30:00

Virus-like particles (VLPs) mimic the structural skeleton of real viruses, but are non-infectious as they lack the viral genome. They are typically stable over wide ranges of pH, temperature, and solvent and have repetitive surfaces that can display a high density of intrinsic or attached molecules. Because of these unique features, VLPs are particularly attractive as vaccines as they offer in vivo stability, trafficking to lymph nodes, and stimulation of B and T cell responses to displayed epitopes.

In this study, we describe *Escherichia coli*-based cell-free protein synthesis (CFPS) as a method to rapidly produce Hepatitis B core protein (HBc) VLPs that display unnatural amino acids as unique conjugation sites. However, repeated observations indicated that these nanoparticles were at least partially disassembled by the click conjugation conditions. We therefore took advantage of the repetitive and modular nature of VLPs to introduce 240 new disulfide bonds to totally stabilize the nanoparticles with covalent connections. VLPs have been successfully produced in different expression host systems (bacteria, yeast, insect, plant and mammalian), but CFPS enables rapid production of VLPs at high yields (~0.5 g/L) in a few hours. This CFPS technology has been shown to be scalable from the microliter to the liter scale. Taking advantage of the open CFPS environment and post-expression processing, the virus capsid coat proteins self-assemble into stable VLPs under reducing conditions and can then be cross-linked using a mild oxidizing agent.

(794i) Nanoscale Patterning of Membrane-Bound Proteins Formed Through Curvature-Induced Partitioning of Phase-Specific Receptor Lipids

Ogunyankin, M. O., University of California, Davis
Longo, M. L., University of California, Davis
Sasaki, D., Sandia National Laboratories
Originally presented on: 11/8/2013 10:30:00 - 10:45:00

Lipid mixtures self-assemble to form planar bilayers containing coexisting lipid phases in many compositional regimes. Coexisting lipid phases each have distinct compositions that can selectively partition optically active dyes, ligands, and membrane proteins. Therefore, precise arrangements of these phases in continuous lipid bilayers could be used to create displays of proteins, DNA strands, optically, chemically, or electrically active molecules, or nanoparticles. Curvature-based sorting of functionalized lipids that comprise or partition into specific lipid phases could serve as a way of templating active molecules. A variety of methods have been developed to attach functional molecules to lipid headgroups. These include covalent bonding and linkers such as single-stranded DNA, biotin, and glycan-phosphatidyl inositol. Lipid headgroups can also be functionalized with a metal chelate group such as iminodiacetic acid (IDA) and nitrilotriacetic acid (NTA). These chelate groups coordinate to divalent transition-metal ions such as Cu^{2+}, Ni^{2+}, and Zn^{2+} and sequester proteins rich with histidine moieties. This work describes a technique for forming high-density arrays and patterns of membrane-bound proteins through binding to a curvature-organized compositional pattern of metal-chelating lipids (Cu^{2+}-DOIDA or Cu^{2+}-DSIDA). In this bottom-up approach, the underlying support is a poly(methyl methacrilate) (PMMA) e-beam formed, square lattice pattern of hemispheres. This curvature pattern sorts Cu^{2+}-DOIDA to the 200 nm hemispherical lattice sites of a 600 nm ′ 600 nm unit cell in L_d - L_o phase separated lipid multibilayers. Binding of histidine-tagged green fluorescent protein (His-GFP) creates a high density array of His-GFP-bound pixels localized to the square lattice sites. In comparison, the negative pixel pattern is created by sorting Cu^{2+}-DSIDA in L_d - L_b phase separated lipid multibilayers to the flat grid between the lattice sites followed by binding to His-GFP. Lattice defects in the His-GFP pattern lead to interesting features such as pattern circularity. We also observe defect-free arrays of His-GFP that demonstrate perfect arrays can be formed by this method suggesting the possibility of using this approach for the localization of various active molecules to form protein, DNA, or optically active molecular arrays.

(794j) Reaction-Limited Fusion Mechanism of Zwitterionic Phospholipid Nanodiscs

Xia, Y.,
Nieh, M. P., University of Connecticut
Originally presented on: 11/8/2013 10:45:00 - 11:00:00

Lipid-based bilayered nanodiscs can be self-assembled by a mixture of phospholipids in aqueous solutions. It has been reported that the nanodiscs grow (fuse) upon dilution via a small angle neutron scattering (SANS) study. In this presentation, we will evaluate the kinetics of zwitterionic nanodiscs made of 1,2-dimyristoyl-*sn*-glycero-3-phosphocholine (DMPC) and 1,2 –dihexanoyl-*sn*-glycero-3-phophocholine (DHPC) at a molar ratio of 3.2 upon dilution by free DHPC (pre-micellation) solutions of different concentrations using time-resolved dynamics light scattering (DLS). The result shows that the growth rate of nanodisc induced by dilution strongly depends on the DHPC concentration, suggesting that the disc fusion process is controlled by the surrounding DHPC concentration in solutions. This result provides the direct evidence that the fusion of nanodiscs is attributed to disturbance of equilibration between the concentrations of DHPC in solution and nanodisc upon dilution. The equilibrium constant is also investigated to identify whether it is thermodynamically controlled. This result will be compared with that of charged nanodiscs consisting of DMPC, DHPC, and small amount of charged 1,2-dimyristoyl-sn-glycero-3-phospho-(1'-rac-glycerol) (DMPG) to further understand the charge effect on the fusion mechanism.

(810a) Breakdown in the Wetting Transparency of Graphene

Shih, C. J., Massachusetts Institute of Technology
Wang, Q. H., Massachusetts Institute of Technology
Lin, S., Massachusetts Institute of Technology
Park, K. C., Massachusetts Institute of Technology
Strano, M. S., Massachusetts Institute of Technology
Blankschtein, D., Massachusetts Institute of Technology
Originally presented on: 11/8/2013 12:35:00 - 12:53:00

We develop a theory to model the van der Waals interactions between liquid and graphene, including quantifying the wetting behavior of a graphene-coated surface. Molecular dynamics simulations and contact angle measurements were also carried out to test the theory. We show that graphene is only partially transparent to wetting and that the predicted highest attainable contact angle of water on a graphene-coated surface is 96. Our findings reveal a more complex picture of wetting on graphene than what has been reported recently as complete "wetting transparency."

(810b) Confinement of a DNA Dodecamer Onto Pristine Carbon Nanotubes: Stability of the Canonical B Form

Cruz, F. J. A. L., Universidade Nova de Lisboa

Mota, J. P. B., Universidade Nova de Lisboa

de Pablo, J. J., University of Wisconsin-Madison

Originally presented on: 11/8/2013 12:53:00 - 13:11:00

Confinement of a DNA Dodecamer onto Pristine Carbon Nanotubes and the Stability of the Canonical B Form

Fernando J.A.L. Cruz[1,2], Jose P. B. Mota[1] and Juan J. de Pablo[2,3]

[1] Requimte/CQFB, Dept. Chemistry, Universidade Nova de Lisboa, Caparica, Portugal. [2] Dept. Chem. Bio. Eng., University of Wisconsin-Madison, Madison, WI, USA. [3] Institute of Molecular Engineering, University of Chicago, Chicago, Illinois, USA.

fj.cruz@fct.unl.pt

<>I. Introduction

Deoxyribonucleic acid (DNA) and single-walled carbon nanotubes (SWCNTs) are prototypical one-dimensional structures; the former plays a central role in chemical biology and the latter holds promise in nanotechnology applications.[1-3] From the point of view of biological purposes and DNA manipulation, carbon nanotubes have been proposed as templates for DNA encapsulation, intracellular penetration via endocytosis and delivery of biological payloads and ultrafast nucleotide sequencing. While structure in its natural form and environment is well established (*e.g.* B-DNA in aqueous solution), their interactions have been the subject of intense investigation, nonetheless, the corresponding molecular-level phenomena remain rather unexplored. Is confinement spontaneous from a thermodynamical point of view (free-energy)? How important are the conformational properties of the double strand upon confinement (entropy)? What is the effect of the media electrical charge density on the negatively charged DNA backbone (ionic strength)? How does the confinement process depend on intrinsic properties of the nanotube such as diameter (D)? Previous theoretical and experimental work has focused almost exclusively on exoadsorption of DNA at the SWCNTs external walls, overlooking the possibility of endohedral confinement; the energetics of the latter is one of the main purposes of the present work. Confinement driven adsorption is a phenomenon that is utterly unmapped, and most of the previous work has been focused in systems at temperatures away from the physiological value and/or ionic strength, thus precluding extrapolation of results to *in vivo* conditions. We'll address this issue using molecular dynamics and enhanced sampling techniques (metadynamics, umbrella sampling) to probe the encapsulation mechanism of an atomistically detailed DNA dodecamer, onto SWCNTs of different diameters, employing precise physiological conditions, 310 K and [NaCl] = 134 mM.

II. Results and Discussion

Confinement of DNA onto a (51,0) topology, $D = 4$nm, is fast and becomes complete before 16 ns of observation time (Fig.1A); initially ($t = 0 - 2$ ns), the double strand is in the bulk, and as it diffuses towards the SWCNT undergoes structural rearrangements leading to minor increases in pitch length, P, and end-to-end distance, L. After 2 ns the dodecamer is already at the nanopore entrance, experiencing strong van der Waals attractions towards the solid, leading to complete encapsulation at 15.42 ns, after which it relaxes and reassumes its unconfined structure ($P = 3.4$ nm, $L = 4.1$ nm) consistent with the geometrical characteristics of B-DNA in bulk.[4] It is interesting to observe that confinement is unalterable, and the biomolecule never returns back to the bulk solution during the observation time window, always maintaining direct contact with the solid at a distance of closest approach of *ca.* 0.26 nm; nonetheless, the confined molecule retains its translational mobility diffusing along the nanopore main axis (Fig. 1B). In contrast, the narrowness of a (40,0) topology, $D = 3$ nm, compared to the DNA skeletal diameter, inhibits encapsulation even for an observation time of 0.1 μs. We determined the characteristic lengths, radius of gyration (R_g) and its projection along the nanotube main axis (R_g^z), as well as the root-mean squared deviation (*RMSD*) compared to the B-DNA dodecamer used as starting configuration; results are plotted in Fig. 1C. Because the *RMSD* compares the structure at any time t with the original DNA structure ($t = 0$), the blue curve recorded in Fig. 1C suggests a minor relaxation of the double-strand from the crystal structure to accommodate liquid state flexibility, whilst maintaining the relative average distance between each atom in the double-strand. After 20 ns, results have smoothly converged to average values of $RMSD = 0.36 \pm 1.9 \times 10^{-3}$ nm and R_g^z = $1.02 \pm 1.5 \times 10^{-3}$ nm, with the latter indicating an alignment of the biomolecule with the nanopore main axis. The canonical Watson-Crick H-bonds network is roughly maintained throughout confinement, exhibiting probability distributions corresponding to more than 75 %; the R_g^z = 0.8 nm depression observed at 56.6 ns is transient and matches a total number of canonical H bonds of 28.

The thermodynamical stability of encapsulation is probed by the free-energy (F) differences associated with the process, using well-tempered metadynamics.[5] An inspection of the resulting three-dimensional surface (Fig.2) reveals the existence of five distinct free-energy minima, sharing in common the fact that all are located at discrete positions along the nanopore internal volume, $\xi_1 < 1.8$ nm; the absolute minimum at $\xi_1 = 0.117$ nm evidences the SWCNT center as the energetically most stable region subsequent to encapsulation. In order to escape from those deep free-energy valleys, $F \sim -40$ kJ/mol^{-1}, the DNA molecule has to overcome large energetic barriers, rendering the exit process towards bulk solution a thermodynamically expensive one. The number density maps depicted in Fig.3, obtained from atomically detailed mass histograms, corroborate the finding that free-energy is minimum at the nanopore center; Fig.3 also reveals a thin and cylindrically-shaped exclusion volume close to the hydrophobic solid walls, arising from the repulsion between the latter and the heavily charged phosphate groups. It is remarkable to observe that the exact position of the free-energy minima oscillate little around $\xi_2 = 4.1$ nm, coherent with $L = 4.1$ nm characteristic of a B–DNA structure (Fig.1).

As far as we are aware these observations are the first of their kind, and they come to pave the way for the design of smart nanotube based devices for *in vivo* DNA encapsulation.

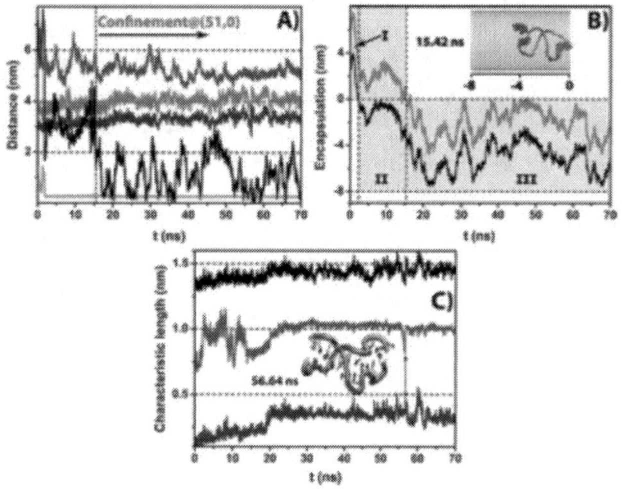

FIGURE 1 - Encapsulation of DNA onto a (51,0) SWCNT. A) Kinetics. Encapsulation is complete after 15.42 ns and the double-strand never returns back to the bulk solution during the observation time window; DNA maintains its translational mobility within the nanopore. Notice that a 2 nm c.o.m. distance between the biomolecule and the solid corresponds to the threshold bellow which complete encapsulation occurs: *black)* distance between centres of mass of DNA and (51,0) SWCNT, *dark red)* distance between centres of mass of DNA and (40,0) SWCNT, *green)* DNA end-to-end length, *blue)* DNA pitch length, *grey)* minimum distance between any DNA atom and the SWCNT. **B) State diagram.** Distances between terminal nucleobase pairs and the SWCNT entrance, projected along the nanotube main axis: *black)* terminus 1 and *red)* terminus 2. The horizontal dashed lines correspond to the SWCNT boundaries and the inset magnification depicts the first instant just after complete encapsulation, representing the terminus 1 atoms in black and the terminus 2 atoms in red. The encapsulation mechanism can be described by a three-step kinetics: *I)* fast diffusion of bulk DNA towards the nanopore (0 – 2 ns), *II)* strong van der Waals attraction towards the solid, leading to confinement of terminus 1 at 2.33 ns, and finally *III)* penetration of terminus 2 onto the nanopore volume resulting in complete encapsulation of DNA. Note the confined molecule maintains its translational mobility along the nanopore axis. **C) Characteristic lengths.** *black)* radius of gyration, R_g, *grey)* radius of gyration z-component, R_g^z, and *blue)* root mean-squared deviation, *RMSD*. After confinement, molecular configuration deviates minimally from the canonical B-DNA form, $RMSD = 0.36 \pm 1.9 \times 10^{-3}$ nm, and maintains a quasi-linearity about the nanopore main axis.

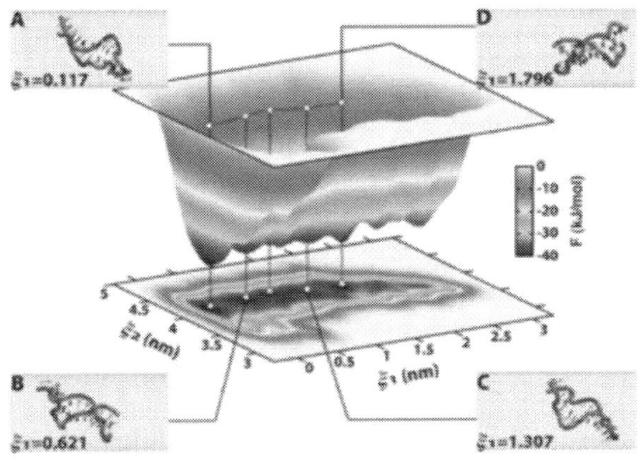

FIGURE 2 - Free energy landscape of DNA@(51,0)SWCNT hybrid. ξ_1 is the distance between centres of mass of the DNA and SWCNT, projected along the nanopore main axis, and ξ_2 is the absolute distance between (GC) termini on opposite sides of the double-strand, equivalent to the DNA end-to-end length. The several free-energy minima along ξ_1 demonstrate that the molecule is relatively mobile to translocate along the nanotube, however, the absolute minimum at $\xi_1 = 0.117$ nm indicates that the nanopore center is the energetically favoured region. Interestingly, all the ξ_1 minima are located along a *quasi*-linear path defined by ξ_2 Å 4.1 nm highlighting the enhanced thermodynamical stability corresponding to the canonical B form. Snapshots were taken at (ξ_1, ξ_2) nm = *A)* (0.117, 4.112), *B)* (0.621, 4.164), *C)* (1.307, 4.164) and *D)* (1.796, 4.115).

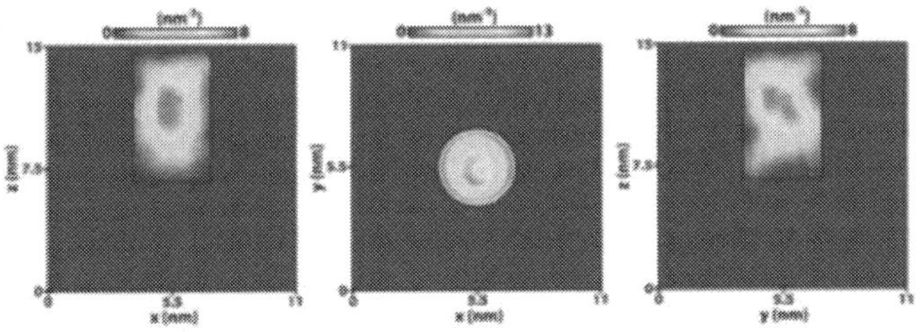

FIGURE 3 – Density maps of DNA@(51,0)SWCNT. The existence of a cylindrical exclusion volume close to the walls, r Å 0, is the direct consequence of direct repulsion between the heavily charged phosphate groups and the hydrophobic solid. The dashed lines indicate the boundaries of the nanotube.

References

(1) H. Kumar *et al*, *Soft Matter* **2011**, *7*, 5898.

(2) B.M. Venkatesan, R. Bashir, *Nature Nano.* **2011**, *6*.

(3) A. D. Franklin *et al*, *Nano Lett.* **2012**, *12*, 758-762.

(4) J.M. Vargason *et al*, *Proc. Nat. Acad. Sci.* **2001**, *98*, 7265.

(5) A. Barducci *et al*, *Phys. Rev. Let.* **2008**, *100*, 020603.

(810c) On the Intrinsic Wettability of Graphene

Li, L.,
Li, Z.,
Wang, Y.,
Kozbial, A.,
Zhou, F.,
Liu, H.,
Originally presented on: 11/8/2013 13:11:00 - 13:29:00

On the Intrinsic Wettability of Graphene

Zhiting Li[2], Yongjin Wang[1], Andrew Kozbial[1], Feng Zhou[2], Lei Li[1]and Haitao Liu[2]

[1]Department of Chemical and Petroleum Engineering and [2]Department of Chemistry, University of Pittsburgh, Pittsburgh, PA, 15261, U.S.A.

The water wettability of graphene plays a key role in many important applications, not only defining the interaction between graphene and its environment but also directly impacting many of its properties, such as charge doping, carrier mobility, wetting, and adhesion. Here we report the effect of airborne hydrocarbon contaminants on the wettability of supported CVD graphene. We show that the water contact angle (WCA) of a freshly prepared supported graphene surface increases upon exposure to ambient air. Infrared (IR) spectroscopy and X-ray photoelectron spectroscopy (XPS) showed that airborne hydrocarbon adsorbed onto graphitic surface during this process. Both thermal annealing and controlled UV/O_3 treatment could partially remove the hydrocarbon contaminant and resulted in a concurrent decrease in the WCA. Our findings indicate that graphitic surfaces are more hydrophilic than previously believed and suggest that the previously reported hydrophobic nature of graphitic surface is due to unintentional hydrocarbon contamination from ambient air. The theoretical calculation results also support the experimental findings.

(810d) The Structure and Rheology of Carbon Nanotubes Adsorbed At An Air-Water Interface

Vora, S., University of Connecticut
Patanwala, H., University of Connecticut
Guo, Y., University of Connecticut
Ma, A. W. K., University of Connecticut
Originally presented on: 11/8/2013 13:29:00 - 13:47:00

Particles of appropriate size and wetability adsorb strongly at gas-liquid and liquid-liquid interfaces and lower the interfacial energy. The ability of particles to stabilize foams and emulsions was well documented by Ramsden and Pickering more than a century ago. The effect of particle shape has, however, remained largely unexplored and unexploited. We hypothesize that the particle shape matters and that rod-like particles will increase the number of particle-particle contacts, leading to the formation of a stronger layer at the interface relative to spheres. To examine this hypothesize, we have chosen carbon nanotubes (CNTs) as a model system given their high aspect ratio (>1000). In this presentation, we will report our recent findings on the microstructure and interfacial rheology of a CNT-decorated interface. The latter is directly related to the mechanical integrity and stability of the interface. Our findings indicate that CNTs adsorb strongly at the air-water interface, leading to a significant increase in surface pressure. The highly compressible nature of an interface also enables the creation of two-dimensional CNT "liquid crystals" by compressing a CNT-decorated interface.

(810e) Understanding the Interaction of Surfactants and Other Adsorbents With Single Walled Carbon Nanotubes

Xu, J., University of Florida
Clar, J. G., University of Florida
Bonzongo, J. C. J., University of Florida
Ziegler, K. J., University of Florida
Originally presented on: 11/8/2013 13:47:00 - 14:05:00

The ability to understand and control the surface structure of nanomaterials is a difficult barrier that must be overcome for their practical application in biology and medicine. Single walled carbon nanotubes (SWCNTs) have a large specific surface area and, thus, a high adsorption capacity. The high adsorption capacity can be utilized for practical applications in drug loading but also has important implications on their fate and toxicity as the materials enter natural systems during production and disposal. The changes that occur to the localized region surrounding the SWCNT can be tracked by photoluminescence (PL), which has been proven to be a quick and reliable probe of SWCNT surfaces. In this approach, the SWCNTs act like a chemical sensor that qualitatively detects the adsorption of molecules around itself, such as drugs, chemicals, or environmental species. While PL provides information about direct contact with the SWCNT, Proton NMR is another probe that we have started to use to track the specific interaction of surfactants with SWCNTs and other molecules adsorbed onto their surface. Changes to the NMR chemical shift of the surfactant molecules provides information about their interaction with either the SWCNTs or other molecules. These changes are shown to correspond to molecular structural changes. By combining PL and Proton NMR analysis, we aim to understand the formation and morphology of both surfactant and solvent structures around SWCNTs. We find characteristic trends in the PL and Proton NMR changes that help elucidate the different interactions occurring between SWCNTs, surfactants, and adsorbents. The ability to understand and control the structure around SWCNTs while interacting with different adsorbents has novel applications in drug delivery and will also improve our understanding of the fate and transport of SWCNTs in natural systems.

(810f) Energetics, Structure, and Dynamics of Graphene-Based Nanoparticles

McNutt, N., University of Tennessee
Wang, Q., University of Tennessee Knoxville
Rios, O., Oak Ridge National Laboratory
Keffer, D., University of Tennessee
Originally presented on: 11/8/2013 14:05:00 - 14:23:00

The understanding of the behavior of carbon materials that include nanoparticles composed of a few graphene domains would be greatly aided by the availability of basic properties of the isolated crystalline nanoparticles. In this work, the energetics, structure, and dynamics of three isolated graphitic nanoparticles of different sizes were studied using Molecular Dynamics (MD) simulations. Results show that the bonded potential energies (stretching, bending, and torsion) and in-plane nonbonded potential energy become less favorable with increasing temperature and particle size, corresponding to increased out of plane ripples. The interplanar binding energy becomes less favorable with increasing temperature and decreasing particle size leading to structural energetics that are more like freestanding graphene. The heat capacity varies little with either nanoparticle size or temperature. The interplanar spacing increases as the binding energy between the graphene layers is relaxed with decreasing nanoparticle size, in agreement with experimental observations. Three types of planar motion (breathing, sliding, and rocking) are characterized across a range of temperatures. The fundamental frequencies were calculated through Fourier transforms of the corresponding motions. The frequency for sliding is close to that seen for AB-stacking graphite. Two major frequencies are also observed for the breathing motion. These frequencies show weak dependence on temperature in the temperature range studied. The fundamental information on the energetics, structure, and dynamics of these graphitic nanoparticles will help explain such processes as Li-ion diffusion in these kinds of materials in the future.

(810g) High Capacity Adsorption of Dyes By "Super Sand"

Sheath, P., Monash University
Hill, M. R., CSIRO
Majumder, M., Rice University
Originally presented on: 11/8/2013 14:23:00 - 14:41:00

Recent reports have shown that graphene oxide (GO), synthesised from the oxidation of graphite, has a great capacity for adsorption of model pollutants (usually dyes) from water on account of a multitude of oxygen functionalities on its surface. These same oxygen functionalities render GO soluble in water, making its removal from a solution a challenge and thus hindering its usability as a water purification tool. RGO, derived from chemical or thermal reduction of GO, has also been investigated as a potential water purification material, its surface usually retains a number of oxygen functional groups whilst being insoluble and therefore easily filtered out of water, however, typically the adsorption capacities of RGO materials have not been particularly promising.

Herein we explore GO and RGO adsorption using three dyes of different charge and size and a $NaBH_4$-reduced RGO which has equivalent adsorption capacity to GO (~600mg/g for methylene blue) in batch experiments. Further, we explore a facile chemically derived modification with copper which has an enhanced capacity for adsorption of basic dyes with capacities up to 600mg/g for anionic Acid Blue 9 dye, and the use of RGO as a coating over sand (RGO-sand) (following from [1]) whereby a three order-of-magnitude improvement over GO-coated sand is observed for cationic methylene blue dye in a fixed-bed-column. The superior adsorption capacity of RGO-sand is explained in terms of a water-induced swelling of the RGO coating which is not observed for the GO-coating. Batch kinetics and isotherms are presented with adsorption mechanisms confirmed by correlation between the different adsorption character of each dye and a pH-study. This work makes strides towards determining the feasibility of using GO/RGO for water purification purposes.

[1] Engineered Graphite Oxide Materials for Application in Water Purification, Wei Gao, Mainak Majumder, Lawrence B. Alemany, Tharangattu N. Narayanan, Miguel A. Ibarra, Bhabendra K. Pradhan, and Pulickel M. Ajayan, *ACS Applied Materials & Interfaces* 2011 *3* (6), 1821-1826

(810h) Development of Single-Wall Carbon Nanotube Adsorbent Systems for Removal of Organic Compounds From Aqueous Environments

Rogers, R. E. Jr., Rochester Institute of Technology
Sherwood, T. J., Rochester Institute of Technology
Benton-Smith, J.,
Weinstein, S. J., Rochester Institute of Technology
Landi, B. J., Rochester Institute of Technology
Originally presented on: 11/8/2013 14:41:00 - 14:59:00

Carbon nanotubes show promise as effective adsorbents for environmental applications. Both single-wall carbon nanotubes (SWCNT) and multi-walled carbon nanotubes (MWCNT) have demonstrated strong uptake of various organic compounds, including herbicides and toxins such as trichloroethylene (TCE). In this presentation, a previously-developed semi-empirical model for the adsorption of a model polyaromatic compound is reviewed. Using the model as a framework, the time-dependent and equilibrium adsorption of 2,4-dicholorophenoxyacetic acid (2,4-D) on SWCNT is examined. This compound is of particular interest in that it is typically used in fertilizing systems and can lead to contamination of groundwater. Using UV spectroscopy techniques, adsorption of 2,4-D is measured to obtain kinetic and equilibrium parameters and compared against model predictions. Comparison between activated carbon, graphene, and SWCNTs show that SWCNTs have a much higher affinity for adsorption of 2,4-D and better adsorption capacity over time. A practical perspective on the use of SWCNTs for environmental remediation is discussed.

(813a) Towards a Biologically Inspired Nano-Factory for Molecular Assembly, Capture, Transport, and Characterization

Mahajan, K. D., The Ohio State University
Vieira, G., The Ohio State University
Ruan, G., The Ohio State University
Bouxsein, N., Sandia National Labs
DorcÃ©na, C. J., The Ohio State University
Lustberg, M., The Ohio State University
Bachand, G., Sandia National Labs
Chalmers, J., The Ohio State University
Sooryakumar, R., The Ohio State University
Winter, J. O., The Ohio State University
Originally presented on: 11/8/2013 12:30:00 - 12:55:00

Controlled manipulation of matter at the nanoscale is a common goal of nanotechnology and molecular diagnostics. A more ambitious goal is to build a nanofactory that incorporates functions such as molecular or nanoparticle assembly, capture, transport and characterization at the same time on the same platform. Various technologies have been developed to achieve these individual functionalities but an integrated platform combining all these functionalities is lacking. Here we describe a nano-engineered platform capable of building, manipulating and capturing multiple nanostructures in a massively parallel array. Nanostructures are formed by a two step hierarchical self assembly process. In the first step, fluorescent and magnetic nanoparticles are encapsulated in a self-assembled micellar structure. In the second level of self-assembly, the micellar structures are further complexed using protein-protein or complementary DNA interactions into higher order multifunctional nanoassemblies. The second level of assembly is shown to be concentration dependent and possesses fluorescent and magnetic properties of component nanoparticles. The magnetic functionality of the nanoassembly can be used to capture and transport aggregates on a magnetic microchip. Programmable traps can be created on the magnetic microchip, which consists of micro-patterned magnetic arrays and external electromagnets that control x-y and z directional magnetic fields. Transport of the nanoassemblies can be visualized in real time using the fluorescence functionality. The additive nature of fluorescence also helps characterize the nanostructures as they are being transported. We have shown assembly of quantum dots and superparamagnetic iron oxide nanoparticles in ampiphillic block copolymer micelles, followed by linker concentration dependent assembly in higher order structures; these structures are then transported along a nanoconveyor belt platform on magnetic microchip. We have demonstrated applications of this nanofactory in ultrasensitive detection and separation of protein, and DNA, in a diagnostic platform for detection and separation of circulating tumor cells. By inorporating molecular motors as molecular shuttles in the assembly we have achieved higher throughput leading to scale up. This platform can also be used for biophysical characterization of single molecules; as a proof of concept we have demonstrated preliminary experiments for determining bending strain values of microtubule-kinesin motor protein system.

(813b) Ultrasensitive Detection of Enteric Fever Using Magnetic Nanoparticles: Evanescent Wave Optical Illumination Studies

Upadhyay, M., Indian Institute of Technology, New Delhi

Originally presented on: 11/8/2013 12:55:00 - 13:20:00

An ultrasensitive method for early stage detection of *Salmonella* is demonstrated using total internal reflection fluorescence (TIRF) microscopy. *Salmonella* is captured from the milliliters volumes of spiked serum samples using fluorescent magnetic nanoparticles conjugated with anti-*Salmonella*. The labeled cells are separated from the unbound nanoparticles by simple gravity-based settling in confined volumes, circumventing the use of bulky ultracentrifuges. Thus the fluorescent magnetic nanoparticle conjugates serves both as pathogen preconcentrators and biosensing reagents. The fluorescently-labeled cells are selectively imaged at the bottom of the substrate by evanescent wave illumination using a TIRF microscope. Simultaneously the amount of cells captured was calculated by using plating method in order to validate the captured cells observed under the microscope. This technique allows rapid visualization of both live and dead bacterial cells unaffected by the antimicrobial properties of fresh blood and the presence of antimicrobial agent in the blood (in case of pretreatment with antibiotics). The current known methods for enteric fever diagnosis using cell culture often have low sensitivity and specificity, and long turnaround times. Thus, because of high sensitivity and rapid detection of pathogens the proposed approach could be very promising in the field of diagnostics, unlocking many biomedical opportunities.

(813c) Thermal Modeling of Wirelessly Heated Tissue Mimics

Coffel, J.,
Nuxoll, E., University of Iowa
Originally presented on: 11/8/2013 13:20:00 - 13:45:00

Roughly four percent of the 3.6 million medical devices implanted annually in the U.S. become infected and must be treated. Despite current prophylactic techniques, infection rates have decreased little over the past decade and existing treatments are limited to surgical explantation and replacement. One solution to treat and eliminate these infections is the use of thermal sterilization. Using heat inside the body to deactivate surface colonized bacteria is often overlooked because of the intense power requirements needed to induce effective thermal gradients and the risk of damaging surrounding, healthy tissue. In this work, magnetic hyperthermia is used to wirelessly deliver heat on implanted device surfaces with enough power to target deactivation time and temperature protocols. Remote surface heating is generated via a superparamagnetic iron oxide nanoparticle (SPION), poly(vinyl alcohol) composite coating exposed to an alternating magnetic field. Specific heating protocols using the coating are achieved by varying the SPION loading and the magnetic field strength. Additionally, experimental and theoretical thermal transport models are constructed to target a heating protocol that will eliminate infection while minimizing tissue damage. In the present work, these models quantify the thermal profiles generated from the SPION composite coating. Experimental heat transport models are produced through composite solid/liquid structures mimicking physiological constructs. These models provide results for the most extreme temperature gradients; including heat transport through mimicked bulk tissue, blood flow, and codependent bulk tissue/blood flow models. Experimental temperature profiles for both conductive and convective transport phenomena are measured using an in-house fabricated flow chamber and microscale thermistors.

(813d) Wirelessly Controlled Microfluidic Actuators Using Radiofrequency Electromagnetic Induction

Syed, W.,

Wu, W. I., McMaster University

Campbell, S. B., McMaster University

Hoare, T. R., McMaster University

Selvaganapathy, P. R., McMaster University

Originally presented on: 11/8/2013 13:45:00 - 14:10:00

Wirelessly Controlled Microfluidic Actuators using Radiofrequency Electromagnetic Induction

Wasi Syed*, Wen-I Wu*, Scott Campbell**, Todd Hoare** and P. Ravi Selvaganapathy*

* Department of Mechanical Engineering, McMaster University, CANADA

** Department of Chemical Engineering, McMaster University, CANADA

E-mail: campbesb@mcmaster.ca | hoaretr@mcmaster.ca

INTRODUCTION: The biomedical applications of microfluidic systems have enormous potential, particularly for their use in handling biomolecules.[1] However, the development of commercializable integrated microfluidic systems have been limited by the lack of reliable microfluidic components, such as micropumps and microvalves. A promising type of microvalve for these systems are thermopneumatic actuators, which operate based on the temperature-induced volume expansion of a phase-change material to block off a segment of a microfluidic channel to control flow. These actuators can provide large displacement forces and have facile fabrication as compared to electrostatic, magnetic, piezoelectric, and bimorph actuators, leading to their widespread use in microfluidics.[2]

Conventional thermopneumatic actuators are controlled by application of an electrical current to microheaters that produce heat due to resistance losses. However, in some instances, such as in therapeutic implants and microsurgical tools, it is not practical to have wired connection and preferable to have wireless actuation.[3] Recently, hyperthermia treatments induced using superparamagnetic iron oxide nanoparticles (SPIONs) and external oscillating magnetic fields (OMF) has also been recognized as a promising therapy for cancers by safely and remotely delivering heat to a local area inside the body.[4] In this work we plan to dissipate the heat generated by similar nanoparticles upon the application of an external OMF to the surrounding phase-change material in order to control microfluidic flow rates around the actuator. The phase-change material used herein is polyethylene glycol (PEG, MW ~1,000), which has a melting temperature of 39°C and a corresponding volume change of 25%. Subsequently, when PEG is combined with SPIONs this volume expansion can used to control micro pumps and valves wirelessly using an OMF.

EXPERIMENTAL: The SPIONs are synthesized using a coprecipitation method[5] with $Fe(III)Cl_3$ and $Fe(II)Cl_2$ followed by PEG (MW ~8,000) surface functionalization. The resulting iron oxide nanoparticles are 10-20 nm and aggregate into larger clusters, as shown in Figure 1.

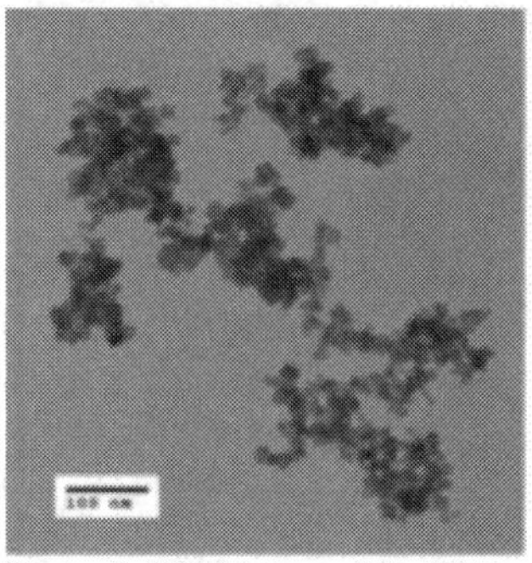

Figure 1: TEM image of the PEG-coated magnetic SPIONs.

The nanoparticles are mixed with molten PEG at concentrations of 0.5, 1.0 and 1.5 mg/mL and cast as thin sheets (2.5 mm thick) at room temperature. Small disks (∅6 mm) are punched and then placed onto spin-coated PDMS membranes (100 μm thick) as shown in Figure 2.

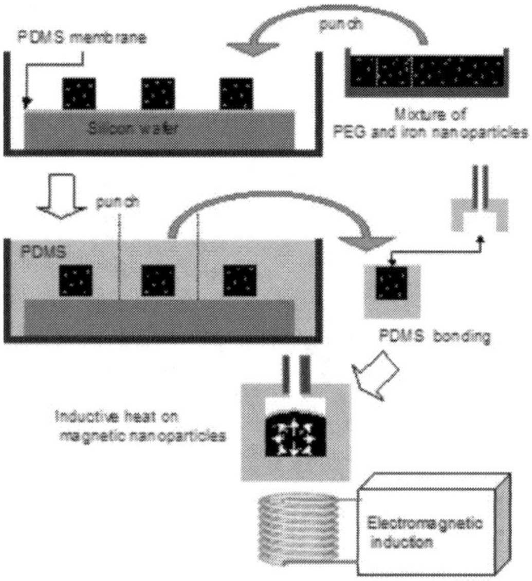

Figure 2: Process flow of microactuators with magnetic nanoparticles and phase-change material.

Uncured PDMS pre-polymer is casted on top to seal the disks. The actuator blocks are diced from this composite after a 24 hour curing process at room temperature. Subsequently, these actuator blocks are integrated with microchambers and microchannels by plasma treatment to form microfluidic pumps and valves respectively, both shown in Figure 3.

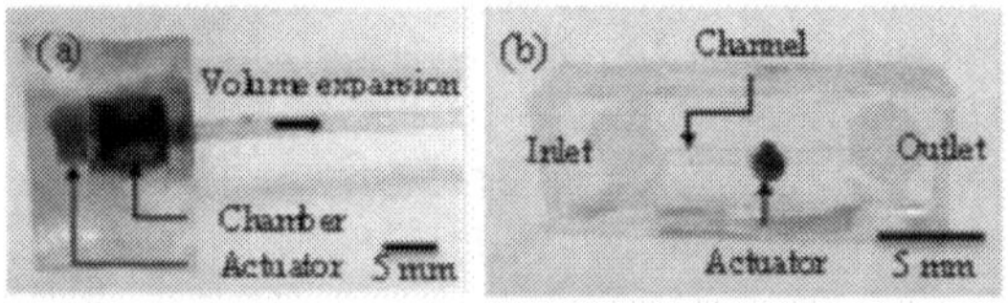

Figure 3: A prepared (a) micro pump and (b) micro valve.

The OMF used to induce heating of the SPIONs has three coils 4 cm in diameter, and operates at a frequency of 220 kHz to produce a maximum manetic flux of 4.5 mT for all experiments.

RESULTS: Initial experiments were performed on OMF induced micropumps and microvalves. In the case of micropumps, the microchamber of the pump is filled with dye for visualization and a camera is used to track the volume expansion as a change in the height of the liquid in the outlet column, as shown in Figure 4a.

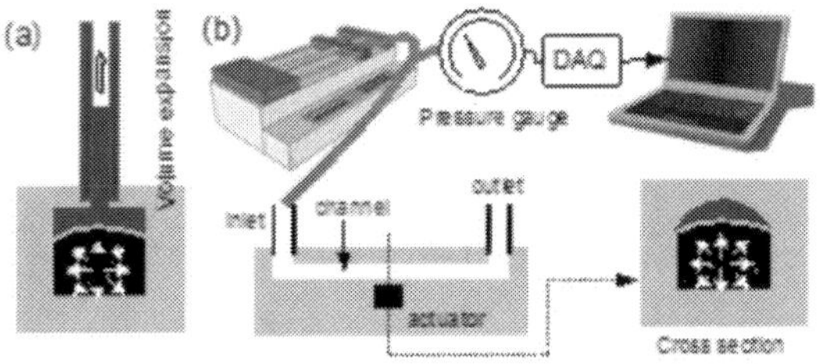

Figure 4: Experimental setup for (a) micro pump and (b) micro valve.

It was observed that volumetric expansions of ~25% are possible (similar to that of the literature[6]) and the rate of volumetric expansion is proportional to the concentration of the SPION nanoparticles, as shown in Figure 5. Notably, the micropumps were shown to be capable of providing high output volumes (>18 mm^3), which could further increase its potential applications.

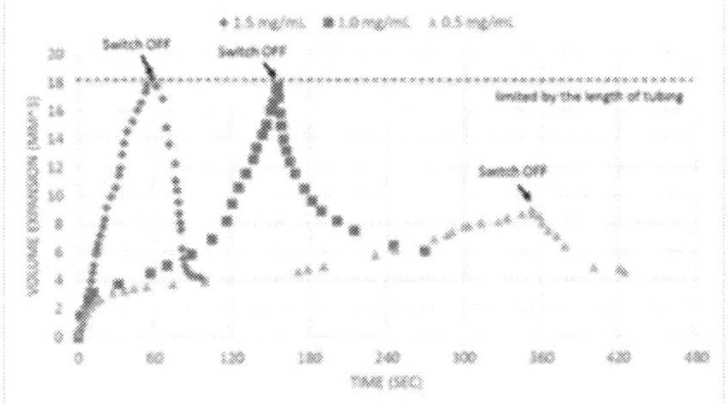

Figure 5: Volume expansion from micro pumps with 0.5, 1.0, and 1.5 mg/mL nanoparticle concentrations in PEG. Pumps were started upon an external magnetic field of 4.5 mT (220 kHz) and stopped immediately when switched off.

In the case of the valve, a flow rate of 1 mL/min (DI water) is infused by a syringe pump through a microfluidic channel with rounded cross-sectional shape (20 mm length, 1 mm width, 0.5mm height) and the upstream pressure is monitored by a pressure sensor (as depicted in Figure 4b). Figure 6 shows the valve is effectively closed against a pressure of 10 psi as long as the magnetic field is on. When the magnetic field is turned off, the PEG within the valve solidifies and the valves are opened within three minutes.

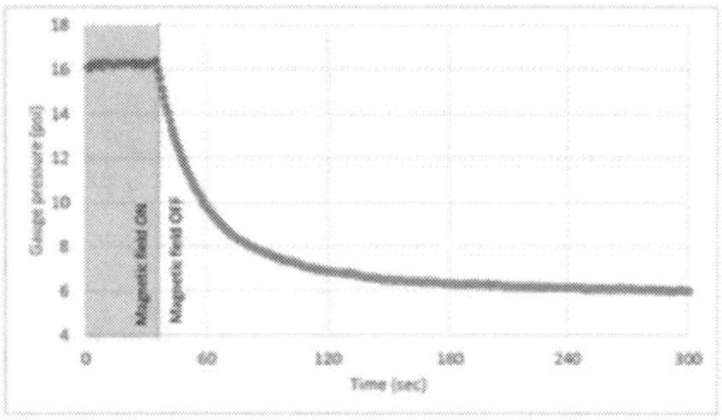

Figure 6: Response of the upstream pressure during valve opening. Valves were closed against 10 psi upon an external magnetic field of 4.5 mT (220 kHz) and opened when switched off.

The slow response to fully open the valve after OMF application is due to the time needed to cool the molten PEG, which can be further reduced by increasing its surface-to-volume ratio (ie. smaller PEG volume). The effects of the dimensions of the actuator, the SPION concentration in the PEG material, and the controlled actuation of multiple will also be presented.

CONCLUSIONS: Novel magnetic microfluidic actuators that can be controlled by an external OMF are capable of both providing high output volumes as micropumps and can withstand high pressures as microvalves, which could potentially permit their use in integrated microfluidic systems or as implanted devices for medical therapy applications.

REFERENCES: (1) Oh, K. W.; Ahn, C. H. *Journal of Micromechanics and Microengineering* **2006**, *16*, R13–R39. (2) Gravesen, P.; Branebjerg, J.; Jensen, O. S. *Journal of Micromechanics and Microengineering* **1993**, *3*, 168–182. (3) Smith, T. J.; Coyne, P. J.; Smith, W. R.; Roberts, J. D.; Smith, V. *American journal of hematology* **2005**, *78*, 153–154. (4) Mornet, S.; Vasseur, S.; Grasset, F.; Duguet, E. *Journal of Materials Chemistry* **2004**, *14*, 2161–2175. (5) Hoare, T.; Santamaria, J.; Goya, G. F.; Irusta, S.; Lin, D.; Lau, S.; Padera, R.; Langer, R.; Kohane, D. S. *Nano letters* **2009**, *9*, 3651–3657. (6) Song, W. H.; Kwan, J.; Kaigala, G. V; Hoang, V. N.; Backhouse, C. J. *Journal of Micromechanics and Microengineering* **2008**, *18*, 045009.

ACKNOWLEDGEMENTS: This research is funded by the Natural Sciences and Engineering Research Council of Canada (NSERC), the Ontario Graduate Fellowship, and the J.P. Bickell Foundation (Medical Research Grant).

(813e) Surface Functionalization of Iron Oxide Nanoparticles for Biological Encapsidation Applications

Ko, B. Y.,
Swartz, J. R., Stanford University
Originally presented on: 11/8/2013 14:10:00 - 14:35:00

Super-paramagnetic iron oxide nanoparticles (SPIO) have been utilized for more than two decades as contrast agents for magnetic resonance imaging (MRI), but recent years in particular have seen a dramatic increase in the number of potential applications for these nanoparticles. SPIO and SPIO-based constructs are being studied for a range of uses, including tumor imaging, drug delivery, and tissue repair. However, many of these applications require that the SPIO be packaged within or otherwise fused to a larger biological construct while maintaining stability in a physiological environment. Current approaches for functionalizing water-soluble SPIO generally result in relatively thick surface coatings that limit the size of SPIO that can be encapsidated; this, in turn, decreases the potential transverse relaxivities of the SPIO, rendering them less effective contrast agents. Here we demonstrate a novel biphasic method for stabilizing organic SPIO in an aqueous environment. This method minimizes the thickness of the nanoparticle surface coating to facilitate encapsidation within a biological construct of limited size and also maximizes transverse relaxivity. Briefly, sonication creates interfacial surface area between an aqueous SPIO-receiving phase and an organic phase containing both the hydrophobic oleic acid-coated SPIO and amphipathic molecules that intercalate into the oleic acid layer to produce a new hydrophilic surface coating. Following sonication, the resultant emulsion is phase-separated into its component layers via centrifugation, and the now-colored aqueous layer contains SPIO that have undergone successful phase transfer. Various parameters including sonication time and intensity, aqueous buffer and composition, and different amphipathic molecules, were studied to optimize SPIO phase transfer while maximizing relaxivities. This technique yields monodisperse, water-soluble SPIO with minimal coating thickness (< 3 nm) as measured by dynamic light scattering (DLS) and high transverse relaxivities ($r_2 \geq 200\ mM_{Fe}^{-1}s^{-1}$) via ^1H NMR. Furthermore, the amphipathic coating molecules can be selected based on the desired SPIO surface charge and functional group (e.g. amine or carboxylic acid). Thus, this functionalization technique, with fine control over the surface chemistry while minimizing the coating thickness, offers the ability to produce effective SPIO with the flexibility of being encapsidated within various biological constructs.

(813f) Structure-Relaxivity Relationships of Well-Defined Magnetite Clusters for Sensitive Magnetic Resonance Imaging

Davis, R. M., Virginia Tech
Balasubramaniam, S., Virginia Tech
Kayandan, S., Virginia Tech
Pothayee, N., Virginia Tech
Hu, N.,
Lin, Y., Virginia Tech
Pothayee, N., Virginia Tech
Koretsky, A.,
House, M., The University of Western Australia
Woodward, R., The University of Western Australia
St. Pierre, T. G., The University of Western Australia
Riffle, J. S., Virginia Tech
Originally presented on: 11/8/2013 14:35:00 - 15:00:00

In the past decade, multifunctional magnetite nanocarriers that integrate therapeutic agents into a single system have attracted considerable interest in theranostics integrating drug delivery and drug biodistribution. Magnetic nanoparticles have long been recognized as T_2-contrast agents for MRI. We report colloidally-stable magnetite clusters with hydrophobic and hydrophilic cores, and controlled sizes and compositions. Discrete ~ 8 nm nanoparticles (M_s = 65 A.m^2/kg) were synthesized by reducing iron (III) acetylacetonate in benzyl alcohol and their surfaces were subsequently modified to be hydrophobic by coating with oleic acid. A poly(ethylene glycol)-*b*-poly(D,L-lactide) block copolymer [mPEG(5 kDa)-*b*-PDLLA(10 kDa)] was synthesized by ring-opening polymerization of D,L-lactide in the presence of mPEG using stannous octoate as the catalyst. Hydrophobic-core clusters were prepared by rapid nanoprecipitation using an amphiphilic poly(*D*,*L*-lactide)-poly(ethylene oxide) diblock copolymer stabilizer. Nanoparticles with hydrodynamic diameters in the range of 100-150 nm and polydispersity indices less than 0.2 were obtained, as measured by dynamic light scattering. The transverse and longitudinal NMR relaxivities of the magnetite-loaded nanoparticles were ~ 200 s^{-1}.mM Fe^{-1} and ~ 1 s^{-1}.mM Fe^{-1} respectively. The transverse relaxivities enabled sensitive T_2-weighted MRI and the low longitudinal relaxivities are consistent with the hydrophobic nature of the cores of the nanoparticle clusters.

For the hydrophilic-core clusters, magnetite-poly(ethylene oxide-*b*-acrylate) complexes were synthesized, followed by crosslinking the amine tips of the H_2N-PEO corona with PEO diacrylate. The transverse relaxivities of the two classes of clusters correlated with particle sizes and compositions. Interestingly, the hydrophilic-core clusters had transverse relaxivities approaching the static dephasing limit of ~ 600 s^{-1}.mM Fe^{-1} and were more than twice that of the hydrophobic-core clusters at similar sizes and compositions. We posit that the high magnetic field gradients accessible to water protons by diffusion within the core of the hydrophilic clusters lead to the higher relaxivities.

(818a) Fluorescent Pentablock Conjugated Polymer Nanoparticles for Diagnostics and Therapeutics

Ahmed, E., University of Washington
Swager, T., Massachusetts Institute of Technology
Originally presented on: 11/8/2013 12:30:00 - 12:52:00

Nanoparticle-based materials are envisioned to play a critical role in the future of medicine, particularly in the areas of target-specific drug/gene delivery, early diagnosis, and disease treatments. Conjugated polymer nanoparticles (NPs) represent a novel class of promising alternative nanomaterials for diagnostics and therapeutics due to their high extinction coefficients, high fluorescence quantum yields, robust photostability, and ease of functionalization with target ligands. This talk we will highlight our recent efforts in developing a new class of multicomponent nanostructured materials and a modular platform that combines fluorescent probe, targeting ligands, and therapeutics, using simple synthetic chemistry, for detection of biological analytes and *in vivo* targeted imaging for early diagnostics and treatments of cancer. The nanoparticles consist of pentablock copolymers that include a far-red emitting dye made of a π-conjugated polymer for monitoring cellular uptake, an oligo(ethylene glycol) block that promotes stealth-like persistence in the blood stream and avoids protein opsonization and macrophage uptake in the body, and a targeting block containing folate for tumor targeting. Our results demonstrate the modulity of this platform that can allow for independent tuning of optical properties, targeting agents, and nanoparticle size.

(818b) Tracking Single Molecules in Living Cells and Whole Organisms With Carbon Nanotubes

Fakhri, N., University of Gottingen
Butkevich, E., University of Gottingen
Wessel, A. D.,
Willms, C.,
Klopfenstein, D. R.,
Pasquali, M., Rice University
MacKintosh, F. C., Vrije Universiteit
Schmidt, C. F., University of Gottingen
Originally presented on: 11/8/2013 12:52:00 - 13:14:00

Despite rapid improvements of fluorescent probes, observing individual proteins in living cells or whole organisms remains virtually impossible because of probe instability and autofluorescence of cells. Novel nanomaterials open up approaches that can overcome these problems. Single-walled carbon nanotubes (SWNTs) exhibit highly stable, near-infrared luminescence. Here we specifically target SWNTs to kinesin-1 motor proteins expressed in COS-7 cells as well as in C. elegans neurons. The combination of photostability and high signal-to-noise ratio makes it possible to accurately track individual motors traversing the entire cell. Remarkably, we distinguish the motor's motion along its microtubule (MT) track from the track's underlying random non-thermal fluctuations. Visualizing intracellular single-molecule dynamics within whole living organisms will be extremely powerful to explore the molecular basis of fundamental biological processes, including intracellular transport, in situ.

(818c) Engineering Biological Photoreceptors As Fluorescent Molecular Probes for Bioimaging

Mukherjee, A., University of Illinois at Urbana-Champaign
Schroeder, C. M., University of Illinois at Urbana-Champaign
Originally presented on: 11/8/2013 13:14:00 - 13:36:00

We report on the development of naturally occurring biological photoreceptors as an advanced class of fluorescent imaging probes with broad applications in synthetic biology, biological engineering, biomedical, and industrial bioprocess platforms. The elegant photophysical properties of natural photoreceptors were recently harnessed to develop genetically encoded imaging probes known as flavin-based fluorescent proteins (FbFPs) [1,2]. FbFPs are characterized by two key advantages relative to the widely used green fluorescent protein (GFP): namely, small size (≈ 50 % of GFP) and oxygen-independent maturation of fluorescence [1,2]. However, FbFPs are at a nascent stage of development and a complete understanding of their performance and properties as a practical set of biological probes is lacking. Importantly, FbFPs have modest brightness, which further limits pervasive applications of FbFPs for biological imaging.

With a view to advancing FbFP-based imaging technology, we systematically addressed the aforementioned issues. **First**, we extensively characterized the existing set of FbFPs in terms of key biophysical parameters, such as brightness (quantum yield), oligomeric state, maturation time, fraction of (fluorescent) holoprotein, pH tolerance, redox sensitivity, and thermal stability. We demonstrated that FbFPs show important advantages as broad-spectrum imaging probes, including broad pH tolerance (pH 4-11) and thermal stability (up to 60°C) as well as rapid maturation of fluorescence (≈ 3 min. *vs.* ≈ 30 min. in case of GFP) [3]. **Second,** we applied protein engineering using directed evolution to develop bright mutants of FbFPs. Specifically, we constructed and screened a library of approximately 30,000 FbFP variants and isolated two FbFP mutants (F37S and F37T) that displayed a 2-fold increase in overall brightness of fluorescence emission [4]. Our work demonstrated that enhancements in FbFP probe brightness can be engineered through improvements in quantum yield as well as increasing the strength of association between the protein and its fluorescent flavin cofactor. In addition, we demonstrated that aromatic amino acids in the fluorophore (flavin) binding pocket of FbFPs can potentially quench probe fluorescence. **Finally,** we validated the application of FbFPs as stable and noninvasive fluorescent reporters of gene expression *in vivo* by constructing a series of FbFP-based transcriptional fusions to probe promoter dynamics in *Escherichia coli* under different conditions of growth [3].

Overall, our results identify FbFPs as an emerging class of fluorescent protein probes that are characterized by stable cellular expression and robust performance under a wide range of environmental conditions, such as lack of oxygen and extremes of pH and temperature. Importantly, we develop a molecular engineering framework to identify and engineer new and improved FbFP-variants using directed evolution based on site saturation mutagenesis. In summary, we anticipate that our work will encourage the continued development of FbFPs and enable their broad application as fluorescent reporters of gene expression, protein localization, and whole cell microbial sensors in conditions where existing imaging probes fail to perform optimally (*e.g.*, fast time-scale cell processes, extremophilic microbes, anaerobic bioprocesses such as fermentation).

References

1. T. Drepper, **et al**. *Nature Biotechnology*, 25: 443:445 (2007).

2. S. Chapman, **et al**. *PNAS*, 105(50):20038:20043 (2008)

3. A. Mukherjee, **et al**. PLOS ONE, accepted (2013)

4. A. Mukherjee, **et al**. Journal of Biological Engineering, 6(1):20 (2012)

(818d) Using Luciferase Expression to Quantify the Accumulation of Genetically Engineered Mesenchymal Stem Cells in Murine Tumors

Jampol, R. P., Georgia Institute of Technology
Ghosh, D., Georgia Institute of Technology
Rodriguez, K., Georgia Institute of Technology
Dawson, M. R., Georgia Institute of Technology
Originally presented on: 11/8/2013 13:36:00 - 13:58:00

Mesenchymal stem cells (MSCs) are bone marrow derived multipotent cells that can be expanded rapidly and genetically engineered to express therapeutic proteins. MSCs have a natural ability to home to tumors after systemic injection, but reduced homing capabilities arise after extended culture and genetic manipulation. Treatment strategies that increase the migration of genetically engineered cells are essential in the development of MSC-based gene delivery systems that home to tumors after infusion.

MSCs isolated from C57BL/6 mice were genetically engineered with a tetracycline inducible vector to express the luciferase reporter gene upon treatment with doxycycline. The inducibility of luciferase expression was characterized in vitro. Engineered MSCs and B16F1 tumor cells were co-implanted in C57BL/6 mice. Mice given control diet or doxycycline diet were imaged using the IVIS bioluminescent imaging system. Luciferase was easily visualized at the primary tumor site as well as in the lungs or tail of mice infused with genetically engineered MSCs. Luciferase expression will be used to track the location of systemically injected MSCs and to determine if in vitro treatment with soluble growth factors may be useful for improving the homing efficiency of genetically engineered MSCs.

Stagg J (2008) Mesenchymal stem cells in cancer. *Stem Cell Rev* 4(2):119-124.

(818e) Extending Bioavailability of Polymersomes for Enhanced Tumor Detection

Lai, M. H., University of Illinois at Urbana-Champaign
Lee, S., Kyung Hee University
Kim, K., Korea Institute of Science and Technology
Kong, H., University of Illinois
Originally presented on: 11/8/2013 13:58:00 - 14:20:00

Nanoparticles have emerged as a novel tool enabling targeting and imaging of cancer because they can diffuse through the leaky tumor vasculature and accumulate in the extravascular tissue effectively via the enhanced permeation and retention (EPR) effect. Successful use of nanoparticles in this application greatly relies on a structural stability of nanoparticles in circulation; however, according to previous studies, nanoparticles which are often used for clinical diagnosis and treatments are plagued by their shorter lifetime in circulation.

To address this problem, we synthesized a poly(*N*-hydroxyethyl-*DL*-aspartamide) modified with pedant alkyl chains and methacrylate groups, termed PHEA-g-C_{18}-MA, which can form a polymersome with chemically cross-linked shell via sequential self-assembly and photo-induced cross-linking reaction of the bilayer. The resulting polymer and its self-assembled morphology of PHEA-g-C_{18}-MA were investigated by nuclear magnetic resonance and tramission electron microscopy. The resulting PHEA-g-C_{18}-MA polymersomes with hardened shells exhibited the better *in vitro* stability than a control, the same polymersome without cross-linked shells. The polymersome was further functionalized to present a capability of imaging target tumor vasculature by labeling the particle with a near-infrared (NIRF) dye. The polymersome carrying NIRF dye was injected into mice with squamous cell carcinoma tumors through the tail vein, and its bio-distribution was monitored using a whole-body imaging system. Interestingly, PHEA-g-C_{18}-MA polymersomes with hardened shells displayed superior capabilities to accumulate in tumor sites and show entire tumor tissue, compared to controls including polymersomes with softer shell.

We propose that this result should clearly address the important role of structural stabilities of nanoparticles on improving their capabilities to target and image cancer tissue. In addition, this study serves to provide an invaluable biomedical tool that can significant enhance the quality of the cancer diagnosis or treatment.

(818f) The Use of Hyperspectral Imaging for the Detection of Gold Nanorods in An In Vitro System

Comfort, D., University of Dayton
Stacy, B., University of Dayton Research Institute
Comfort, K., University of Dayton
Hussain, S., Air Force Research Laboratory
Originally presented on: 11/8/2013 14:20:00 - 14:42:00

Due to their unique plasmonic and optical properties, gold nanorods (GNR) have shown tremendous potential for nano-based applications extending into a variety of fields including advanced bioimaging techniques. These distinctive, shape-specific properties are strongly dependent upon the GNR aspect ratio, thus producing the ability to be targeted for an application by fine-tuning their physical parameters. It is owing to their characteristic spectral signature, which is vastly different from that of a cellular setting, that GNRs are emerging as an ideal candidate for nano-based imaging applications. However, one challenge that has emerged in the field of bioimaging is the need to account for the observed plasmon coupling effect that arises from GNR agglomeration in a physiological environment. In this study, GNRs with aspect ratios of 2.5 and 6.0 were actively identified in an in vitro setting through a hyperspectral imaging (HSI) analysis; which successfully recognized and separated the light scattering pattern of these particles from that of the surrounding cells. Through inclusion of agglomerated GNR spectral patterns in the HSI library, this imaging technique was able to overcome the complication of plasmon coupling, though to varying degrees. Results indicated that the GNR transverse peak is not as impacted by plasmon coupling effect as the longitudinal, identifying it as a preferable target for imaging purposes. These results demonstrate the tremendous potential of GNRs coupled with HSI analysis to advance the field of nano-based sensing and imaging mechanisms.

(818g) Nano BaSO4 : A Novel Means to Create Antimicrobial Radiopaque Thermoplastics

Aninwene, G. II, Northeastern University
Stout, D. A., Brown University
Yang, Z., Brown University
Webster, T. J., Northeastern University
Originally presented on: 11/8/2013 14:42:00 - 15:04:00

Abstract:

Hospital acquired infections remain a major costly problem. This study seeks to understand how incorporating nano-barium sulfate into pellethane composites affect the physical properties and antimicrobial nature of the resulting polymers. The results of this study showed that the incorporation of nano-barium sulfate into pellethane polymers yielded polymers which had enhanced antimicrobial properties, yet had similar hydrodynamic properties and were still radiopaque.

Introduction: Hospital acquired infections remain a costly problem. Each year there are approximately 2 million hospital acquired infections, 90,000 of which are fatal [1]. Catheter-related bloodstream infections alone represent a potential burden on the healthcare system of at least $35billion [2]. The main challenge is to prevent bacteria growth early before biofilm production takes place, since once the bacterial biofilm matrix forms, bacterial infections can become profoundly more resistant to the host defenses as well as antibiotic treatments [3]. Barium sulfate ($BaSO_4$) is a common agent used to make medical tubing radiopaque; however, in addition to this, $BaSO_4$ polymeric formulations have been shown to exhibit antimicrobial activity[4]. Additional studies have shown that the use of nano structures on surfaces can lead to surfaces that are antimicrobial or resistant to bacterial proliferation [5]. The goal of this study was to investigate if nano-$BaSO_4$ pellethane composites are able to effectively act as surfaces which prevent initial bacterial adhesion and proliferation, while still remaining radiopaque.

Materials and Methods: Foster Biomedical Polymers and Compounds (Putnam, CT, USA) extruded pellethane tapes with various weight percentages of $BaSO_4$ powder and nano- $BaSO_4$. Seven different sample groups made with varying weight percentages of $BaSO_4$ (0% $BaSO_4$, 20% nano-$BaSO_4$, 30% nano-$BaSO_4$, 40% nano- $BaSO_4$, 20% $BaSO_4$, 30% $BaSO_4$, and 40% $BaSO_4$). Tapes were then cut into disks that fit into 12-well plates (approximately 22mm in diameter). These disks were then sterilized with EtOH and UV light treatment prior to use.

Contact angle measurements were made on a Krüss Easy Drop contact angle instrument (Krüss, Germany) connected to an image analysis program (Drop Shape Analysis (Version 1.8)). The Krüss Easy Drop apparatus was used to measure the contact angles that resulted when a 10μL drop of either H_2O, glycerol, or ethylene glycol was placed on the surface of a sample disk [6].

Polymer samples were labeled and radiograph images were taken using a infinity XMA HF-30AP, set to Manual technique mode with an exposure time of 0.016 seconds and MAS@6.1 and 70KV. Images were taken of each sample individually and s-values, numeric value of exposure received by the receptors in the digital system, were recorded [7]. For analysis the s-value for the 0% BaSO4 sample was used as a base and subtracted from all the other values to normalize the results.

Stock solutions of *Staphylococcus aureus* (*S. aureus*) (ATCC# 25923) and *Pseudomonas aeruginosa (Schroeter) Migula* (*P. aeruginosa*) (ATCC# 27853) were diluted and frozen. Bacteria from stock solutions were streaked out for isolation on agar plates. Bacteria was cultured in sterile tryptic soy broth (TSB) (Sigma Aldrich) then diluted to a density of 1×10^7 bacteria/mL (as estimated by the McFarland scale which corresponded to an optical density of 0.52 at 562 nm then further diluted at a ratio of 1:100) [8]. The sterile polymer disks were placed in separately labeled wells of a sterile 12 well tissue culture plate. They were then covered in 1ml of the bacteria in. The plates were sealed with parafilm and then incubated at 37°C while shaking at 200rpm for 1.5 hours. Next, 10μL samples of each solution were obtained and diluted in fresh TSB. Finally, colony counts were taken to determine the number of colony forming bacteria units in each well.

Results and Discussion: Contact angle trials yielded no significant differences. Radiopacity trials indicated that the nano-$BaSO_4$ samples were still radiopaque. Figure 1 displays the growth curve of *S. aureus* at room temperature (line) paired with bacterial growth on polymer samples (bars) The bacteria results indicated a significant decrease in bacteria proliferation at certain concentrations of nano $BaSO_4$. In the case of *S. aureus*, significant decreases were observed when 20% and 40% nano $BaSO_4$ polymers were used. In the case of *P. aeruginosa*, significant decreases verse control were observed when 0% $BaSO_4$ as well as 30% and 40% nano $BaSO_4$ polymers were used. Additionally the 40% nano $BaSO_4$ led to a significant decrease in P. aeruginosa compared to the 40% $BaSO_4$ polymer (results not shown). The results of these trials indicated that adding nano-$BaSO_4$ to the extrusion process of pellethane was able to change the surface dynamics and microbial surface interactions of the resulting polymer sample. From the results, it appears that the nano 20% and the nano 40% polymer blends yielded a marked reduction in *S. aureus* proliferation, while the 30% and 40% blends showed mark decreases in P. aeruginosa proliferation. These results indicated that longer antimicrobial trials should be conducted to investigate the potency and longevity of the antimicrobial effects seen.

Conclusion: These trials indicated that although the nano-$BaSO_4$ did not change not change the hydrodynamic nature of the samples there was a significant change in bacteria function when nano-$BaSO_4$ was present in the polymer. The result was a reduction in bacterial proliferation. Further trails need to be completed to better understand the mechanisms of the trends observed in the present study as well as determine bacteria growth on samples over prolonged periods of time.

Acknowledgements: The authors would like to thank Adriana Noemí Santiago, and Gozde Durmus for aid in performing trials. Special thanks to Foster Biomedical Polymers and Compounds for providing samples.

References:

1. Burke JP. Infection Control — A Problem for Patient Safety. *New England Journal of Medicine.* 2003;348(7):651-656.

2. Fears R, van der Meer JWM, ter Meulen V. The Changing Burden of Infectious Disease in Europe. *Science translational medicine.* Oct 2011;3(103).

3. Hall-Stoodley L, Costerton JW, Stoodley P. Bacterial biofilms: From the natural environment to infectious diseases. *Nat. Rev. Microbiol.* Feb 2004;2(2):95-108.

4. LaBrec BC. Brown University Biomedical Research Sponsored By Foster Corporation. In: Webster T, ed. Email Comunication between Foster and Brown university. ed2011.

5. Taylor E, Webster TJ. Reducing infections through nanotechnology and nanoparticles. *Int. J. Nanomed.* 2011;6:1463-1473.

6. Puckett SD, Lee PP, Ciombor DM, Aaron RK, Webster TJ. Nanotextured titanium surfaces for enhancing skin growth on transcutaneous osseointegrated devices. *Acta Biomaterialia.*6(6):2352-2362.

7. Sprawls P. S-Values - Digital Radiology. 2012; http://www.acrrt.com/index.php/11-articles/29-s-values-digital-radiology. Accessed 11-7-12, 2012.

8. Puckett SD, Taylor E, Raimondo T, Webster TJ. The relationship between the nanostructure of titanium surfaces and bacterial attachment. *Biomaterials.* 2010;31(4):706-713.

AUTHOR INDEX

AUTHOR INDEX

AUTHOR INDEX

AUTHOR INDEX

AUTHOR INDEX

AUTHOR INDEX

AUTHOR INDEX

AUTHOR INDEX

AUTHOR INDEX

AUTHOR INDEX

AUTHOR INDEX

AUTHOR INDEX